HISTORY OF CARCINOLOGY

CRUSTACEAN ISSUES

8

General editor:

FREDERICK R. SCHRAM

Instituut voor Systematiek en Populatiebiologie, Amsterdam

A.A. BALKEMA / ROTTERDAM / BROOKFIELD / 1993

HISTORY OF CARCINOLOGY

Edited by
FRANK TRUESDALE
Louisiana State University, Baton Rouge

A.A. BALKEMA / ROTTERDAM / BROOKFIELD/ 1993

Published by
A.A. Balkema, P.O. Box 1675, 3000 BR Rotterdam, Netherlands
A.A. Balkema Publishers, Old Post Road, Brookfield, VT 05036, USA

ISSN 0168-6356
ISBN 90 5410 137 7

Table of contents

Preface

In recent years, a renaissance has occurred in the study of the history of biology. The founding of the Division of History and Philosophy of Science within the American Society of Zoologists helped this in large part. In fact, this volume grows out of a co-sponsored symposium presented at the 1990 annual meeting in San Antonio, Texas of the ASZ and The Crustacean Society. This symposium commemorated the tenth anniversary of the foundation of The Crustacean Society. The contents of this book differ from the traditional substance of a Crustacean Issues volume. This does not mean, however, that we should give it less consideration than we would a volume on hard science.

The papers here take several forms, from strict chronologies to detailed historical analyses. Thus we see some contributions that in whole or part confine themselves to providing a 'roll call' of crustacean workers in particular places or times. Other papers present slightly fuller listings of the efforts of particular workers, sort of 'hats off' accolades, but leave unexamined the relationship of these contributions to the development of carcinology as a whole. Still other chapters relate the researchers and their work to the times in which they lived and analyze the effects workers had on each other. We should expect such a spread of approaches given the still nascent stage that history of science as a formal discipline finds itself in today.

The particular presentation chosen by an author, however, can create inaccurate perceptions in the mind of the reader. Because one author might present only a roll call should not imply that true success in carcinology occurred only in other spheres. Likewise, if another author gives a person or group of people only a short accolade does not mean that a more detailed examination of archives somewhere will not produce some significant and interesting new insights.

Nor should we rely on making traditionally viewed 'great' persons the sole focus of our studies and interpreting them as the sources of all progress in our science. I believe we already have an unfortunate perception of some historical figures as sort of 'eternal understudies' in the fellowship of carcinology. We tend to forget, however, that success often results merely from good luck, working in the right place at the right time, which does not necessarily reflect innate talents. After all, what could a 'lesser figure' like William Stimpson have achieved if he had lived longer, or if he had not lost almost everything in the Great Chicago Fire? And what would 'the great' Sidnie Manton have accomplished if, instead of living in London and working at the University of London and the British Museum, she had lived alone next to the Great Salt Lake and had only branchiopods to look at? In addition, great things in any human endeavor come often as the result of shear volume of work. Given time and an open field devoid of little or no previous research, 'greatness' seems inevitable. I do not denigrate anyone's achievements; I merely suggest that we keep anyone's achievements in perspective.

Finally, we must remember that we make our own history. As historians we create history merely by either choosing to analyze particular events or people, or defining success in limited and isolated terms. When we do this, we make it very difficult to weave a coherent tale from all the threads of human activity that actually could present themselves to us if we would just but look. Much in this volume can form the basis of further, more detailed, and analytical studies. I would hope these inquiries will occur in time. And when they do emerge, we should remember that to ignore new findings just to retain old, familiar, comfortable icons will risk losing a chance to gain new insights into how the history of our science unfolded.

A serious study of history derives from a need to understand our roots, and in doing that we seek to understand why we believe the things that we do. To enshrine particular workers or to retain outdated ideas because we find them comfortable deflects history into the path of mythology. As scientists we should not fear re-examining, redefining, and reinterpreting what we learned as students. We should not hesitate to seek in the past the neglected threads of thought that, if picked up again, could lead to conceptual advances.

So Frank Truesdale and I offer this volume for both your entertainment and your stimulation. If it does the former, our satisfaction will result; if it causes the latter, moving you to pursue further study and reading, we will experience delight.

Frederick R. Schram
Amsterdam

Introduction

In 1986 Fred Schram and Paul Haefner, President and Secretary, respectively, of The Crustacean Society (TCS), suggested that the Society celebrate its 10th anniversary with a symposium on the history of carcinology, and then publish the delivered papers, along with others not given at the symposium, in a 'souvenir' volume. Because I too am, as Tony Rice describes himself, a 'bit of a history freak,' I was moved to answer a help wanted ad in the *Ecdysiast* (TCS newsletter) for volunteers to aid in the proposed celebration. Surprisingly, I was quickly informed that I had made the shortlist for symposium organizer and chairperson, and when the selection committee deadlocked on the two frontrunners, I got the job!

Of the 19 papers in this volume, 11 are from the 12 presented at the symposium, including those of three non-US scientists (Christiansen, Harada, and Wolff) who graciously helped me realize my desire to make the symposium an international effort. I was particularly honored in that Harada Eiji was making his first visit to the US

Many of those who accepted my invitation to participate in the symposium and this volume had published on the history of carcinology and zoology, and therefore I generally allowed all to select and develop their own topics. I did ask Patsy McLaughlin to write on women in carcinology, and when I first contacted her I said I hoped her response would not be the same as Libbie Hyman gave Waldo Schmitt. It seems (Blackwelder 1979), Schmitt thought an all-women symposium on the lophophorates was in order for an annual meeting of the Society of Systematic Zoology, and he asked Hyman to be chair. Hyman replied that it was a 'damn fool idea.' Well McLaughlin's response was much more flattering, and the presentation on women in carcinology, delivered by her co-author Sandra Gilchrist, drew the largest audience of the symposium. (Oh, Libbie Hyman did go ahead and chair the lophophorate symposium, anyway). Whether the topic was suggested or not, its obvious that the contributions reflect that what is history varies with the author, and 'that for some, history is literature; for others, facts; for some delving in archives; for others, interpretations of the sources; for some an art; for others, a science; for some, drudgery; for others, a romance; for some, an explanation of the present; for others, a revelation and a realization of the past' (Thorndike 1937, as quoted in Barzun & Graff 1977).

If to some, like me, history is 'the story of past facts' (Barzun & Graff 1977), the barest facts in this volume are the birth and death dates for a legion of individuals who have somehow influenced the field of carcinology. The long productive lives of such legends as G.O. Sars (see Christiansen) and Mary Rathbun (See McLaughlin & Gilchrist; Bishop) may lead us to muse: What if the incredibly productive lives of Georg Marcgraf (see Tavares), William Stimpson (see

1

Manning), and Giuseppe Nobili (see Rodriguez) had not been so tragically short? Gudger (1912) said: 'Had [Marcgraf] lived a few years longer to have put into shape his Brazilian collections and observations, [he] certainly would have raised himself to the rank of first natural historian of his time, and possibly the greatest since Aristotle.' Stimpson, too, needed only a few more years to publish works that would be the culmination of his 'remarkably energetic life' (Mayer 1918), when he lost everything in the Chicago fire of 1871 and died soon after. Nobili was but 31-years-old when he died, but had already published 53 papers on Crustacea, including a monographic work on the stomatopods and decapods of the Red Sea (Rathbun 1909).

Let me string together some more facts from this volume and make a story or two.

Spencer Fullerton Baird (1823-1887) was not a carcinologist, but as several of the papers in this volume will attest, he had a profound effect on North American carcinology. As the first US Commissioner of Fish and Fisheries not only did he provide for seaside facilities for summer research, including the then (1885) state-of-the-art Fisheries Station at Woods Hole, Massachusetts, used by many prominent US aquatic zoologists of the late 1800s (Allard 1990), but he was responsible, almost singlehandedly, for the US government building the survey vessels Fish Hawk and Albatross. Fish Hawk, in commission 1880-1926 (Galtsoff 1962, Reintjes 1973), was designed primarily as a floating fish hatchery, and was not seakindly in the open coastal waters of New England (Linton 1915) where it was used for numerous dredging and trawling surveys. Albatross, in commission 1883-1921 (Hedgepeth & Schmitt 1945), was approved by Congress, due to Baird's lobbying, as an exploratory fishing vessel, but it was designed and equipped for scientific research in deep waters beyond the range of most commercial species (Allard 1990). Albatross was well equipped and comfortable, as documented by the comments of Alexander Agassiz who conducted three expeditions on the vessel and who was prone to *mal de mer*: 'The Albatross is an excellent sea boat and she rides the sea wonderfully well, and really much better than many a large ocean steamer I have been on....The accommodations for work and for taking care of the collections are excellent' (G.R. Agassiz 1913). In 1945 Hedgepeth & Schmitt wrote: 'It is practically impossible to name an eminent American zoologist who has not been associated in some way with the Albatross. Nor has this influence ceased [as] there still remain a large number of unsorted collections [and] it will be many years before the last Albatross report is published.' Fenner Chace (b.1908) of the US National Museum of Natural History has recently added to the continuing list of Albatross reports (e.g., 1988). Writing of the caridean shrimps of the Albatross Philippine Expedition, 1907-1910, Chace (1983) said: 'I came to realize that having been collected by dedicated professionals, they are in far better condition than much of the material received by museums today.' Chace's earlier involvement with an Albatross specimen, also from the Philippine Expedition, further documents the professionalism of the Albatross staff, and the continuing influence of Albatross collections on carcinology.

Preliminary sorting of Philippine Expedition materials by National Museum staff in the early 1970's had yielded a crustacean unknown to Chace, and which seemed to be lacking its major claws. 'I had everybody on the lookout for those lost claws,' Chace said, and 'when visitors came in I'd frequently pick up this [specimen] and show it to them to see if they had any ideas about it. The last time was when Jacques Forest [b.1920] was over here from Paris. I showed it to him, and his eyes lit up....He spotted [it] as resembling the glypheids, which were presumably long since extinct. Certainly, eventually it proved to be that.'[1] Later Forest lead a French expedition to try and collect more specimens (see McLaughlin & Gilchrist's biography of St. Laurent, this volume). Chace continued: 'Lo and behold, they went out to the same spot

[Albatross] found it in...and got nine more specimens! Of course those clinched the matter as far as the glypheids are concerned; they proved to be glypheids. These missing claws weren't claws at all; they were just additional legs. But all of them that came up, apparently, threw off that pair of legs....so this was a typical reaction to capture....All of these nine specimens Forest took were taken within a half mile of the original [Albatross] collecting site.'[2]

Carcinologists afield and other collectors of crustaceans have been amply documented in this volume by name, dates, expeditions, and collections, but stories beyond the confines of specimen label and formal paper, for the most part, remain untold. Horton Hobbs Jr. (b. 1914) of the US National Museum of Natural History, expert on freshwater decapod crustaceans, especially crayfishes, relates a story of the joys of collecting ones own specimens, with the help of a good graduate student or two. 'One afternoon we'd been out [Hobbs, a faculty colleague at the University of Florida, and two graduate students], it was getting late....Dr. Sherman was driving, and I said, "Dr. Sherman, here's a little place right close to the road. Stop, it won't take Billy [W.M. McLane] and me a moment to have a look." So we rushed over. It was nothing but a depression, but on our way back to the car there was a perfectly cylindrical chimney that went right down...between 50 and 70 feet. I peeked over the edge of it and saw that there was a little water in the bottom of it which excited me....McLane and I went down into the hole and the other two stayed up above. We got down to the bottom and I saw white crayfish on the bottom. We looked around and there was a little opening into the side, about 2 1/2 feet in diameter; and we crawled into this opening, and that led into a fissure that was about 4 feet wide and some 60 or 70 feet long, with no floor, but the entire thing with water under us. It was shallow at one end, then it dropped off rather quickly....I saw the white crayfish down there and I turned...and said, "Billy, if you'll catch one of those crayfishes and it's a new one, I'll name it for you." Well I knew what they were so I was perfectly safe. So Billy jumped into the water with a dip net and started scurrying around, but he was missing them. I said, "Let's go." The water had gotten so cloudy you couldn't do anything....When we reached the top of the cave Billy turned to me and said, "Well, Doc, I didn't get any of the big ones but I got this little one." He handed me a vial with a little tiny crawfish in it, the most amazing animal I had ever seen in my life [*Troglocambarus maclanei* Hobbs, 1942].[3]

William Stimpson's visit to Japan during the North Pacific Exploring Expedition of 1853-1856 (see Manning) had no direct bearing on the westernization of Japanese zoology (see Harada). However, in 1877 his friend and colleague E.S. Morse (see Manning), visiting Japan to study brachiopods, accepted a position at the newly formed University of Tokyo to teach zoology and evolution (Watanabe 1990). Morse, a student of Louis Agassiz and who split from the master in the 'Salem secession' (see Manning), probably did not become an 'avid Darwinian' until 1870-1873 (Winsor 1991), but he is credited with presenting (6 October 1877) the first lecture on Darwinism ever given in Japan (Watanabe 1990). Although the stated purpose of the North Pacific Exploring Expedition was to 'extend the empire of commerce and of science....' (US Sec. of the Navy J.P. Kennedy, quoted in Cole 1947), Stimpson's collecting was not facilitated by his Navy supervisors (see Manning), and the squadron's surveying of Japanese waters was not welcomed by the Japanese Government (see Cole 1947). A half-century later (1900) Albatross steamed into into Yokohama discharging Alexander Agassiz and his assistant Alfred Goldsborough Mayer (1868-1922), at the end of the 'second Agassiz' or South Sea expedition. Before beginning its homeward leg, Albatross sampled in Japanese waters, sometimes with students and professors from the University of Tokyo onboard (Hedgpeth & Schmitt 1945). Since Stimpson's visit Japan had undergone (1868) a reformation, the Meiji Revolution (see Harada), and was rapidly assimilating Western knowledge, including modern science, at an astonishing rate (Watanabe 1990). Western scientists such as Morse, who was succeeded at the University of Tokyo by C.O. Whitman (1842-1910), were invited to teach at Japanese universities, and numerous students, including Watase Shozaburo (1862-1929) who received his doctorate at Johns Hopkins under W.K. Brooks (see Benson), were sent to study at US and European universities. David Starr Jordan, ichthyologist, educator, sometimes-decapod-collector (see Rodriguez; Rathbun 1902), and a US scientist revered in Japan for contributions to development of modern Japanese zoology (Kaburaki 1926), arrived for his first visit to Japan, in the summer of 1900 (Jordon 1922), while Albatross was just leaving Japanese waters (Hedgpeth & Schmitt 1945). [Albatross would return on the Japanese Expedition of 1906, when according to Hedgpeth (1947) the most extensive survey of Japanese waters by a foreign vessel prior to World War II was made.]

Rice's mention (this volume) of Sandoz and her co-workers offers an opportunity to acknowledge several little-known researchers whose names, ironically, have become commonplace in the literature of certain areas of carcinology. Mildred Sandoz (dates not available) worked at the Virginia Fisheries Laboratory, Yorktown, Virginia (now Virginia Institute of Marine Science, Gloucester Point), during the early to middle 1940's. Among her co-workers were Rosalie Rogers (Rosalie Rogers-Talbert, b. 1922) and Sewell H. Hopkins (1906-1984). Williams mentions Sandoz & Hopkins' 'Early life history of the oyster crab, *Pinnotheres ostreum* (Say)', and although William's paper is a historical review, the work by Sandoz & Hopkins is still frequently cited in the research literature (see *Science Citation Index* [*SCI*]). However, that by Sandoz & Rogers – 'The effect of environmental factors on the hatching, moulting, and survival of zoea larvae of the blue crab, *Callinectes sapidus* Rathbun' – may be a *Citation Classic* (cf. Garfield 1982). Since 1955, I counted 55 citations to this paper in *SCI*, including several in the 1990's, the fifth decade after publication.

Schram's paper on the correspondence between Libbie Hyman and the young carcinologist Martin Burkenroad winds up giving greater insight into Hyman, a non-carcinologist, than into Burkenroad. (This is, in part, because Burkenroad's actual letters to Hyman have been lost.) However, most carcinologists of my generation grew up on the various volumes of Hyman's treatise on invertebrates, but with little, or no, accurate information on Hyman herself. Also, Hyman planned a final volume, of the four originally projected, on the Arthropoda (*The Invertebrates* Vol.I, 1940:v), and no doubt if she had not run out of time (cf. *The Invertebrates* Vol.V, 1959:v) that volume, too, would have been a classic. Although Burkenroad used the facilities of the Louisiana State University Field Laboratory at Grand Isle (also called Louisiana State University Marine Laboratory) during research for his 1934 paper on the Penaeidae of Louisiana, he does not seem to have mentioned it as a seaside laboratory Hyman might visit in her quest for the ideal summer sojourn. Interestingly, the Grand Isle facility had been established by Hyman's fellow assistant in zoology at the University of Chicago, Ellinor H. Behre (1886-1982). Behre, who directed the Marine Laboratory from (1928-1945), wrote a couple of papers on Crustacea and compiled an important checklist of the fauna of the Grand Isle region (1950). Manning & Felder (1989) named *Pinnixa behreae* in Behre's honor, calling her a 'pioneer of marine science studies in the Gulf of Mexico.'

One other carcinologist who used the Grand Isle facility was the copepodologist and founding editor of the *Journal of Crustacean Biology*, Arthur G. Humes (b.1916), who collected *Callinectes sapidus* there in summers of 1939 and 1940 for his monograph on *Carcinonemertes* (1942).

Enough of my stories!

My association with this project has been a revelation in more ways than I dare mention. I still believe that the history of carcinology offers a rich area of research for historians of science as well for carcinologists themselves. This volume will document that the fields of evolutionary biology, embryology, genetics, morphology, and biogeography, among others, have benefited from the efforts of zoologists who for at least part of their careers were carcinologists. I also hope it will show, especially to graduate students, that in most areas of carcinological research a historical perspective is essential to a complete understanding of previous research.

ACKNOWLEDGEMENTS

All participants in the symposium and proceedings contributed more than just their papers to the success of this project. I thank all of them for their support, encouragement, and advice. To this group I add the names of others who generously gave me information, advice, and support: R. Brusca, D. Damkaer, D. Felder, J. Martin, G. Poore, and S. Wang. I warmly thank D. Frey for presenting a paper at the symposium and for the friendship extended by him and his wife to my wife and me. I also thank R. Brusca, D. Felder, D. Frey, J. Martin, and F. Schram for reviewing manuscripts; R. Condrey, W. Kelso, S. Morris, A. Rutherford, and R.E. Turner for help with problems of 'electronic manuscripts;' and L. Bonvillain for retyping certain tables and manuscripts. I thank D. Walker and W.A. Van Engle for supplying information on women carcinologists at the Virginia Fisheries Laboratory; Harada Eiji for providing biographical information on Watase Shozaburo; and W. Deiss, Smithsonian Institution, for supplying transcripts of the Oral History Project Interviews with Fenner Chace and Horton H. Hobbs Jr. Special thanks to my friend R. Bauer for his patient listening and advice as I endlessly recounted the vicissitudes of 4 years of work on the history of carcinology. And, last but not least, my deepest thanks to my wife, Meredith, and our children, Aimée and Frank, for their support during this project, and my sincerest apologies to them for my many deviations from the normal during the ordeal.

NOTES

1   Transcript of Oral History Project Interviews (1977) with Dr. Fenner A. Chace Jr., Smithsonian Institution Archives, SIA9514:69.
2   Ibid.:70.
3   Transcript of Oral History Project Interview (1976) with Dr. Horton H. Hobbs Jr., Smithsonian Institution Archives, SIA9509:13-14.

REFERENCES

Agassiz, G.R. (ed.) 1913. *Letters and Recollections of Alexander Agassiz*. Boston: Houghton Mifflin.
Allard, D.C. 1990. The Fish Commission laboratory and its influence on the founding of the Marine Biological Laboratory. *J. Hist. Biol.* 23: 251-270.
Behre, E.H. 1950. Annotated list of the fauna of the Grand Isle region, 1928-1946. *Occas. Pap. Mar. Lab., La. St. Univ.* 6: 1-66.
Blackwelder, R.E. 1979. *The Zest for Life: The Life of Waldo LaSalle Schmitt*. Lawrence: Allen Press
Barzun, J. & H.F. Graff 1977. *The Modern Researcher*. 3rd.ed. New York: Harcourt Brace Jovanovich.
Chace, F.A. Jr. 1983. The caridean shrimps (Crustacea: Decapoda) of the Albatross Philippine Expedition, 1907-1910, Part 1: Family Stylodactylidae. *Smith. Contr. Zool.* 381:1-21.

Chace, F.A. Jr. 1988. The caridean shrimps (Crustacea: Decapoda) of the Albatross Philippine Expedition, 1907-1910. Part 5: Family Alpheidae. *Smith. Contr. Zool.* 466: 1-99.

Cole, A.B. (ed.) 1947. *Yankee Surveyors in the Shogun's Seas.* Reprint 1968. New York: Greenwood Press.

Galtsoff, P.S. 1962. The story of the Bureau of Commercial Fisheries Biological Laboratory Woods Hole, Massachusetts. *US Fish. Wildl. Serv. Circ.* 145: 1-119.

Garfield, E. 1982. *Citation Classics*--four years of the human side of science. *Essays of an Information Scientist* 5:123-134.

Gudger, E.W. 1912. George Marcgrave, the first student of American natural history. *Pop. Sci. Month.* Sept.: 250-274.

Hedgpeth, J.W. 1947. The steamer Albatross. *Sci. Month.* 65: 17-22.

Hedgpeth, J.W. & W.L. Schmitt 1945. The United States Fish Commission Steamer Albatross. *Amer. Neptune* 5:5-26.

Hobbs, H.H. Jr. 1942. A generic revision of the crayfishes of the subfamily Cambarinae (Decapoda, Astacidae) with the description of a new genus and species. *Amer. Midl. Nat.* 28: 334-357.

Humes, A.G. 1942. The morphology, taxonomy, and bionomics of the nemertine genus *Carcinonemertes. Illinois. Biol. Monogr.* 18 (4): 1-105.

Hyman, L.H. 1940. *The Invertebrates: Protozoa through Ctenophora,* Vol.I. New York: McGraw-Hill.

Hyman, L.H. 1959. *The Invertebrates: Smaller Coelomate Groups,* Vol.V. New York: McGraw-Hill.

Jordan, D.S. 1922. *The Days of a Man,* Vol.2. New York: World Book.

Kaburaki, T. 1926. On the fauna of Japan. In Shinjo, S. (ed.), *Scientific Japan, Past and Present*: 105-135. Kyoto.

Linton, E. 1915. Reminiscences of the Woods Hole laboratory of the Bureau of Fisheries, 1882-89. *Science* n.s.41: 737-753.

Manning, R.B. & D.L. Felder 1989. The *Pinnixa cristata* complex in the Western Atlantic, with a description of two new species (Crustacea: Decapoda: Pinnotheridae). *Smith. Contr. Zool.* 473: 1-26.

Mayer, A.G. 1918. Biographical memoir of William Stimpson. *Biog. Mem. Nat. Acad. Sci.* 8:419-433.

Rathbun, M.J. 1902. Japanese stalk-eyed crustaceans. *Proc. US Nat Mus.* 26:23-55.

Rathbun, M.J. 1909. Dr. Giuseppe Nobili. *Science* n.s. 29: 249.

Reintjes, J.W. 1973. Fisheries research steamer *Fish Hawk. Mar. Fish. Rev.* 35 (11):14-17.

Sandoz, M. & S.H. Hopkins 1947. Early life history of the oyster crab, *Pinnotheres ostreum* (Say). *Biol. Bull.* 93:250-258.

Sandoz, M. & R. Rogers 1944. The effect of environmental factors on the hatching, moulting, and survival of zoea larvae of the blue crab, *Callinectes sapidus* Rathbun. *Ecology* 25:216-228.

Thorndike, L. 1937. [Book review: *The Human Comedy: As Devised and Directed by Mankind Itself* by H.E. Barnes] *J. Mod. Hist.* 9:367-369.

Watanabe, M. 1990. *The Japanese and Western Science.* Philadelphia: Univ. Penn. Press.

Winsor, M.P. 1991. *Reading the Shape of Nature.* Chicago: Univ. Chicago Press.

Frank M. Truesdale
School of Forestry, Wildlife, and Fisheries
Louisiana State University
Agricultural Center
Louisiana Agricultural Experiment Station
Baton Rouge
June 1992

Toward the history of pre-Linnean carcinology in Brazil

Marcos S. Tavares

Universidade Santa Ursula, Rio de Janeiro and Museum National d'Histoire Naturelle, Paris

1 INTRODUCTION

The dipterist and historian Nelson Papavero entrusted to me for study a collection of color transparencies of crustaceans figured in the *Theatrum Rerum Naturalium Brasiliae*, a collection of oil paintings and watercolors produced during the Nassau-Siegen Mission, 1637-1644, and presumed destroyed during World War II. This motivated me to piece together the extensive data I had compiled on pre-Linnean carcinology in Brazil. Examination of the relevant literature showed that there was very little information, despite the richness of pre-Linnean texts on the Brazilian fauna. My investigations revealed, however, that there was a wealth of information in unexplored texts.

The present contribution is divided into four parts: 1) Narratives of voyages of early explorers and missionaries including the letters exchanged with Europe; 2) Dutch explorers in Brazil during the Nassau-Siegen Mission; 3) the influence of literature produced in Brazil on Linnaeus' *Systema Naturae*; 4) influence of Marcgraf's *Historia Naturalis Brasiliae* on Sebas' *Thesauri*. Pre-Linnean authors are discussed in chronological order with biographical notes and a brief review of their work. Space constraints necessitated the omission of some details concerning the Dutch period in Brazil, and only relevant bibliographical details are given.

The works reviewed herein represent the most important documents concerning pre-Linnean carcinology in Brazil. However, it is possible that additional information will come to light from paintings of the period, other than those of the artists of Nassau-Siegen. For example, as a result of the activities carried out by the Dutch West India Company in the New World, the exhibition of American species (mainly birds) in Europe was rather common. The painting by Dutch artist Frans Snyders (1579-1657), 'The parrots and other birds' (Grenoble Museum), illustrates two endemic species of parrots from Brazil, *Ara ararauna* (Linnaeus 1758) and *Ara chloroptera* Gray, 1859, but Snyders apparently never visited Brazil. He probably saw an exhibition of American species and painted two of them. Albertin (1985) and Whitehead & Boeseman (1989) provided a list of further iconographic sources concerning the period and I looked at those which could possibly relate to carcinology. They are as follows: 1) Kupferstich-Kabinett, Dresden: a) Anonymous drawings, 18th century, 1 vol. (some copied from Z. Wagner), b) Samuel Niedenthal's drawings, 3 vols. (some of them copied from Z. Wagner), c) Abbé Joseph Lebisch's drawings 1 vol. (some of them copied from Z. Wagner); 2) Sächsische Landesbibliothek, Dresden: Jacob Grieb's Naturalien-Buch, 1 vol. (some of them copied from

Z. Wagner plus some *Theatri* paintings); 3) Forschungsbibliothek, Gotha: Caspar Schmalkal-
den's drawings, 1 vol. (? copied from *Theatrum*).

2   EARLY EXPLORERS AND MISSIONARIES

Brazilian carcinology was begun by the explorers and missionaries who came to Brazil very
early in the 16th century and published the first descriptions of the Brazilian fauna. These
documents mainly took the form of letters exchanged with Europe and narratives of voyages.
There was neither an intention to produce an inventory of the fauna, nor a catalogue. The plants
and animals were usually referred to according to their utility (edible, medicinal, noxious, etc.).

Often these early texts contained fanciful notions. Jean de Léry's (1578) voyage narrative
refers to the tapir as a hybrid between a cow and an ass. He also described an Indian religious
ceremony in which imaginary animals and beasts, part human, part animal, appeared. Yves
d'Evreux (1615) was also unable to separate fantasy and reality. He accepted the existence of
birds which 'ont le bec faict comme ces couteaux qui se replient dans leur manche, qu'on
appelle communement Iambettes & Rasoir: ainsi leurs becs sont inutiles àles pouvoir d'aucune
norriture, & leurs becs trenchans ne servent d'autre chose qu'à leur donner du passetemps, lors
qu'ils se promenent és rivages de la.Mer, rencontrans en leur chemin quelque Poisson courant
au bord, ils le découpent en deux, ainsi qu'avec un couteau, & se contentent de cela.' These
fantasies reflect European attitudes of that time toward nature . Still, under the strong influence
of the cosmogony established during the Middle Ages (according to Voltaire 'the age of faith',
meaning a period of superstitions), people often failed to distinguish between fantasy and
reality and yet despite this the accuracy of some drawings and texts is surprising.

In some works of the pre-Linnean period, the fauna of the New World (mainly South
America) is already clearly recognized as being different from that of the other continents, e.g.,
sloths, opossums, armadillos, and giant anteaters. More than 3 centuries later this region appeared in Sclater's (1858) biogeographical classification of the world (based upon the distributions of birds) as a separate faunal region, and some years later Wallace (1876) stated that such
faunistic peculiarities might apply not only to birds but also to other animals.

2.1   *Pero Vaz de Caminha*

Pero Vaz de Caminha was the scribe of the Portuguese fleet which, under the command of Pedro
Alvares Cabral, reached Porto Seguro, Bahia, on the northeastern Brazilian coast in April 1500.
Just after its discovery, he described the new land of Brazil in a letter sent to Dom Manuel,
Emperor of Portugal (Cortesão 1943). This is the first document concerning Brazil. From an
ethnological point of view this letter is very instructive because it describes the Indians' physical
aspects, dress, habits, food, etc., but his comments about the fauna and flora are rather vague.
The absence of details makes identification very difficult. His only reference to crustaceans is
the mention of the catch of some robust and short shrimp. Leitão (1941) suggested these shrimp
could have been either the freshwater *Macrobrachium acanthurus* (Wiegmann 1836), or the
marine *Squilla* Fabricius 1787, or *Lysiosquilla* Dana 1852. However, according to Caminha's
letter they landed beside the mouth of a small river where they stayed for some time. In such
areas, the most abundant and conspicuous shrimps are those of the genus *Macrobrachium* Bate
1868. Gabriel Soares de Sousa was in Bahia just 69 years later, and like Caminha he employed
the term 'grosso' (robust) to describe a shrimp called by the Indians 'potiuaçu' (Sousa, 1987). It

seems likely that both Gabriel Sousa and Caminha encountered the same species of shrimp. Among the color plates of the *Icones Aquatilium* (see below) there is an illustration of a shrimp called 'potiguaçu' (='potiuaçu'), differences in spelling are due to the difficulties inherent in translating the Indians' speech sounds), which corresponds very well with *Macrobrachium carcinus* (Linnaeus 1758). *M. carcinus* has the tip of the rostrum somewhat curved upwards, bearing in its upper margin 11 to 14 teeth, and on its lower margin 3 to 4 teeth (Holthuis 1952). The rostrum of the 'potiguaçu' in the illustration is somewhat curved upwards bearing 11 teeth on its upper margin and 4 teeth on its lower margin. According to Holthuis (1952) the best characters for identification of members of the genus *Macrobrachium* are the shape of the rostrum and second legs. Unfortunately it is not possible to discern any details of the second leg on the illustration of the 'potiguaçu'. However, the color pattern on the illustration of the 'potiguaçu', dark brown with lighter spots along the sides of the carapace and abdomen, is very much the same as that of *M. carcinus* (for a complete description of the color of this species see Holthuis 1952). Further support for the identification of the shrimp as *M. carcinus* is that the area visited by Pero Vaz de Caminha is in the distributional range of *M. carcinus*, fresh and brackish waters from Florida to southern Brazil.

2.2   *José de Anchieta*

Many explorers who visited Brazil for only a short time found it difficult to interpret the native languages, and the spelling of native names often differed among these visitors (see for example Claude d'Abbeville). However, due to the monumental work of José de Anchieta the spelling of the native names of species mentioned by pre-Linnean authors does not differ very much from one author to another. Each Indian group had its own language or dialect and the problem in communication made the missionaries task of converting them to Christianity difficult. Consequently, Anchieta organized a general language called Tupi, which was not the language of a particular Indian group but a synthesis of several dialects into a single grammar. This language was imposed by Jesuits throughout the country and, as it enabled different tribes of Indians, explorers, and missionaries to communicate with each other, Tupi consequently became a general language. Afterwards there was a hybridization between Tupi and the Guarani language (largely spoken in Paraguay, Uruguay, and Argentina), resulting in a wide-spread language called Tupi-Guarani. For this reason numerous geographical names of rivers, plants, and animals employed by pre-Linnean authors are still in use (for example: 'aratu', 'guanhamu').

José de Anchieta was born on 19 March 1534, in São Cristóvão de Laguna, Canary Islands. He was 16 years old when he entered the University of Coimbra, and one year later he was ordained a Jesuit friar. On 8 May 1553, Anchieta embarked for Brazil with seven other missionaries arriving on 15 July. He lived in Brazil for 44 years, where he died on 9 July 1597. In a letter written in May 1560 (first published in 1799) to his superior, he briefly described the natural history of São Vicente (now São Paulo). Anchieta (from the 1933 translation) reported on climatology and described some species of mammals, birds, reptiles, fishes, crustaceans, insects, and plants. His comments about crustaceans are very brief. He recognized a variety of crabs, but only four species were described: Portunidae sp. (swimming paddles), *Uca* sp. (chelipedes, and burrow), *Ucides cordatus* (Linnaeus 1763) (color, and burrow) and *Cardiosoma guanhumi* Latreille, 1828 (size, color, and burrow). For the latter two species these are the first published descriptions.

2.3   *First attempt at settlement by France*

Vice-admiral Nicolas Durant de Villegagnon was an enthusiastic supporter of the reformation and was extremely eager to establish a French settlement in Brazil. He obtained support from Henry II for an expedition, and in May 1555 he sailed to Brazil, reaching Rio de Janerio 6 months later after a series of difficulties. Among the expeditioners were the friars André Thevet and Jean de Léry, whose narrative of the voyage is one of the earliest documents about Brazil. André Thevet (1502-1592) spent 3 months collecting information on ethnology and natural history, but no information about crustaceans is contained in his work published in 1558.

2.3.1   *Jean de Léry*

Jean de Léry (1534-1611) was born in La Margelle, neighboring the abbey of Saint-Seine de Bourgogne. Also an enthusiastic follower of the reformation movement he went to Geneva at the age of 18, attending a course on theology by Calvin. In 1557, Calvin received a letter form the vice-admiral Nicolas Durant de Villegagnon, a colleague at the University of Paris, offering an opportunity to spread the cause of Calvinism in the New World. Calvin accepted Villegagnon's invitation and sent an official party of 14, headed by P. Richier and G. Chartier. Among them was Jean de Léry, future narrator of the mission. Later in 1557, Jean de Léry embarked aboard the Grande Roberge, one of the three ships under the command of Bois-Le-Comte, nephew of Villegagnon, landing at Rio de Janeiro on 26 February 1557. After 11 months in Brazil, Léry returned to France and then to Geneva to finish his theological studies. Initially, he did not publish his notes on the voyage, but encouraged by colleagues he gave a manuscript to a friend in 1563. It was unfortunately lost, but based on his own drafts, Léry started a second version of the voyage narrative. In 1572, disturbed by religious conflicts, he left the city of La Charité-sur-Loire and took refuge in Sancerre, and during this time the second manuscript was also lost. In 1576, the first version of the narrative was found and finally published, 18 years after Léry's voyage. Léry died in Berna in 1611. His narrative is largely devoted to aspects of the life of the Indians and to religious questions, and his accounts of the fauna and flora are scanty. He referred to some birds, bats, bees, and flies. The only crustacean mentioned by Léry (1578) is the 'ussa' (='uça', name still used in Brazil) which corresponds to the ocypodid crab *Ucides cordatus* (Linnaeus). Léry remarked that this species appeared in great numbers on the mangroves. His observation that burrows of this species are usually dug next to mangrove trees has been confirmed by Tavares & Albuquerque (1989).

2.4   *Gabriel Soares de Sousa*

Gabriel Soares de Sousa was born around 1540, most probably in Lisbon. In 1569, the expedition of Francisco Barreto, on its way to Mozambique, visited Brazil. Some of the expeditioners decided to stay, among them Gabriel de Sousa who established a farm in Bahia and lived prosperously for 17 years. His brother, João Coelho de Sousa, travelled for 3 years in the interior of Brazil finding gold and precious stones and mapping his discoveries. When João Coelho de Sousa died, Gabriel Soares became the owner of the map of his brother's discoveries. This inheritance produced a deep change in his life and influenced enormously the study of Brazilian natural history. At the end of August 1584 Gabriel de Sousa left Bahia for Spain to ask Philip II, at that time sovereign of Portugal, to support an expedition to the interior of Brazil. It took a full 6 years before his request was granted. During this period de Sousa wrote his monumental *Tratado descritivo do Brasil em 1587* (literally, Descriptive treatise of Brazil in 1587) and on 1 May 1587 gave it to the influential politician Dom Cristovão de Moura. This work (first published in 1851) was intended mainly to bring to the attention of the authorities the richness of the land, the risks of its occupation by foreigners, and consequently the benefits of an expedition to better explore the country. In December 1590, de Sousa finally succeeded in obtaining support from Phillip II and on 7 April 1591 sailed from Lisbon on board the Grifo Dourado accompanied by 360 men. The ship sank in Vasa-Barril, a sheltered locality near Salvador, Bahia, but de Sousa and most of the expedition members survived. Arriving in Salvador de Sousa received some support from Dom Francisco de Souza and started the inland exploration. Gabriel de Sousa died of malaria before finding the mines discovered by his brother.

Gabriel de Sousa's book was the result of 17 years spent in the Brazilian territory. It contains detailed information on several subjects such as physical geography, mineralogy, ethnology, zoology, and botany. The crustaceans referred to by de Sousa (1987) are as follows: On page 288 he mentions the 'potipema' which seems to be the same 'potipema' of Marcgraf (1648), and which belongs to the species *Macrobrachium carcinus* (Linnaeus).

Two species are mentioned on page 289: The spiny lobster *Panulirus echinatus* Smith, 1869, referred to as 'potiquiquiá,' which appears in Marcgraf's work as 'Potiquiquiya,' and *Ucides cordatus* mentioned under its native name 'uça.' He also mentioned that about 300 crabs, caught by only one fisherman, were cooked every day to feed the slaves on the farms. Far more interesting is the correlation that he made between the molting and the reproductive season. He pointed out that females are ovigerous once a year and molt around this time.

Crabs called 'siris' are described on page 290 as having squared chelipedes with long fingers. The native name 'siri' is still commonly employed in Brazil for the swimming crabs of the genus *Callinectes* Stimpson, 1860. It means literally 'one who slides,' an allusion, in this case, to the way the crab swims. Among the seven species recorded from Brazil, *C. danae* Smith, 1869, is the most common in this area. Three other genera of portunids occur in Bahia: *Portunus* Weber, 1795, with seven species known under 'siri-candeia;' *Arenaeus* Dana, 1851, with only one species largely known as 'siri-chita;' and *Cronius* Stimpson, 1860, with two species, for which there is no native name, as far as I know.

Gabriel de Sousa also commented on two common species in mangrove areas. He described 'aratus' [*Aratus pisonii* (H. Milne Edwards 1837)] as crabs with a rather soft carapace, which live on mangrove trees, eating the leaves. He also discussed the crab *Goniopsis cruentata* (Latreille 1803), under the native name 'guaiararas,' which in his words is a 'full colored species living in brackish waters.'

He mentioned another species, the 'guaias,' saying only it has short walking legs and chelipedes, and a very nice taste. Although such vague descriptions are difficult to interpret, the local fisherman apply the name 'guaiá' to the xanthid *Menippe nodifrons* Stimpson, 1859, rather than to any other species. The last crustacean referred to on page 290 is the 'guaiauças,' described as a white crab which digs deep burrows on sand beaches in the intertidal zone, clearly the ghost crab *Ocypoda quadrata* (Fabricius 1787). According to de Sousa this species was used by the Indians as bait for fishing. In a note on the 'sururu' on page 292, one of the species of mollusks of the genus *Mytilus* Linnaeus, 1758, occurring in Brazil, he mentions some very small crabs living inside, probably peacrabs of the family Pinnotheridae. Four genera and seven species of pinnotherids are recorded from Bahia and any of them could be the peacrab mentioned by de Sousa.

Four freshwater shrimps are recorded on page 297: 'poti', 'aratus', 'araturé' (whose descriptions leave no doubt that they belong to the genus *Macrobrachium*) and 'potiuaçu,' which is the same as Marcgraf's 'potiguaçu,' *Macrobrachium carcinus* (Linnaeus). Finally on page 298 he described the land crab *Cardiosoma guanhumi* Latreille: 'This land crab the Indians call 'guoanhamu' lives in wet soils not too far from the sea...it is large, blue in color with the carapace and walking legs bright; males have large chelipeds and are bigger than females...its burrows are deep, and from February through March the females are ovigerous.'

2.5   *Second attempt at settlement by France*

The first attempt to establish a French settlement in southeastern Brazil resulted in the works of André Thevet and Jean de Léry. More than 50 years later the French made a second attempt at settlement which was also unsuccessful but similarly resulted in two useful publications on Brazilian natural history.

At the end of the 16th century, a French fleet under the command of Jaques Rifault, when departing Brazil, left part of its crew behind, because two of the three ships of the fleet sank around Santa Ana Island, Maranhão. Some of the crew were captured by the Portuguese governor Feliciao Coelho. Among these was Charles des Vaux, who upon his return to France tried to convince the French authorities of the benefits of a settlement in Brazil. Daniel de la Touche and Jean Moguet were organizing an expedition to settle Cayenne, where they had visited in 1604; however, after hearing from Charles des Vaux, they changed their minds and decided to go to northern Brazil. Such an expedition was prepared, financed, and military support obtained. To spread Christianity in the region, four Capuchin friars joined the expedition: Yves d'Evreux, as the superior of the religious group, Claude d'Abbeville, Arsène de Paris, and Ambroise d'Amiens. D'Evreux (1570-1650) stayed 2 years in Brazil, and his narrative is chiefly devoted to ethnology and religious subjects. He showed little interest in native names and descriptions of species. Crustaceans are only very vaguely cited: '...cet oyseau choisit une viande singuliere et speciale, à laquele seul il tend son vol, sçavoir est, des Crabes, ou Escrevisses de mer.' (D'Evreux 1615). Claude d'Abbeville however made a more significant contribution.

2.5.1   *Claude d'Abbeville*
Claude Foullon was born in Abbeville, and in 1593 was ordained a Capuchin friar. On 19 March 1612, he embarked for northeastern Brazil on board the Regente, the flagship of a fleet of three. After a series of difficulties the fleet reached Fernando de Noronha Island, off the coast of Brazil, where they stayed for 15 days. On 21 July, nearly 5 months after sailing from the French harbor of Cancale, the expedition finally reached Santa Ana Island, on the coast of Maranhão. Claude d'Abbeville stayed in Brazil for about 4 months, sailing back to Paris on 7 December onboard the Regente. His narrative was first published in 1614 and has been re-issued seven times in French and Portuguese.

Claude d'Abbeville's work (from 1975 translation) contains mainly the description of his experiences among the Indians; observations on their physical form and appearance; and short descriptions of local zoology, botany, and geography. His work contains a surprising volume of information for having been compiled in only a few short months. He devoted four chapters to the fauna of Maranhão: one on birds; a second on fishes (among which he put the crustaceans, crocodiles, two species of lizards and one species of mammal); a third about 'terrestrial animals' (mammals, reptiles and amphibians); and in the last chapter, 'imperfect animals of Maranhão,' he describe some flies, mosquitoes, ants, fleas, and crickets.

Claude d'Abbeville mentioned seven species of crustaceans, all brachyuran crabs. As he was not familiar with the Tupi-Guarani language, his spelling of native names for these species is very peculiar and sometimes unintelligible. Two species are mentioned on page 197: *Cardiosoma guanhumi* Latreille, referred to as 'oégnomoin' (instead of 'guanhamu'); and *Ucides cordatus* (Linnaeus) as 'oussa' (instead of 'uçá'). Of the first species he gives a brief description of the coloration and burrowing behavior, while the latter species, a mangrove crab, is described by its red and hairy walking legs. On page 198, he goes on to describe five other species of crabs. The 'ouia ouässou' (normally spelled 'guaiá-açú) is probably *Menippe nodifrons* Stimpson. The 'aratou' (instead of 'aratu') is described as marine, yellow and blue in color, and said to be little smaller than the preceding species – by this color pattern it should be *Eriphia gonagra* (Fabricius 1781). Curiously Claude d'Abbeville applied the native name 'aratu,' typically used for *Aratus pisonii* (H. Milne Edwards), to this species. *Callinectes* sp. is described under 'siry' (instead of 'siri') as a marine, blue and white crab. The 'aouära-oussa' (the appropriate spelling is 'grauçá') is described as a white crab which digs its burrows on sand beaches, clearly the ghost crab *Ocypode quadrata* (Fabricius). Claude d'Abbeville showed he was a careful observer by mentioning that the ghost crab brings fragments of stone or shells to the burrow.

The last two references to crustaceans are difficult to interpret. The first is the 'ouraroup' (it was not possible to find a corresponding term within the Tupi-Guarani language), which according to d'Abbeville is bigger than a fist, lives only in freshwater (perhaps a species of the fresh water family Trichodactylidae) and is preyed upon by the 'oussapeve' (=uçá-peva, a term which literally means flattened crab). It is possible that d'Abbeville's flattened crab was *Callinectes sapidus* Rathbun, 1895, which can occur in freshwater far from the sea.

2.6   *Ambrósio Fernandes Brandão*

Ambrósio Fernandes Brandão is reputed to be the author of the *Dialogues of the Attributes of Brazil* written in 1618. Unfortunately little is known about him except that he was Portuguese and in 1583 lived in the state of Pernambuco. Before 1613, he lived in the state of Paraba where the *Dialogues* were probably written. He was an infantry captain and fought against Indian tribes and the French. It is not known when he died, but it was before the Dutch arrived in Brazil in 1624.

Brandão chose a rather original style to describe the natural history of northeastern Brazil. He wrote six dialogues between the characters 'Alvino' and 'Brandônio.' Alvino symbolized a newly arrived foreigner who could well have been Nuno Alvares, Brandão's colleague. Brandônio, the main interlocutor of the dialogues, was in Brazil in 1583 and was probably Brandão himself. The zoology is contained in the fifth dialogue, pages 201-238 of the fourth edition (1977), and here the fauna is separated into four categories according to the Aristotelian elements: fire, air, water, and earth. The fire was said to be sterile despite the 'story about the lizard called salamander, which is thought to live in it.' He described several species of birds, mentioned the migration northward of a species of butterfly, referred to marine and freshwater fishes, turtles, crocodiles, mollusks, and crustaceans, and finally described some species of land mammals and snakes.

The crustaceans were briefly described. He referred (p. 214) to the abundance of marine (?Penaeidae) and freshwater shrimps (Palaemonidae); mentioned (p. 218) certain kinds of 'lagostins' which 'can be easily obtained on the reefs during the low tide' (according to Garcia 1977, such lobsters belong to the family Scyllaridae, but it is likely that it is the same reef lobster referred to by Gabriel de Sousa under 'potiquiquiá', i.e., *Panulirus echinatus* Smith). Finally he listed (page 220) seven species of brachyuran crabs, remarking that they were utilized as food by the Indians and slaves. The species were referred to by their native names: 'uçá' [*Ucides cordatus* (Linnaeus)]; 'siri,' (*Callinectes* sp.); 'goajá' (instead of 'guajá or 'guaiá'), which is probably *Menippe nodifrons* Stimpson; 'guoazaranha' ('guoa'= crab; 'aranha'= spider), perhaps a spider crab, family Majidae; 'aratu,' (*Aratus pisonii* (H. Milne Edwards)); 'garauçá' (instead of 'graçá'), for *Ocypode quadrata* (Fabricius); and 'guanhamus', (*Cardiosoma guanhumi* Latreille).

2.7 *Cristóvão de Lisboa*

Cristóvão Severin was born on 25 July 1583 in Lisbon. At an early age he went to the University of Evora, and at 19 was ordained a Capuchin friar. On 25 March 1624, he sailed to Brazil to spread Christianity, arriving in the State of Pernambuco on 2 May and moving later to the State of Maranhão. He left Brazil for Lisbon in 1635 where he died at the age of 69.

Between 1624 and 1627 he wrote a four-volume manuscript about the social and natural history of Maranhão, which was lost during the Lisbon earthquake in 1755. The volume pertaining to the fauna and flora was found in 1933 and published for the first time in 1967 (Walter 1967). It contains 157 numbered plates (plates 29, 30, 31, 35, 61, 70 missing) with 260 poorly prepared drawings of animals and plants (24 mammals, 67 birds, 7 reptiles, 105 fishes, 3 crustaceans and 54 plants). No descriptions accompany the illustrations of the three crustaceans on plates 25 ('poty'), 26 ('ciry'), and 42 (not named) (Plates 1a-c). The figure of the 'poty' is inadequate for identification, however, de Lisboa's 'ciry' surely belongs to the genus *Callinectes* Stimpson 1860. I tentatively assign the shrimp figured on Lisboa's plate 42, to the genus *Macrobrachium* Bate, 1868. The specimen shows a strong and chelate second leg, and no chelae on the third leg; however, the pleura of the second abdominal somite does not overlap those of the first and third segments. (see Almaça this volume)

3   DUTCH EXPLORERS IN BRAZIL

The Dutch occupation of Brazil was primarily a result of the wider consolidation of their mercantile interests. Dutch trade and colonization had been mainly directed towards the Orient where they had the East India Company. However, they started expanding westward in 1621, founding the Dutch West India Company, which unlike the East India Company was strongly political and formed primarily to harass the settlements of Spain and Portugal in the New World. In 1624, they expelled the Portuguese from the Cape Verde Islands and went on to the New World, occupying some Caribbean islands, Suriname, and the northeastern coast of Brazil.

Johan Maurits (or Johann Moritz) van Nassau-Siegen, the 13th of 25 children of the two marriages of Johann der Mittleren, was born in 1604 in the castle of Dillenburg. He was descended in an almost direct line from the Emperor Charlemagne. At the age of 12 he went to university, first in Basel and then in Geneva, where he was educated in Latin, French, and Italian while also studying law and economics. In 1613, Johann des Mittleren died leaving one-third of the small territory of Siegen to Johan Maurits and the remanding two-thirds to the other two sons of his first marriage. The oldest son was upset over not inheriting more of the estate and tried to implicate Johan Maurits in a Protestant plot against the Emperor. Anticipating that serious problems could develop, Johan Maurits moved to Haia where he met several influential people in the court of Queen Elizabeth. In 1636, the Dutch possessions in the New World were entrusted to him as Governor and 1 year later he went to Pernambuco, taking with him a team of artists and naturalists. Johan Maurits van Nassau-Siegen governed northeastern Brazil from

1637 to 1644, a period in which much was learned of the natural history of Brazil, mainly by the contributions of the naturalists Georg Marcgraf and William Piso, and the artists Albert Eckhout, Frans Post, and the amateur Zacharias Wagener. The aims of these Dutch investigators in Brazil were in all ways completely different from those of earlier explorers and missionaries, as they intended to produce a detailed document on the local natural history. This was to be a catalog of the plants, animals, diseases, local medicines, ethnology, astronomy, topography, scenery, and native peoples 96 perhaps for use by the West India Company, since the existing information, when available, was considered inadequate. However, in 1640 the Dutch settlement began to deteriorate. Coincidentally, Portugal again became independent from Spain and concentrated much attention on Brazil. In 1644, Nassau-Siegen failed to obtain additional military and financial aid from the Board of Directors of the West India Company, and he returned to Holland. In 1654, Portugal resumed control of Pernambuco and in 1661 the West India Company finally left Brazil. Count Johan Maurits van Nassau-Siegen died in Cleves, Germany, in 1679.

3.1 *Johan de Laet*

Johan de Laet (1595-1649) was the managing director of the Dutch West India Company. His interest in the New World led him to publish the *Novus Orbis* (1633) in which he described its geographical limits and political boundaries. Laet's work is subdivided into 18 volumes, Brazil being covered in volume 15 (chapters I-XXVII). His descriptions and illustrations of Brazil were largely based on Thevet's (1558) and Léry's (1578) narratives, as well as on a certain 'authorem lusitanum.' Crustaceans, fishes, mangrove trees, and sea birds were mentioned together in chapter 13. Only three species of Crustacea were mentioned: 'Uzae' ('Uzas' or 'Ucaa,' Laet 1641), *Ucides cordatus* (Linnaeus); 'Guainumu', *Cardiosoma guanhumi* Latreille; and 'Aratu', *Aratus pisonii* (H. Milne Edwards).

3.2 *Theatrum Rerum Naturalium Brasiliae*

In the 19th century, the seven volumes of watercolors and oil paintings produced by the artists of the Nassau-Siegen Mission were combined with a large collection of paintings called *Libri Picturati* making 144 volumes in total. The paintings concerning Brazil, which depict nearly 2000 Brazilian plants and animals, were from then on referred to as *Libri Picturati A 32-38*. They are separated into three parts as follows: 1) The *Libri Picturati A 32-35* (also called *Theatrum*, mainly oil paintings) which is subdivided into four volumes: *Icones Aquatilium, Icones Volatilium, Icones Animalium*, and *Icones Vegetalibum*, these volumes forming the *Theatrum Rerum Natualium Brasiliae*, attributed to Eckhout; 2) the *Libri Picturati A 36-37* (two 'Handbooks' of watercolors), also known as *Libri principis* because they are supposed to have been handled by Prince John Maurits van Nassau-Siegen and have some notes attributed to him (the drawings of the second part have been attributed to Georg Marcgraf and to Zacharias Wagener); and 3) *Liber Picturatus A 38* concerning only Brazil, probably by the hand of Eckhout, included accidentally in the *Miscellanea Cleyeri* between 1660-1664 by Christian Mentzel, physician of the Elector of Brandenburg. The *Miscellanea Cleyeri* was a collection of drawings by Andreas Cleyer of Kassel, a colleague of Mentzel, who had worked as physician to the East India Company.

Not long after Nassau-Siegen's term as Governor of Brazil, these collections were dispersed. In 1644, Nassau-Siegen returned to Holland to live in The Hague at the 'Mauritshuis' (Maurits' house) specially constructed for him, but in 1647 he moved to the principality of Cleves, Germany, as its administrator. In 1652, perhaps seeking to improve his political status, Nassau-Siegen gave a large part of his collection to Friedrich I, Elector of Brandenburg. Not long after, Nassau-Siegen obtained the position of 'Herrenmeister' and later in the same year the title of Prince. He also received a large amount of money which enabled him to buy the castle of Freudenberg. The gifts to the Elector of Brandenburg were quoted in a list written in September 1652, and included the collections of drawings and manuscripts about Brazil's fauna and flora accumulated during Nassau-Siegen's sojourn in Brazil.

In 1653, Friedrich III of Denmark decorated Nassau-Siegen with the order of the White Elephant – one of the most important Protestant distinctions of that time. One year later Nassau-Siegen returned the favor by giving the King a collection of paintings, most of them supposedly painted by Eckhout.

The third phase of dispersal of Nassau-Siegen's collections occurred in 1679 when he offered a large collection of paintings to Louis XIV of France, expecting in exchange a large amount of money to supply his financial needs. However, Nassau-Siegen died in Cleves that same year before receiving anything.

Three private collectors also received part of Nassau-Siegen's collections. These were Albert Seba and Friedrich Ruijsch of Amsterdam, and Olaus Worm of Denmark, whose collection was later incorporated in the collection of Friedrich III. In 1717, Seba's and Ruijsch's collections were bought by Czar Peter the Great and now are in Leningrad.

About 1717 the *Theatrum* and the *Libri Principis* were found in the Königlichen Bibliothek of Berlin, and the *Miscellanea Cleyeri* was found in the Berlin Library. Around 1828 the *Theatrum*, the *Libri Principis* and the *Miscellanea Cleyeri* were combined to form the *Libri Picturati*. During the early part of World War II the *Libri Picturati* was kept in the Preussische Staatsbibliothek, but when the library was damaged by British bombings the paintings were transferred to 24 sites in what would later be 'Eastern Europe,' and to 5 in 'Western Europe.' In 1941, as part of the plan to evacuate the manuscripts, the *Libri Picturati A 32-38* was transported to the castle of Fürstenstein in Silesia (now Poland). Two years later, when the Germans occupied the castle, the manuscripts were sent to the Benedictine monastery of Grüssau (now Krzeszó, neighboring Kracow). After the war, the material in the West was deposited (1957) in the Neue Staatsblibliothek Preussischer Kulturbesitz, West Berlin. The manuscripts sent to the monastery of Grüssau were thought to have been destroyed by a great fire there and remained lost until recently. Nevertheless, Brazil, the United States, and England made efforts to find the missing manuscripts. The Brazilian government asked the Soviet Union about them. In the United States, the director of the National Arts Foundation, Carleton Smith, who was also searching for original works of Bach, Beethoven, Haydn, and Mozart (later revealed to be housed with the Brazilian materials), organized an intensive search and, in 1974, discovered and examined the missing collections in Poland. He was permitted to examine them under the condition that he did not reveal where they were housed. Peter Whitehead (British Museum of Natural History) also learned that the collections moved to Silesia in 1941 were not destroyed in the fire of Grüssau, and were probably somewhere in Poland. In 1973, he began negotiations with the Polish government to search for the collections, and in 1977 he finally received official notification that they were safe in the Jagiellon Library, Kracow. In 1979, he was able to examine the drawings (Whitehead 1976, Albertin 1985). Whitehead (1976) outlined how such an important collection, over 300 years old, had remained almost inaccessible for study: 'Linnaeus never saw them, Bloch (1787) saw and used only the 'Handbooks', and Lichtenstein (1818, 1819, 1822, 1829), considered both the 'Handbooks' and the *Theatrum* but dealt only with (part of) the animals (mammals, birds, fishes, and amphibians). Martius (1853) claimed to have had exact copies made of the plants but these cannot be found (Whitehead 1973). The only studies made in the modern period are those of Ehrenreich (1894) on the ethnology and Schneider (1938) on the birds.' In February 1980, the historian Petronella J. Albertin went to Kracow to study the collection of illustrations concerning Brazil, gathering a complete series of photos of them. The photos pertaining to crustaceans provide the basis for the following study.

3.2.1 *Icones Aquatilium*

The *Icones Aquatilium* (*Liber Picturatus A 32*) includes 44 fish, 25 crustacean illustrations, and 1 human figure (Plates 2A-J; 3A-H; 4A-F) The other three volumes of the *Theatrum* are organized as follows: *Icones Volatilium*, comprising 111 bird illustrations; *Icones Animalium*, with 52 drawings of mammals, reptiles, insects, and spiders, and 13 illustrations of human figures; and the *Icones Vegetalibum*, with 172 illustrations of plants, flowers, and fruits. A detailed description of the *Libri Picturati A 32-38* was provided by Albertin (1985). The results of the study of the 25 crustacean color plates included in the *Icones Aquatilium* are summarized in Table 1.

Twenty five illustrations of crustaceans, representing 16 species (13 decapods, 1 stoma-topod, 2 barnacles), are included. Only two species, *Carpilius corallinus* (Herbst 1783), and *Plagusia depressa* (Fabricius 1775), of which there are two illustrations of each in the *Icones Aquatilium*, were not referred to in the *Historia Naturalis Brasiliae* (see below). Among the 17 illustrations of animals appearing in the frontispiece of the *Icones Aquatilium*, 2 are crustaceans:

Table 1. List of the crustaceans depicted in the *Icones Aquatilium* with original designations and corresponding binomials. Hb. = 'Handbooks;' Th. = 'Theatri;' Nn. = Not named.

Icones aquatilium	Remarks	Identification
Guaiâ apára	Hb., p.326	*Calappa ocellata* Holthuis 1958
?Guaiâ apára	Th., p.341	*Calappa ocellata* Holthuis 1958
Guaiâ	Th., p.339	*Calappa ocellata* Holthuis 1958
N.n.	Hb., p.328	*Persephona punctata* (Linnaeus 1758)
Guaiâ	Hb., p.338	*Mithrax (Mithrax) hispidus* (Herbst 1790)
çiri	Hb., p.352	*Callinectes danae* Smith 1869
Guaiâmû	Th., p.355	*Cardiosoma guanhumi* Latreille 1828
N.n.	Hb., p.320	*Cardiosoma guanhumi* Latreille 1828
N.n.	Hb., p.348	*Goniopsis cruentata* (Latreille 1803)
Potîquiquiya	Hb., p.384	*Panulirus echinatus* Smith 1869
Poticucuma	Th., p.319	*Panulirus echinatus* Smith 1869
çiriayeima	Th., p.335	*Parribacus antarcticus* (Lund 1793)
Potiquiquyixe	Th., p.335	*Parribacus antarcticus* (Lund 1793)
Tamarû guaçu	Hb., p.324	*Lysiosquilla scabricauda* (Latreille 1818)
Tamarû	Th., p.311	*Lysiosquilla scabricauda* (Latreille 1818)
Guaricurû	Th., p.331	*Atya scabra* (Leach 1815)
Potipema	Th., p.323	*Macrobrachium carcinus* (Linnaeus 1758)
Potiguaçu	Th., p.329	*Macrobrachium carcinus* (Linnaeus 1758)
Potîatinga	Th., p.329	*Macrobrachium carcinus* (Linnaeus 1758)
N.n.	Hb., p.314	*Lepas hillii* (Leach 1818)
N.n.	Hb., p.366	*Plagusia depressa* (Fabricius 1775)
N.n.	Hb., p.366	*Plagusia depressa* (Fabricius 1775)
Reñapiya (Punaru)	Th., p.15	*Balanus tintinnabulum* (Linnaeus 1758) and Pinnotheridae sp.
Illegible	Th., p.337	*Carpilius corallinus* (Herbst 1783)
N.n.	Th., p.337	*Carpilius corallinus* (Herbst 1783)

'Guaja,' *Carpilius corallinus* (Herbst), and 'Tamaru,' *Lysiosquilla scabricauda* (Lamarck); only 6 correspond to the illustrations of the *Theatrum*: 'Guaja,' 'Iaciatataguaçu,' 'Iacarepetim-buaba,' 'Camaruru,' 'Ajereba,' and 'Tamaru' (Plates 5A-B).

3.3   *Leningrad drawings*

The Leningrad drawings consist of a collection of 145 folio sheets presently housed in the Leningrad Academy of Science, depicting 283 Brazilian animals and 7 sheets of text (Albertin 1985; Whitehead & Boeseman 1989; Boeseman et al. 1990). Almost nothing is known about the early history of the Leningrad drawings, only that the collection was executed not before 1650 (the Nassau-Siegen Mission ended in 1644). After nearly 3 centuries the evidence of their existence surfaced in a brief account by Soloviev (1934). The drawings were found in 1832 by J.F. Brandt (1802-1879), director of the Zoological Institute, and sent to Johann Horkel (1763-1846), who after comparing them with the *Libri Picturati*, split them into two series: Series A (drawings matching the watercolors of the 'Handbooks': 21 folios), and Series B (drawings matching the oil paintings of the *Theatrum*: 124 folios). Afterwards they were sent back to Brandt and included in two portfolios erroneously labelled 'Lehmaniana,' and then fell again into obscurity. As outlined by Whitehead & Boeseman (1989) and Boeseman et al. (1990), the portfolios were transferred from the Zoological Institute to the Archives of the Academy of Sciences of the USSR, where they were found again by Dom Clemente da Silva Nigra in 1965 and exhibited in Rio de Janeiro in 1968.

A detailed description of the 145 folios of the Leningrad drawings is given by Whitehead & Boeseman (1989) and Boeseman et al. (1990). The folios contain 283 drawings of mammals, birds, fishes, reptiles, crustaceans, spiders, and insects. Albertin (1985), Whitehead & Boeseman (1989), and Boeseman et al. (1990) believe the Leningrad drawings to be poor copies of the *Libri Picturati A32-38*. Indeed, the quality of some of the drawings of the Leningrad collection is far inferior to the originals. All the crustaceans from the *Icones Aquatilium* were copied except for *Persephona punctata* (Linnaeus 1758). On the other hand, *Callinectes sapidus* Rathbun, and *Conchoderma virgatum* (Spendler 1790) are not represented in the *Theatrum* and 'Handbooks'. A preliminary account of the crustaceans depicted in the Leningrad collection was given by Hoogmoed et al. (1990) and Holthuis & Boeseman (1990), here I only add comments that arise from a comparison of some of the poorly executed outline drawings and oil paintings of the Leningrad series with the originals in the *Theatrum*. Hoogmoed et al. (1900) interpreted the outline drawing 'ciri' (folio 19, numbered 352 by Johann Horkel) as belonging to the species *Cronius ruber* (Lamarck, 1818); however, the detailed oil painting of the 'çiri' on page 352 of the 'Handbook' and also Marcgraf's 'ciri apoa' have exactly the same carapace shape, and the legs and chelipedes are arranged in the same way. This crab clearly belongs to the genus *Callinectes*, and is most probably *Callinectes danae* Smith, 1869. The crab figured on folio 43 was interpreted as *Eurypanopeus* sp.; however, the proximity of the eyes lead me to believe it is Pinnotheridae sp. Finally, according to my interpretation of the 'Potiguaçû' [*Macrobrachium acanthurus* (Wiegmann), see Holthuis & Boeseman 1990] and of the 'Potiatinga' [*Palaemon pandaliformes* (Stimpson 1871), see Holthuis & Boeseman 1990], both are the same species, *Macrobrachium carcinus* (Linnaeus).

3.4   *Georg Marcgraf*

Georg Marcgraf (also Marcgrave or Markgrave) was born on 20 September 1610 in Liebstadt, near Dresden. At an early age he learned Latin and Greek from his father and his maternal grandfather, who also took care to have Marcgraf instructed in music and the art of painting. He was 17 years old when he left his father's house for Germany to study mathematics, botany, chemistry, and medicine. In Stettin, he studied astronomy with Laurence of Eichstadt. Then he went to Leiden, spending two years working under the botanist Adolphus Vorstius and the astronomer Jacob Golius. After hearing many stories about Brazil, Marcgraf asked Johan de Laet, who lived in Amsterdam and was the managing director of the Dutch West India Company (which at that time controlled the northeast coast of Brazil), for an opportunity to go to Brazil. He was engaged as astronomer and left Holland on 1 January 1638, reaching Brazil 2 months later. During his 6 or 7 years in Brazil he explored the zoology, botany, astronomy, cartography, and meteorology. In 1644, both Nassau-Siegen and Marcgraf were preparing to return to Holland, but unexpectedly Marcgraf was sent to Angola where he died. According to Gudger (1912), had Marcgraf 'lived but a few years longer to have put into shape his Brazilian collections and observations, he would certainly have raised himself to the rank of the first natural historian of his time, and possibly that of the greatest since Aristotle.' Marcgraf's manuscript for his *Historia Naturalis Brasiliae* was entrusted to Johan de Laet, who published it in 1648.

3.4.1 *Historia Naturalis Brasiliae*

Marcgraf's *Historia Naturalis Brasiliae* (1648) is a 'work which for a 150 years served as the naturalist's *vade mecum* for Brazil, and indeed for much of the Neotropical region...and is one of those pre-Linnean works that still demand the attention of the taxonomist' (Whitehead 1979).

The crustaceans are in volume IV, chapters XIX-XXII. Many specialists have attempted to identify the crustaceans including: H. Milne Edwards (1837), Moreira (1900), Rathbun (1918, 1930, 1937), Sawaya (1942), Holthuis (1952, 1956, 1958, 1959a), and Lemos de Castro (1962). Curiously, Marcgraf's work was illustrated by poorly prepared wooden engravings instead of the outstanding illustrations attributed to Eckhout. The artist responsible for the engravings is not known, but it is clear that he used many oil paintings and watercolors from the *Theatrum* and the 'Handbooks' as the basis for the woodcuts. Twenty eight specimens of crustaceans, belonging to 24 species (21 decapods, 1 stomatopod, and 2 barnacles) were briefly described and referred to under native names, but only 18 illustrated. As noted by Lichtenstein (1818, 1819, 1822, 1829) and Lemos de Castro (1962), sometimes Marcgraf's descriptions do not conform with the illustrations, and this has caused misidentifications. Such non-correspondence between illustration and text is probably due to de Laet's editing, but this in itself is also surprising as de Laet had had previous experience with Brazilian fauna. Some of the Brazilian vernacular names employed by Marcgraf in the *Historia Naturalis Brasiliae* have subsequently been introduced into the zoological nomenclature: 'uçá,' *Cancer uca* Linnaeus, 1767, and *Uca* Leach, 1814; 'guaia,' *Guaia* H. Milne Edwards, 1837; 'aratu,' *Aratus* A. Milne Edwards, 1853; 'maracoani,' *Uca maracoani* (Latreille, 1802-1803); 'guanhamu,' *Cardiosoma guanhumi* Latreille, 1828.

The comparison between the woodcut illustrations of Marcgraf's work (Plates 5C-H, 6A-J) and the color plates of the *Libri Picturati* was very useful in clarifying the identity of some species for which illustrations were poor or lacking. A few of the woodcut figures have no corresponding color picture. My comparisons are summarized in Table 2. Further comments are restricted to species which presented particular problems of identification.

The first attempt to identify the 'Guaia alia species' was that of Sawaya (1942), who suggested that it could be a crab of the family Cancridae. On the other hand, Lemos de Castro (1962) remarked that the carapace, walking legs, and chelipeds resemble more a crab of the genus

Table 2. List of the crustaceans species referred to in the *Historia Naturalis Braziliae* (1648) with original spellings and binomial assignments by listed authors.

Marcgraf (1648)	H. Milne Edwards (1837)	Moreira (1900)	Rathbun (1918*, 1930**, 1937***)
'Guaia apara'	*Calappa marmorata* Latreille (1802-1803)	*Calappa flamaea* (Herbst 1793)	x
'Guaia alia species'	*Guaia punctata* (Linnaeus 1758)	*Persephone punctata* (Browne 1769)	*Persephona punctata punctata* (Linnaeus 1758)***
'Guaia alia species'	x	x	x
'Guaia miri'	x	x	x
'Aratu Peba'	x	x	x
'Ciri Apoa'	x	*Cronius ruber* (Lamarck 1818)	*Callinectes danae* (Smith 1869)**
'Ciri Obi'	x	x	*Arenaeus cribrarius* (Lamarck 1818)
'Maracoani'	x	*Uca maracoani* (Latreille 1803)	*Uca maracoani* (Latreille, 1802-1803)*
'Carara Una'	x	x	x
'Aguara Uca'	x	x	x
'Uca Una'	x	*Oedipleura cordata* (Linnaeus 1767)	*Ucides cordatus* (Linnaeus 1763)*
'Uca Guacu'	*Ocypode rhombea* (Fabricius 1798)	x	x
'Cunuru'	x	x	x
'Guanhumi'	*Cardiosoma guanhumi* Latreille 1825	*Cardiosoma guanhumi* Latreille 1825	*Cardiosoma guanhumi* Latreille, 1825*
'Aratu and Aratu pinima'	x	x	*Aratus pisonii* (H. Milne Edwards 1837)*
'Ciecie Ete'	x	x	*Uca thayeri* (Rathbun 1900)*
'Ciecie Panema'	x	x	x
'Potiquiquiya'	x	x	x
'Potiquiquyixe'	x	x	x
'Tamaru Guacu'	*Squilla scabricauda* (Lamarck 1818)	x	x
'Guaricuru'	x	x	x
'Carara Pinima'	x	x	x
'Potipema'	x	*Palaemon jamaicenses* (Herbst 1796)	x
'Paranacare'	x	x	x
'Poti Atinga'	x	x	x
'Potiguau'	x	x	x
'Reri Apiya'	x	x	x v
'Reri Apiya'	x	x	x

Table 2 (continued).

Sawaia (1942)	Lemos de Castro (1962)	Present contribution
Calappa flammea (Herbst 1784)	*Calappa ocellata* (Holthuis 1958)	*Calappa ocellata* (Holthuis 1958)
Persephona punctata punctata (Linnaeus 1758)	*Persephona punctata punctata* (Linnaeus 1758)	*Persephona punctata* (Linnaeus 1758)
Cancer sp.	*Mithrax (Mithrax) hispidus* (Herbst 1790)	*Mithrax (Mithrax) hispidus* (Herbst 1790)
Panopeus sp.	*Panopeus* sp.	*Xanthidae* sp.
Ovalipes sp.	*Ovalipes punctatus* (de Haan 1833)	*Ovalipes trimaculatus* (de Haan 1833)
x	*Callinectes danae* (Smith 1869)	*Callinectes danae* (Smith 1869) v
x	*Arenaeus cribrarius* (Lamarck 1818)	*Arenaeus cribrarius* (Lamarck 1818)
Uca maracoani (Latreille 1802-1803)	*Uca maracoani* (Latreille 1802-1803)	*Uca maracoani* (Latreille 1802-1803)
Uca sp.	*Sesarma* (Holometopus) rectum (Randall 1840)	*Sesarma (Holometopus)* rectum (Randall 1840)
Ocypode albicans (Bosc 1801-1802)	*Ocypode quadrata* (Fabricius 1787)	*Ocypode quadrata* (Fabricius 1787)
Ucides cordatus (Linnaeus 1763)	*Ucides cordatus* (Linnaeus 1763)	*Ucides cordatus* (Linnaeus 1763)
Ocypode albicans (Bosc 1801-1802)	*Ucides cordatus* (Linnaeus 1763)	*Ucides cordatus* (Linnaeus 1763)
Ocypodidae sp.	*Ucides cordatus* (Linnaeus 1763)	*Ucides cordatus* (Linnaeus 1763)
Cardiosoma guanhumi Latreille 1825	*Cardiosoma guanhumi* Latreille 1825	*Cardiosoma guanhumi* Latreille 1828
Aratus pisonii (H. Milne Edwards 1837)	*Goniopsis cruentata* (Latreille 1803)	*Goniopsis cruentata* (Latreille 1803)
Uca sp.	*Uca* sp.	*Uca* sp.
x	*Panulirus* sp.	*Panulirus echinatus* (Smith 1839)
Scyllaridae sp.	*Parribacus antarcticus* (Lund 1793)	*Parribacus antarcticus* (Lund 1793)
Squillidae sp.	*Lysiosquilla scabricauda* (Lamarck 1818)	*Lysiosquilla scabricauda* (Lamarck 1818)
Atya sp.	*Atya scabra* (Leach 1815)	*Atya scabra* (Leach 1815)
Aratus pisonii (H. Milne Edwards 1837)	*Aratus pisonii* (H. Milne Edwards 1837)	*Aratus pisonii* (H. Milne Edwards 1837)
x	*Macrobrachium carcinus* (Linnaeus 1758)	*Macrobrachium carcinus* (Linnaeus 1758)
Paguridae sp.	*Paguridae* sp. *Paguridae* sp.	
x	*Penaeus* sp.	*Macrobrachium carcinus* (Linnaeus 1758)
x	*Penaeus* sp.	*Macrobrachium carcinus* (Linnaeus 1758)
Lepas sp.	*Lepas* sp.	*Lepas hillii* (Leach 1818)
Balanus sp.	*Balanus* sp.	*Balanus tintinnabulum* (Linnaeus 1758)

Mithrax Latreille, 1817, probably *Mithrax (Mithrax) hispidus* (Herbst 1790). Examination of the color illustration supports Castro's identification. A number of details are clarified including shape and exact number of antero-lateral spines, ornamentation of the carapace, and color pattern – all omitted on the poor woodcut.

Unfortunately there is no color equivalent for the very poor woodcut of the unidentifiable xanthid 'Guaia miri.' Also, there is no color plate for Marcgraf's 'Carara Una;' however, as noted by Lemos de Castro (1962), the shape of the carapace in the woodcut is very much the same as that in the genus *Sesarma* Say, 1817, and the absence of a lateral tooth behind the outer tooth, is characteristic of the subgenus *Holometopus*. These characters along with Marcgraf's description suggest that the 'Carara Una' is *Sesarma (Holometopus) rectum* Randall, 1840.

After studying the illustration that accompanies the description of 'Aratu' and 'Aratu pinima,' Rathbun (1930) and Sawaya (1942) identified it as *Aratus pisonii* (H. Milne Edwards). However, as noted by Lemos de Castro (1962), Marcgraf's description does not conform with the illustration. The description refers to *Goniopsis cruentata* (Latreille 1803), while the illustration corresponds to the tree crab *A. pisonii* (H. Milne Edwards).

Rathbun (1930) considered the figure of 'Ciecie Ete' to be the fiddler crab *Uca thayeri* Rathbun, 1900. While the figure is clearly of a *Uca* species, in the absence of a more detailed illustration, it is difficult to be more precise.

Lemos de Castro (1962) considered that because of the poor illustration and description of the 'Potiquiquiya' it is only possible to affirm that it is a spiny lobster, genus *Panulirus* White, 1847. Among the color plates of the *Icones Aquatilium*, there are two illustrations of lobsters, one named 'Potiquiquiya,' and the other 'Poticucuma,' both representing the same species illustrated in Marcgraf's work. Study of these color plates shows the presence of transverse incomplete grooves on the abdominal segments, and light striped pereiopds, both characteristic of the brown spiny lobster *Panulirus echinatus* Smith, 1869.

Examination of the color plate of the 'Potipema' from the *Icones Aquatilium*, confirms Holthuis' (1952) and Lemos de Castro's (1962) interpretations, based only on the description given by Marcgraf, that it is *Macrobrachium carcinus* (Linnaeus). Sawaya (1942) referred to the 'Potiquiquiyxe' as Scyllaridae sp., but I agree with the interpretations of Holthuis (1956) and Lemos de Castro (1962) that it is *Parribacus antarcticus* (Lund 1793).

Marcgraf's descriptions of the 'Poti Atinga' and 'Potiguaçu' (there are no figures) lead Lemos de Castro (1962) to suggest that these are shrimps of the genus *Penaeus*. However, in the collection of the *Icones Aquatilium* there are two color plates, one illustrating the 'potiatinga' and other illustrating the 'potiguaçu' but both belonging to the same species, *Macrobrachium carcinus* (Linnaeus).

3.5 *Willem Piso*

Willem Piso (or Pies) was born in 1611, in Leiden. His father was a musician from Cleves, Germany. At the age of 12 he went to Caen (Normandy), where in 1634 he obtained the title of 'Doctor Medicinae.' He was 27 years old when he joined the Nassau-Siegen Mission in 1638, and went to Brazil to replace Willem de Milaenen, who had died shortly after his arrival in Brazil. With the end of the Nassau-Siegen Mission in 1644, Piso left Brazil for Leiden (1645) and afterwards moved to Amsterdam where he died on 28 November 1678. For 7 years Piso collected information about tropical diseases and Indian medicine, publishing the results of his research (1648) in the De *Medicina Brasilien*, along with Marcgraf's *Historia Natualis Brasiliae*. In 1658, Piso published another work on New World medicine and included a chapter on

Brazil's fauna and flora. This work is mentioned here as it contains nine figures of crustaceans copied from Marcgraf's *Historia Naturalis Brasiliae*. The crustaceans referred to by Piso (1658: 75-78) are: 'Guáia-Guaçú' (Marcgraf's 'Guaia alia species,' *Mithrax hispidus* (Herbst)); 'Guáia-Apara' (*Calappa ocellata* Holthuis); 'çiri' (Marcgraf's 'Ciri Apoa,' *Callinectes danae* Smith); 'Uça-úna' (*Ucides cordatus* (Linnaeus)); 'Cunurú' (Marcgraf's 'Aguara Uca,' *Ocypode quadrata* (Fabricius)); 'Maracoani' (*Uca maracoani* (Latreille)); 'Guanhúmi' (*Cardisoma guanhumi* Latreille); 'Potiquiquíya' (*Panulirus echinatus* Smith); and 'Poti' (Marcgraf's 'Paranacare,' Paguridae sp.).

3.6   *The Nassau-Siegen Mission artists*

In a letter written to the Marquis de Pomponne, Johan Maurits mentioned that 'dans mon service les temps de ma demeure au Brésil, six peintres, dont chacun acurieusement peint à quoy il estoit le plus capable...;' however, the 'Lijste van de domistiquen ant hoff van Sijn Extie Johann Mauritz van Nassau...op den 1 April 1643 genietende de vrije taefel' (in English: List of the domestics who had access to the table of his Excellency Johan Maurits van Nassau in April 1643) only cites the names of Albert Eckhout and Frans Post (Larsen 1962). Such a discrepancy makes the exact number and identity of the artists who worked for Johan Maurits a subject for debate.

Very little is known about the life of Albert Eckhout. He was probably born in Groningen in 1610 and joined the Nassau-Siegen Mission in 1636, going to Brazil in the company of Nassau-Siegen. Eckhout's work in Brazil was mainly devoted to pictorially recording the natives, and the fauna and flora. Several of his paintings are dated 1641 and 1643. In 1644, Eckhout returned to Holland, and in 1653, under the direction of Nassau-Siegen, he moved to Dresden with his wife and daughter to work in the Court of Johann Georg II, where he stayed probably until 1653. Eckhout died in Groningen in 1654. The illustrations included in the *Theatrum Rerum Naturalium Brasiliae* and in the *Liber Picturatus A 38* are attributed to him. Additional information relevant to carcinology also comes from the painting 'Tupi male Indian' (269 × 170 cm, Copenhagen Museum) (Plate 7A) which includes an illustration of the land crab *Cardisoma guanhumi* Latreille, accurately representing its color and the rows of spines on the dactyli of the pereiopods. This painting was probably a part of the collection of paintings offered by Nassau-Siegen to Freidrich III of Denmark. In 1686, Eckhout cartoons, presented in 1679 to Louis XIV by Nassau-Siegen, were portrayed on sets of tapestries known as 'Anciennes Indes' which were made at the Gobelins factory in Paris. In 1730, two other series of tapestries were made based upon the same cartoons, 'les Petites Indes' and 'les Grandes Indes,' causing damage to the originals. In 1735, the artist François Desportes (1661-1743) replaced the damaged originals with a series of paintings, which were utilized as models for a series of tapestries called 'les Nouvelles Indes' (Pinault 1990). Desportes' 'Chasseur indien' (Marseille Museum) is the only tapestry in which some crustaceans appear. It is largely based upon Eckhout's work and does not contain any original information. Finally, six species of crabs (some of them used for 'les Nouvelles Indes') are represented in Desportes' oil painting 'Fishes hanging, also crabs and tortoise' (Sèvres): *Cardisoma guanhumi* Latreille, *Calappa ocellata* Holthuis, *Mithrax hispidus* (Herbst), *Goniopsis cruentata* (Latreille), *Persephona punctata* (Linnaeus), and *Carpilius corallinus* (Herbst).

Frans Post was born in 1612 in Haarlem and it seems that he was already an artist of renown when he joined the Nassau-Siegen Mission in 1636. His paintings deal mainly with the landscape of Brazil, but he sometimes added plants and animals. In 1644, he left Brazil for Europe, and in 1646 he returned to Haarlem, where he continued working. Post died in 1680.

Zacharias Wagener was born in 1614 in Dresden. He was 19 when he travelled to Hamburg then to Amsterdam, staying for one year at the house of the famous librarian Wilhelm Janson Blau. On 18 July 1634, on the advice of Mr. Blau, he accepted employment as a soldier on board the Amsterdam and sailed to Brazil. Johan Maurits, perhaps seeing Wagener's talents, made him his special secretary. Wagener spent 7 years in northeastern Brazil and during this time had access to several Dutch natural history texts about Brazil and, as a result, decided to make a kind of report to his friends and family in Germany by describing and illustrating the Indians and the local fauna and flora in a 'Thierbuch.' The manuscript, presently housed in the Kupferstich Kabinet Staatliche Kunstsammlung, Dresden, contains more than 100 illustrations of animals, plants, human types, and landscapes, to which Wagener added short descriptions. The botanical and zoological specimens he figured were given to him by the Indians. It is likely that for the illustrations of the Indian's, Wagener utilized some of the drawings made by Eckhout but in the process of copying corrected some of the imprecise observations in the originals. On 1 April 1641, he departed from Brazil on board the Grosse Tiger travelling to Texel. He did not stay long in Europe, sailing on 29 September 1642 to East India, where he made a series of local expeditions. In July 1668, Zacharias Wagener returned to Amsterdam, where he died on 1 October 1668.

Six crustaceans were illustrated and described by Wagener (1964; Figs 23 to 27, last species not numbered). The quality of his illustrations enable the immediate recognition of several species (Plates 7B-G): The mantis shrimp *Lysiosquilla scabricauda* (Lamarck), referred to under the native name 'Tamalu asu' (instead of 'tamaru-açu'), for which he provided a brief description of the raptorial claws and its aggressive behavior; the goose-necked barnacle *Lepas hillii* (Leach), for which he did not know the native term, providing only the Dutch name 'Langk Hälse' (which literally means long neck), and mentioning the blue color of its stalk. Four species of brachyuran crabs were mentioned in the text under 'crangejo:' the land crab *Cardisoma guanhumi* Latreille; the box crab *Calappa ocellata* Holthuis; the mud crab *Ucides cordatus* (Linnaeus) (the illustration of this species lacks the characteristic long hairs on the lower surface of the walking legs because the specimen used by Wagener had partially decayed); and the mangrove crab *Goniopsis cruentata* (Latreille).

4   CAROLUS LINNAEUS AND THE SYSTEMA NATURAE

Carolus Linnaeus (also known as Carl Linné) was born on 23 May 1707 in Raashult, Smaaland, Sweden. In 1727, he went to the University of Lund and soon afterwards to Uppsala. In 1735, financed by Johan Moraeus, Linnaeus went to Harderwijk, Holland, where he was awarded the title of 'Doctor Medicinae' on 24 June of the same year. Later, Linnaeus entered the University of Leiden, where he completed the first edition of his *Systema Naturae*. In 1738, he returned to Sweden and in 1740 entered the University of Uppsala. Linnaeus assembled under him a number of collaborators and students, whose travelling and collections of natural history speci-mens were used to augment the subsequent editions of the *Systema Naturae*. From South America he received assistance from three collaborators who sent collections to Europe ena-bling him to add new material to his work (Carl Gustaf Dahlberg 1721-1781, Daniel Rolander 1725-1793 and Pehr Loefling 1729-1756). Only Dahlberg is important here, as Rolander and Loefling sent no crustaceans to Linnaeus. Born in Nykoping, Sweden, Dahlberg went to Suriname in 1746 as a corporal in the Dutch forces. Interested in natural history, he built up a

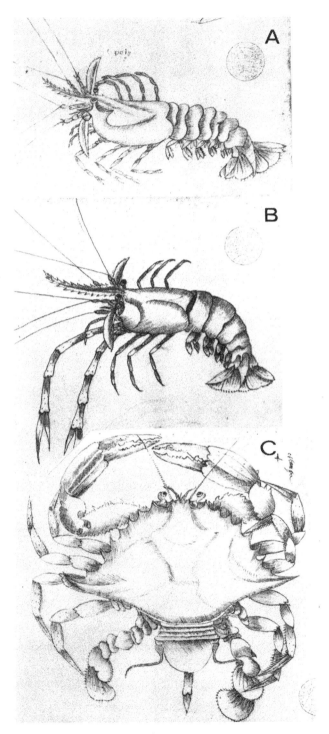

Plate 1. Crustaceans depicted in the *História dos animais e árvores do Maranhão* by Cristóvão de Lisboa. A. 'Poty' B. *Macrobrachium* sp. C. 'Ciry,' *Callinectes* sp.

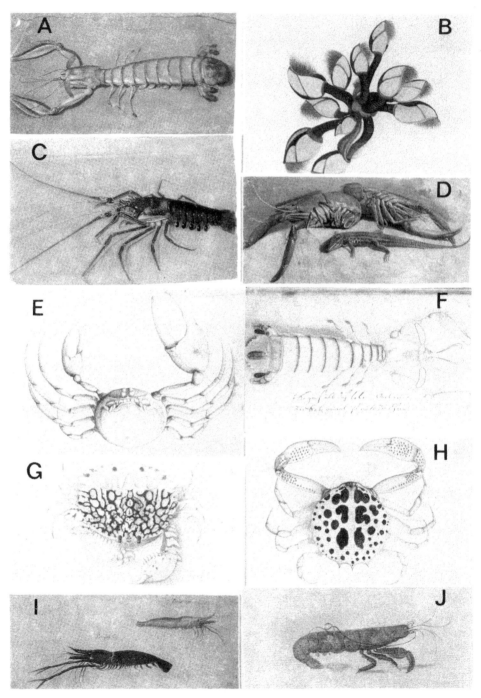

Plate 2. Crustaceans depicted in the *Icones Aquatilium*. A. 'Tamarû guaçu,' *Lysiosquilla scabricauda* (Lamarck). B. *Lepas hillii* (Leach). C. 'Poticucuma,' *Panulirus echinatus* Smith. D. 'Potipema,' *Macrobrachium carcinus* (Linnaeus). E. *Cardisoma guanhumi* Latreille. F. 'Tamarû,' *Lysiosquilla scabricauda* (Lamarck). G. 'Guaiâ apá ra,'*Calappa ocellata* Holthuis. H. *Persephona punctata* (Linnaeus). I. 'Potiguaçu' and 'Potîatinga,' *Macrobrachium carcinus* (Linnaeus). J. 'Guaricurû,'*Atya scabra* (Leach).

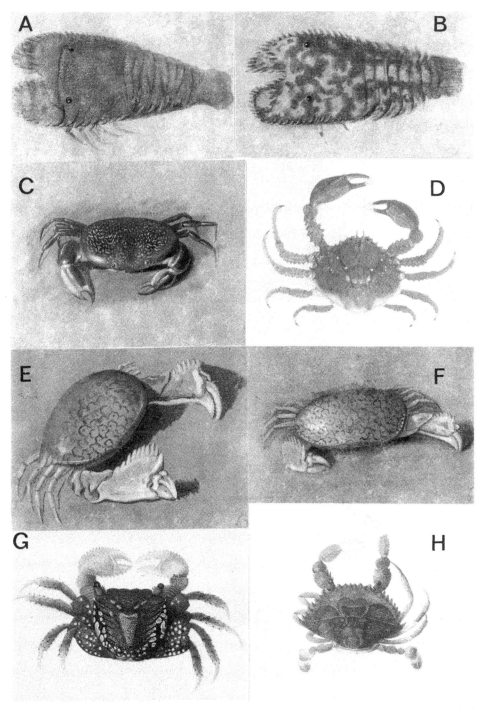

Plate 3. Crustaceans depicted in the *Icones Aquatilium*. A. 'çîrayieima,' *Parribacus antarcticus* (Lund). B. 'Potiquiquyixe,' *Parribacus antarcticus* (Lund). C. *Carpilius corallinus* (Herbst). D. 'Guaiâ,' *Mithrax (Mithrax) hispidus* (Herbst). E. 'Guaiâ,' *Calappa ocellata* Holthuis. F. '? Guaiâ apára,' *Calappa ocellata* Holthuis. G. *Goniopsis cruentata* (Latreille). H. 'çiri,' *Callinectes danae* Smith.

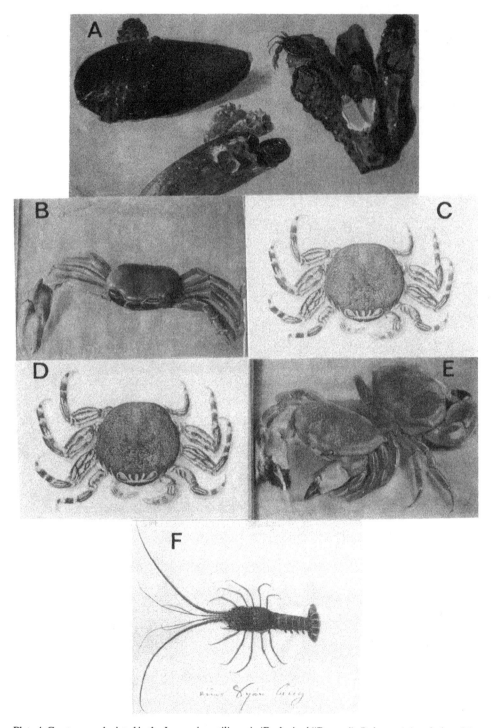

Plate 4. Crustaceans depicted in the *Icones Aquatilium*. A. 'Reñapiya' ('Punaru'), *Balanus tininnabulum* (Linnaeus). and Pinnotheridae sp. B. 'Guaiâmû,' *Cardisoma guanhumi* Latreille. C. *Plagusia depressa* (Fabricius). D. *Plagusia depressa* (Fabricius). E. *Carpilius corallinus* (Herbst). F. Potîquiquiya,' *Panulirus echinatus* (Smith).

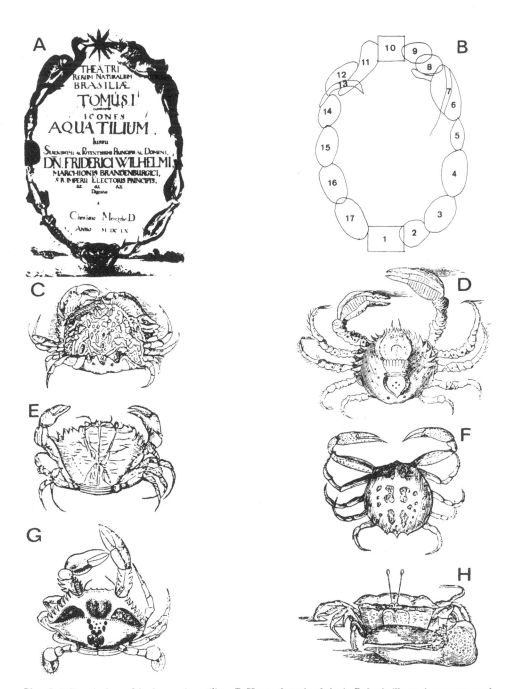

Plate 5. A. Frontispiece of the *Icones Aquatilium*. B. Key to the animals in A. Only six illustrations correspond with those of the *Theatrum*: 1. 'Guaja,' *Carpilius corallinus* (Herbst); 10. 'Iaciatataguaçu'; 11. 'Iacarepetim-buaba'; 13. 'Camaruru'; 15. 'Ajereba'; 17. 'Tamaru,' *Lysiosquilla scabricauda* (Lamarck), A-B after Albertin (1985) C-H. Crustaceans depicted in the *Historia Naturalis Braziliae* (1648): C. 'Guaia apara,' *Calappa ocellata* Holthuis. D. 'Guaia alia species,' *Mithrax (Mithrax) hispidus* (Herbst). E. 'Guaia miri,' Xanthidae sp. F. 'Guaia alia species,' *Persephona punctata* (Linnaeus). G. 'Ciri Apoa,' *Callinectes danae* Smith. H. 'Maracoani,' *Uca maracoani* (Latreille).

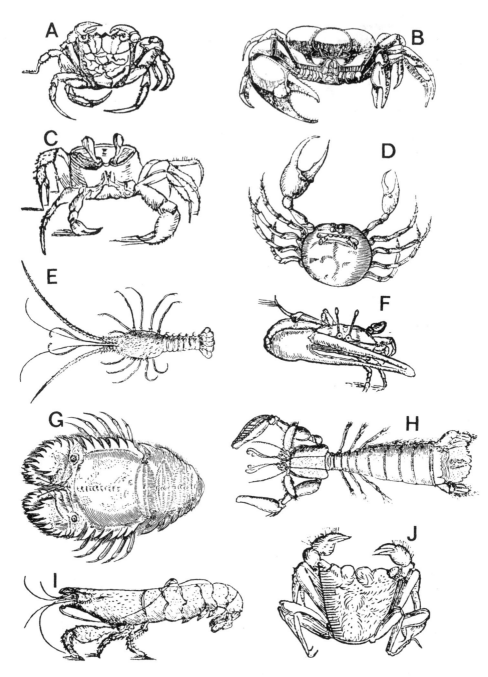

Plate 6. Crustaceans depicted in the *Historia Naturalis Braziliae* (1648). A. 'Carara Una,' *Sesarma (Holometopus) rectum* Randall. B. 'Uca Una,' *Ucides cordatus* (Linnaeus). C. 'Aguara Uca,' *Ocypode quadrata* (Fabricius). D. 'Guanhumi,' *Cardisoma guanhumi* Latreille. E. 'Potiquiquiya,' *Panulirus echinatus* (Smith). F. 'Ciecie Ete,' *Uca* sp. G. 'Potiquiquyixe,' *Parribacus antarcticus* (Lund). H. 'Tamaru Guacu,' *Lysiosquilla scabricauda* (Lamarck). I. 'Guaricuru,' *Atya scabra* (Leach). J. 'Carara Pinima,' *Aratus pisonii* (H. Milne Edwards).

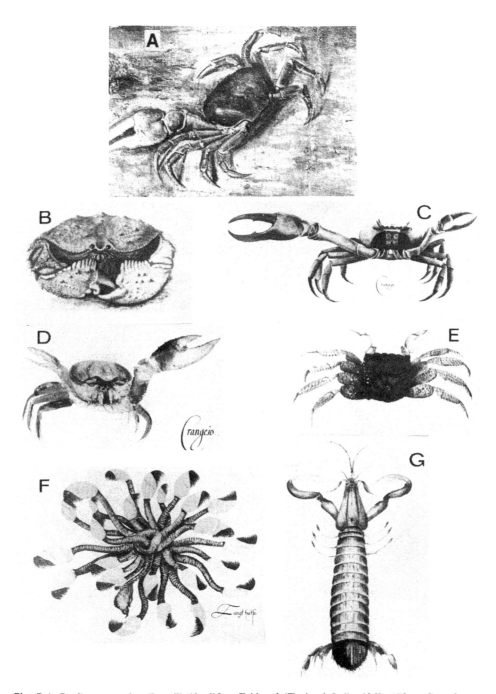

Plate 7. A. *Cardisoma guanhumi* Latreille (detail from Eckhout's 'Tupi male Indian,' 269×170 cm, Copenhagen Museum). B-G. Crustaceans depicted in Wagner's 'Thierbuch.' B. 'Crangejo,' *Calappa ocellata* Holthuis. C. 'Crangejo,' *Cardisoma guanhumi* Latreille. D. 'Crangejo,' *Ucides cordatus* (Linnaeus). E. 'Crangejo,' *Goniopsis cruentata* (Latreille). F. 'Langk hälse,' *Lepas hillii* (Leach). G. 'Tamalu asu,' *Lysiosquilla scabricauda* (Lamarck).

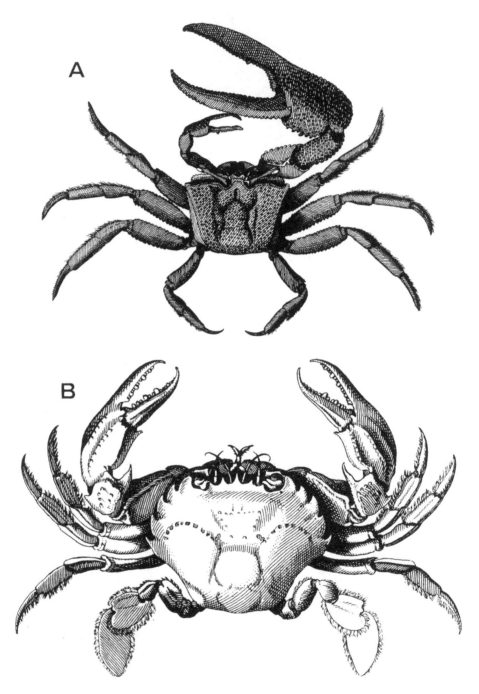

Plate 8. Crustaceans species from Seba's *Thesauri* to which were applied Marcgraf's nomenclature. A. 'Uka una,' *Uca major* (Herbst). B. 'Ciri apoa or Aratus pinima,' *Ovalipes trimaculatus* (de Haan).

collection of plants and animals, offering them to King Adolf Fredrik of Sweden during his first visit back to Europe in 1754. On 25 April 1761, he travelled again back to Europe in order to visit Holland and Sweden, and brought new collections to Sweden, probably including the type material of *Cancer cordatus* Linnaeus, 1763 (*Ucides cordatus*), which he gave to King Gustaf III. This collection is presently in the Zoological Institute of Uppsala University. Dahlberg died on 6 September 1781 in Paramaribo (Holthuis 1959b; Papavero 1971). Linnaeus died on 10 January 1778 and his collections (excluding the library and herbarium which remained with his son) were shipped to Mr. James Edward Smith, and later bought by the Linnean Society of London. Linnaeus' *Systema Naturae* was not only influenced by the crustacean collections from the New World but also by the works of Marcgraf (1648), Sloane (1725), Catesby (1734), and Seba (1759). The influence of Marcgraf's work was such that many of the animals and plants of Brazil named by Linnaeus (and subsequent binomials by other authors) were wholly or partly based on Marcgraf's descriptions and figures, and Linnaeus perpetuated native names of species, including *Cancer uca* Linnaeus, 1767 (*Ucides cordatus*) and *Cancer vocans* Linnaeus, 1767, (Marcgraf's 'maracoani,' *Uca maracoani* (Latreille); *non Cancer vocans* Linnaeus, 1758, which was based on Rumphius' material.)

5 ALBERTUS SEBA'S THESAURI

Albertus Seba was born on 2 May 1665 in Etzel, East Friesland, now Germany. He was an apothecary in Amsterdam who made two large collections of natural history objects mainly by purchasing curiosities from travellers returning from the Indies, America, and Africa. When Seba's first collection was purchased in 1717 by Peter the Great and transferred to St. Petersburg, he started a second collection, which was later dispersed by sale at auction in 1752. These two collections form the basis of Seba's *Locupletissimi Rerum Naturalium Thesauri*, also known as Seba's *Thesaurus*. It comprises 4 folio-size volumes with 449 well executed illustrations but poor descriptions of the animals in Latin and French. Volumes one and two were issued in 1734 and 1735, before Seba's death on 2 May 1736, and volumes three and four were edited by Arnout Vosmaer in 1759 and 1765 (Holthuis 1969). The crustaceans are included in volume three.

Seba (1759) is admittedly post-Linnean, but like Holthuis (1959a) I consider that since Seba ignored the binomial nomenclature his work falls into the same category as pre-Linnean works. The importance of Seba's work in the context of pre-Linnean carcinology in Brazil comes from the fact that Seba adopted Marcgraf's nomenclature for two species of crabs. In fact, a mistake by Seba is responsible for the introduction into the zoological literature of the generic name *Uca* for fiddler crabs, instead of mangrove crabs. Marcgraf misspelled 'uçá una,' a name in common usage for mangrove crabs, as 'uca una.' Seba copied the term from Marcgraf's work, but applied it to a fiddler crab instead of a mangrove crab ('Uka una' vol. 3: 44; pl. XVIII, fig. 8). Seba's illustration of the 'Uka una' (Plate 8A) was the basis for Herbst's description of *Cancer vocans major* Herbst, 1782, type species of the genus *Uca* Leach, 1814. In order to correct this mistake, Latreille (1817) created the genus *Gelasimus* for the fiddler crabs and proposed that the name *Uca* Leach should be applied only to the mangrove crabs. As Latreille's procedure was invalid, Rathbun (1897) and Ortmann (1897), established respectively the genera *Ucides* and *Oedipleura* for mangrove crabs, retaining *Uca* Leach for the fiddler crabs. *Oedipleura* Ortmann fell into the synonymy of *Ucides* Rathbun by 42 days (Oliveira 1939, Tavares 1990). The mangrove crab *Ucides cordatus* (Linnaeus), is referred to by Seba under 'Cancer fluviatilis, sive Gammarus Americanus' (vol. 3: 43, pl. XX, fig. 4). Seba's description of the 'Ciri apoa or Aratus pinima' (native names copied from Marcgraf) is inaccurate, and it could well be any species of portunid crab from Brazil; however, the illustration (Plate 8B) probably belongs to the species *Ovalipes trimaculatus* (De Haan 1833).

6 CONCLUSIONS

Knowledge of the Brazilian crustaceans in the pre-Linnean period is spread over the publications of 13 authors who studied mainly northeastern Brazil (chiefly the States of Maranhão, Paraíba, Pernambuco, and Bahia), and on only few occasions explored the south (Rio de Janeiro and São Paulo). The pre-Linnean carcinolgy of Brazil can be subdivided into two distinct phases: one in which the species were usually poorly described in narratives of voyages, or letters (from 1500 to 1627); and a second phase characterized by an intention to catalogue the fauna, carried out during the governorship of Nassau-Siegen (1637-1644). Thirty one littoral species of crustaceans were referred to in the pre-Linnean texts (27 decapods, 1 stomatopod, 3 barnacles), 23 illustrated, and 3 included in Linnaeus' *Systema Naturae*. The main contribution in the 17th century was that of the Nassau-Siegen Mission, during which far more scientific activity was carried out than in the preceding 138 years. The most important early account of Brazilian zoology is the *Historia Naturalis Brasiliae* (1648) which contained 24 species (21 decapods, 1 stomatopod and 2 barnacles), of which 18 were illustrated.

After the West India Company left Brazil, the Brazilian harbors remained closed to visitors for more than 100 years. During the 18th century very little exploration was carried out. However the transfer of the court of the Emperor Dom João VI from Portugal to Brazil inaugurated a new phase for the study of Brazilian natural history. In 1808, Brazil opened its harbors enabling scientific expeditions to enter the country. From then until 1817, about 20 scientific expeditions were carried out, and the amount of information accumulated was so large that it led the ornithologist O. des Murs (1855) to mention that at that time Brazil was 'le pays du Globe, après l'Europe, sans doute le mieux exploité sous le rapport de ses productions dans les trois règnes de la nature.'

ACKNOWLEDGEMENTS

I am greatly indebted to Nelson Papavero (Museu de Zoologia, Universidade de São Paulo) and Petronella J. Albertin (São Paulo) for entrusting to me for study the collection of color transparences of pictures of the crustaceans in the *Theatrum Rerum Naturalium Brasilliae*; Carlos Diniz Freitas (Universidade Santa Ursula, Rio de Janeiro) for opening his personal library; Colin L. McLay (University of Canterbury, Christchurch) and Peter Davie (Queensland Museum, Brisbane) for checking the English of this paper; Elaine F. Albuquerque (Universidade Santa Ursula, Rio de Janeiro) and Claudia Mendes Tavares, for helping to locate many obscure references.

The many discussions I had with Michèle de Saint Laurent (Muséum National d'Histoire Naturelle, Paris) on the subject helped me to improve the manuscript. I also received the kind cooperation of Alain Crosnier and Jacques Forest (Muséum National d'Histoire Naturelle, Paris), and Johann Becker (Museu Nacional do Rio de Janeiro). Gary C.B. Poore (Museum of Victoria, Australia) informed me of the proposed volume, *The History of Carcinology*, and is responsible for me contacting the editor Frank M. Truesdale. Jacques Rebière (Muséum

National d'Histoire Naturelle, Paris) kindly took the photographs of the drawings in the *Historia Naturalis Brasiliae* and Wagner's 'Thierbuch.'

I am also grateful to Daniele Guinot (Muséum National d'Histoire Naturelle, Paris) for critically reading the manuscript.

This study was supported by Muséum National d'Histoire Naturelle, and by the Brazilian Council for Scientific Research (CNPq) under grant no. 202252/89.2.

REFERENCES

Abbeville, C. (Foullon, C.). 1614. *Histoire de la mission des près capucins en l'Isle de Maragnan et terres circonvoisines*. Paris. (not seen)

Abbeville, C. (Foullon, C.). 1975. *História da missão dos padres capuchinhos na ilha do Maranhão e terras circunvizinhas*. Itatiaia: Belo Horizonte. (Portuguese translation of Abbeville 1614)

Albertin, P. J. 1985. Arte e ciência no Brasil Holandês - Theatri Rerum Naturalium Brasiliae: Um estudo dos desenhos. *Revta Bras. Zool.* 3: 249-326.

Anchieta, J. 1799. *Epistola quam Plurimarum Rerum Naturalium quae S. Vicenti (nunc. S. Pauli) Provinciam Incolunt Sistens Descriptionem*. Lisbon: Typis Academiae. (not seen)

Anchieta, J. 1933. *Cartas Jesuíticas*. Rio de Janeiro: Civilizaão Brasileira S. A. (Portuguese translation of Anchieta 1799)

Bloch, M. E. 1787. *Naturgeschichte der ausländischen Fische*. Berlin.

Boeseman, M., L.B. Holthuis, M.S. Hoogmoed & C. Smeenk 1990. Seventeenth century drawings of Brazilian animals in Leningrad. *Zool. Verh.* 267: 1-189.

Brandão, A. F. 1977. *Diálogos das Grandesas do Brasil*. São Paulo: Melhoramentos. (first integral edition 1930)

Catesby, M. 1734. *The Natural History of Carolina, Florida, and the Bahama Islands: Containing the Figures of Birds, Beasts, Fishes, Serpents, Insects and Plants, Particularly the Forest-Trees, Shrubs, and other Plants, not Hitherto described, or Very Incorrectly Figured by Authors, Together with Their Descriptions in English and French, To Which Are Added, Observations on the Air, Soil, and Waters: with Remarks Upon Agriculture, Grain, Pulse, Root, Etc*. Two volumes. London.

Cortesão, J. 1943. *A Carta de Pero Vaz de Caminha*. Rio de Janeiro: Livraria Editora Livros de Portugal.

Ehrenreich, P. 1894. Über einige ältere Bildnisse Südamericanische Indianer. *Globus, Ill. Zeitschr. Länder Völkerkunde Bunswick*. 66(6): 81-90.

Evreux, Y. 1615. *Suite de l'Histoire des Choses Plus Memorables Advenües en Maragnan en Années 1613 et 1614. Second traité*. Paris: F. Huby.

Garcia, R. 1977. Notas. In A. F. Brandão (ed.), *Diálogos das grandesas do Brasil*:pp. 240-241. São Paulo: Melhoramentos.

Gudger, E. W. 1912. George Marcgrave, the first student of American natural history. *Pop. Sci. Monthly* 81: 250-274.

Holthuis, L. B. 1952. The Subfamily Palaemoninae. A General Revision of the Palaemonidae (Crustacea, Decapoda, Natantia) of the Americas. II. *Occ. Pap. Allan Hancock Found.* 12: 1-396.

Holthuis, L. B. 1956. Proposed addition to the 'Official List of Generic names in Zoology' of the names of twenty-five genera of Macrura Reptantia (Class Crustacea, Order Decapoda), including proposals for the use of the plenary powers (a) to validate the spelling *Cherax* as the valid original spelling for the generic name published as *Cherax* and *Cheraps* by Erichson in 1846, (b) to suppress the specific name *goudotii* Guirin-Méneville, 1839, as published in the combination *Astacoides goudotii*, and (c) to validate the emendation to *Palinurus* of the generic name *Pallinurus* Weber, 1795. *Bull. Zool. Nomencl.* 12: 107-119.

Holthuis, L. B. 1958. West Indian Crabs of the Genus *Calappa* with a description of three new species. *Studies Fauna Curaçao Caribb. Is.* 8. (34): 146-186.

Holthuis, L. B. 1959a. The Crustacea Decapoda of Suriname (Dutch Guiana). *Zool. Verh. Leiden* 44: 1-296.

Holthuis, L. B. 1959b. Notes on pre-Linnean carcinology (including the study of Xiphosura) of the Malay Archipelago. In (H. C. de Wit (ed.), *Rumphius Memorial Volume*: pp. 63-125. Baarn.

Holthuis, L. B. 1969. Albertus Seba's 'Locupletissimi Rerum Naturalium Thesauri....' (1734-1765) and the 'Planches de Seba' (1827-1831). *Zool. Med.* 43: 239-252.

Holthuis, L. B. & M. Boeseman, 1990. Crustaceans and further invertebrates. In Seventeenth century drawings of Brazilian animals in Leningrad. *Zool. Verh. Leiden* 267: 1-189.

Hoogmoed, M. S., M. Boeseman & L. B. Holthuis 1990 Reptiles, fishes and invertebrates. In Seventeenth century drawings of Brazilian animals in Leningrad. *Zool. Verh. Leiden* 267: 1-189.

Laet, J. de. 1633. *Novus Orbis seu Descriptionis Indiae Occidentalis Libri XVIII. Authore Ioanne de Laet Antuerp. Novis Tabulis Geographicis et Variis Animalium, Plantarum, Fructuumque Iconibus Illustrati.* Lugduni Batavorum: Elzevirios.

Laet, J. de. 1641. *L'Histoire du Noveau Monde ou Description des Indes Occidentales, Contenant Dix-huit Liures Par le Sieur Iean de Laet d'Anuers; des Animaux, Plantes et FruitsD. Leiden: Bonaventure & Elsevirios.*

Larsen, E. 1962. *Frans Post, Interprète du Brésil.* Amsterdam & Rio de Janeiro: Colibris Editora.

Latreille, P. A. 1817. Gélasime, Gelasimus (Buffon). In *Nouveau Dictionnaire d'Histoire Naturelle, Appliquée aux Arts, à l'Agriculture, à l'économie Rurale et Domestique, à la Mdecine, Etc.* p. 517-520.

Leitão, C. M. 1941. *História das expedições científicas no Brasil.* São Paulo: Editora Nacional.

Lemos de Castro, A. 1962. Sobre os crustáceos referidos por Marcgrave em sua Historia Naturalis Brasiliae (1648). *Arch. Mus. Nac. Rio de Janeiro* 52: 37-51

Lery. J. de. 1578. *Histoire d'un Voyage Fait en la Terre du Brésil, autrement dite Amerique. Contenant la Navigation, & Choses Remarquables, Vuës sur Mer par l'Aucteur: Le Comportement de Villegagnon, en ce Pais là. Les Meurs & Façons de Vivre Estranges des Sauvages Ameriquains: avec un Colloque de Leur Langage. Ensemble de la Description de Plusieurs Animaux, Arbres, Herbes & Autres Choses Singulieres, & du Tout Inconues par Deça, Dont on Verra les Sommaires des Chapitres au Commencement du Livre. Non Encore Mis en Lumiere, pour les Causes Contenues en la Preface. Le Tout Recueilli sur le Lieux par Jean de Léry Natif de la Margelle, Terre Sainct Sene au Duché de Bourgogne. Seigneur, ie te Celebreray entre les Peuples, & te Diray Pseaumes entre les Nations.* La Rochelle.

Lichtenstein, K. M. H. 1818. Die werke von Marcgrave und Piso über die Naturgeschichte Brasiliens, erläutert aus den wieder anfgefunderen Originalzeichnungen. *Abh. Kön. Akad. wiss. Berlin.* 5: 201-222.

Lichtenstein, K. M. H. 1819. Die werke von Marcgrave und Piso über die Naturgeschichte Brasiliens, erläutert aus den wieder anfgefunderen Originalzeichnungen. II, Vögel. *Abh. Kön. Akad. wiss. Berlin.* 6: 155-178.

Lichtenstein, K. M. H. 1822. Die werke von Marcgrave und Piso über die Naturgeschichte Brasiliens, erläutert aus den wieder anfgefunderen Originalzeichnungen. III, Amphibien. *Abh. Kön. Akad. wiss. Berlin.* 8: 237-257.

Lichtenstein, K. M. H. 1829. Die werke von Marcgrave und Piso über die Naturgeschichte Brasiliens, erläutert aus den wieder anfgefunderen Originalzeichnungen. IV, Fische. *Abh. Kön. Akad. wiss. Berlin.* 8: 267-288.

Linnaeus, C. 1758. *Systema Naturae per Regna tria Naturae, Secundum Classes, Ordines, Genera, Species, cum Characteribus, Differentii, Synonymis, Locis.* Ed. 10. Holmiae.

Linnaeus, C. 1763. Centuria insectorum, quam, presisidae D. D. Car. von Linné, proposuit Boas Joahansson, Calmariensis. In C. Linnaeus, *Amoenitates Academicae; seu Dissertationes Variae, Physicae, Medicae, Botanicae, Antehac Seorsim Editae, nunc Collectae & Auctae,* 6: 21-35. Stockholm: Salvius

Marcgraf, G. 1648. Historia Rerum Naturalium Brasiliae, Libri octo: Quorum tres priores agunt de plantis. Quartus de piscibus. Quintus de avibus. Sextus de quadrupedibus et serpentibus. Septimus de insectis. Octavus de ipsa regione, et illus incolis. Cum appendice de Tapuyis, et Chilensibus. In G. Piso & G. Marcgraf, *Historia Naturalis Brasiliae. Auspicio et Beneficio Illustriss. I. Mauritii Com. Nassau illius Provinciae et Maris summi Praefecti adornata, in qua non tantum Plantae et Animalia, sed et Indigenarum morbi, ingenia et mores describuntur et Iconibus supra quingentas illustrantur:*pp. 1-293. Leiden & Amsterdam.

Martius, C. F. P. 1853. Versuch eines commentars über die pflanzen in den werken von Marcgrav und Piso über Brasilien. *Abh. Bayer Akad. Wiss.* 7: 181-238.

Milne Edwards, H. 1837. *Histoire Naturelle des Crustacés, Comprenant l'Anatomie, la Physiologie et la Classification de ces Animaux.* Paris: Librairie Encyclopédique de Roret.

Moreira, C. 1900. Crustaceos do Brasil. Contribuições para o conhecimento da fauna Brasileira. *Arch. Mus. Nac. Rio de Janeiro.* 11: 1-151.

Murs, O. des 1855. Oiseaux. In F. Castelnau, *Animaux Noveaux ou Rares Recueillis pendant l'Expédition dans les Parties Centrales de l'Amérique du Sud, de Rio de Janeiro a Lima, et Lima au Para; Exécute par Ordre du Gouvernement Francais pendant les Années 1843 a 1847 sous la Direction du Cmte Francis de Castelnau:* p. 61.

Oliveira, L. P. H. de 1939. Contribuição ao conhecimento dos crustáceos do Rio de Janeiro. Genero *Uca* (Decapoda, Ocypodidae). *Mem. Inst. Osw. Cruz.* 34: 115-148.

Ortmann, A. E. 1897. Carcinologisch studien. *Zool. Jb. (Syst.)* 10: 258-372.

Papavero, N. 1971. *Essays on the History of Neotropical Dipterology, with Special References to Collectors (1750-1905).* Two volumes. São Paulo: Museu de Zoologia Universidade de São Paulo.

Pinault, M. 1990. *Le Peintre et l' Histoire Naturelle.* Flammarion, Paris.

Piso, G. & G. Marcgraf. 1648. *Historia Naturalis Brasiliae, Auspicio et Beneficio Illustriss. I. Mauritii Com. Nassau Illius Provinciae et Maris Summi Praefecti Adornata, in qua non Tantum Plantae et Animalia, Sed et Indigenarum Morbi, Ingenia et Mores Describuntur et Iconibus Supra Quingentas Illustrantur.* Leiden & Amsterdam.

Piso, G. 1658. *De India utriusque re naturali et medica Libri quatuordecim, Quorum contenta pagina sequens exhibet.* Amsterdam.

Rathbun, M. J. 1897. A revision of the nomenclature of the Brachyura. *Proc. Biol. Soc. Wash.* 11: 153-167.

Rathbun, M. J. 1918. The grapsoid crabs of America. *Bull. U. S. Nat. Mus.* 97: 1-461.

Rathbun, M. J. 1930. The crancroid crabs of America of the families Euryalidae, Portunidae, Atelecyclidae, Cancridae and Xanthidae. *Bull. U. S. Nat. Mus.* 152: 1-609.

Rathbun, M. J. 1937. The Oxystomatous and allied crabs of America. *Bull. U. S. Nat. Mus.* 116: 1-278.

Sawaya, P. 1942. Comentários sobre os crustáceos, moluscos e equinodermas. Caps. XIX-XXII do livro IV da Historia Naturalis Brasiliae de Jorge Marcgrave. In J. P. de Magalhães (trans.), *Historia Natural do Brasil*: pp. 61-65. São Paulo

Schneider, A. 1938. Die Vögelbider zur Historia Naturalis Brasiliae des Georg Marcgrave. *J. Orn. Lpz.* 86 (1): 74-106.

Sclater, P. L. 1858. On the general geographical distribution of the members of the Class Aves. *J. Linn. Soc. Zool.* 2: 130-145.

Seba, A. 1759. *Locupletissimi Rerum Naturalim Thesauri Accurata Descriptio et Iconibys Artificiosissimis Expressio per Universam Physices Historiam. Opus, Cui, in hoc Rerum Genere, Nullum par Exstitit. Ex Toto Terrarum Orbe Collegit, Digessit, Descripsit, et Depingendum Curavit.* Amsterdam: H. K. Arksteum & H. Merkum, et Petrum Schouten.

Sloane, H. 1725. *A Voyage to the Islands Madera, Barbadoes, Nieves, St Christophers, and Jamaica, with the Natural History of the Herbs and Trees, Four-footed Beasts, Fishes, Birds, Insects, Reptiles, Etc. of the Last of those Islands, To which is prefix'd, an Introduction, wherein an Account of the Inhabitants, Air, Waters, Diseases, Trade, Etc. of That Place; with Some Relations concerning the Neighbouring Continent, and Islands of America.* Two volumes. London.

Soloviev, M. 1934. Materiali ekspeditii Moritza Nassauskogo v Brasili (1636-1643) v Zoologiczeskom Institute Akademii Nauk SSSR. *Trudi Inst. Istorii Nauki i Techniki Leningrad* (1) 2: 217-225.

Sousa, G. S. de. 1987. *Tratado descritivo do Brasil em 1587.* São Paulo: Companhia Editora Nacional. (first integral edition 1851)

Tavares, M. S. & E. F. Albuquerque 1989. Levantamento taxonomico preliminar dos Brachyura (Crustacea: Decapoda) da Lagoa de Itaipu, Rio de Janeiro, Brasil. *Atlântica.* 11(1): 101-108.

Tavares, M. S. 1990. Considérations sur la position systématique du genre *Ucides* Rathbun, 1897 (Crustacea, Decapoda, Ocypodidae). *Bol. Mus. Nac., N. S., Zool.* 342: 1-8.

Thevet, A. 1558. *Les Singularités de la France Antarctique, Autrement Nommée Amérique, et de Plusieurs Terres et Isles Découvertes de Notre Temps.* Paris.

Wagener, Z. 1964. *Zoobiblion. Livro dos Animais do Brasil.* São Paulo: Brasiliensia Documenta 4.

Wallace, A. R. 1876. *The Geographical Distribution of Animals.* Two volumes. London: Macmillan.

Walter, J. 1967. *Frei Crisóvão de Lisboa. História dos animais e órvores do Maranhão.* Lisboa: Arquivo Histórico Ultramarino e Centro de Estudos Históricos Ultramarinos.

Whitehead, P. J. P. 1973. The clupeoid fishes of the Guianas. *Bull. Br. Mus. Nat. Hist. (zool.).* 5: 1-227.

Whitehead, P. J. P. 1976. The original drawings for the Historia Naturalis Brasiliae of Piso and Marcgrave. *J. Soc. Bibliogr. Nat. Hist.* 7: 409-422.

Whitehead, P. J. P. 1979. The biography of Georg Marcgraf (1610-1643/44) by his brother Christian, translated by James Petivier. *J. Soc. Bibliogr. Nat. Hist.* 9: 301-314.

Whitehead, P. J. P. & M. Boeseman 1989. *A Portrait of Dutch 17th Century Brazil. Animals, Plants and People by the Artists of Johann Maurits of Nassau.* Amsterdam: North-Holland,

The beginning of Portuguese carcinology

Carlos Almaça

*Museu Bocage, Departamento de Zoologia e Antropologia and Centro de Fauna Portuguesa (INIC)
Faculdade de Ciencias, Lisboa, Portugal*

1 INTRODUCTION

The development of natural history in Portugal was a consequence of the Portuguese voyages of discovery in the 15th and 16th century. In the 12th and 13th century, centers of learning had been in church schools (e.g. Lisbon, Braga, Coimbra) close to cathedrals and in monasteries (e.g. Alcobaça, Santa Cruz de Coimbra). A few of these centers had remarkable libraries. The library of the monastery of Alcobaça was renowned, including Aristotle's works and medieval encyclopedias in its holdings. Other notable libraries in Portugal during the Middle Ages were court libraries, particularly those of King Deniz (reigned 1279-1325) and King João I (reigned 1385-1433), which are described as having many natural history manuscripts.

A university was founded in Lisbon in 1290, moving more than once from Lisbon to Coimbra and back again, until permanently established at Coimbra in 1537. Like other medieval universities this one was devoted mainly to theological studies. However, medicine was also included in the curriculum, the treatises of Avicenna (980-1037) being followed, among others.

Natural history research was not encouraged by the Church and what little research there was, was conducted under the king's authority and restricted to applied subjects like falconry. A remarkable book containing many original observations on birds of prey, *Livro de Falcoaria*, was written by Pero Menino, King Fernando's falconer, in the 14th century. Even the natural philosophy of Aristotle was not taught at the university until 1431.

Therefore, in the 15th and 16th century, when Portuguese settled overseas in the newly-discovered territories, the few learned men accompanying the colonizers were not schooled in natural history. Nevertheless, their unsophisticated observations are interesting and certainly among the first recorded from the then unknown tropics.

2 THE PORTUGUESE IN BRAZIL

The Jesuit missionaries José de Anchieta (1533-1597), Fernão Cardim (1540-1625), and Gaspar Afonso (c. 1548-1618); the 'capitão-mor' Gabriel Soares de Sousa (c. 1540-1591); the humanist Pero de Magalhaes Gandavo (16th century, dates of birth and death unknown); and the Franciscan Frei Cristvão de Lisbon (?-1652) were the first educated men to write about

Brazilian natural history. Large mammals and brilliantly-colored birds impressed them most, but at least two of them also mentioned Crustacea.

José de Anchieta, who was born at Tenerife (Canary Islands) and came to Portugal as a child, wrote his observations in a letter of May 31, 1560, to the 'Padre Geral' of the Society of Jesus at São Vicente, and in the 'Informação da Provincia do Brasil para o Nosso Padre,' in 1585. In the letter, Anchieta simply mentions that shellfish are most abundant over all the coast, mentioning lobsters, crabs, and shrimps, some of which he says are as large as nine inches.

However, in the letter of May 1560, Anchieta writes of terrestrial and aquatic crabs, briefly describing their morphology and lifestyles. The terrestrial crabs he described as abundant, greenish, much larger than the aquatic crabs, and as digging the holes they lived in. He seems to be describing *Cardiosoma*. The aquatic crabs were more diverse, and in Anchieta's words, 'They live always under water; the last legs are flat for swimming; others dig holes in the mangroves; of these, some have red legs and black body; others are blueish and hairy; still others have two mouths, one of them nearly as large as all the body, the other proportional to the body.' (By 'mouths,' Anchieta means chelipedes and in the same letter, referring to the scorpions, he also writes, 'they have two mouths like the crabs....') Several species are apparently included in these descriptions. The first probably refers to *Callinectes*; common name in Tupi dialect is 'siri,' meaning that which runs or slides.

Later, in a manuscript written by Frei Cristòvão de Lisbon, *Historia dos Animais e Arvores do Maranhão*, which was apparently completed in 1627 (Frade 1966), two species of decapod crustaceans are described and figured: now recognized as *Macrobrachium acanthurus* (Wiegmann) and *Callinectes bocourti* A. Milne Edwards. *Macrobrachium acanthurus* was named 'opoti' (shrimp) and is figured on plates 25 and 42 of Frei Cristòvão's manuscript and described as, '...flesh whitish colored, with red antennae, very abundant and very good for eating.' *Callinectes bocourti*, the 'siri,' is figured on plate 26. In Frei Cristòvão's words, 'It is like a crab, very wide, grey whitish with blue and white legs and blue spines on the carapace; it is also very good for eating.' (see Tavares, page..., Fig. 1)

Observations on Brazilian Crustacea by Anchieta and Frei Cristòvão de Lisbon preceded Marcgraf's *Historia Rerum Naturalium Brasiliae* (1648). Unfortunately, publication of their observations was delayed for centuries: Anchieta's letter was first published in 1799 and Frei Cristòvão's manuscript in 1967!

3   MODERN NATURAL HISTORY STUDIES IN PORTUGAL: THE LINNEAN PHASE

The University of Coimbra was reorganized in 1772, and scientific studies, including the study of natural history, were introduced into the curriculum. Natural history was now under the influence of Linnaeus's systematic approach to nature and his standardization of nomenclature. In 1722, Dr. Domingos Vandelli (ca.1735-1816), an Italian naturalist, was invited to teach natural history and to organize the University's museum of natural history. In that same year, Real Museu e Jardim Botânico da Ajuda was founded in Lisbon, also under the direction of Vandelli. The Real Museu was initially created by the Prime Minister, the Marquis of Pombal, for the education of the grandchildren of King Jose I. However, with the encouragement of enlightened members of government and the scientific patronage of Vandelli, the Real Museu also prepared naturalists to be sent to Portuguese colonies in America, Asia, and Africa.

Vandelli, along with a few associates and supporters, helped establish a scientific milieu in

Portugal. Through correspondence with Vandelli, other European scientists expressed interest in Portuguese fauna and flora, poorly known at that time. For example, Thomas Pennant (English zoologist, 1726-1798) wrote to Vandelli expressing interest in getting a systematic catalog (list) of specimens of 'crustaceous insects' of the Portuguese coast. In the same letter Pennant offered to exchange specimens with Vandelli (Fig. 1).

Vandelli's teaching at the University of Coimbra prepared the first professional Portuguese naturalists. With the founding of the Academy of Sciences in 1778, regular publication of scientific memoirs was facilitated. Furthermore, the Academy intended to organize its own museum of natural history, publishing, in 1781, instructions for preparing and sending specimens to the museum (Fig. 2).

The preparation of naturalists for scientific expeditions in overseas territories included some research in Portugal under Vandelli's direct supervision. In 1783, one of his students, Manoel Dias Baptista, prepared the *Faunae Conimbricencis Rudimentum*, published in 1789, in which two species of isopods, *Oniscus asellus* and *O. armadillo*, are referenced (Fig. 3). This is the first paper on Portuguese fauna using Linnean nomenclature.

Vandelli himself published (1797) a list of Portuguese fauna and flora, which also included Brazilian species. He recorded 19 species of Portuguese decapod and isopod Crustacea (Fig. 4). One puzzling problem posed by that list is that *Cancer astacus* is recorded with the vernacular name 'lagosta,' the common name for *Palinurus elephas* (Fabricius 1787). I do not know now whether Vandelli misnamed *P. elephas* or actually was referring to *Astacus pallipes* Lereboullet, 1858, the crayfish, which is only supposed to have been introduced into Portugal in the late 19th century (Almaca 1990).

The Portuguese began overseas scientific expeditions in 1783. In that year four naturalists left Lisbon, each to explore a separate Portuguese territory: Alexandre Rodrigues Ferreira to Brazil, João da Silva Feijò to the Cape Verde Islands, Joaquim José da Silva to Angola, and Manuel Galvão da Silva to Mozambique. Some months later, shipments of specimens began to

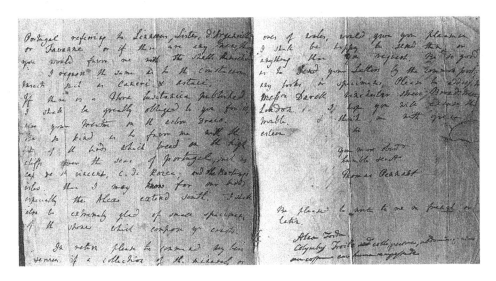

Figure 1. Letter (2 June 1786) from Thomas Pennant to Domingos Vandelli asking for specimens of Crustacea and other groups of the Portuguese fauna (Document CE/P-36, historical archives, Museu Bocage, Lisbon).

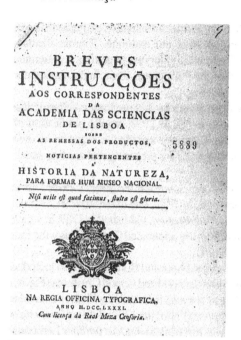

Figure 2. Cover from Lisbon Academy of Science instructions (1781) for collection of specimens, and a page of instructions for collection of crustaceans.

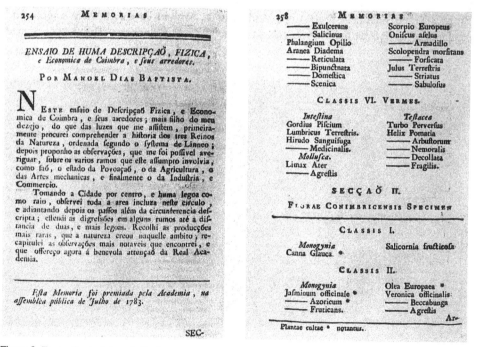

Figure 3. *Fauna Conimbricensis Rudimentum* by Manoel Dias Baptista (1789), containing the first listing of Portuguese Crustacea according to the Linnean system.

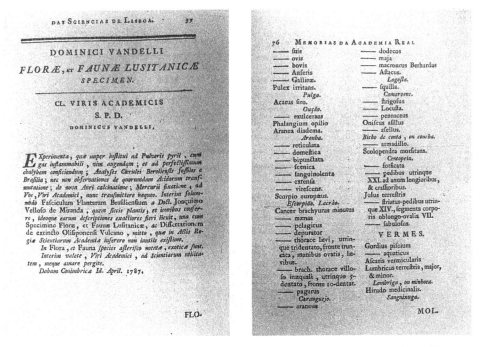

Figure 4. List of Portuguese fauna showing crustacean entries including *Cancer astacus*. From *Florae et Faunae Lusitanicae Specimen* (Vandelli 1797).

arrive in Lisbon, and the collections of the Real Museu da Ajuda became more and more diverse, particularly in Brazilian items. Specimens were also sent to the Museum of the Academy of Sciences and the Museum of the University of Coimbra. By the end of the 18th century, the Real Museu da Ajuda was described as not very large but with remarkable collections, especially of birds and shells (Simon 1983).

How large the crustacean collection of the Real Museu da Ajuda was is not known, but indirect evidence suggests it was modest. In fact, when (1808) Geoffroy Saint-Hilaire (French zoologist, 1772-1844) removed substantial collections (894 animal species represented by 1593 specimens) from the Museu da Ajuda to the Musum d'Histoire naturelle (Paris), only 12 specimens of Crustacea, representing 5 species were included (Fig. 5). Furthermore when the remnants of the Ajuda collections were incorporated in to the Museum of Lisbon (1858), there were no more than 30 to 40 species of Crustacea (Bocage 1862).

Following the Napoleonic invasions and the removal of collections by Geoffroy Saint-Hilaire, the Real Museu da Ajuda was completely disorganized. In 1836, Queen Maria II decreed the transfer of what remained of its collections to the Museum of the Academy of Sciences. The Escola Politécnica was founded in 1837, and in 1858 claimed the Museum of the Academy of Sciences to support its chairs of natural history.

4 THE ORGANIZATION OF TAXONOMIC RESEARCH

The Chair of Comparative Anatomy and Physiology, and Zoology of the Escola Politécnica, Barbosa du Bocage (1823-1907), was also the director of the zoological section of the Museum

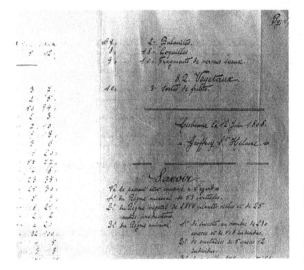

Figure 5. Portion of list, with Saint-Hilaire's signature, showing species of Crustacea and number of specimens removed from the collections of the Museu da Ajuda (Document D IV-16c, historical archives, Museu Bocage, Lisbon).

of Natural History, which from 1862-1905 would be named Museu Nacional de Lisbon. With unusual energy, du Bocage dedicated himself for nearly 50 years to work as a taxonomist, professor, and museologist. He assembled important zoological collections obtained through purchase and royal donations, and also through compensation from the Musum d'Histoire Naturelle (Paris) for those collections requisitioned by Saint-Hilaire. However, the greatest contributions to the collections came from persons living in, or exploring, Portuguese Africa and India, whom Bocage persuaded to make collections. For these correspondents, Bocage prepared instructions for collection, preparation, and shipping of zoological specimens (Fig. 6, left).

Carcinology has been one of the main concerns of the Museu de Lisbon. Felix de Brito Capello (1828-1879), naturalist at the Museu de Lisbon, published nine papers on Portuguese and African Crustacea, between 1865 and 1870 (Fig. 6, right). Besides his research on crustaceans, Capello (Fig. 7, left) also studied the taxonomy of Portuguese fishes and African arachnids.

Some years after Capello's death, carcinological research in the Museu de Lisbon was revived by Balthasar Osorio (1855-1926). Like Capello, Osorio (Fig. 7, right) dedicated himself to carcinological and ichthyological research. Between 1887 and 1923, he published 17

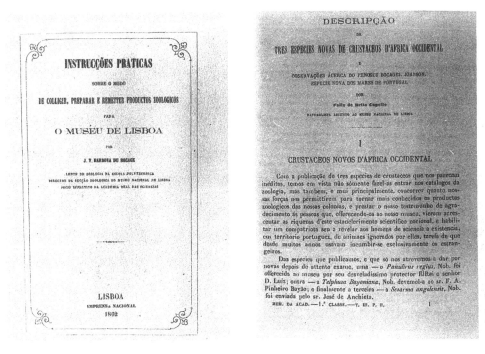

Figure 6. Left: Cover from Bocage's (1862) instructions for collecting and shipping zoological specimens to Museu de Lisboa. Right: First page from Capello's (1865) publication on West African Crustacea.

Figure 7. Left: Felix de Brito Capello. Right: Baltazar Osorio.

Figure 8. Augusto Nobre.

papers on Portuguese and African Crustacea. From 1921 to 1925, Osorio was director of the Museu Bocage (the name of the Zoological and Anthropological Section of the Museu Nacional de Lisbon since 1905).

At the Zoological Museum of the University of Coimbra, Manuel Paulino d'Oliveira (1837-1899) made a brief appearance in carcinology with the description of *Maja goltziana* in 1888. Still in the 19th century, Augusto Nobre (1865-1946) started his work in carcinology at the zoological museum of the University of Porto. Nobre (Fig. 8) organized the Zoological Museum and between 1896 and 1946 published 15 papers on Portuguese Crustacea (copepods, cirripedes, isopods, amphipods, schizopods, decapods, and stomatopods) and one book on Portuguese Decapoda and Stomatopoda (1931, first edition; 1936, second edition). Today, his papers on marine fauna are still fundamental to the study of the Portuguese biota.

5 SUMMARY

Three phases are recognized in the development of carcinology in Portugal. The first phase, or unsophisticated phase, was characterized by educated men, not specifically trained in natural history, reporting their observations on the Brazilian fauna. The second phase, or Linnean phase, was conducted by trained naturalists who lead overseas expeditions for the Real Museu e Jardim Botânico da Ajuda in the late 18th century. The third phase, or methodical phase, started in the mid-19th century when the National Museum of Lisbon, now Museu Bocage, was organized. Such researchers of the period as du Bocage, Capello, Osario, and Nobre are the founders of modern Portuguese carcinology.

ACKNOWLEDGEMENTS

I am deeply grateful to Dr. L.B. Holthuis for the determination of the species *Macrobrachium acanthurus* and *Callinectes bocourti* as those crustaceans figured in Frei Cristòvão's manuscript and to Dr. Frank Truesdale for revising my manuscript.

REFERENCES

Almaça, C. 1990. *Recursos animais e sua conservação. As populações Portuguesas do lagostim-de-rio, Astacus pallipes Lereboullet, 1858*. Lisboa: Museu Nacional de Historia Natural.

Anchieta, J. 1560. (1933). Ao Padre Geral, de São Vicente, ao ultimo de Maio de 1560. In *Cartas Jesuiticas III*: pp. 103-124. Rio de Janeiro: Civilização Brasileira.

Anchieta, J. 1585. (1933). Informação da Provincia do Brasil para o nosso Padre – 1585. In *Cartas Jesuiticas III*: pp.409-436. Rio de Janeiro: Civilização Brasileira.

Baptista, M.D. 1789. Ensaio de huma descripção fizica, e economica de Coimbra, e seus arredores. *Mem. Econ. da Acad. Real das Sci. de Lisboa* 1:254-298.

Bocage, J.V.B. 1862. *Instrucções praticas sobre o modo de colligir, preparar e remetter productos zoologicos para o Museu de Lisbon*. Lisboa: Imprensa Nacional.

Cristòvão de Lisbon, Frei 1627 (1967). *Historia dos animais e arvores do Maranhão*. Lisboa: Arquivo Historico Ultramarino.

Frade, F. 1966. Comentàrio zoològico relativo à Història dos animais e árvores do Maranhão (1625-1631), de Frei Cristòvão de Lisbon. *Garcia de Orta* 14(3) 343-350.

Marcgraf, G. 1648. Crustacei Pisces. In *Historiae Rerum Naturalium Brasiliae*: pp.182-189. Amsterdam: L. Batavorum.

Oliveira, M.P. 1888. Nouveau oxyrhynque du Portugal. *Instituto* 36:78-79.

Simon, W.J. 1983. *Scientific Expeditions in the Portuguese Overseas Territories (1783-1808)*. Lisboa: Instituto de Investigação Cientifica Tropical.

Vandelli, D. 1787. Florae, et faunae Lusitanicae specimen. *Mem. da Acad. Real das Sci. de Lisboa* 1:37-79.

From Oviedo to Rathbun: The development of brachyuran crab taxonomy in the Neotropics (1535-1937)

Gilberto Rodriguez
Instituto Venezolano de Investigaciones Científicas, Caracas, Venezuela

1 INTRODUCTION

The Neotropical Region extends from Mexico to Chilean Patagonia, and although originally defined by Sclater (1858) on the basis of the avifauna, 'Neotropical Region' is used here so that all Latin American countries and Caribbean islands may be considered in a common historical context. The process of describing and recording the biota of this vast region, after the establishment of the first European settlements, has been slow and not extraneous to the larger historical processes of each epoch.

Knowledge of the brachyurans of the Neotropical Region steadily progressed through the efforts of pre-Linnean naturalists, the French entomologists, and many other 18th and 19th century taxonomists, helped by a host of often forgotten collectors. By 1937 brachyuran species recorded from the Neotropics, and considered valid today, included 748 marine and estuarine species, and 61 freshwater species.

To deal with this large number of taxa, I compiled a data base from all major works dealing with Neotropical crabs, with species names cross-referenced with authors, dates, localities, collectors, collection dates, repository museums, and other miscellaneous data. Analysis of this data base allowed the historical development of taxonomy of Neotropical Brachyura to be traced.

2 THE PRE-LINNEAN BEGINNINGS

2.1 *Oviedo – Just after the discovery*

The first description of a crab from the New World was published in 1535, in Seville, by the Spanish chronicler Gonzalo Oviedo (b. Madrid 1478, d. Santo Domingo 1557). During his extensive travels through the Caribbean, Oviedo came to know intimately the natural history of the New World. The first volume of his monumental work *The Natural and General History of the Indies*, containing the chapters on the flora and fauna, was published in 1535, but the rest of the work, dealing with anthropology and political events, as well as the illustrations, remained in manuscript until the Royal Academy of History began to publish it in 1851 (Oviedo 1851-1855). Writing about the animals of Hispaniola and the Spanish Main, Oviedo said: 'The crabs are terrestrial animals which come out of burrows they make on the ground, and their head and body are rounded like a hawk's hood, with four legs on each side, and two pincer-like mouths of unequal size; ... their shells and other parts are smooth and thin, like eggshell, but harder. Their color is brown, or white, or purple, approaching blue, and they walk sidewards; they are good to eat, and the Indians are very fond of this dish; and when the Christians go inland, it is a quick food roasted over live coal. Finally, their shape is like that of the Sign of Cancer. In Andalucia, near the seashore, and in the Guadalquivir, where this river enters the sea, and in other places, there are also crabs, but these live in the water, whereas the present crabs live on the land' (Oviedo 1535).

The color, terrestrial habitat, and edibility of the crabs described by Oviedo identify them as *Cardisoma guanhumi*, a species formally described 3 centuries later by Latreille.

2.2   *Marcgraf and the Natural History of Brazil*

For the next 100 years, several chroniclers reported on natural history of the New World including Anchaeta, de Lry and de Sousa, who explored in Brazil (See Tavares, this volume), but only cursory mention was made of the invertebrates. However, towards the mid-17th century a fortuitous event brought an able and competent naturalist to South American shores, and with him knowledge of the Brachyura really began. Georg Marcgraf (b. 1610, d. 1644) arrived in Brazil in 1638 as part of the cultural mission of Johan Maurits van Nassau-Siegen, the first and last governor-general of Dutch Brazil. For the next 6 years Marcgraf devoted himself to astronomy, geography, and natural history, exploring the northeastern section of the country, which comprises the present states of Pernambuco, Parahyba, and Rio Grande do Norte. In 1644, the young naturalist was expecting to return to Europe, but instead he was sent to Angola, Africa, then under Dutch rule, where he died (Chardon 1949, Tavares, this volume).

When Johan Maurits returned to the Netherlands he gave all of Marcgraf's writings and scientific observations from Brazil to Johan de Laet (b. 1595, d. 1649), scholar and Prefect of the Dutch West India Company, who patiently organized the manuscripts and deciphered those written in code. Willem Piso (or Pies) (b. 1611, d. 1678), who had been Johan Maurits' personal physician in Brazil and another scientist with Marcgraf in the cultural mission, published *Historia Naturalis Brasiliae* in Amsterdam in 1648. The first part of this work, devoted to diseases and other medical topics, was written by Piso. The other two sections, 'Historiae Plantarum,' and 'Rerum Naturalium Historiae,' were the work of Marcgraf. In this last section is a list of crabs with their native names (Marcgraf 1648), some of which can be identified (Rathbun 1918, 1930, 1937) as follows:

Guaiâ alia species = *Persephona punctata*
Aratu pinima = *Aratus pisonii*
Granhumi = *Cardisoma guanhumi*
Uca una = *Ucides cordatus*
Maracoani = *Uca maracoani*
Ciecie Ete = *Uca thayeri*
Ciri Apoa = ?*Callinectes danae*
Ciri Obi = ?*Arenaeus cribrarius*

2.3   *Naturalists in the British possessions in the Caribbean*

The next step in our knowledge of Neotropical Brachyura came 50 years later, in the Caribbean, from Sir Hans Sloane (b. 1660, d. 1753). Sloane was a renowned physician and a naturalist who described a few species, but who is best known for amassing collections studied by others.

When Christopher Monck, second duke of Albermarle, was appointed governor of Jamaica, Sloane accompanied him as physician, sailing on 12 September 1687. Sloane stayed in Jamaica less than 2 years, and returned to England in 1689 to pursue a successful medical career (de Beer 1975).

Sloane's description of the voyage and the observations on the inhabitants, diseases, plants, animals, and meteorology of the West Indies were published between 1707 and 1725, in *A Voyage to the Islands of Madera, Barbados, Nieves, St. Christophers and Jamaica*. Two species of crabs from Jamaica mentioned by Sloane (1725) can be identified with those bearing binomials (Rathbun 1918). His 'Cancer terrestris cuniculos sub terra agens' is *Gecarcinus ruricola* and 'Cancer palustris cuniculos sub terra agens' is *Uca maracoani*.

In addition to his personal collections, Sloane bought other collections of natural history and ethnography, as well as antiquities, paintings by famous artists, coins, and medals. These collections, together with his library of 50000 volumes and 3500 manuscripts, were purchased by the British government and formed, with the Harleian collection and the Cotton library, the basis of the British Museum (de Beer 1975).

Patrick Browne (b. 1720, d. 1790), who studied natural sciences in Paris and graduated in medicine at Leiden, was another physician-naturalist who collected in Jamaica. Browne had been in the Caribbean in his early youth, and in 1746 returned and established himself as physician in Jamaica. During this stay he extensively collected specimens all over the island, particularly plants of which he gathered 1200 species, 400 more than Sloane. He returned to Britain in 1755, and published *The Civil History of Jamaica* (1756). Browne made still another trip to the Caribbean, visiting Montserrat and Antigua, and then retired to Ireland. Of the Brachyura recorded by Browne (1756), the 'Three thorned Crab' is *Persephona punctata*, 'Cancer 4' (or 'larger long-shanked crab') is *Lupella forceps*, and 'Cancer 9' is *Macrocoeloma trispinosum*. The 'Cancer 1' or 'Oyster Crab' is perhaps *Pinnotheres geddesi* (see Rathbun 1918).

2.4   *Hughes and 'The natural history of Barbados'*

The Natural History of Barbados was published (1750) in London 3 years before Linnaeus's *Species Plantarum*, and was a well illustrated treatise by Griffith Hughes (b.ca.1707 d.?), Master of Arts, Fellow of the Royal Society, and Rector of St. Lucy Parish in Barbados (Chardon 1949). The work dealt in 10 successive books (sections) with the geography, materia medica, health, land animals, flora, and marine animals. In the 9th book Hughes lists 15 kinds of crabs, but the descriptions are so brief and the names so trivial that it is impossible to determine most of the species with any great degree of certainty. Only two of the brachyuran crabs are figured; one, the Horned Crab is an undoubted *Stenocionops furcata*, the other, the Lazy Crab, is a composite, having the chelipedes and ambulatory legs of *Mithrax spinossisimus* attached to the body of a parthenopid crab, *Daldorfia horrida* (=*Parthenope horrida* of authors), an Indo-Pacific species' (Rathbun 1921).

3   LINNAEUS AND THE ENTOMOLOGISTS, 1758-1836

Among a few score of pre-Linnean names for Neotropical crabs published before the middle of the 18th century, 14 species can be discerned, as follows: Species from mangroves – (1) 'Aratv pinima' = *Aratus pisonii* (2) 'Vca vna' = *Ucides cordatus* (3) 'Maracoani' or 'Cancer palustris cuniculos sub terra agens' = *Uca maracoani* (4) 'Ciecie Ete' = *Uca thayeri* (5) 'Oyster Crab' = ?*Pinnotheres geddesi*; semiterrestrial habitats (6) 'Granhumi' = *Cardisoma guanhumi* (7) 'Cancer terrestris cuniculos sub terra agens' = *Gecarcinus ruricola*; littoral lagoons and the immediate sandy or rocky sublittoral (8) 'Ciri Apoa' = ?*Callinectes danae* (9) 'Three thorned Crab' = *Persephona punctata* (10) 'Ciri Obi' = ?*Arenaeus cribrarius* (11) 'Cancer 4' = *Lupella forceps* (12) 'Horned Crab' = *Stenocionops furcata* (13) 'Lazy Crab' = part. *Mithrax spinossisimus* (14) 'Cancer 9' = *Macrocoeloma trispinosum*.

Up to now, not only were the names of Neotropical crabs largely indeterminate, but their descriptions were the product of traveler-naturalists who worked without museums and permanent collections. Even after publication of the 10th edition of Linnaeus' *Systema Naturae* some descriptions of American crabs were made under colloquial names. Father Louis Nicholson (no dates available), a French Dominican who lived for 4 years on Hispaniola, listed the dromiid *Hypoconcha sabulosa* as 'Faux Bernard l'Hermite' in the glossary to his book *Essai sur L'histoire Naturelle de Saint-Domingue* (Nicholson 1776). Antonio Parra (no dates available), commissioned by the Spanish government and the Botanical Garden of Madrid to collect in Cuba for the Royal Cabinet of Natural History (Parra 1787; Chardon 1949), described under Spanish names the common Cuban crabs *Cardisoma guanhumi* ('Cangrejos terrestres'), *Lupella forceps* ('Xaiva de horquilla'), *Stenorhynchus seticornis* ('Arana'), *Libinia rhomboidea* ('Cangrejo peludo'), *Mithrax spinossisimus* ('Cangrejo santoya'), ?*Mithrax verrucosus* ('Cangrejo denton'), *Mithrax cornutus* ('Cangrejo espinoso'), and *Stenocionops furcata* ('Cangrejo cornudo'). Similarly, Albert Seba (b.1665 d.1736) in his posthumous *Locupletissimi Rerum Naturalium Thesauri* (Seba 1761) gives descriptive Latin names for the following species : 'Cancer Pagurus, hirsutus, Americanus, pronus' (= *Ucides cordatus*); 'Cancer sulcatus, terrestris, sive montanus, Americanus' (= *Gecarcinus ruricola*); 'Cancer Uka una, Brasiliensis' (= *Uca heterochelos*); 'Cancer marinis, scutiformis' (= *Portunus sebae*); 'Aranaeus marinus' (= *Mithrax sculptus*).

In the 10th edition of *Systema Naturae* (1758) Linnaeus (b. 1707, d. 1778) gave binomial names to three species of American crabs, *Cancer grapsus* (= *Grapsus grapsus*), *C. ruricola* (= *Gecarcinus ruricola*), and *C. punctatus* (= *Persephona punctata*), and in *Amoenitates Academici* (1763) he named *C. cordatus* (= *Ucides cordatus*) and *C. epheliticus* (= *Hepatus epheliticus*). The range of *Cancer chabrus* (= *Plagusia chabrus*), described originally from the Indian Sea (Linnaeus 1764) was extended by later authors to Chile, and *C. minutus* (= *Planes minutus*), described as 'Pelagi Fuco natante' in 1758, was also recorded from America.

After the system of binomial nomenclature was established, the development of carcinology came under the influence of entomologists, particularly those in France and Germany. This was the consequence of Crustacea being included as a class of insects in *Systema naturae*. Lamarck writes: 'Dan mon cours de l'an VII (1799) j'ai établi la classe des crustacés. Alors M. Cuvier, dans son Tableau des Animaux, p. 451, comprenait encore les crustacs parmi les insectes; quoique cette classe soit esentiellement distincte, ce ne fut néanmoins que six ou sept ans aprés que quelques naturalistes consentirent a l'adopter' (Lamarck 1809).

Johan Christian Fabricius (b. 1745, d. 1808), the foremost entomologist of the 18th century, was a pupil of Linnaeus for 2 years in Upsala. After some years of travel abroad he was appointed professor in Copenhagen (1770), and then in Kiel (1775) (Tuxen 1967). Fabricius followed Linnaeus' system and adapted it for the insects in a series of works that began in 1775 with *Systema Entomologiae*. In this work and in others (1781, 1787, 1792-1794) Fabricius described 10 species of Neotropical crabs including the common beach crab *Ocypode quad-*

rata. Coetaneous with Fabricius was J. F. W. Herbst (b. 1743, d. 1807), a German priest, but also an able entomologist who contributed particularly to the systematics of Coleoptera (Lindroth 1973). Herbst named 11 species of Neotropical crabs in the three volumes of *Versuch einer Naturgeschichte der Krabben und Krebse* (Herbst 1781-1804), among them the arrow crab, *Stenorhynchus seticornis*.

Guillaume Antoine Olivier (b. 1756, d. 1814), studied medicine at Montpellier, but became interested in natural history, collecting insects in England and Holland, providing material for his *Encyclopédie Méthodique* (1791). In the sixth volume of this work, he described *Stenocionops furcata* (Olivier 1880, Papavero 1971-1972).

The Paris Museum or the Muséum National d'Histoire Naturelle was founded in 1793, during the French revolution, from the reform of the Jardin des Plantes. The museum soon gathered some of the most distinguished entomologists in Europe, among them, Lamarck, Bosc, and Latreille.

Jean-Baptiste-Pierre-Antoine de Monet, de Lamarck (b. 1744, d. 1829) was a precise naturalist as well as a philosopher. Destined to the priesthood, young Lamarck preferred a military career, which he followed until 1765 when he became a clerk in a Paris bank. At 26 he began his studies of medicine and natural history, becoming a skilled botanist under the guidance of Bernard de Jussieu (b. 1699, d.1777). In 1793, Lamarck was appointed a professor in the Paris Museum, not in botany but in invertebrate zoology (Guyénot 1857). During this period he published *Système des animaux sans vertébres* (1801), and *Histoire naturelle des animaux sans vertébres* (1818), in which he described 7 species of Neotropical crabs.

Louis Agustin Guillaume Bosc (b. 1759, d. 1828), a close friend of Fabricius, studied natural sciences in Paris and later served the Directoire as French Consul in New York. In America he collected insects and other organisms. In 1803 he was appointed inspector of the Gardens of Versailles, and in 1806 professor in the Paris Museum (Papavero 1971-1972). In 1801-1802 he published the first volume of *Histoire naturelle des Crustacés*, in which he described *Sesarma cinereum*, collected during his visit to the Carolinas in 1798-1800.

Pierre-André Latreille (b. 1762, d. 1833) was the leading French entomologist of this period. Fabricius, who was his friend and sometimes also his antagonist, called him 'entomologorum nostri temporis omnium princeps' (the foremost entomologist of our time), an opinion largely shared by other contemporaries (Lindroth 1973).

Latreille, an ordained priest, was arrested during the French revolution, sentenced to be deported to Cayenne, but saved by Georges-Marie Bory de Saint-Vincent (b. 1778, d. 1846). Latreille entered the Paris Museum in 1798 as 'aide-naturalist,' through the influence of his protector Lamarck, where he published the 14 volumes of *Histoire Naturelle, Génrale et Particulière, des Crustacés et des Insectes* (1802-1805), *Genera Crustaceorum et Insectorum* (1806-1809), the *Considérations générales sur L'ordre Naturel des Animaux Compossant les Classes des Crustacés, des Arachnides et des Insectes* (1810), the chapter on the insects of Cuvier's *Règne Animal*, and the *Mémoires pour Servir a L'histoire des Insectes* (1819). He continued as 'aide' until 1820 when he replaced Lamarck, then almost entirely blind, as professor. With the death of Lamarck in 1829, the chair in invertebrate zoology was divided to create a new one of entomology, which Latreille occupied from 1830 until his death in 1833 (Papavero 1971-1972).

Latreille described seven Neotropical species in his contributions of 1803, 1819, and 1825. Some of Latreille's materials were obtained by two collectors from the Paris Museum: The French botanist Auguste Plée (b. 1787, d. 1825) who visited Puerto Rico, the Lesser Antilles, and the coast of Venezuela, from 1820 to 1823 (Papavero 1971-1972); and Alexander Ricord

(b. 1798, d. ?), a native of Baltimore, who traveled extensively in the Antilles (Chardon 1949).

During this period Thomas Say (b. 1787, d. 1834) published descriptions of several North American species of brachyurans whose ranges were later extended to include the Neotropics. At first Say was an apothecary in his native Philadelphia, but when he was 25 years old devoted himself entirely to the study of natural history. When the Academy of Natural Sciences of Philadelphia was founded in 1812, he was one of its first members. In 1825 he left Philadelphia to join in the Utopian experiment at New Harmony, Indiana, where he remained until his death. Say is best known for his entomological and malacological work, but he also contributed 5 papers on the Crustacea of the United States. These were published in 12 parts, in 1817 and 1818, in the first volume of the *Journal of the Academy of Natural Sciences, Philadelphia* (Holthuis 1969). In these papers Say brought together all available information and added descriptions of many new species. The following are Say's new species of crabs (with their present generic associations) later recorded from the Neotropics: *Pinnotheres depressus, P. maculatus, Fabia byssomiae, Eurytium limosum, Menippe mercenaria, Pilumnus sayi* and *Metoporaphis calcarata.*

During this period other species were described from localities outside the Neotropics but later recorded there. These included *Latreillia elegans*, described from the Mediterranean by Polydore Roux (b. 1792, d. 1833) in his *Crustacés de la Méditerranée et de Son Littoral* (1828-1830); and *Ovalipes punctatus*, described by W. de Haan (b. 1801, d. 1855) from Japan in the Crustacea section of P. F. von Siebold's *Fauna Japonica* (1833). Felix Edouard Guérin-Méneville (b. 1799, d. 1874), one of the most prolific writers in entomology and who studied the insects collected during the circumnavigation of *La Coquille* (1822-1825), published (1828) a memoir on *Eurypodius latreilli* from the Falkland Islands, a species whose range was later extended to temperate South America.

The last of the great French entomologists to publish on Neotropical crabs during this period was Henry Milne Edwards (b. 1800, d. 1885). After medical and zoological studies in Paris he joined, in 1832, the cole Centrale des Arts et Manufactures as professor of hygiene and natural history. His researches on invertebrates earned him the Academy of Sciences prize in experimental physiology in 1828 and his election to the zoology section of the Academy in 1838. In 1841 he was appointed to the chair of entomology at the Paris Museum where he had long occupied a laboratory. In 1861, he was transferred to the chair of mammalogy left vacant by the death of Geoffroy Saint-Hilaire (b. 1772, d. 1844) (Anthony 1974).

Milne Edwards's early investigations on the internal anatomy of crustaceans, published with Jean Audouin (b. 1797, d. 1841) who preceeded him as the chair of entomology at the Museum, were the basis for his classic *Histoire naturelle des crustacés*, in three volumes (1834-1840). In the Histoire and in three other miscellaneous papers (1832, 1833, 1835) he described 17 new species of Neotropical crabs.

From 1758 to 1834, 75 species of Neotropical brachyurans were described. Over the next decade two other French entomologists added minor contributions: Christophe de Freminville (b. 1787, d. ?), zoologist and naval officer who described *Gecarcinus lateralis* (1835), and Hyppolyte Lucas (b. 1814, d. 1899), entomologist and arachnologist, who published, with H. Milne Edwards (1843), descriptions of several Neotropical crabs.

4   THE DEVELOPMENT OF CARCINOLOGY IN CHILE

Knowledge of the crab fauna of the Pacific coast of South America during the 19th century came mostly from Chile. This was because the civil and cultural development of Chile was ahead of that of other South American countries, and Valparaiso was a main port-of-call.

The first carcinologist born in the New World was Juan Ignacio Molina (b. Talca, Chile, 1740, d. Bologna, Italy 1829). He studied languages and natural sciences in the Jesuit college at Concepcion, where he entered the Jesuit order and was made librarian of the college. In 1768 he left Chile because the Jesuits had been expelled from Spanish dominions, and his notes were lost during the exodus to Italy. In 1774, Molina was appointed professor of natural sciences at the Institute of Bologna, where he wrote most of his works from memory, with the help of the travel narratives of Frezier, Anson, and Feuillet (Guerra 1974).

Molina's work *Compendio della Storia Geografica, Naturalle e Civile del Regno di Chile* (Bologna 1776) was published anonymously, and subsequently greatly improved and enlarged as *Saggio sulla storia Naturale del Chili* (1782). In the *Saggio*, Molina gave short diagnoses of the following Chilean crabs: (1) *Cancer talicuna* (2) *C. apancora* (3) *C. setosus* (= *C. polyodon*) (4) *C. xaiva* (?*Taliepus dentatus*) (5) *C. coronatus*. His *Cancer talicuna* and *C. apancora* cannot be recognized, and *C. coronatus* seems to be an imaginary species (Philippi 1894), or perhaps conspecific with *C. plebejus* (see Rathbun 1930). Molina also described the lithodid *Cancer santolla* (= *Lithodes santolla*), and the palaemonid prawn *Cancer caementarius* (= *Cryphiops caementarius*.)

Half a century passed before Chilean crabs were mentioned again in the literature. Milne Edwards (1833, 1834) described four species from Chile; one of these was perhaps conspecific with Molina's *Cancer jaiva*, and another, *Epialtus bituberculatus*, was an Atlantic species erroneously assigned to Chile.

In 1835, the English zoologist Thomas Bell (b. 1792, d. 1880), professor of Zoology in King's College, London, and president of the Linnean Society, described four species of *Cancer* collected near Valparaiso: (1) *Cancer irroratus* (= *C. plebejus* Poeppig, not *C. irroratus* Say); (2) *C. dentatus* (= *C. polyodon* Poeppig, not *C. dentatus* Herbst); (3) *C. edwardsii*; (4) *C. longipes* (= *C. porteri* Rathbun). Bell (1836) also described new species of Majidae from Pacific America; 13 were from the Galapagos Islands, 1 from Peru, 3 from Central America, and the syntypes of another were collected in the Galapagos and at Valparaiso. Most of the material described by Bell in 1835 and 1836 came from the collections of Hugh Cuming (see below).

Another important contribution to knowledge of the brachyuran fauna of Chile was made in 1836 by the German traveler-naturalist Eduard Friederich Poeppig (b. 1798, d. 1866). He listed nine species of Brachyuran crabs he had collected in Valparaiso, including five new species (Poeppig 1835-1836). But, the largest contribution to Chilean carcinology during the 19th century was that of H. Milne Edwards and Lucas in the fifth volume of D'Orbigny's *Voyage dans l'Amérique Méridionale* (1843). They listed 40 species from Chile, and gave illustrations of almost all of them. Additionally they recorded 1 species from Guayaquil and 3 others from El Callao; 14 of the Chilean species were new to science. However, their *Pseudothelphusa chilensis* (= *Hypolobocera chilensis*) was not a Chilean species, but a freshwater crab from Lima.

In 1837, the Chilean Government commissioned Claude Gay (b. 1800, d. 1873), a naturalist of French origin but resident in Chile from 1828 until his death (Papavero 1971-1972), to write a physical description of the country, including its natural history. Gay's *Historia Fisica y Politica de Chile* was published in 1849, with the section on Crustacea written by the French entomologist Hercule Nicolet (b. ?, d. 1872), a specialist in Thysanura and Collembola (Lindroth 1973). Nicolet (1849) added nine species to the list of Milne Edwards & Lucas (1843). However of these nine, only *Pinnotheres bipunctatus* is considered valid today. For example,

Trichodactylus granarius is not a Trichodactylidae but a synonym of *Hemigrapsus crenulatus* and *Liriopea leacheii* and *L. lucasii* are synonyms of *Halicarcinus planatus*.

To close this section on the development of Chilean carcinology two Chilean zoologists must be briefly mentioned. Rudolph Amandus Philippi (b. Charlottenburg, Germany, 1808, d. Santiago Chile, 1904) settled in Chile in 1851, and was professor of botany and zoology in the University of Chile and director of the National Museum for 43 years. Philippi published more than 414 papers (Papapvero 1972), but only 7 were on decapods (Balss & Gruner 1961). Carlos Emilio Porter (b. Valparaiso 1867, d. Santiago, 1942) was director of the Valparaiso Museum from 1897 to 1906 and curator in the invertebrate section of the Museo Nacional in Santiago from 1911 to 1928 (Anonymous 1980). Porter published 23 papers on decapods from 1903 to 1937 (Balss & Gruner 1961), mostly giving new records for species, but also describing *Paromola rathbuni* in 1908.

5   WHITE'S LIST OF CRUSTACEA OF THE BRITISH MUSEUM

In a booklet published in 1847, Adam White (b. 1817, d. 1879) gave 'a complete list of the specimens of Crustacea contained in the collections of the British Museum' (White 1847a). Of the 466 species of Brachyura on the list, 70 species were already known from Latin America, and 10 of 11 new species from this area were nomina nuda; only *Valdivia serrata* was validated with a short description published elsewhere (White 1847b). The material for the list came from various sources. As stated in the introduction by John Edward Gray (b. 1800, d. 1875), British naturalist and keeper of the zoological collections at the British Museum, 'the specimens collected by ... Dr. W. E. Leach, may be considered as the nucleus of this collection.' Possibly six Latin American species came from this old collection. The types of species from the United States described by Say in 1817-1818, were also deposited at the British Museum; White's record of *Panopeus herbstii* is of a Say species that ranges to the Caribbean.

Some of the old Sloane collection are represented on White's list by specimens of *Persephona latreillii* (= *P. punctataD*) which Leach had used in 1817 in describing the species. The specimens under *Persephona lamarckii* (= *P. punctataD*) also came from the Sloane collection, since Thomas Bell (1855), referring to the type specimens of the two Leach species, states: 'It is remarkable that Leach should have been unaware that those specimens were originally in the Sloanian Collection, and therefore brought from the West Indies.' Many collectors or donors which appear in White's list are well known 19th century traveler-naturalists. Specimens of West Indian species were presented by Rev. Lansdowne Guilding, H. B. Hillier, and Lieutenant E. Redman, or purchased from Mr. Scrivener and Mr. Gosse. Philip Henry Gosse (b. 1810, d. 1888), the well known British naturalist, studied bird and plant life in Jamaica in his early years (Gosse 1840).

The Reverend Lansdowne Guilding (b. Kingstown St. Vincent, 1797, d. Jamaica (?), 1831) studied at Oxford and returned to St. Vincent in 1817, where he developed his interest in natural history (Guilding 1828). In a paper on some Crustacea from the West Indies, published in 1825, he described *Leptopodia ornata* (= *Stenorhynchus seticornis*); the specimen mentioned by White is probably Guilding's type from St. Vincent.

The South American species of Crustacea in White's list were represented by specimens donated by Lord Stuart de Rothesay, W. Swainson, and P. George Smith and his wife, from Brazil; Sir R. Schomburgk, from British Guiana; Capt. Sir James C. Ross, Dr. Joseph Hooker, W. Wright, and Charles Darwin, from the South coast of America (i.e., Patagonia and adjacent islands); T. Bell from Chile; Dr. Andrew Sinclair, from Guayaquil (Ecuador).

William Swainson (b. Liverpool, 1786, d. New Zealand 1855) traveled to Brazil in 1816 and explored there for 2 years. J. P. George Smith, mentioned with his wife, was the son-in-law of J. E. Gray, keeper of zoology in the British Museum (see above), and seems to have collected in Para and Pernambuco (Brazil) around 1844.

Sir Robert Hermann Schomburgk (b. 1804, d. 1864), explored the British and Venezuelan Guianas between 1835 and 1839, and described his travels in a book (Schomburgk 1841) and several articles. In 1848 he was British consul to Haiti. His brother Richard also explored in British Guiana (1840-1844). In White's List, Schomburgk's name only appears as the collector of a new species of pseudothelphusid crab, *Potamocarcinus schomburgkii* (= *Kingsleya latifrons*).

The names of Captain James Clark Ross (b. 1777, d. 1856) and the botanist Sir Joseph Dalton Hooker (b. 1817, d. 1911) appear in connection with the British Antarctic Expedition (1839-1843). Two South American crabs came from this expedition.

In some cases it is not clear from White's list who were collectors and who were donors of collections made by others. M. Dufresne (Pierre Dufresne, botanist from Geneva, b. 1786, d. 1836) probably collected the specimen of *Atelecyclus chilensis* reported from Chile, but George B. Sowerby, English conchologist (b. 1788, d. 1834), transmitted collections made by others. This is also the case with Thomas Bell who is mentioned in the list as describing 11 species of crabs in 1835 and 1836. Although not indicated by White, most of the specimens attributed to Bell were collected by Cuming. Hugh Cuming, British conchologist (b. 1791, d. 1865) was also a sailmaker who established a successful business in Valparaiso, Chile, in 1820. In 1826, he retired from sailmaking and built a yacht, Discoverer, fitted expressly for the collecting and storing of natural history objects. With this yacht he made two long collecting expeditions. The first took him, by way of Juan Fernandez Islands, to Easter Island and many islands of Polynesia, including Tahiti, returning to Valparaiso in June 1828. The second, from 1828 to 1830, took him to many localities on the coast of Chile, Peru, Ecuador, Colombia, Panama, Costa Rica, Nicaragua, Honduras, and in the Galapagos Islands. Having put his collections in order, he left Valparaiso in May 1831, and returned to England. On 1 May 1832 Cuming was recognized by the Linnean Society of London as Fellow, an honor he repaid by donating collections of animals and plants to the Society's museum (Dance 1966).

Curiously, only one crustacean collected by Darwin during the voyage of the Beagle (1831-1836) is mentioned in the list, the isopod *Serolis orbignyi* from the Falkland Islands. The bulk of Darwin's crustaceans, other than the barnacles, went to Thomas Bell and although Bell was a contemporary of White and associated with the British Museum, the Beagle crustaceans were apparently kept in Bell's private collection and never worked up (D. Porter 1985). Bell's zoological collections were purchased in 1863 by Professor John Obadiah Westwood (b. 1805, d. 1893) to enrich the Hope Entomological collections at Oxford University, and were transferred in 1899 to the Oxford University Museum (Chancellor et al. 1988). Among these collections, Chancellor et al. (1988) were able to identify 21 species of Brachyura from South America collected by Darwin. It is interesting to note that six of these species were described many years after the return of the Beagle, by Dana (1852), Stimpson (1859b), A. Milne Edwards (1875), and Rathbun (1898a, 1914), using materials from other collections.

Some of the other collectors mentioned in White's list, like Lt E. Redman (Jamaica) and Dr. Andrew Sinclair (Guayaquil), were Royal Navy personnel onboard ships in the mentioned localities. The largest collection in the list, from the West Indies, was 'purchased from Mr. Scrivener.' Unfortunately, I have been unable to identify this person among the several prominent persons of that name who figured in English history during the first half of the 19th century.

6 AMERICAN CARCINOLOGISTS AND THE DEEP-SEA FAUNA

6.1 *Beginnings of the American influence in carcinology*

From 1839 to 1858, 79 species of crabs from the Neotropics were described, some of them by Bell (1835, 1836); Milne Edwards & Lucas (1843); White (1847b); Nicolet (1849); Adolf Gerstaecker (b. 1828, d. 1895), a German zoologist and professor in Greifswald, who added 3 new species from the Galapagos, Mexico, and Venezuela in 1856; and Henri de Saussure (b. 1829, d. 1905), from the Geneva Museum, who described 6 new species from Mexico (1853, 1857, 1858), as a result of his 1853 trip to that country. But the beginnings of U.S. influence on brachyuran taxonomy also began during this period. J.W. Randall (b. 1813, d. 1892), a Boston physician (Holthuis 1959a), described (1840) *Sesarma rectum*, from material collected in Surinam by Dr C. Hering (b. 1800, d. 1880), and four other species of crabs from the United States whose ranges were later extended to the Neotropics. Augustus Addison Gould (b. 1805, d. 1866), also a Boston physician, described *Rhithropanopeus harrisii* in *Report on the Invertebrata of Massachusetts* (1840), a species later found in Mexico and South America. Gould is best known for his studies of Mollusca, of which he described 1100 species (Gifford 1972). Also, Lewis R. Gibbes (b. 1810, d. 1894), a naturalist and physician from Charleston, South Carolina, described, in 1848 and 1850, five North American species of Brachyura later recorded from tropical America.

6.2 *United States Exploring Expedition (1838-1842)*

The US Exploring Expedition was the first great scientific mission undertaken by the United States, and also the first by the Depot of Charts and Instruments, forerunner of the U.S. Hydrographic Office. The expedition was carried out with a fleet of five vessels under the command of Lieutenant Charles Wilkes (b. 1798, d. 1877), head of the Depot since 1832, on the sloop-of-war Vincennes; the squadron also included the vessels Peacock, Porpoise, Flying Fish, and Sea Gull. The scientific personnel were Dr. Charles Pickering (b. 1805, d. 1878), James D. Dana (b. 1813, d. 1895), J.P. Courthouy (b. 1808, d. 1864), and Titian R. Peale (b. 1800, d. 1885), zoologists; William Rich (b. 1800, d. 1864), and J.D. Breckenridge (b. 1810, d. 1893), botanists; and Horatio Hale (b. ? 1818, d. 1896), philologist.

The expedition sailed from Norfolk, Virginia, on 18 August 1838, and arrived back at New York City on 10 June 1842, after circumnavigating the world. On its South American leg the expedition visited Rio de Janeiro, from 23 November 1838 to 6 January 1839; and Rio de la Plata from 25 January to 3 February 1839; and then proceeded to Argentinian Patagonia; rounded Cape Horn; and entered Orange Harbor, New Island, Tierra del Fuego, on 17 February 1839. After a month-long expedition to Antarctica, the squadron departed for Valparaiso, Chile, arriving in mid-May and staying until the beginning of June. On the 20 June 1839 the squadron gathered at San Lorenzo Island, near Callao, Peru, and later, from Callao, the scientific party made trips to Lima and the Cordillera. On 13 July 1839 the expedition left the coast of South America for Polynesia (Wilkes 1852).

James Dwight Dana, later a professor at Yale University (1850-1890) wrote the reports on Crustacea, zoophytes and geology. Although best known as a geologist, Dana was also an important carcinologist. The following new species of crabs collected by the US Exploring Expedition in South America were described by Dana in 1851 and 1852: *Epialtus brasiliensis* (Rio de Janeiro), *Sesarma angustipes* (South America, probably Rio de Janeiro), *Cyrtograpsus angulatus* (Rio Negro, Patagonia) and *Cyclograpsus cinereus* (Valparaiso).

6.3   *European oceanographic expeditions*

Although several exploratory expeditions were organized by the French Government, one of the few to collect crabs in the Neotropics was that of the corvette *L'Uranie* (1817-1820), under command of Captain Louis Claude des Saulses de Frecinet (b. 1779, d. 1824). Charles Gaudi-chaud-Beaupèr (b. 1789, d. 1854), was the naturalist onboard and was assisted in his collecting by the ship's surgeon Jean Rene Constantin Quoy (b. 1790, d. 1869) and Joseph Paul Gaimard (b. 1790, d. 1858). The expedition visited Rio de Janeiro in 1817, and again on its way home in 1820, after the L'Uranie was lost in the Falklands and the expedition transferred to an American ship rechristened La Physicienne. In Rio de Janeiro, Quoy and Gaimard collected the holotype of *Pilumnus quoyi* described by H. Milne Edwards in 1834.

The Austrian frigate Novara circumnavigated the globe (1857-1859) and made collections in Chile and Ecuador. The collections were deposited at the Vienna Museum, and were reported on in 1865 by Camil Heller (b. 1823, d. 1917). They included a new species *Pachygrapsus pubescens* from Chile. Heller (1862) had already dealt with Neotropical decapods, describing two of the most common freshwater prawns of South America, *Macrobrachium amazonicum* and *M. brasiliensis*, from material collected by Johan Natterer (b. 1787, d. 1843) for the Vienna Museum, during the latter's explorations (1817-1836) in Brazil (Holthuis 1959a).

The most famous European expedition was the Challenger Expedition, which circumnavigated the world from 7 December 1872 to 24 May 1876. Neotropical crabs were not particularly abundant in the Challenger materials due to limited collecting in South America and the depth of the dredgings, usually over 1000 fathoms (Thomson 1877). The Challenger worked in the Neotropics on three occasions. From 15-25 March 1873 the vessel was in the Caribbean around the islands of Sombrero and St. Thomas (390-460 fathoms). From September 1st to September 14th, the expedition surveyed the Brazilian littoral, dredging at depths less than 30 to 1600 fathoms (Campbell 1877), off Fernando Noronha, Cape San Roque, Pernambuco, Parahyba, Bahia, Maceio, and the San Francisco River. Finally, on the trip home, Challenger approached South America from the Pacific, surveyed the Juan Fernandez Islands (13 November 1875) and the coast in the vicinity of Valparaiso, Chile (19 November 11 December 1875). From Chiloe and the Gulf of Penas, Challenger sailed through the Strait of Magellan and arrived at the Falkland Islands on 18 January 1876. From the Falkland Islands, Challenger made a last stop in South American at Montevideo.

The crabs of the Challenger Expedition, reported by Miers in 1886, included 6 new species: *Macrocoeloma concavum*, Fernando Noronha, 7-20 fathoms; *Notolopas brasiliensis*, Bahia, 7-20 fathoms; *Libinia smithii* (= *Libidoclea smithii*), Chile, 245 fathoms; *Libinia gracilipes* (= *L. granaria*), Chiloe, 45 fathoms; *Picroceroides tubularis*, Fernando Noronha and Bahia, shallow water; *Eurypodius longirostris*, Chile, 175 fathoms; and *Iliacantha intermedia*, Bahia.

6.4   *The development of museums in the United States*

At the beginning of the 19th century, museums associated with local scientific societies were already centers of activity for naturalists like Thomas Say, who was associated with the Philadelphia Academy of Arts and Sciences. The Charleston Museum, the first public museum in the United States (1773), was the center of activity for the physician-naturalist Lewis R. Gibbes (see above). But the major development of museums in the United States dates from about the middle of the 19th century (Coleman 1939). Three museums of this era came to be important repositories of Neotropical crabs during the second half of the 19th century and into the 20th century: The National Museum in Washington, D.C.; the Museum of Comparative Zoology in Cambridge, Massachusetts; and the Peabody Museum of Natural History, in New Haven, Connecticut.

In 1826, the Englishman Smithson's bequest of an endowment to the United States paved the way for the National Museum, but the first unit of the museum dates only from 1850 (Coleman 1939). Collections were poorly stored in those early days, leading to the loss of many specimens. In 1925, Mary Rathbun wrote: 'Of the Crustacea obtained by the United States Exploring Expedition in 1838-1842 ... very little remains owing to the inadequate housing of the former collection before the existence of a National Museum building.' The present building dates from 1911. In the beginning, Richard Rathbun Assistant Commissioner of the US Fish Commission was also in charge of crustacean collection in the National Museum; later the collections of the Commission were transferred to the National Museum, and since then specimens collected by the Commission and other bureaus of the US Government have been regularly deposited there (R.Rathbun 1883).

In 1852, Harvard bought Louis Agassiz's collections, forming the nucleus of the Museum of Comparative Zoology (1859). Louis Agassiz (b. 1807, d. 1873), the foremost museum collector of his day, also contributed specimens of Neotropical crabs from his trip to Brazil (1865-1866; see below) and from his 1871 expedition with the US Coast Survey steamship Hassler. Alexander Agassiz (b. 1835, d. 1910) continued his father's work with trips to the Gulf of Mexico, the Caribbean, and waters off the Carolinas, in the Coast Survey steamship Blake (1877-1880) and with three expeditions to the Pacific in the Fish Commission's Albatross.

The Peabody Museum of Natural History, at Yale University was established in 1866 from benefactions of George Peabody. Sidney I. Smith described Neotropical species of Brachyura from the Peabody collections, and Mary Rathbun used crab collections from the Peabody as a source for her monographs.

6.5   *Sydney Smith and William Stimpson: 1859-1871*

Sydney Irving Smith (b.1843 d.1926), whose first work in natural history was a collection of insects that Louis Agassiz found so comprehensive he purchased it for Harvard, entered Yale in 1864, where he worked with Addison E. Verril. After graduation in 1867, he stayed at Yale as assistant, and in 1875 was appointed first professor of comparative anatomy in the department of zoology, a post he held until his retirement in 1906. From 1864 to 1870 Smith was involved in several expeditions with Verrill and other zoologists. In 1871, he was zoologist to the US Lake Survey, and the next year joined the US Coast Survey and became a member of the US Fish Commission, working on Fish Hawk and Albatross. His collections were deposited in the Peabody Museum of Natural History and the National Museum of Natural History (Simpkins 1975). Smith is well known in carcinology for his studies of the development of the American lobster, but he also described 27 new species of Neotropical crabs between 1869 and 1883.

Even more productive than Smith was William Stimpson (b. 1832, d. 1872). Stimpson's contributions to knowledge of American Brachyura can be compared only with those of Alphonse Milne Edwards and Mary Rathbun. His contributions to Neotropical carcinology began with the first part of his 'Notes on American Crustacea' published in 1859 and 1860. In this paper he described 65 new species from the south eastern coast of the United States (later recorded also from the Caribbean, Central America, and the Pacific coast of Mexico. From Mexico he described 24 new species from material collected by Janos (or John) Xantus (b. 1825, d. 1894) at Cape San Lucas, the tip of the Baja California Peninsula. Eleven years later,

Stimpson (1871) continued to publish on this same material, adding eight new species from Cape San Lucas; six from Manzanillo, Mexico, also collected by Xantus; seven from Panama collected by Alexander Agassiz in 1860; two from St. Thomas; and one from Barbados.

The same year, Stimpson described 36 new species dredged in the Gulf Stream by Louis de Pourtalés (b.1824-d.1880) in cruises arranged by the US Coast and Geodetic Survey in 1867 and 1868. Pourtalés, a Swiss-American student of Louis Agassiz and who had accompanied the latter to America. Previously, knowledge of American crabs was almost entirely restricted to those of littoral and shallow waters. With the Pourtalés material, the description of deep-sea species began, a task Alphonse Milne Edwards would continue. During his lifetime Stimpson described 124 species of crabs that are now part of the Neotropical fauna (See Manning this volume).

6.6   *The US Coast and Geodetic Survey*

After the pioneer work of Wilkes in the US Exploring Expedition, a more systematic approach to scientific exploration of the sea was undertaken by the US Coast and Geodetic Survey. For example, beginning in 1844 Survey Director Alexander Bache (b.1806, d.1867) arranged for bottom samples to be taken with soundings and for competent study of such collections (Coker 1962).

The first ship of the US Coast and Geodetic Survey to carry out extensive dredgings in the Neotropics was the Steamer Hassler, which was transferred to California from the East Coast in 1871 and conducted deep-sea research during its circumnavigation of South America (Mills 1983). The dredging started in Barbados in December 1871, continuing to Bahia (Brazil) and then, following the South American littoral, to Rio de Janeiro, Rio de La Plata, Point San Antonio, Bahia Blanca (Bermejo, Rio Colorado) and the Gulf of San Matias. After passing through the Strait of Magellan, Hassler followed the Chilean coast, stopping at Chiloe, Talcahuano, Valparaiso, and the Juan Fernandez Islands and working some localities in northern Chile and Peru before crossing to the Galapagos. From the Galapagos Islands Hassler proceeded to Acapulco and Lower California, arriving there in mid-August 1872. Sixty two species of crabs were later identified among the Hassler collections; 14 of them were new to science.

In April 1872, William Stimpson onboard the US Coast Survey Steamer Bache dredged from shallow water to more than 125 fathoms in West Florida (Sand Key, Sombrero), the Yucatan Channel (Contoy and Mujeres islands) and Cuba. Five years later the Steamer Blake was commissioned to carry out an extensive survey of the deep-sea fauna of the Caribbean under the scientific direction of Alexander Agassiz. During 1877 and 1878 the Blake dredged on the Yucatan Bank, in the Gulf of Mexico, and off Cuba. From January 5 to March 10, 1879, the Blake explored the Lesser Antilles, starting in St. Thomas, through the Flannegan Passage to St. Croix, Saba, St. Christopher, Nevis, and Montserrat, and then continuing in the Windward Islands up to Grenada, and then to Barbados.

6.7   *The French scientific mission to Mexico*

The French intervention in Mexico (1862) placed Maximilian on a shaky throne. The Austrian frigate Novara which brought Maximilian to Mexico was the same ship of the Novara Expedition (1857-1859), whose results were reported by Heller (1865). The French decided to send a cultural mission to the new Mexican Empire, and Marie-Fermin Bocourt (b.1819 d.1904), taxidermist at the Paris Museum and collector in Siam during 1861-1862, was chosen naturalist. He embarked late in 1864, but when he arrived in Mexico armed rebellion made it too dangerous for his work, and instead the French government allowed him to explore Guatemala and, to a lesser extent, other parts of Central America. Bocourt went to Belize and visited Lake Isabel, Alta Verapaz, and the mountains of Solola and Totonicapam, arriving on the Pacific coast at the mouth of the River Nagualate. He went by boat along the coast to Panama, stopping at La Libertad, La Union, Realejo, and Punta Arenas (Vaillant 1904). Bocourt returned to France in 1867 and spent most of the rest of his life working with Dumeril on the reptiles he collected (Vanzolini 1977).

Once the Bocourt collections arrived in France, the Ministry of Public Instruction entrusted the crabs to Alphonse Milne Edwards (b. 1835, d. 1900), Director of the Paris Museum. However during the shelling of Paris in 1870, in the last stages of the Franco-Prussian war, a projectile exploded in the room where the largest portion of the Bocourt collections was stored, destroying almost all the crab material. ('Tout y a été brissé, réduit en poussière,' according to A. Milne Edwards 1881). The monograph produced by Milne Edwards from 1873-1881, from study of specimens borrowed from the Smithsonian and the Museum of Comparative Zoology, specimens sent by Sydney Smith, and with extant collections of the Paris Museum, became one on crabs of the entire Neotropical region and much more extensive than study of the Bocourt collections alone would have been. He excluded from the monograph only those genera without tropical representatives, and covered 141 species, or a number slightly smaller if synonyms are considered. Of the new species, 15 were from the collections of the Paris Museum, 1 from the British Museum, 9 were collected by the Blake, 11 by the Hassler, and 5 by Stimpson on the Bache. Of the few crab specimens collected by the Bocourt Expedition, two from Belize are considered the syntypes of *Callinectes bocourti* (see Rathbun 1930).

6.8   *The deep-sea fauna*

Up to the middle of the 19th century the majority of crabs known from the Neotropics were shallow water and supralittoral forms, particularly those from mangroves and rocky shores. Some sublittoral forms were known because they were found entangled in fishermen's nets and traps, as was the case of *Stenocionops furcata*, recorded by Hughes, Parra, and other early naturalists. The study of the sublittoral and deep-sea crabs began with the systematic dredgings of Stimpson and with the cruises of the US Coast and Geodetic Survey vessels Hassler, Bache, and Blake.

Perhaps Stimpson's premature death (1872) led Alexander Agassiz to ask for the help of Alphonse Milne Edwards to study the crab collections from the Blake, which were then in the Museum of Comparative Zoology. Agassiz sent the collections of Bache and Hassler, along with those of the Blake, to the Paris Museum. As mentioned above, A. Milne Edwards included in his monograph (1873-1881) many species of crabs of the Mexican region (Cyclometopa and Oxyrhyncha), and he also prepared a preliminary report on the deep-sea crabs and other Crustacea from the Blake. This report, published in 1880, included 35 new species: 27 collected by the Blake, 5 by Stimpson aboard the Bache, and 3 by the Hassler. The area covered was mostly the Caribbean, except for two species from Hassler collections, one in Brazil and one in Patagonia.

In 1902, a more detailed monograph on the dromiids and oxystomes was published by Milne Edwards and Bouvier. But it was only in 1923, after the first world war ('le moment des Gothas et des Berthes', according to Bouvier), and long after the return of the collections to Cambridge, that Eugne Louis Bouvier (b. 1856, d. 1944) published a monograph on the other groups of crabs, with A. Milne Edwards as co-author, although the latter had died many years before.

7 ANTILLIAN CARCINOLOGY IN THE 19TH CENTURY

7.1 *Authors and collectors*

Between the publication of Sloane's list of Jamaican crabs (1725) and Henry Milne Edwards' *Histoire Naturelle des Crustaces* (1834-1837) knowledge of the Antillean fauna had increased steadily, and was augmented during the second half of the 19th century by the deep-sea dredgings of the Blake and the Hassler, and the work of several individual collectors. Henri de Saussure, in 1857 and 1858, described 7 species of Brachyura he collected from Guadeloupe and others of the Antilles, and Stimpson, from 1859 to 1871, recorded 3 species from Barbados and 14 from St. Thomas collected by A. Riise and others. Another minor contribution (1872) was that of T.H. Streets (b. 1847, d. ?), who listed several crabs from St. Martin, including the new species *Gelasimus affinis* (= *Uca mordax*?), collected by Hendrik Eling van Rijgersma (b. 1834, d. 1877), a Dutch physician on St. Martin and a correspondent of the Academy of Natural Sciences of Philadelphia (Holthuis 1959b).

Early in the 19th century the Muséum d'Histoire Naturelle supported Plée and Ricord (see above) as collectors in the Antilles, and later (1838-1841) employed Louis Daniel Beauperthuy (b. 1807, d. 1871), a physician born in Guadaloupe, as a traveling naturalist in Guadaloupe and Venezuela. Paul Serre (no dates available), an associate of the Muséum d'Histoire Naturelle, acting as French consular agent, made numerous collections of Crustacea in Puerto Rico (1907), where he collected the holotype of *Pinnotheres serrei*; Cuba (1909-1910); Bahia (1912-1913); and Trinidad (1914-1915). Individual collectors added new species from the Antillian crab fauna, but detailed knowledge of the distribution of these species came only from systematic lists developed for some of the islands by Schramm (1867), Gundlach (1887), and Rathbun (1902b).

7.2 *Desbonne and Schramm's list of Crustacea from Guadaloupe*

Isis Desbonne (no dates available) was a French physician, in Moule, a minor port near Pointe-a-Pitre, Guadeloupe. For more than 10 years, until his death (ca. 1866), he studied and recorded the crabs of the island. In his manuscript he described 63 species, 35 of which he considered new. After his death, his notes and collections were given to the government by his widow. Alphonse Schramm (b. ?, d.1875) a resident naturalist on Guadeloupe, was entrusted with the publication of Desbonne's manuscript. He added 17 other records to those of Desbonne (Schramm 1867), either from the Desbonne's collection or from Saussure's records from Guadaloupe, and later made some corrections to his 1867 report (Schramm 1874). Later authors have retained seven species and one subspecies originally described by Desbonne, including the common *Pitho lherminieri* and *Macrocoeloma trispinosum nodipes*, and the rare *Pinnotheres barbata*, which lives in the stomach of the gastropod *Cytharium pica*. Several specimens from Desbonne's collections were sent to the 1867 International Exposition in Paris (Schramm 1867).

7.3   Gundlach's list of Puerto Rican Crustacea

Juan Gundlach (b. 1810, d. 1896), was a curator at the University of Marburg and later at the museum of Frankfurt am Mein. In 1839, he went with the expedition of Pfeiffer and Otto to Cuba, and stayed on the island after the other two naturalists departed. Gundlach carried out extensive explorations of the Cuban fauna, and in 1867 the Municipality of Havana commissioned him to carry specimens from his collections to Paris for the Universal Exposition. In 1892 the Spanish Government bought his collections for 8000 gold pesos and deposited them at the Instituto de La Habana (Chardon 1949). Gundlach visited Puerto Rico twice, invited by Leopold Krugg, German consul at Mayagüez, who had sent collections of the reptiles and batrachians of the island to the Berlin Museum. The iteneraries for his visits in June 1873, and in September 1875 were summarized by Evermann (1902) in the introduction to the results of the Fish Hawk expedition to Puerto Rico. Several of the crabs collected by Gundlach in Puerto Rico and Cuba were sent to the Berlin Museum and identified by Edward von Martens (b. 1831, d. 1904). Gundlach published (1878-1887) the results of his work in Puerto Rico under the general title *Apuntes para la fauna Puerto Riquena*. His list of Crustacea contains 52 species of decapods, of which 37 are Brachyura and 8 are Macrura (Gundlach 1887). The crabs listed were species fairly common in the Caribbean, and the identifications were correctly done, except in a few instances, like the misidentification, by von Martens, of *Epilobocera cubensis* and the listing of a composite species, *Lupa diacantha*. From Gundlach's collections in Cuba, von Martens (1872) described two new species of crabs, *Pitho anisodon* and *Pachygrapsus corrugatus*. Gundlach's notes on his collections of Crustacea from Cuba, compiled and completed by Jose I. Torralbas (Gundlach 1900-1901), were published in the *Anales de la Academia de la Habana*.

7.4   Explorations of the US Fish Commission Steamer Fish Hawk

The most complete list of Brachyura from the Antilles was published by Rathbun (1902) from the materials gathered by the Fish Hawk. Soon after Puerto Rico became a part of the United States, as an outcome of the Spanish-American war (1898), the US Commissioner of Fish and Fisheries decided that an investigation of the aquatic life of the island was in order. He believed that such an investigation would not only yield important scientific results, but would prove of mutual commercial value to the island and to the United States. For the purpose of future comparisons it was important that the work begin before the island was modified from US influences (Evermann 1902).

The US Fish Commission Steamer Fish Hawk was sent to Puerto Rico with a scientific party of Dr. Barton W. Evermann (b. 1853, d. 1932), naturalist-in-charge; Dr. H. F. Moore (b. 1867, d. 1948), chief naturalist from US Fish Commission Steamer Albatross; Mr. M. C. Marsh; Mr. W. A. Wilcox; and Mr. J. B. Wilson. Fish Hawk reached San Juan Harbor on 2 January 1899, and the scientific party was engaged in investigations for the next 45 days. Evermann and Wilson collected along the shore, particularly in places where they could seine; another party under Marsh collected on the reef. Meanwhile, Fish Hawk trawled and dredged under Moore's direction. Collecting from ship and shore proceeded from San Juan to the west coast of the island, through Aguadilla and Mayagüez. Fish Hawk made extensive collections off the southern coast and around the islands of Vieques and Culebras, while the shore party collected at Ponce and Arroyo. Decapods in the expedition collections, including 162 species of crabs, were dredged from depths up to 225 fathoms, although the majority came from within the 100-fathom line (Rathbun 1902). This total number of species was later increased to 166 with the corrections by Rathbun in her monographs. One genus and 14 species of crabs were new; however, later corrections reduced new species to 12.

8 THE CARCINOLOGICAL WORK OF MARY J. RATHBUN

Mary Jane Rathbun (b. 1860, d. 1943) became a clerk with the US Fish Commission in 1884 and was appointed a copyist in the Department of Marine Invertebrates at the National Museum in 1886. Later she became an aide, and finally assistant curator of that division. After one year, on 31 December 1914, she voluntarily retired from that position to create an opportunity for the appointment of Waldo L. Schmitt (b. 1887, d. 1977). However, Rathbun continued her research work, as an honorary associate, until 1939, 'when she was too feeble to continue commuting daily to the Museum' (Chace 1990).

After A. Milne Edwards's contributions of 1880 and about the time Mary Rathbun joined the National Museum, 478 species of crabs presently included in the Neotropical fauna had been described. (The number would be far greater if synonyms were not taken into account). Still, four more new species were added by Smith in 1881, and Miers in 1886 described eight more in his reports on the Brachyura of the Challenger.

The contributions of Rathbun to Neotropical carcinology began in 1891 when she published, with James E. Benedict (b. 1854, d. 1930) then assistant curator of marine invertebrates, descriptions of eight new species of *Panopeus* mostly from the coasts of the United States. Five of these species are considered valid today, although now distributed among the genera *Panopeus*, *Hexapanopeus*, and *Eurypanopeus*, and their ranges have been extended into the Neotropics. From 1892 to 1935 Rathbun described 210 new species of crabs now included in the Neotropical fauna. During this same period only 50 new species were added by other authors. Among the latter, Walter Faxon (b. 1848, d. 1920), zoologist from the Museum of Comparative Zoology, made the largest contribution, in 1893, when he described 10 new species from West Mexico, Panama, and the Galapagos, in the preliminary report to the 1891 trip of the Albatross. He added still another in 1895 in the final report of that expedition. Other authors added one or two new species: Benedict (1892); De Man (1892); Holmes (1895); Ortmann (1894, 1897); Bouvier (1898); Milne Edwards & Bouvier (1899a, 1899b); Nobili (1901a); Lenz (1902); Verrill (1908); Porter (1908); Borradaile (1916); Stebbing (1914); Balss (1923); Boone (1927); and Glassell (1933, 1935). The materials and information on all these species were incorporated into Rathbun's four famous monographs: *The Grapsoid Crabs of America* (1918), *The Spider Crabs of America* (1925), *The Cancroid Crabs of America* (1930), and *The Oxystomatous and Allied Crabs of America* (1937). Of the species included in these monographs, 748 are recorded from the Neotropics.

The material for the Rathbun monographs came mainly from the collections in the US National Museum, and to a lesser extent from the Museum of Comparative Zoology, the Peabody Museum, and some minor collections. In 1896, Rathbun visited European museums to examine the collections of J. C. Fabricius, in Copenhagen and Kiel; of Herbst, in Berlin; of Saussure, in Geneva; of H. and A. Milne Edwards, in Paris; and of Miers, and others, in the British Museum.

8.1 *Collectors*

Among the 250 collectors mentioned by Rathbun are many ship captains and medical personnel from the US Navy. Several collectors in Panama were connected with the Panama Railroad, which opened in 1855, or were travelers, like Captain J. M. Dow, of the steamship Guatemala, Central American Steamship Company, who beginning in 1856 was in charge of the coasting trade from Panama to San Jose de Guatemala (Otis 1862), and who collected in El Salvador and Panama. Others include Jose Castulo Zeledon (b. 1846, d. 1923), a renowned ornithologist from Costa Rica; the Swiss mining engineer Henry François Pittier (b. 1857, d. 1950) and his fellow countryman Pablo Biolley (b. 1862, d. 1909) (Gomez and Savage 1983); Felipe Poey (b. 1799, d. 1891), the most eminent Cuban naturalist of the 19th century; the American ichthyologist and naturalist David Starr Jordan (b. 1851, d. 1931) who collected in Cuba; and Thomas Hunt Morgan (b. 1866, d. 1945), American geneticist and Nobel laureate (1933), who while working on the regeneration of appendages in Crustacea (Barnhard 1954) gathered 27 species of crabs in Jamaica in 1891. The list of collectors also includes, in Santo Domingo, William More Gabb (b. 1839, d. 1878), an American paleontologist; and in Venezuela, Lieutenant Wirt Robinson, US Army (b. 1864, d. 1929), who had traveled in Colombia in 1892 as an ornithologist (Robinson 1895) and in 1895 collected crabs in La Guaira, Guanta, and on the Island of Margarita. Later, in 1900, he worked with Dr M.W. Lyon (b. 1875, d. 1942) in La Guaira.

Rathbun mentions many collectors from Brazil, a 'classic' haunt of the traveler-naturalist during the 19th century. Among them is Herbert Huntingdon Smith (b. 1851, d. 1919), Curator of the Carnegie Museum, who carried out extensive explorations in Brazil, Mexico, and Colombia. In 1852 Fritz Müller (b. 1821, d. 1897), a German zoologist whose *Für Darwin* was a significant contribution to carcinology as well as a support of Darwin's theory, settled in Brazil and collected for the National Museum of Rio de Janeiro (Möller 1915), contributing several species of crabs he collected in 1866. Hermann von Ihering (b. 1859, d. 1930), another German-Brazilian zoologist, founder and director of the Museu Paulista of Sao Paulo from 1894 to 1916, sent many collections, gathered in Brazil and Uruguay from 1896 to 1919, to the US National Museum.

Two collectors in Argentina are well known outside the field of zoology: Ales Hrdlika (b. 1869, d. 1943), American anthropologist who made collections in 1910 while Curator in the Division of Physical Anthropology, US National Museum (Krogmann 1943); and Luis Federico Leloir (b. 1906), Argentinian biochemist and Nobel Laureate (1970) who collected in 1924. The largest collection of Peruvian crabs was by Robert E. Coker (b. 1876, d. 1967), marine biologist from the University of North Carolina, when he worked for the Peruvian Government as a fisheries expert, from 1906 to 1908.

8.2 *Expeditions*

Specimens for Rathbun's monographs also came from several expeditions during the 19th century and during the first quarter of the 20th century.

8.2.1 *Hartt Explorations in Brazil*
The American geologist Charles F. Hartt (b. 1840, d. 1876), assistant to Louis Agassiz during the Thayer Expedition and later Professor of Geology at Cornell (Wright 1953) was asked by the Brazilian government in 1875 to organize the Geological Commission of the Empire. He brought as assistants the young Americans O. Derby (b. 1851, d. 1915), J. C. Branner (b. 1850, d. 1922), and Richard Rathbun (b. 1852, d. 1918), brother of Mary J. Rathbun (Schwartsman 1979). The majority of the crabs attributed to Thayer's commission were collected by Richard Rathbun in Fernando Noronha, Pernambuco, Rio de Janeiro, Bahia and Rio Vermelo. Twenty-eight of them, all common littoral species, were reported many years later in Mary Rathbun's

monographs. Other members of the commission made smaller collections. O. Derby collected in Grao Para with Powers, and in Maranhao with Wilmot.

8.2.2 *The US Fish Commission Steamer Albatross*
Albatross was designed for marine research, particularly deep-sea dredging, and was in commission from 1882-1921 and for almost 30 years operated in the Pacific, based in San Francisco, California (Hedgpeth & Schmitt 1945). Albatross cruises in the Neotropics are listed as follows: (1) From January 27 to May 1, 1884, dredged in the Greater Antilles; around some of the islands off Venezuela; and at several localities off Colombia, Panama, and Nicaragua (2) from January 17 to March 13, 1885, dredged off Cuba, Yucatan, and in the Gulf of Mexico (3) from November 1887 to May 1888, during transfer of vessel from East Coast to West Coast, dredged from Martinique and St. Lucie to Cape San Roque, Abrolhos Island, and Trinidade in Brazil; and at Rio de La Plata, Gulf San Matias, in Patagonia; and after passing through the Strait of Magellan, for a month explored southern coast of Chile to Concepcion, then proceeded directly to Panama, then to the Galapagos, and finally arriving in San Francisco (4) from January 26 to April 10, 1889, dredged in Gulf of California and down to the Clarion and Socorro Islands of Mexico (5) from February 10 to April 18, 1891, Acapulco, Mexico, to Panama and Costa Rica, then to Cocos, Malpelo, and Galapagos islands, and return to San Francisco by way of Tehuantepec and the Gulf of California (6) From March 12 to April 11, 1899, dredged in the Gulf of California and on the Pacific side of Baja Peninsula (7) From October 14, 1904, to January 8, 1905, Acapulco to Panama, Guatemala, and then to the Galapagos (h) from March 12 to April 30, 1911, dredged around the southern tip of the Baja Peninsula. All crabs from these cruises were studied by Rathbun except those from the 1891 expedition, which were reported on by Faxon in 1893.

8.2.3 *The University of Iowa Expedition to the Florida Keys and the Bahamas*
This expedition gathered a small number of crabs (16 specimens) in Cuba, from 24 May to 3 June 1893, which were reported by Rathbun (1898c). Among these specimens was a new species, *Collodes armatus*.

8.2.4 *The Stanford University Expeditions*
A Stanford University expedition collected 10 species of crabs in the Galapagos from January to May 1899; three of these were new species. Rathbun also mentions a Stanford Expedition to Brazil, in 1911, during which Fred Baker (b.1854 d.1938) collected 3 species of littoral crabs at Ceara and Natal.

8.2.5 *The Branner-Agassiz Expedition*
On this expedition, led by J. C. Branner, who had been on the Hartt expedition of 1875 (see above), and Alexander Agassiz, crabs were collected mostly by A. W. Greeley (no dates available). Collections were made, from June to August 1899, in northwest Brazil, from Natal to Maceio, in mangrove and other coastal environments. Twenty-three species of crabs were collected; two of them, *Uca thayeri* and *Teleophrys cristulipes*, were new to science.

8.2.6 *The British Antarctic (Terra Nova) Expedition*
The Terra Nova Expedition is mentioned by Rathbun (1918) in connection with only one species, *Goneplax hirsuta*, collected in Rio de Janeiro and described by L. A. Borradaile (b. 1872, d. 1945) in 1916.

8.2.7   *The Tomas Barrera Expedition*

The Tomas Barrera was a small (60 ft.) fishing schooner used on an expedition organized by John B. Henderson (b. 1879, d. 1923) and Carlos de La Torre (b. 1858, d. 1950) to collect the flora and fauna, especially the mollusks, on the northwest coast of Cuba. Other scientists onboard included Paul Bartsch (b. 1871, d. 1960), George H. Clapp (b. 1859, d. 1949), J. Rodriguez, Charles T. Simpson (b. 1846, d. 1932) and Manuel Lesmes, a fisheries expert from the Cuban Government. Henderson and Bartsch collected the crustaceans. The exploration comprised dredgings, from May 12 to June 14, 1914, in shallow water along the entire length of the Colorado Reef, from La Esperanza to Cape San Antonio; and collecting in the region about Bahia Honda and Cabanas (Henderson 1916; Turner 1950). Rathbun recorded 58 species from this expedition, 2 of them, *Microphrys interruptus* and *Parapinixa hendersoni*, were new species.

8.2.8   *The University of Iowa Barbados-Antigua Expedition*

Under the leadership of professor Charles C. Nutting   (b. 1858, d. 1927), this expedition extensively collected marine invertebrates in Barbados in May-June 1918, and in Antigua in July 1918. The expedition collected, with dredges and tangles, 93 species of crabs from the up to 118 fathoms, but only one of these was a new species, *Pilumnus barbadensis* (see Rathbun 1921).

8.2.9   *The Harrison Williams Galapagos Expedition (1923)*

This expedition is mentioned by Rathbun (1925) in connection with a record of *Mithrax bellii*.

8.2.10 *The Zaca Expeditions*

The Zaca, a two-masted diesel schooner owned by Templeton Crocker, collected along the Pacific coast of Mexico, from Manzanillo to Sinaloa, and around the tip of the Baja Peninsula, during July and August 1932. Nine species of oxystomatous crabs collected on this cruise and deposited at the California Academy of Sciences were included in Rathbun (1937). Templeton Crocker organized two other expeditions under the scientific direction of William Beebe (b. 1877, d. 1962), Director of the Department of Tropical Research of the New York Zoological Society. The first, in 1936, collected in Lower California and on the Mexican mainland (Beebe 1938); the second, in 1938, collected from Mexico to Colombia (Beebe 1947). The brachyurans collected on the first expedition were studied by Steve A. Glassell (b. 1884, d. 1948) in 1936 and Jocelyn Crane (b. 1909) in 1937; those on the second were studied by Crane (1947).

8.2.11 *The Allan Hancock Pacific Expeditions*

These expeditions were conducted by Captain C. Allan Hancock on his vessels Velero III, beginning in 1931 (Fraser 1943), and Velero IV, in the early 1950's. From January to March 1933, Velero III collected in the Galapagos Islands and at a few stations in Central America, and from January to March 1934, again sampled in the Galapagos, as well as at stations in Mexico (Lower California, Clarion and Socorro islands, Jalisco, Oaxaca), Costa Rica, Panama, Colombia, and Ecuador. The oxystomatous crabs from these two expeditions, including six new species, were dealt with by Rathbun (1933a, 1935) and included in her last monograph (Rathbun 1937). Other crabs from the Allan Hancock Expeditions were monographed by John S. Garth (b. 1909), beginning in 1939.

9   THE HINTERLAND

9.1   *Authors*

Freshwater crabs are represented in the Neotropics by two endemic families, the Trichodactylidae and the Pseudothelphusidae. The first description of a species of Trichodactylidae, *Cancer orbicularis*, was published by Meuschen (b. 1719, d. ? 1800) in 1781. But since his *Index Zoophylacium Gronovianum* is not accepted as a precedent in binomial nomenclature, the junior synonym of *C. orbicularisD*, *Cancer septemdentatus*, described by Herbst in 1783, is considered the first valid name proposed for a species of this family. In 1825, Latreille published the first description of a pseudothelphusid crab, *Thelphusa dentata* (= *Guinotia dentata*), from Guadeloupe, and the description of a second species of a trichodactylid, *Trichodactylus fluviatilis*, from Rio de Janeiro. There followed from 1839 to 1901, descriptions of 27 new species of Trichodactylidae and 56 of Pseudothelphusidae. Several of the new names proposed, however, are now considered synonyms.

Randall (1840) described two common freshwater crabs, the trichodactylid *Orthostoma dentata* (= *Dilocarcinus dentatus*) and the pseudothelphusid *Potamia latifrons* (= *Kingsleya latifrons*). White (1847a) described *Valdivia serrata*, and Stimpson (1861) described *Dilocarcinus pagei*, two trichodactylids widely distributed in the Amazon basin. But, two of the most prolific describers were Henry Milne Edwards, who described six trichodactlyds (1854) and one pseudothelphusid (H. Milne Edwards & Lucas 1843); and Alphonse Milne Edwards, who in 1869 described four trichodactlyds and three pseudothelphusids. One of the foremost specialists in freshwater decapods during this period was Arnold Edward Ortmann (b. 1863, d. 1927), a German zoologist who settled in the US in 1894, where he became curator at Princeton University, and later curator at the Carnegie Institution and professor at Pittsburg University. He described two trichodactylids and two pseudothelphusids (Ortmann 1893, 1897), but also contributed important revisions of the systematics and biogeography of the two families (Ortmann 1903). The Italian zoologist Giuseppe Nobili (b. 1877, d. 1908) described two new species of pseudothelphusids collected by Enrico Festa (b. 1868, d. 1939) in Ecuador, and two new Trichodactylidae collected by Alfredo Borelli (b. 1858, d. 1943) in Argentina. Minor contributions were made to the knowledge of Neotropical freshwater crabs by Eydoux & Souleyet (1842), Gerstaecker (1856), Saussure (1857), von Martens (1869), Goeldi (1886), Smith (1870), Pearse (1911), and Moreira (1912).

In 1889, Reginald I. Pocock (b. 1863, d. 1947), an English zoologist associated with the Zoological Garden of London described *Boscia tenuipes*, from Dominica, a junior synonym of *Guinotia dentata*. John Sterling Kingsley (b. 1854, d. 1929), who was a professor of zoology at several American universities and Curator of the Peabody Academy of Sciences from 1876 to 1878 (Holthuis 1959a), described *Dilocarcinus spinifrons* (= *Sylviocarcinus devillei*) (Pocock 1880).

The largest number of American freshwater crabs was described by Rathbun in several papers (1893c, 1896c, 1897e, 1898d), and in her monograph on the freshwater crabs of the world (Rathbun 1905, 1906). The monograph was based largely on specimens she examined in the Paris Museum (see above). Rathbun described 9 new trichodactylids and 49 new pseudothelphusids. Of these numbers, 4 trichodactylids and 27 pseudothelphusids are presently considered valid species.

The main reason for taxonomic uncertainty in the Trichodactylidae and the Pseudothelphusidae is that in most cases the only specific diagnostic characters are found in the first male gonopod, an appendage disregarded by all the early taxonomists except Rathbun who recognized its significance in brachyuran systematics. In her monograph (Rathbun 1905, 1906) she illustrated first male gonopods of 17 pseudothelphusids; however, she ignored the gonopod in describing the remainder of pseudothelphusids and in describing all trichodactylids. In fact, she used female holotypes for several of her new species. After her mongraph, Rathbun described nine additional pseudothelphusids (1912, 1915a, 1919, 1933c); six of these are considered valid today.

In 1920, Colosi described a new pseudothelphusid, and in 1939 Coiffmann (1939) added two more, but then followed an inactive period in the taxonomic study of American freshwater crabs. Then in the late 1950's several authors added new species to both families, and today the number of species considered valid are 42 for the Trichodactylidae, and 168 for the Pseudothelphusidae (Rodriguez 1982, 1992).

9.2   *Collectors*

In her monograph Rathbun (1905, 1906) gives the names of 80 collectors of American freshwater crabs, including: C F. Baker, Beauperthuy, Biolley, Gabb, Garman, Goldman, Guérin, Gundlach, Meek, Krøyer, McNiel, Nelson, Capt. Page, Pittier, Palmer, Lt Robinson, Saussure, H. H. Smith, Tristan, Xantus, and Zeledon. Of some of these 80 collectors little is known, as in the case of Gollmer who is described by Gerstaecker (1856) as the 'apotheker' who sent him crabs from Caracas, possibly around 1850. However, some of the collectors were also naturalists who explored the hinterland, for example: F. Geay; Auguste Sallé (b. 1820, d. 1896); the German geologist Wilhelm Reiss (b. 1838, d. 1908), who explored the cordilleras of Colombia, Ecuador, and Peru, and several localities in Brazil, from 1868 to 1875; and Eugéne Simon (b. 1838, d. 1924) who during extensive explorations in Venezuela collected the holotypes of *Pseudothelphusa simoni* and *P. venezuelensis* (now in the genera *Neopseudothelphusa* and *Orthothelphusa*, respectively).

9.3   *Land expeditions*

Other important sources of materials were land expeditions to South America during the 19th century. The Scientific Mission to Mexico (see above) collected abundant material of the Pseudothelphusidae in the region of Alta Verapaz, Guatemala. Rathbun described several new species from this material, but not all are valid today.

One of the first to collect freshwater crabs in tropical America was the French naturalist Alcides Dessalines D'Orbigny (b. 1802, d. 1857), sent by the Paris Museum. He arrived in Rio de Janeiro on 24 September 1826, collected in the vicinity for a short time, then departed for Argentina where he spent three years (February 1827 – December 1829) exploring the Parana River, the northern provinces, and Patagonia. In early 1830 he sailed to Chile and proceeded to Bolivia where he collected from May 1830 to June 1833. He visited Peru in the vicinities of Callao and Lima in mid-1833 and sailed from Valparaiso, Chile, on 18 October 1833, back to France. The Crustacea collected by D'Orbigny were studied by H. Milne Edwards and Lucas, and reported by them in D'Orbigny's *Voyage a l'Amérique Meridional*, a monumental work published in nine volumes between 1843 and 1847. H. Milne Edwards and Lucas described a new species of Pseudothelphusidae collected by D'Orbigny in Lima (Peru), *Potamia chilensis* (= *Hypolobocera chilensis*).

Francis Louis Nompart de Caumont de la Porte de Castelnau (b. 1810, d. 1880) was sent by the French Government to explore South America. With Castelnau were Eugene d'Osery, mining engineer; Hugh Algerson Weddell (b. 1819, d. 1877), an English botanist; and Emile

Deville, an employee of the Muséum National d'Histoire Naturelle. The party arrived in Rio de Janeiro on 17 June 1843 and spent almost 4 years exploring Brazil, Bolivia, and Peru. The report of the expedition was published by Castelnau in 15 volumes (Castelnau 1850-1859). The crabs, however, were treated separately by H. Milne Edwards, in a contribution of the *Archives du Muséum* (1853) in which he described five new species of trichodactylids, four of which are valid today. *Silviocarcinus devillei* and *Dilocarcinus castelnaui* were collected north of Goiás (Brazil), and *D. pictus* (= *Sylviocarcinus pictus*) and *D. emarginatus* (= *Zilchiopsis emarginatus*) in Loreto (Peru), on the banks of the Amazon. H. Milne Edwards also described the pseudothelphusid *Boscia macropa* (= *Neostrengeria macropa*), the holotype said to have been collected by Weddell in Bolivia. This locality, however, is an error, since the species is restricted to the neighborhood of Bogota (Colombia).

The Thayer Expedition, one of the main sources of freshwater crabs for Rathbun's monograph, was led by Louis Agassiz and named after Nathaniel Thayer, a friend of Agassiz and a trustee of the Harvard Museum, who covered the expenses. The party included Agassiz's wife, Elizabeth Cabott (Cary) Agassiz (b. 1822, d. 1907), who wrote a diary of the expedition (Agassiz & Agassiz 1868); Frederick C. Hartt and Oreste St. John, geologists; Joel A. Allen (b. 1838, d. 1921), ornithologist; John G. Anthony (b. 1804, d. 1877), conchologist; James Burkhardt, artist; George Sceva, preparator; and a number of volunteers. The expedition arrived at Rio de Janeiro on 23 April 1865, and under the patronage of the Brazilian Emperor Dom Pedro II, who provided ships and field assistants, Agassiz and his party collected in the vicinity of Rio, and in the valleys of the San Francisco, Parnahyba, Tocantins, and Amazon rivers until 2 July 1866.

Many specimens of *Trichodactylus fluviatilis* were obtained in Rio de Janeiro, in the Rio dos Macacos, and at Rio Parahyba. In the Amazonas, at Para, Agassiz and Bourget collected the holotype of *Pseudothelphusa agassizi* (= *Fredius reflexifrons*), and Agassiz obtained several species of Trichodactylidae from the Amazon. At Tabatinga, Bourget collected the holotype of *Trichodactylus* (*Valdivia*) *bourgeti* (= *Valdivia harttii*). Other trichodactylids were collected in the river Poty by O.H. St. John, and at Bahia possibly by N. Dexter or S.V.R. Thayer.

Another important expedition was that of Jean Chaffanjon (b 1854 d.1913), a French naturalist, who was entrusted by the Ministry of Public Instruction and Fine Arts, in 1884, to explore the Orinoco River basin, Venezuela. Chaffanjon made two separate trips upriver, the first from January to June 1885 and the second from April to December 1986. During his second trip, he claimed to have reached the source of the Orinoco, but he actually only went a little beyond the Raudal de los Guaharibos, approximately 150 km from the source. Chaffanjon wrote a narrative of his travels (Chaffanjon 1889), and was given the golden medal of the Geographical Society of France in 1880. Rathbun described two new species collected by Chaffanjon in the Orinoco, *Trichodactylus* (*Valdivia*) *venezuelensis* (= *Forsteria venezuelensis*), and the pseudothelphusid *Potamocarcinus chaffanjoni* (= *Fredius chaffanjoni*).

Toward the close of the 19th century several Italian naturalists were active in South America. Enrico Festa (b. 1868, d. 1939), an assistant in the Museum of the Institute of Zoology, Turin University, arrived in La Guaira, Venezuela, 18 May 1895 and proceeded to Panama where he explored along several river courses. In September 1895, he went to Guayaquil, and began exploring in Ecuador. At the end of March 1898 he was back in Genoa (Festa 1909).

The crab collections gathered by Festa in Panama and Ecuador were reported on by Nobili (1901b). At Gualaquiza (Equador) and during his explorations of the Rio Santiago (Equador), Festa collected the holotypes of *Pseudothelphusa conradi* and *P. henrici* (now in the genus *Hypolobocera*). The status of three other species, described by Colosi (1920), from Festa's collections is uncertain.

10 CONCLUSIONS

This chronology places development of the taxonomy of Neotropical brachyurans into a general historical context and emphasizes that this development has been dependent on the activities of collectors and zoologists from different national and social backgrounds.

Although some of the early Spanish chroniclers, like Oviedo, identified the most common plants and animals found in the New World with related species from their original homeland, in general the Spanish empire, largely involved in mineral exploitation of its colonies, did not promote the study of natural history in the New World. Other European powers, whose economies relied heavily on international trade in living resources, were more interested in the 'natural productions' of potential overseas colonies. For instance, Dutch interest in Brazil during the governorship of van Nassau-Siegen (1638-1644) resulted in the work of Marcgraf and Piso. Similar interest by the British was shown through early studies of West Indies biota by Sloane, Browne, Hughes, Guilding, and many others whose collections in the British Museum were listed by White (1842).

Spain's lack of interest in natural history of the New World during the colonial period also carried over into the new republics created after independence, except most importantly in Chile where the social and geographical conditions promoted natural history study. But for the most part, after the establishment of the Linnean nomenclature, most knowledge of Neotropical biota came from a few learning centers in Europe. At the beginning of the 19th century, the British had many collectors in the Neotropics, but did not have enough carcinologists to study their collections. This task fell to German zoologists, mostly entomologists, and even more to French zoologists, and these, all heirs of the encyclopedists and pursuing knowledge for its own sake rather than for economic reasons, laid the foundations of brachyuran taxonomy in the Neotropics. During this period and for some time, the more accessible shallow water along the coasts was the most intensively studied, for instance in the Antilles, where crab fauna was listed by Desbonne (1867), Gundlach (1887) and Rathbun (1902b), and in Brazil, where many collectors worked for the Paris and British museums.

In the latter half of the 19th century, the United States established several federal agencies whose responsibilities included the survey of natural resources. The vessels of these agencies – *Blake, Hassler, Albatross* – explored the continental waters of Mexico, the Caribbean, and South America as an extension of their domestic duties in US waters. These vessels were equipped for deep-sea dredging, and as a result, collections of crabs from the continental shelf became available to zoologists for the first time. Several carcinologists then gave a second great impetus to the study of Neotropical brachyurans; Mary J. Rathbun represents the most important of these. After the publication of her last monograph in 1937, little was left for other systematists to describe of the marine and estuarine Brachyura of the Neotropics. However, freshwater crabs from the hinterland of Latin America were still poorly known due to the vastness of this area and the inaccessibility of some regions. The process of describing these freshwater Brachyura continues.

ACKNOWLEDGEMENTS

Many thanks are due Frank Truesdale for critiquing and editing the manuscript. I am particularly grateful to Doris Oliva, from the Universidad Catolica de Chile, for providing invaluable data on Chilean carcinology, and to Elisa Martinez and Isabel Rodriguez-Diaz for compiling the extensive data bases used in the present contribution. The editor and I thank L.B. Holthuis for generously providing biographical information on some carcinologists and collectors.

REFERENCES

Agassiz, L. & E. Agassiz 1868. *A Journey in BrazilD. Boston: Ticknor & Fields.*

Anonymous 1980. *Biobiografias, 1830-1980.* Museo de Historia Natural de Chile, Santiago de Chile.

Anthony, J. 1974. Henry Milne Edwards. In C. C. Gillispie (ed.), *Dictionary of Scientific Biography.* Vol. 9: pp. 407-409. New York: Charles Scribner's Sons.

Barnhard, C.L. (ed.) 1954. *The New Century Cyclopedia of names.* New York: Appleton-Century-Croft.

Balss, H. 1923. Decapoden von Juan Fernandez. *The Natural History of Juan Fernandez and Easter Island* 3:329-340.

Balss, H. & H. Gruner 1961. Schriftenverzeichnis. In H. Balss, W. v. Buddenbrock, H.E. Gruner & E. Korschelt (eds), *Dr. H.G. Bronns Klassen und Ordnungen des Tierreichs* 5. Band I. Abteilung 7. Buch Decapoda: pp. 1823-2061.

Beebe, W. 1938. *Zaca Venture.* New York: Harcourt, Brace.

Beebe, W. 1947. *Book of Bays.* London: The Bodley Head.

Beer, G. de 1975. Sir Hans Sloane. In C.C. Gillispie (ed.), *Dictionary of Scientific Biography.* Vol. 12: pp. 456-459. New York: Charles Scribner's Sons.

Bell, T. 1835. Observations on the genus *Cancer* Leach (*Platycarcinus* Latreille) with descriptions of three new species. *Proc. Zool. Soc. London* 3:86-88; *Trans. zool. Soc. London.* 1:335-342, pl. 43-47.

Bell, T. 1836. Some account of the Crustacea of the coasts of South America, with descriptions of new genera and species. I Oxyrhynchi. *Proc. Zool. Soc. London.* 3:169-173; *Proc. zool. Soc. London.* 2:39-66, pl. 8-13.

Bell, T. 1855. Horae carcinologiceae, or notices of Crustacea. I A monograph of the Leucosiidae. *Trans. Linn. Soc. London.* 21:277-314.

Benedict, J. & M.J. Rathbun 1891. The genus *Panopeus. Proc. US Natl. Mus.* 26:889-895.

Benedict, J. 1892. Decapod Crustacea of Kingston Harbour. *John Hopkins Univ. Circ.* 11:77.

Boone, L. 1927. The littoral crustacean fauna of the Galapagos Islands. Part I Brachyura. *Zoologica, New York.* 8:127-288.

Borradaile, L.A. 1916. Crustacea. Part I. Decapoda. *British Antarctic ('Terra Nova') Expedition, 1910, Natural History Reports, Zoology.* 3(2):75-110.

Bosc, L.A.G. 1801-1802. *Histoire naturelle des Crustacés, contenant leur Description et leur Moeurs.* Vol. 1, pp. 1-258, pls. 1-8. Paris.

Bouvier, E.L. 1898. *Lithadia digueti,* nouveau Crustacé brachyure de la famille des Leucosiid. *Bull. Soc. ent. France.* 1898:330-331.

Browne, P. 1756. *The Civil and Natural History of Jamaica.*

Campbell, G. 1877. *Log Letters from 'The Challenger.'* 5th ed. London: Macmillan.

Castelnau, F. de 1850-1859. *Expédition dans les parties centrales de l'Amérique du Sud, de Rio de Janeiro Lima, et de Lima au Para; exécuté par ordre du gouvernement français pendant les années 1843 a 1847, sous la direction de Francis de Castelnau.* Paris: P. Bertrand.

Chace Jr., F.A. 1990. Mary J. Rathbun (1860-1943). *J. Crust. Biol.* 10:165-167.

Chaffanjon, J. 1889. *L'Orénoque et le Caura.* Paris: Hachette.

Chancellor, G.A. DiMauro, R. Ingle & G. King. 1988. Charles Darwin's 'Beagle' collections in the Oxford University Museum. *Arch. Nat. Hist.* 15:197-231.

Chardon, C.E. 1949. *Los Naturalistas en la America Latina.* Santo Domingo: Secretaria de Estado de Agricultura, Pecuaria y Colonizacion.

Coker, R.E. 1961. *This Great and Wide Sea.* Chapel Hill: The University of North Carolina Press.

Coifmann, I. 1939. Potamonidi della Guiana Inglese raccolti da Prof. Nello Beccari. *Arch. Zool. Ital.* 27:93-116.

Coleman, L.V. 1939. *The Museum in America. A Critical Study.* Washington, D.C.: The American Association of Museums.

Colosi, G. 1920. I Potamonidi dell R. Museo Zoologico di Torino. *Boll. Mus. Zool. Anat. Torino.* 35(734):1-39.

Crane, J. 1937a. The Templeton Crocker Expedition. III Brachygnathus crabs from the Gulf of California and the west coast of lower California. *Zoologica, New York.* 22:47-78.

Crane, J. 1937b. The Templeton Crocker Expedition. VI Oxystomatous and dromiaceous crabs from the Gulf of California and the west coast of lower California. *Zoologica, New York.* 22:97-108

Crane, J. 1947. Eastern Pacific expeditions of the New York Zoological Society. XXXVII Intertidal brachygnathous crabs from the west coast of tropical America with special reference to ecology. *Zoologica, New York.* 32: 69-95.

Dana, James D. 1851. Crustacea Grapsoidea, (Cyclometopa, Edwardsii): Conspectus crustaceorum quae in orbis terrarum circumnavigatione Carolo Wilkes e classe republicae foederatae duce lexit et descripsit. *Proc. Acad. Nat. Sci. Phila.* 5:247-254.

Dana, James D. 1852. Crustacea, Part 1. In *United States Exploring Expedition during the years 1838, 1839, 1840, 1841, 1842, under the Command of Charles Wilkes, USN* 13:1-659, Atlas (1855):1-27, pls. 1-96. Philadelphia.

Dance, S.P. 1966. *Shell Collecting.* Berkeley: Univ. of Calif. Press.

Darwin, F. (ed.) 1887. *The Life and Letters of Charles Darwin*, Vol. 1. London: John Murray.

Evermann, B.W. 1902. Summary of the scientific results of the Fish Commission Expedition to Porto Rico. *Bull. US Fish Comm.* 20 (1):3-26.

Eydoux, F., & L.F.A. Souleyet 1842. Crustacés. *Voyage autour du monde execute pendant les annes 1836 et 1837 sur la corvette La Bonite comande par M. Vaillant, capitaine de vaisseau. Zoologie.* 1:219-272, 5 pls. Paris.

Fabricius, J.C. 1775. *Systema Entomologiae, sistens Insectorum Classes, Ordines, genera, Species, adiectis Synonymis, locis, Descriptionibus, Observationibus.* Flensburgi-Lipsiae.

Fabricius, J.C. 1781. *Species Insectorum exhibentes eorum Differentias Specificas, Synonyma Auctorum, Loca natalia, Metamorphosin adjectis Observationibus, Descriptionibus.* Hamburgi-Kilonii.

Fabricius, J.C. 1787. *Mantissa Insectorum sistens eorum Species nuper detectas adjectis Characteribus Genericis, Diferentiis Specificis, Emendationibus, Observationibus.* Hafniae.

Fabricius, J.C. 1792-1794. *Entomologia systematica, emendata et aucta, secundum Classes, Ordines, Genera, Species, adjectis Synonimis, Locis, Observationibus, Descriptionibus.* Hafniae.

Faxon, W. 1893. Reports on the dredging operations off the West Coast of Central America to the Galapagos by the 'Albatross.' VI Preliminary descriptions of new species of Crustacea. *Bull. Mus. comp. Zool. Harvard.* 24:149-220.

Festa, E. 1909. *Nell Darien e nell' Ecuador, Diario di Viaggio di un Naturalista.* Torino: Tipografia Editrice Torinese.

Fraser, C.M. 1943. General account of the scientific work of the 'Velero III' in the Eastern Pacific, 1931-41. *Allan Hancock Pacific Exped.* 1:1-432.

Freminville, C. de 1835. Notice sur les Torlouroux ou crabes de terre des Antilles. *Ann. Sci. Nat. (Zool.).* 3: 213-224.

Gerstaeker, A. 1856. Carcinologische Beitrage. *Arch. Naturg.* 22(1):101-162, pl. 4-6.

Gibbes, L.R. 1848. Catalogue of the fauna of South Carolina. Appendix to: M. T. *Report on the Geology of South Carolina.* Columbia, South Carolina.

Gibbes, L.R. 1850. On the carcinological Collections of the cabinets of Natural History in the United States. With an enumeration of the species contained therein, and descriptions of new species. *Proc. Amer. Assc. Sci. Charleston.* 3:165-201.

Gifford Jr, G.E. 1972. Augustus Addison Gould. In C.C. Gillispie (ed.), *Dictionary of Scientific Biography*, Vol.5: pp. 477-479. New York: Charles Scribner's Sons.

Glassell, S.A. 1933. Descriptions of five new species of Brachyura, collected in the west coast of Mexico. *Trans. San Diego Soc. Nat. Hist.* 7:331-344, pl. 22-26.

Glassell, S.A. 1935a. Three new species of *Pinnixa* from the Gulf of California. *Trans. San Diego Soc. Nat. Hist.* 8:91-106, pl. 9-16.

Glassell, S.A. 1935b. New and little known crabs from the Pacific coast of northern Mexico. *Trans. San Diego Soc. Nat. Hist.* 8:91-106, pl. 9-16.

Glassell, S.A. 1936. The Templeton Crocker Expedition I. Six new brachyuran crabs from the Gulf of California. *Zoologica, New York.* 21: 213-218.

Goeldi, E.A. 1886. Studien über neue und weniger bekannte Podophthalmen Brasiliens. Beiträge zur kenntnis der sübwasser genera *Trichodactylus, Dilocarcinus, Sylviocarcinus* und der marinen genera *Leptopodia, Stenorhynchus Arch. f. Naturg.* 52:19-46, pl. II, III.

Gomez, L.D. and J.M. Savage. 1983. Searchers on that rich coast: Costa Rican field biology, 1400-1980. In D.H. Janzen (ed.), *Costa Rican Natural History.* Chicago: The University of Chicago Press.

Gosse, E. 1890. *The Life of Phillip Henry Gosse.* London. Gosse, P.H. 1840. *A Naturalist Sojourn in Jamaica.* London: Longmans, Brown, Green, Longmans.

Gould, A.A. 1841. *Report on the Invertebrata of Massachussets, comprising the Mollusca, Crustacea, Annelida, and Radiata.* Cambridge, Mass.

Guérin-Méneville, F.E. 1828. Mémoire sur l'Eurypodius, nouveau genre de Crustacé décapode brachyure. *Mém. Mus. His. Nat. Paris.* 16:345-355, pl. 14.

Guerra, F. 1974. Juan Ignacio Molina. In C.C. Gillispie (ed.), *Dictionary of Scientific Biography.* Vol. 9: pp. 458. New York: Charles Scribner's Sons.

Guilding, L. 1825. An account on some rare West-Indian Crustacea. *Trans. Lin. Soc. London.* 14:334-338.

Guilding, L. 1828. Observations on the zoology of the Caribbean Islands. *Zool. J.* 3:403-408, 527-544; 4:164-175.

Gundlach, J. 1887. Apuntes para la fauna Puerto Riqueña, VI Crustáceos. *Anales de la Sociedad española de Historia Natural.* 16:115-133.

Gundlach, J. 1900-1901. Contribución al estudio de los Crustáceos de Cuba. Notas compiladas y completadas por el Dr. J.I. Torralbas. *Anales Acad. La Habana.* 36:292, 305, 326-332, 362-374; 37:51-65, 148-160, 5 pl.

Guyénot, é. 1957. *L'évolution de La Pensée Scientifique. Les Sciences de La Vie au XVIIe et XVIIIe Siécles. L'Idee D'évolution.* Paris: Albin Michell.

Haan, W de 1833-1850. Crustacea. In *P.F. von Siebold, Fauna Japonica sive Descriptio Animalium, quae in Itinere per Japoniam, Jussu et Auspiciis Superiorum, qui Summum in India Batava Imperium Tenent, Suscepto, Annis 1823-1830 Collegit, Notis, Observationibus et Adumbrationibus* Leiden: Lugduni-Batavorum.

Hedgpeth, J.W. & W.L. Schmitt. 1945. The United States Fish Commission Steamer Albatross. *The American Neptune.* 5:1-25.

Heller, C. 1862. Beiträge zur näheren Kentniss der Macrouren. *S. B. Akad. Wiss. Wien.* 45:389-426, pl. 1, 2.

Heller, C. 1865. *Reise der österreichischen Fregatte 'Novara' um die Erde in Crustaceen den In Jahren 1857, 1858, 1859.* Zoologischer theil 2. Vienna: Kaiserlich-Königlichen Hof- und Staatsdruckerei.

Henderson, J.B. 1916. *The Cruise of the Tomas Barrera.* New York: Putnam and Sons.

Herbst, J.F.W. 1782-1804. *Versuch einer Naturgeschichte der Krabben und Krebs nebst einer systematischen Beschreibung ihrer verschidenen Arten.* 3 volumes. Zurich, Berlin, and Stralsund.

Holmes, S.J. 1895. Notes on West American Crustacea. *Proc. Calif. Acad. Sci.* (2) 4:563-588, 2 pl.

Holthuis, L.B. 1959a. The Crustacea Decapoda of Suriname (Dutch Guiana). *Zool. Verhand. (Leiden)* 44:1-296, pl. i-xvi.

Holthuis, L.B. 1959b. H.E. van Rijgersma - A little-known naturalist of St. Martin (Netherland Antilles). *Stud. Fauna Curaao and other Carib. Is.* 9:69-78.

Holthuis, L.B. 1969. Thomas Say as a carcinologist. In Thomas Say, *An Account of the Crustacea of the United States.* Reprint 1969. Lehre: J. Cramer.

Holthuis, L.B. 1979. H. Milne Edwards' 'Histoire Naturelle des Crustacs' (1834-1840) and its dates of publication. *Zool. Medel.* 53:286-296.

Hughes, G. 1750. *The Natural History of Barbados.* Book ix. Of Crustaceous Animals, pp. 261-266, pl. XXV.

Kingsley, J.S. 1880. Carcinological notes No. 1. *Proc. Acad. Nat. Sci. Phila.* 1880:34-37.

Krogmann, W.M. 1943. Obituary Ales Hrdlika. *Science* 98:254-255.

Lamarck, J.B. 1801. *Système des Animaux sans Vertèbres.* Paris.

Lamarck, J.B. 1809. *Philosophie Zoologique.* Paris.

Lamarck, J.B. 1818. *Histoire Naturelle des Animaux sans Vertèbres.* Vol. 5, Decapoda S.200-273. Paris.

Latreille, P.A. 1803. *Histoire Naturelle, générale et particulière des Crustacés et des Insectes.* 6:1-391, pl. 44-57. Paris.

Latreille, P.A. 1819. *Nouveaux Dictionaire d'Histoire Naturelle.* 28:1-570.

Latreille, P.A. 1825. In *Genre de Crustacés. Encyclopedie méthodique. Histoire Naturelle. Entomologie, ou histoire naturelle de Crustacés, des Arachnides et des Insectes.* 10:1-832.

Leach, W.E. 1817. *The Zoological Miscellany, Being Descriptions of New or Interesting Animals.* Vol. 3. London.

Lemoine, W. & M.M. Suarez 1984. *Beauperthuy: De Cumana a la Academia de Ciencias de Paris.* Caracas: Cromotip.

Lenz, H. 1902. Die Crustaceen der Sammlung Plate (Decapoda und Stomatopoda). *Zool. Jb. Syst. Suppl.* 5 (Fauna Chilensis 2):731-772, pl. 23.

Lindroth, C.H. 1973. Systematics specializes between Fabricius and Darwin: 1800-1859. In R.F. Smith, T.E. Mittler, & C.N. Smith (eds), *History of Entomology.* Palo Alto, California: Annual Reviews Inc.

Linnaeus, C. 1758. *Systema Naturae per Regna Tria Naturae, Secundum Classes, Ordines, Genera, Species, cum Characteribus, Differentiis, Synonymis, Locis.* Edition 10. Vol. 1. Holmiae.

Linnaeus, C. 1763. Centuria Insectorum, Quam, Praeside D.D. Car. von Linne, Proposuit Boas Johnson, Calmariensis. In C. Linnaeus, *Amoenitatis Academicae; seu Dissertationes variae, physicae, medicae, botanicae, Antehac seorsim editae, nunc collectae & auctae.* Vol. 6: pp. 384-415.

Linnaeus, C. 1764. *Museum S:ae R:ae M:tis Ludovicae Ulricae* Holmiae.

Man, J. G. De. 1892. Carcinological Studies in the Leyden Museum. No. 6. *Not. Leyden Mus.* 14:225-264.

Marcgraf, G. 1648. Historiae Rerum Naturalium Brasiliae, Libri octo: Quorum tres priores agunt de Plantis. Quartus de Piscibus. Quintus de Avibus . Sextus de Quadrupedibus, et Serpentibus. Septimus de Insectis. Octavus de ipsa Regione, et illus Incolis. Cum Appendice de Tapuyis, et Chilensibus. In *Historia Naturalis Brasiliae, Auspicio et Beneficio Illustriss. I. Mauritii Com. Nassau illius Provinciae et Maris summi Praefecti adornata, in qua non tantum Plantae et Animalia, sed et Indigenarum morbi, ingenia et mores describuntur et Iconibus supra quingentas illustrantur.* Pt. 2. Leiden & Amsterdam.

Martens, E. von 1869. Südbrasilische Süs-und Brackwasser-Crustaceen nach den Sammlungen des Dr. Reinh. Hensel. *Arch. f. Naturg.* 35:1-37, Pl. I, II.

Martens, E. von 1872. Über Cubanische Crustaceen nach den Sammlungen Dr. J. Gundlach's. *Arch. f. Naturg.* 38/I:77-147; 257-258, pl. 4, 5.

Meuschen, F.C. 1781. Index continens Nomina Generica Specierum propia, trivalia ut et synonyma. In L.T. Gronovius, *Zoophylacium Gronovianum, exhibens Animalia Quadrupeda, Amphibia, Pisces, Insecta, Vermes, Mollusca, Testacea, et Zoophyta, quae in Museu suo adservavit, examini subjecit, systematice disposuit atque descripsit.*

Miers, E.J. 1886. Report on the Brachyura collected by H.M.S. Challenger during the years 1873-76. *Reports on the Voyager of H.M.S. 'Challenger.'* 17 (2) (part 49), L + 362, pl. 1-29.

Milne Edwards, A. 1869. Révision des genres *Trichodactylus, Sylviocarcinus* et *Dilocarcinus* et description de quelques espéces nouveau qui s'y ratttachent. *Ann. Soc. Entom. France.* (4) 9:170-178.

Milne Edwards, A. 1873-1881. Études sur les Xiphosures et les Crustacés de la région Mexicaine. In *Recherches zoologiques pour servir a l'histoire naturelle de l'Amérique Centrale et du Mexique.* Cinquiéme partie, tome premier. Paris: Imprimerie National.

Milne Edwards, A. 1880. Études préliminaires sur les Crustacés. In Reports on the results of dredging under the supervision of Alexander Agassiz, in the Gulf of Mexico, and the Caribbean Sea, 1877, '78, '79, by the US Coast Survey Steamer 'Blake.' *Bull. Mus. comp. Zool. Harvard.* 8:1-68.

Milne Edwards, A. & E.L. Bouvier 1899a. Espèces nouvelles du genre *Palicus* Phil. (*Cymopolia* Roux) recueillis par le 'Blake' dans le Mer des Antilles et dans le Golfe du Mexique. *Bull. Mus. Hist Nat. Paris.* 5:122-125.

Milne Edwards, A. & E.L. Bouvier. 1899b. Dorippidés nouveau recuiellis par le 'Blake' dans le Mer des Antilles et dans le Golfe du Mexique. *Bull. Mus. Hist Nat. Paris.* 5:384-387.

Milne Edwards, A. & E.L. Bouvier. 1902. Les Dromiacés et les Oxystomes. In Reports on the results of dredging by the United States Survey Steamer 'Blake.' *Mem. Mus. comp. Zool. Harvard.* 27:1-127, pl. 1-25.

Milne Edwards, A. & E.L. Bouvier. 1923. Les Porcellanides et des Brachyures. In Reports on the results of dredging by the United States Survey Steamer 'Blake.' *Mem. Mus. comp. Zool. Harvard.* 47:281-395, pl. 1-12.

Milne Edwards, H. 1832. Observations sur les Crustacés du genre *Mithrax. Mag. Zool.* 2 (classe 7):1-16, pl. 1-5.

Milne Edwards, H. 1833. Description du genre *Leucippa*, établi d'après un Crustacé nouveau de la classe des Décapodes. *Ann. Soc. entom. France.* 2:512-517, 1 pl.

Milne Edwards, H. 1834. *Histoire naturelle des Crustacés, comprenant l'anatomie, la physiologie et la classification de ces animaux.* Vol. 1. Paris.

Milne Edwards, H. 1835. Footnote, p. 218. In C. de Freminville, 1835. Notice sur les Tourlouroux ou Crabes de terre des Antilles. *Ann. Sci. Nat. (Zool.).* (2) 3: 213-224.

Milne Edwards, H. 1837. *Histoire Naturelle des Crustacés, comprenant l'Anatomie, la Physiologie et la Classification de ces Animaux.* Vol. 2, Atlas (1834, 1837, 1840). Paris. Milne Edwards, H. 1854. Notes sur quelques Crustacés nouveaux ou peu connus conservés dans la collection du Muséum d'Histoire naturelle. *Arch. Mus. Hist. Nat. Paris.* 7:145-192, pl. 9-16.

Milne Edwards, H. & H. Lucas. 1843. Crustacés. In A. d'Orbigny, *Voyage dans l'Amérique méridionale (le Brésil, la république orientale de l'Uruguay, la république Argentine, la Patagonie, la république de Chili, la république de Bolivia, la république du Pérou), exécuté pendant les années 1826, 1827, 1828, 1829, 1830, 1831, 1832 et 1833.* 6(1). Strasbourg.

Molina, J.I. 1782. *Saggio Sulla Storia Naturale del Chili.* Bologna.

Moreira, C. 1912. Crustacés du Brésil. *Mem. Soc. Zool. France.* 25:145-154, pl. 3-6.

Müller, A. 1915. *Fritz Müller, Werke, Briefe und Leben.* Jena.

Nicholson, l. 1776. *Essai sur l'Histoire Naturelle de Saint-Domingue.* Paris.

Nicolet, H. 1849. Crustaceos y Arachnidos. In C. Gay (ed.), *Historia fisica y politica de Chile. Zoologia* Vol. 3: pp. 120-220. Paris.

Nobili, G. 1896. Crostacei Decapodi. Viaggio del Dott. A. Borelli nel Chaco Boliviano e nella Repubblica Argentina e nel Paraguay, . *Boll. Mus. Zool. Anat. Comp. R. Univ. Torino.* 11(265):1-3.

Nobili, G. 1897. Decapodi terrestri e d'acqua dolce. Viaggio del Dr. Enrico Festa nella Repubblica dell'Ecuador e regioni vicini. *Boll. Mus. Zool. Anat. Comp. R. Univ. Torino.* 12(275):1-6.

Nobili, G. 1901a. Decapodi raccolti dal Dr. Filipo Silvestri nell' America meridionale. *Boll. Mus. Zool. Anat. Torino.* 16(402):1-16.

Nobili, G. 1901b. Decapodi e Stomatopodi del viagio del Dr. Enrico Festa nella Repubblica dell'Ecuador e regione vicine. *Boll. Mus. Zool. Anat. Torino.* 16(415):1-58.

Olivier, G.A. 1791. Crabe. Cancer, Lin Geoff, Fab. In *Encyclopedie methodique, Zoology*, 6 (Histoire Naturelle, Insectes): pp. 142-182. Paris.

Olivier, E. 1880. *G.A. Olivier, membre de l'Institut de France, sa Vie, ses Travaux, ses Voyages.* Paris: Moulins.

Ortmann, A.E. 1893. Die Dekapoden-Krebse des Strassburger Museum. VII. *Zool. Jb. Syst.* 7:411-495, pl. XVII.

Ortmann, A.E. 1894. Die Dekapodenkrebse des Strassburger Museum. VIII. Abteilung Brachyura I. Unterabteilung Cancroidea. 2. Section: Cancrinea. 2. Gruppe: Catometopa. *Zool. Jb. Syst.* 7:683-772.

Ortmann, A.E. 1897. Carcinologische Studien. *Zool. Jb. Syst.* 10:258--372, pl. 17.

Ortmann, A.E. 1903. The geographical distribution of the fresh-water decapods and its bearing on ancient geography. *Proc. Amer. philos. Soc.* 41:267-400.

Otis, F.N. 1862. *Illustrated History of the Panama Railroad.* New York: Harpers & Brothers, (reprinted 1971, Pasadena, California: Socio-Technical Books.)

Oviedo y Valdés, G.F. de 1535. *La Historia General de las Indias Islas y Tierra Ferma del Mar Oceano.* Sevilla: Juan Cromberger.

Oviedo y Valdés, G.F. de 1851-1855. *Historia General y Natural de las Indias, Islas y Tierra-Firme del Mar Oceano.* 4 vols. Madrid: Real Academia de la Historia.

Papavero, N. 1971-1972. *Essays on the History of Neotropical Dipterology with Special Reference to Collectors (1750-1905).* São Paulo: Universidade de São Paulo, (Vol. 1, 1971; Vol. 2 1972).

Parra, A. 1787. *Descripcion de Diferentes Piezas de Historia Natural, las Mas del Ramo Maritimo, Representadas en Setenta y Cinco Laminas.* Havana.

Pearse, A.S. 1911. Report on the crustacea collected by the University of Michigan-Walker expedition in the state of Vera Cruz, Mexico. *Rep. Michigan Acad. Sci.* 13:108-113.

Philippi, R.A. 1894. Dos palabras sobre la sinonimia de los crustaceos decapodos braquiuros o jaivas de Chile. *Anales de la Universidad de Chile.* 87:369-379.

Pocock, R.I. 1889. Contributions to our knowledge of the Crustacea of Dominica. *Ann. Mag. Nat. Hist.* (6)3:6-22, pl. II.

Poeppig, E.F. 1835-1836. *Reise in Chile, Peru und auf dem Amazonenstrom Wahrend der Jahre 1827-1832.* 2 volumes + Atlas of scenery. Leipzig.

Poeppig, E.F. 1836. Crustacea chilensia nova aut minus nota descripsit. *Arch. Nsturg.* 2(I):134-144, 4 pl.

Porter, C.E. 1914. Los Crustaceos decapodos chilenos del Museo nacional. *Bol. Mus. Nac. Chile.* 7:275-277.

Porter. D.M. 1985. The Beagle collector and his collections. In D.S. Kohn (ed.), *The Darwinian Heritage: A Centennial Retrospect*: pp.973-1019. Wellington, New Zealand, Nova Pacifica: Princeton Univ. Press.

Randall, J.W. 1840. Catalogue of the Crustacea brought by Thomas Nuttall and J.K. Townsend, from the West Coast of North America and the Sandwich Islands, with descriptions of such species as are apparently new, among which are included several species of different localities, previously existing in the collection of the Academy. *J. Acad. Nat. Sci. Phila.* 8:106-147, pl. 3-7.

Rathbun, M.J. 1892. Catalogue of the crabs of the family Periceridae in the US National Museum. *Proc. US Nat. Mus..* 15:231-277, pl. 28-40.

Rathbun, M.J. 1893a. Catalogue of the crabs of the family Maiidae in the US National Museum. *Proc. US Nat. Mus.* 16:63-103, pl. 3-8.

Rathbun, M.J. 1893b. Descriptions of new genera and species of crabs from the west coast of North America and the Sandwich Islands. *Proc. US Nat. Mus.* 16:223-260.

Rathbun, M.J. 1893c. Descriptions of new species of American fresh-water crabs. *Proc. US Nat. Mus.* 16:649-661, pl. LXXIII-LXXVII.

Rathbun, M.J. 1894a. Notes on the crabs of the family Inachidae in the United States National Museum. *Proc. US Nat. Mus.* 17:43-75, pl. I.

Rathbun, M.J. 1894b. Description of a new genus and four new species of crabs from the Antillean Region. *Proc. US Nat. Mus.* 17:83-86.

Rathbun, M.J. 1896a. Description of a new genus and four new species of crabs from the West Indies. *Proc. US Nat. Mus.* 17:83-86.

Rathbun, M.J. 1896b. The genus *Callinectes*. *Proc. US Nat. Mus.* 18(1070):349-375, pl. XII-XXVIII.

Rathbun, M.J. 1896c. Descripions of two new species of fresh-water crabs from Costa Rica. *Proc. US Nat. Mus.* 18:377-379, pl. XXIX, XXX.

Rathbun, M.J. 1897a. Synopsis of the American Sesarmae, with description of new species. *Proc. Biol. Soc. Washington.* 11:89-92

Rathbun, M.J. 1897b. Synopsis of the American species of *Palicus philippi* (*Cymopolia* Roux), with descriptions of six new species. *Proc. biol. Soc. Washington.* 11:93-99.

Rathbun, M.J. 1897c. Description of a new species of *Cancer* from Lower California and aditional note on *Sesarma. Proc. Biol. Soc. Washington.* 11:111-112.

Rathbun, M.J. 1897d. List of the Decapod Crustacea of Jamaica. *Ann. Inst. Jamaica.* 1:1-46.

Rathbun, M.J. 1897e. Descriptions de nouvelles espèces de crabes d'eau douce appartenant aux collections du Muséum d'Histoire naturelle de Paris. *Bull. Mus. Hist. Nat. Paris.* 3(2):58-62.

Rathbun, M.J. 1898a. The Brachyura collected by the US Fisheries Commission Steamer 'Albatross' on the voyage from Norfolk, Virginia, to San Francisco, California, 1877-1878. *Proc. US Nat. Mus.* 21:567-616, pl. 41-44.

Rathbun, M.J. 1898b. In W.M. Ranking, The Northrop collection of Crustacea from the Bahamas. *Ann. N. Y. Acad. Sci.* 12:521-545, pl. 17.

Rathbun, M.J. 1898c. The Brachyura of the Biological Expedition to the Florida Keys and the Bahamas in 1893. *Bull. Lab. Nat. Hist. Iowa.* 4:250-294, pl. 1-9.

Rathbun, M.J. 1898d. A contribution to a knowledge of the fresh-water crabs of America. The Pseudothelphusinae. *Proc. US Nat. Mus.* 21:507-537.

Rathbun, M.J. 1900. Synopsis of North American invertebrates. VII. The cyclometopus or cancroid crabs of North America. *Amer. Natural.* 34 (398):131-143.

Rathbun, M.J. 1900. Synopsis of North American Invertebrates. XI. The catometopous or grapsoid crabs of North America. *Amer. Natural.* 34 (403):583-592.

Rathbun, M.J. 1901. Results of the Branner-Agassiz Expedition to Brazil. I. The Decapod and Stomatopod Crustacea. *Proc. Washington Acad. Sci.* 2:153-156, pl. 8.

Rathbun, M.J. 1902a. Papers from the Hopkins Stanford Galapagos Expedition 1898-1899. *Proc. Washington Acad. Sci.* 4:275-292.

Rathbun, M.J. 1902b. The Brachyura and Macrura of Porto Rico. *Bull. US Fish Comm.* 20 (1):3-26.

Rathbun, M.J. 1904. Description of three new species American crabs. *Proc. Washington Acad. Sci.* 17:161-162.

Rathbun, M.J. 1905. Les crabes d'eau douce (Potamonidae). *Nouv. Archs. Mus. Hist. Nat. Paris.* (4) 7:159-321, pl. 13-22

Rathbun, M.J. 1906a. Les crabes d'eau douce (Potamonidae). *Nouv. Archs. Mus. Hist. Nat. Paris.* (4) 8:30-122.

Rathbun, M.J. 1906b. Description of a new crab from Dominica, West Indies. *Proc. Biol. Soc. Washington* 19:91-92.

Rathbun, M.J. 1906c. Description of three new mangrove crabs from Costa Rica. *Proc. Biol. Soc. Washington,* 19:99-100.

Rathbun, M.J. 1907. The Brachyura. Reports on the Scientific results of the expedition to the tropical Pacific, in charge of Alexander Agassiz, Str. 'Albatross,' 1899-1900 and 1904-1905 *Mem. Mus. Comp. Zool. Harvard.* 35:23-74.

Rathbun, M.J. 1909. Description d'une nouvelle espèce de *Pinnotheres* de Porto Rico. *Bull. Mus. Hist. Nat. Paris.* 15:68-70.

Rathbun, M.J. 1910. The stalk-eyed Crustacea of Peru and adjacent coast. *Proc. US Nat. Mus.* 38:531-620, pl. 36-56.

Rathbun, M.J. 1912. New decapod Crustacea from Panama. *Smithson. Misc. Coll.* 59(13):1-3.

Rathbun, M.J. 1914. New genera and species of American Brachyrhynchous crabs. *Proc. US Nat. Mus.* 47:117-129, pl. 1-10.

Rathbun, M.J. 1915a. New fresh-water crabs (*Pseudothelhusa*) from Colombia. *Proc. Biol. Soc. Washington.* 28:95-100.

Rathbun, M.J. 1915b. New species of decapod Crustacea from the Dutch West Indies. *Proc. Biol. Soc. Washington.* 28:117-119.

Rathbun, M.J. 1916. Descriptions of three new species of crabs (*Osachila*) from the eastern coast of North America. *Proc. US Nat. Mus.* 50: 647-652, pl. 36.

Rathbun, M.J. 1918. The grapsoid crabs of America. *US Nat. Mus. Bull.* 97: 1-460, pl. 1-161.

Rathbun, M.J. 1919. Three new South American river crabs. *Proc. Biol. Soc. Washington* 32:5-6.

Rathbun, M.J. 1920. New species of spider crabs from the Straits of Florida and Caribbean Sea. New Species of crabs from Curaçao. *Proc. Biol. Soc. Washington,* 33:23-24.

Rathbun, M.J. 1921. Report on the Brachyura collected by the Barbados-Antigua Expedition from the University of Iowa in 1918. *Univ. Iowa Stud. Nat. Hist.* 9(5):65-90, pl. I-III.

Rathbun, M.J. 1922. New Species of crabs from Curaçao. *Proc. Biol. Soc. Washington.* 35:103-104.

Rathbun, M.J. 1923. New species of American spider crabs. *Proc. Biol. Soc. Washington.* 36:71-73.

Rathbun, M.J. 1923. The Brachyuran crabs collected by the US Steamer 'Albatross' in 1911, chiefly on the west coast of Mexico. *Bull. Amer. Mus. Nat. Hist.* 48:619-637. pl. 26-36.

Rathbun, M.J. 1924. New species and subspecies of spider crabs. *Proc. US Nat. Mus.* 64 (art. 14):1-5.

Rathbun, M.J. 1925. The spider crabs of America. *US Nat. Mus. Bull.* 129: 1-613, pl. 1-283.

Rathbun, M.J. 1930. The cancroid crabs of America of the families Euryalidae, Portunidae, Atelecyclidae, Cancridae and Xanthidae. *US Nat. Mus. Bull.* 152:i-xvi,1-609, pl. 1-230.

Rathbun, M.J. 1931. New crabs from the Gulf of Mexico. *Proc. Washington Acad. Sci.* 21:125-128.

Rathbun, M.J. 1931. A new species of pinnotherid crab from Costa Rica. *Proc. Washington Acad. Sci.* 21:262-263.

Rathbun, M.J. 1931. Two new crabs from the Gulf of Mexico. *Proc. Biol. Soc. Washington.* 44:71-72.

Rathbun, M.J. 1933a. Descriptions of new crabs from the Gulf of California. *Proc. Biol. Soc. Washington.* 46:147-149.

Rathbun, M.J. 1933b. Preliminary description of nine new species of Oxystomatous and allied crabs. *Proc. Biol. Soc. Washington.* 46: 183-186.

Rathbun, M.J. 1933c. A new species of *Pseudothelphusa* from Mexico. *J. Washington Acad. Sci.* 23:360.

Rathbun, M.J. 1935. Preliminary description of seven new species of Oxystomatous and allied crabs. *Proc. Biol. Soc. Washington,* 48:1-4.

Rathbun, M.J. 1937. The oxystomatous and allied crabs of America. *US Nat. Mus. Bull.* 166: 1-278, pl. 1-86.

Rathbun, R. 1883. *Great International Fisheries Exihibition, London, 1883.* Section G. Descriptive catalogue of the collection illustrating the scientific investigation of the sea and fresh waters. Washington: Government Printing Office.

Robinson, W. 1895. *A Flying Trip to the Tropics.* Cambridge: Riverside Press.

Rodriguez, G. 1982. Les crabes d'eau douce d'Amérique. Famille des Pseudothelphusidae. *Faune Tropical (ORSTOM),* 22:1-223.

Rodriguez, G. 1992. Les crabes d'eau douce d'Amérique. Famille des Trichodactylidae with a supplement to the Pseudothelphusidae. *Faune Tropical (ORSTOM)* 31:1-192.

Roux, P. 1828-1830. *Crustacés de la Méditerranée et de son littoral.* (iv+176 pages), pls. 1-45 (pages not numbered). Paris & Marseille.

Saussure, H. de 1853. Description de quelques Crustacés nouveaux de la côte occidentale du Mexique. *Rev. Mag. Zool.* (2) 5:354-368, pl. 12-13.

Saussure, H. de 1857. Diagnoses de quelques Crustacés nouveaux de l'Amérique tropicale. *Rev. Mag. Zool.* (2) 9:304-306, 501-505, pl. 1-6.

Saussure, H. de 1858. Mémoire su divers Crustacés nouveaux des Antilles et du Mexique. *Mém. Soc. Phys. Genève,* 14:417- 496, pl. 1-6.

Say, T. 1817-1818. An Account of the Crustacea of the United States. *J. Acad. Nat. Sci. Phila.* 1(1,2):57-63, 65-80 (pl. 4), 97-101, 155-160, 161-169 (all 1817), 235-253, 313-316, 317-319, 374-380, 381-401, 423-441 (all 1818). (See Holthuis, 1969, for dates of publication.)

Schomburgk, O.A. 1841. *R. H. Schomburgk's Reisen in Guiana und am Orinoko Wahrend der Jahre 1835-1839.* Leipzig: George Wiegand.

Schramm, A. 1867. *Crustacés de la Guadeloupe d'après un Manuscrit du Docteur Isis Desbonne Comparé avec les échantillons de Crustacés de sa Colection et les Dernières Publications de MM. Henri de Saussure et William Stimpson. 1er partie, Brachyures.* Basse-Terre: Imprimerie du Gouvernement.

Schramm, A. 1874. Rectification des noms de quelques Crustacés de la Guadeloupe, publiés d'après les notes manuscrites de Dr. Isis Desbones, et additions d'espèces non comprises au catalogue. *Rev. Mag. Zool.* (3)2:342-344.

Schwartsman, S. 1979. *Formaçao da Comunidade Cientifica no Brasil.* Sao Paulo: Companhia Editora Nacional.

Sclater, P.L. 1858. On the general geographical distribution of the members of the class Aves. *J. Proc. Linn. Soc. London (Zool.).* 2:130-145.

Seba, A. 1761. *Locupletissimi Rerum Naturalium Thesauri Accurata Descriptio et Iconibus Artificiosissimis Expressio per Universam Physices Historiam.* Vol. 3. Amsterdam: Janssonius-Waesberge.

Simpkins, D.M. 1975. Sydney Irving Smith. In C.C. Gillispie (ed.), *Dictionary of Scientific Biography,* Vol. 12: pp. 479-480. New York: Charles Scribner's Sons.

Sloane, H. 1725. *A Voyage to the Islands Madera, Barbadoes, Nieves, St. Christophers, and Jamaica, with the Natural History of the Herbs and Trees, Four-footed Beasts, Fishes, Birds, Insects, Reptiles, etc., of the Last of Those Islands, To which is Prefix'd, an Introduction, wherein is an Account of the Inhabitants, Air, Waters, Diseases, Trade, etc., of That Place, with Some Relations Concerning the Neighbouring Continent, and Islands of America.* Vol. 2. London.

Smith, S.I. 1869a. Notes on new or little known species of American cancroid Crustacea. *Proc. Boston Soc. Nat. Hist.* 12:274-289.

Smith, S.I. 1869b. Notice of the Crustacea collected by Prof. C. F. Hartt on the coast of Brazil in 1867. *Trans. Connect. Acad. Arts Sci.* 2:1-42, pl. 1.

Smith, S.I. 1870. Notes on American Crustacea. I. Ocypodoidea. *Trans. Connect. Acad. Arts Sci.* 2:113-176, pl. 2-5.

Smith, S.I. 1871. List of the Crustacea collected by J. McNiel in Central America. *Ann. Rep. Peabody Acad. Sci. 1869/1870.* Appendix:87-98.

Smith, S.I. 1879. The stalk-eyed Crustaceans of the Atlantic coast of North America north of Cape Cod. *Trans. Connect. Acad. Arts Sci.* 5:27-136, pl. 8-12.

Smith, S.I. 1881. Preliminary notice of the Crustacea dredged, in 64 to 325 fathoms, off the South Coast of New England, by the United States Fish Commission in 1880. *Proc. US Nat. Mus.* 3:413-452.

Smith, S.I. 1883. Preliminary Report on the Brachyura and Anomura dredged in deep water off the South Coast of New England. *Proc. US Nat. Mus.* 6:1-57, pl. 1-6.

Stebbing, T.R.R. 1914. Stalk-eyed Crustacea Malacostraca of the Scottish National Antarctic Expedition. *Trans. R. Soc. Edinb.* 50:253-307, pl. 23-32.

Stimpson, W. 1859a. (*Hapalocarcinus marsupialis*), a remarkable new form of Brachyurous Crustacean on the coral reefs of Hawaii. *Proc. Boston Soc. Nat. Hist.* 6:412-413.

Stimpson, W. 1859b. Notes on American Crustacea, No. 1. *Ann. Lyceum Nat. Hist. New York.* 7:49-93, pl. 1.

Stimpson, W. 1860. Notes on American Crustacea, in the museum of the Smithsonian Institution. No. 2. *Ann. Lyceum Nat. Hist. New York.* 7:176-246, pl. 2, 5.

Stimpson, W. 1861. Notes on certain Decapod Crustacea. *Proc. Acad. Nat. Sci. Phila.* 1861:372-373.

Stimpson, W. 1871a. Notes on American Crustacea, in the museum of the Smithsonian Institution, No. 3. *Ann. Lyceum Nat. Hist. New York.* 10:92-136.

Stimpson, W. 1871b. Brachyura. Preliminary report on the Crustacea dredged in the Gulf stream in the Straits of Florida by L.F. Pourtalès, Assist. US Coast Survey. *Bull. Mus. Comp. Zool. Harvard.* 2:109-160.

Streets, T.H. 1872. Notice on some Crustacea from the Island of St. Martin collected by Dr. van Rijersma. *Proc. Acad. Nat. Sci. Phila.* 1872:131-134.

Thomson, C.W. 1877. *The Voyage of the 'Challenger.' The Atlantic. A Preliminary Account of the General Results of the Exploring Voyage of H.M.S. 'Challenger' During the Year 1873 and the Early Part of the Year 1876.* London: Macmillan.

Turner, R.D. 1950. The voyage of the 'Tomas Barrera.' *Johnsonia* 2(28):220.

Tuxen, L.S. 1967. The entomologist J.C. Fabricius. *Ann. Rev. Entom.* 9:1-16.

Vaillant, L. 1904. Notice nécrologique sur F. Bocourt, garde des galeries honoraire. *Bull. Mus. Hist. Nat. Paris* 10:32-35

Vanzolini, E.P. 1977. *An Annotated Bibliography of the Land and Freshwater Reptiles of South America. Vol. 1, 1758-1900.* Sao Paulo: Universidade de Sao Paulo.

Verrill, A.E. 1908. Decapod Crustacea of Bermuda, Part I, Brachyura and Anomura. *Trans. Conn. Acad. Sci.* 13:299-479, pl. 9-28.

White, A. 1847a. *List of the Specimens of Crustacea in the Collection of the British Museum.* London: Edward Newman.

White, A. 1847b. Short descriptions of some new species of Crustacea in the collection of the British Museum. *Ann. Mag. Nat. Hist.* 20(132):205-207.

Wilkes, C. 1852. *A Narrative of the United States Exploring Expedition During the Years 1838, 1839, 1840, 1841, 1842.* London: Ingram, Cooke & Co.

Wright, A.H. 1953. Letters to C.F. Hartt, first professor of geology at Cornell. A cross section of the Agassiz period. Pre-Cornell and Early Cornell history. II. *Studies in History* 2:1-60.

Studies on decapod crustaceans of the Pacific coast of the United States and Canada

John S. Garth
Allan Hancock Foundation, University of Southern California, Los Angeles, California, USA

Mary K. Wicksten
Department of Biology, Texas A & M University, College Station, Texas, USA

1 RUSSIAN AND BRITISH EXPLORATION – 1815-1841

The first scientific information regarding the decapod crustacean fauna of the northeastern Pacific came as a result of Russian exploration of their far eastern territories. Specimens were collected by expeditions dispatched through the Academy of Science at St. Petersburg. The specimens had to be shipped back overland through Siberia to the Academy where examinations and descriptions were conducted by resident naturalists. The academicians of the time were mostly Germans who were recruited for service to examine 'the nature of the inhabitants, the soil, the vegetable and mineral wealth' of Russia's remote possessions (Thomson 1966; Raeff 1973). The Alaska king crab *Paralithodes camtschatica* (Tilesius 1815) was among the first northern Pacific species described, although, as the specific name implies, it was first reported off Kamchatka. The Napoleonic Wars temporarily halted all exploration and collecting by the Russians or other Europeans in the area.

The British ship Blossom explored the North American coast during 1825-1827. Collections were made by the ship's surgeon, Alexander Collie, and the naturalist, George Lay, at Monterey, California, and north to Point Barrow in the Arctic (Dodge 1971; Papenfuss 1976). Richard Owen, noted professor of comparative anatomy and Curator of the Museum of the Royal College of Surgeons, London, described the specimens (Owen 1839).

Owen recognized three shrimps and a hermit crab from California and the Pacific Northwest. Of these, the hippolytid shrimp *Heptacarpus palpator* and the hermit crab *Labidochirus splendescens* are common (Hart 1982; Wicksten 1986). Of the other two species, *Hippolyte affinis* (now considered to belong to the genus *Spirontocaris*) was reported only once, in 1866, after its original collection. The other, *Hippolyte layi*, has not been reported again (Schmitt 1921).

In 1839, the St. Petersburg Imperial Academy of Sciences sent Ilya Gabrilovic Voznesenski, preparator of the St. Petersburg Zoological Museum, to make collections in the Russian American colonies. During 1840-1841, he traveled and collected from Alaska to California. His material, sent to the Zoological Museum, included not only crustaceans but also botanical specimens and anthropological materials (Papenfuss 1976; Okladnikova 1987). Fedor Brandt (1850a, b, 1851, 1853) published descriptions of the crustaceans taken during this exploratory trip and those collected during the travels of A.T. von Middendorf, including the lithodid crab *Rhinolithodes wosnessenskii* and the isopod *Idotea wosnesenskii*. Brandt's work was the first to document the diversity of the lithodid crabs of the northeastern Pacific. Brandt also described echinoderms from the area, including such well-known species as *Pisaster ochraceus* and *Pycnopodia helianthoides*. Voznesenski also is commemorated by another lithodid crab, *Placentron wosnessenskii* Schalfew. Following this extensive collecting trip, there were no further major works by Russians on northern Pacific decapod crustaceans until the 20th century, when Russian and later Soviet investigators collected and described hippolytid shrimps ranging from the Siberian coast across the Aleutian Islands and into the northeastern Pacific.

2   AMERICAN EXPLORATION – 1836-1856

During 1836, Thomas Nuttall and J.K. Townsend collected in Oregon and in California at Monterey, Santa Barbara, San Pedro, and San Diego. Nuttall, a noted botanist and ornithologist associated with the Philadelphia Academy of Natural Sciences (Spencer 1986; Spamer & Bogan, this volume), collected 15 species of crustaceans, which were sent to the Academy's museum. J.W. Randall (1840) published descriptions of the crabs collected by Nuttall, who is commemorated by the southern kelp crab *Taliepus nuttalli* (Randall).

The United States Exploring Expedition visited the west coast of North America in 1841. During May to August, they visited the Straits of Juan de Fuca and the outer coast of Washington. From August 14 to October 28, members of the expedition travelled and collected in San Francisco Bay and up the Sacramento River. James Dwight Dana, also noted as a geologist, wrote acccounts of the decapods and isopods (1851, 1852) taken by the expedition in the years 1838 to 1842. Later William Stimpson examined other specimens from the expedition (Viola & Margolis 1985; Manning, this volume). The expedition's botanist, William Rich, is commemorated by the kelp crab *Pugettia richii* Dana, and Dana himself is commemorated by the spot prawn *Pandalus danae* Stimpson.

From June 1853 to July 1856, the US Surveying Expedition to the North Pacific explored and collected from China and Japan to the western United States. Most of the time was spent along the Asiatic coast. However, a party was left at the Leniavine Straits, on the coast of Asia near the Bering Strait, from August 1 to September 5, 1855. This party included William Stimpson, the expedition's zoologist. The ships later proceeded to San Francisco after stopping along the Aleutian Islands and at Sitka, Alaska. By October 13, all of the ships of the expedition had arrived in San Francisco (Cole 1947).

William Stimpson wrote extensively on crustaceans of the western United States and northern Pacific (see Manning, this volume). He published a short note in 1856 which provided the first records and living-color notes for common crustaceans of San Francisco Bay. In the note, he commented that 'a respectable number of all orders (of Crustacea), and even a considerable one of Macroura, are now known to exist on these shores.' While director of the Chicago Academy of Sciences, he described crustacean and echinoderm specimens from two expeditions as well as material that was sent to him from fishermen and fish markets. Most of the specimens were lost in the Chicago fire of 1871, but a few duplicate specimens remain at the National Museum of Natural History in Washington, D.C., and the British Museum (Natural History) in London; others may still exist in other collections in the United States and Europe (Deiss & Manning 1981; Spamer & Bogan, this volume). Stimpson's report on the crustaceans of the North Pacific Exploring Expedition was published posthumously (1907). The caridean shrimp *Heptacarpus stimpsoni* Holthuis is named in his honor.

3 THE CALIFORNIA ACADEMY OF SCIENCES AND STUDIES ALONG THE COAST OF CALIFORNIA AND THE PACIFIC NORTHWEST – 1853-1928

The California Academy of Sciences was established in 1853. William Neale Lockington was curator of fishes, reptiles, Crustacea, and 'radiates' from 1875 to 1881. He actively catalogued and described fishes, echinoderms, and crustaceans from 1874 to 1878, particularly decapods in 1874 and 1877. According to Lockington, by 1876 there were '220 species of the highest orders (the stalk-eyed and sessile-eyed) known up to this date along the shores south of Vancouver's Island.' (He was referring to decapods, amphipods and isopods). Some of the specimens he described were collected by William G. Harford, curator of conchology from 1867 to 1869 and from 1873 to 1875, and director of the Academy's museum from 1876 to 1886. Harford collected land plants and marine invertebrates near Santa Barbara and on Santa Rosa Island, as well as land snails from the Sierra Nevada. Harford also described species of isopods and amphipods, although most of these species later fell into synonymy. Both biologists, unfortunately, were involved in arguments regarding the use of a large financial gift from the estate of James Lick to the Academy. Bitter disagreements among members of the Academy resulted in the resignation of Harford in 1886, although he later served as an assistant in the museum from 1899 to 1906. Lockington severed his ties with the Academy completely. Lockington is commemorated by the snapping shrimp *Synalpheus lockingtoni* Coutiére, and Harford by the isopod *Cirolana harfordi* (Lockington) and the visored shrimp *Betaeus harfordi* (Kingsley).

In 1895 and 1900, Samuel J. Holmes published on decapods through the California Academy of Sciences. Holmes first described the genus *Heptacarpus*, the most abundant of the tidepool shrimps in California. He is commemorated by *Spirontocaris holmesi* Holthuis. The crustacean specimens of the California Academy of Sciences were destroyed in the earthquake and fire of 1906. However, syntypes of Lockington's species are known to survive in European museums.

The California Academy of Sciences and other scientific institutions received many specimens from amateur naturalists along the coast. Dr. Henry Hemphill of San Diego was an enthusiastic collector of mollusks and crustaceans at the turn of the century. He is commemorated by the hermit crab *Pagurus hemphilli* Benedict. Another shell collector, Fred Baker of San Diego, is commemorated by the hermit crab *Paguristes bakeri* Holmes. He contributed an early account of the crustaceans of southern California (Baker 1912) and donated a large collection of mollusks to the San Diego Society of Natural History.

The only crustacean biologist on the faculty of Stanford University was Frank W. Weymouth, professor of physiology and a graduate of that university in 1909. He conducted field studies and summer classes at Hopkins Marine Station, and also worked in British Columbia and the Gulf of Mexico. His papers included works on the life history of the Dungeness crab (1914), the behavior of the mole crab *Emerita analoga* (with C.H. Richardson Jr. 1912), and a guide to the brachyuran crabs of Monterey Bay (1910). His interests were not confined to marine biology but also included English literature, optometry, and civil rights. He is commemorated by the pea crab *Pinnixa weymouthi* Rathbun. Weymouth corresponded with N.B. Scofield, head of the California Department of Commercial Fisheries and author of a study of the shrimp fisheries of California (1919).

By the turn of the century, studies on decapods had begun in British Columbia and in Puget Sound, Washington. C.S. Bate (1864), who worked on the crustaceans collected by H.M.S. Challenger, identified species taken by J.K. Lord off Vancouver Island. Other early accounts of decapods, particularly shrimps, of British Columbia were cited by Butler (1980) and included

Whiteaves (1878), Smith (1880), and Newcombe (1898). William A. Herdman, professor of Oceanography at the University of Liverpool, collected decapods in Puget Sound and near Victoria in 1897. Records and descriptions of these specimens were published by C.A. Walker (1898), who named the shrimp *Heptacarpus herdmani* in Herdman's honor. In 1912, George Taylor, first director of the Pacific Biological Station at Nanaimo, listed 60 species of shrimps found in British Columbia.

The Puget Sound Biological Station, forerunner of the Friday Harbor Laboratory of the University of Washington, was established in 1904. Trevor Kincaid, professor of zoology, was instrumental in establishing the station and encouraged studies on the local fauna. Although he was knowledgeable on the decapods, his own studies concerned freshwater copepods. He is commemorated by the shrimp *Heptacarpus kincaidi* (Rathbun) and the isopods *Excirolana kincaidi* (Hatch) and *Ianiropsis kincaidi* Richardson. Belle A. Stevens, who studied with Kincaid and graduated from the University of Washington, published studies on the Callianassidae (1928) and hermit crabs (1925, 1927). She described the tube-dwelling hermit crab genus *Orthopagurus* (1927). The hermit crab *Pagurus stevensae* Hart is named in her honor. Wayne Wells (1928) and Evelyn Dorothy Way (1917) published studies of the Pinnotheridae, and crabs and crab-like anomurans in the publications of the Puget Sound Biological Station.

4 THE SMITHSONIAN INSTITUTION AND THE HARRIMAN ALASKA EXPEDITION – 1878-1937

Many specimens collected from the western United States were sent for study to scientists on the east coast. J.S. Kingsley (1878, 1899) published early works on snapping shrimp (Alpheidae) and other carideans. Mary Jane Rathbun (1899) wrote an account of the crustaceans of the Pribilof Islands as part of a survey of the fur seals and fur seal islands.

The extensive surveys of the US Fisheries steamer Albatross in 1888-1914 from Alaska to the Galapagos Islands added enormously to knowledge of the fauna. Important works resulting from these surveys were those of Walter Faxon (1893, 1895) on mostly deep-sea decapods; J.E. Benedict (1892, 1894, 1902, 1903, 1904) on anomuran crabs, and H. Coutiére (1899, 1909) on snapping shrimp. Faxon's work of 1895 included the first color paintings of Pacific deep-sea crustaceans. Rathbun included many records of specimens from these collections in her monographs on brachyurans (1918, 1925, 1930, 1937). For further information on Rathbun, see McLaughlin & Gilchrist (this volume) and Chace (1990a). Waldo Schmitt, who studied with Rathbun, used many records of the specimens from the Albatross in his work on the decapod crustaceans of California (1921).

In 1899, the ship George W. Elder left Seattle with a complement of scientists, photographers, and the family of its sponsor, Edward Harriman. Three invertebrate zoologists were aboard: Wesley R. Coe of Yale, William E. Ritter of the University of California (and later first director of Scripps Institution of Oceanography), and Trevor Kincaid of the University of Washington. These three collected along the shore and dredged during the voyage from southern Alaska to the Aleutians; the specimens were sent to the Smithsonian Institution (Goetzmann & Sloan 1982). Volume 10 (1904) of the expedition's reports was devoted to crustaceans, and included Harriet Richardson's account of the isopods and Mary Jane Rathbun's extensive and still authoritative account of the decapods. The collections were particularly rich in caridean shrimp, many of which were described for the first time in Rathbun's work. Rathbun's keys still are useful for identifying caridean species.

5 THE UNIVERSITY OF SOUTHERN CALIFORNIA AND OTHER WORK IN
SOUTHERN CALIFORNIA – 1921-1941

World War I halted scientific expeditions, but after the war Schmitt and Rathbun received new specimens from southern California. Percy S. Barnhart, director of the Venice Marine Biological Station, the forerunner of the marine science center of the University of Southern California (USC), was a particularly active collector who sent specimens from the coasts of Los Angeles and Orange counties as well as Santa Catalina Island. He is commemorated by the pinnotherid crab *Pinnixa barnharti*. Many specimens were taken during trawling or dredging by the USC ship Anton Dohrn in 1911-1930. William A. Hilton, director of the Laguna Marine Laboratory, sent specimens from Newport Bay and nearby rocky areas. He edited the *Journal of Entomology and Zoology* of Pomona College, to which he, his students, and other biologists contributed accounts of the local fauna. He is commemorated by the shrimp *Palaemonetes hiltoni* Schmitt. Private collectors also assembled specimens and forwarded them to Barnhart, Hilton, or Schmitt. An account of these early collections with a catalogue of the specimens was given by Wicksten (1984). Many of the records also were incorporated into Schmitt's benchmark work, *The Marine Decapod Crustacea of California* (1921), which resulted from work towards his M.A. degree at the University of California.

The most extensive collecting efforts prior to World War II were those of the Velero III, sponsored by Captain G. Allan Hancock and involving biologists from the University of Southern California and the Smithsonian Institution. Waldo Schmitt was a participant in these cruises, and made detailed color notes on many specimens. Unfortunately, Schmitt never got around to identifying the caridean shrimp that he had intended to catalogue and describe (see Chace 1990b). Records of the palaemonid shrimp were published in 1951 and 1952 by L.B. Holthuis of the Rijksmuseum van Natuurlijke Historie of Leiden, who studied the specimens at the Smithsonian Institution. M.K. Wicksten (1983, 1991), who studied under J.S. Garth at the University of Southern California, examined specimens at the Allan Hancock Foundation and the Smithsonian Institution and included them in accounts of the caridean faunas of the Gulf of California and the Galapagos Islands. As of today, many of the alpheid, hippolytid, and penaeid shrimp of the Hancock collections still have not been examined.

The cruises of the Velero III introduced John Garth to crustacean biology and crustacean biologists. He shared a stateroom with Waldo Schmitt during the cruises and sought his advice during the preparation and publication of his works on the Brachyura of the Galapagos Islands (Garth 1939, 1946). In 1937, Schmitt introduced Garth to Mary Jane Rathbun.

Although most of the cruises of the Velero III were to the south (from Mexico to Peru, and particularly the Galapagos), collections also were made along the mainland and offshore islands of southern California and north to Monterey Bay in 1938. The works by Garth (1958), Garth & Stephenson (1966), and Haig (1960) included records of the crabs taken during these cruises. For information on Janet Haig, see McLaughlin & Gilchrist (this volume).

Steve Glassell is noted as a collector who described new species of pinnotherid crabs (1935, 1938), hermit crabs (1937) and porcelain crabs (1945) from southern California and western Mexico, including specimens taken during the cruises of the Stranger, under Captain Fred E. Lewis. A self-taught carcinologist, Glassell received technical assistance from the paleontologist Ulysses S. Grant III, then with the San Diego Natural History Museum, where Glassell worked in the museum for many years as an unpaid volunteer. He is commemorated by the porcelain crab *Petrolisthes glasselli* Haig. Glassell's talented illustrator, Anker Petersen, later prepared illustrations for John Garth.

George MacGinitie wrote about burrowing invertebrates, including *Upogebia* spp. (1930), and with Nettie MacGinitie made collections of decapods in Newport Bay and co-authored *Natural History of Marine Life* (1949). They are commemorated by the shrimp *Betaeus macginitieae* Hart and the ghost shrimp *Upogebia macginitieorum* Williams.

The collections of the Templeton Crocker Expedition of 1936 were largely conducted in the Gulf of California and at the southern tip of the Baja California peninsula. However, works by Glassell (1937) on hermit crabs; Burkenroad (1937, 1938) on the penaeids; and Chace (1937, 1951) contain descriptions of species that range into southern California. Burkenroad, who studied at Yale and participated in the Yale oceanographic expeditions, later was recognized as an expert on penaeid shrimp (Schram 1986). He visited Garth at the Hancock Foundation in later years prior to publishing (1963, 1981) lengthy considerations of decapod phylogeny. He is commemorated by the deep-sea shrimp *Bentheogennema burkenroadi* Krygier and Wasmer. For many years before his death, Burkenroad, who was financialy independent, was affiliated with the San Diego Natural History Museum as a research associate, and he provided financial support to the Museum for carcinological research programs. F. A. Chace Jr. graduated from Harvard University and served as curator at the Museum of Comparative Zoology before joining the Division of Marine Invertebrates at the United States National Museum in 1946. He has published extensively on carideans, particularly those of the Caribbean and Atlantic, and lately on the Philippine specimens of the Albatross expedition (see Manning 1986).

Research on decapods continued in British Columbia. J.F.L. Hart of the Royal British Columbia Museum published works as early as 1930, and has continued since then to work on larval decapods, new records, and new species. Noteworthy publications include a review of the genus *Betaeus* (1964), new species of *Pagurus* (1971), and a guide to crabs of British Columbia (1982).

6   STUDIES SINCE WORLD WAR II

World War II effectively stopped biological investigations along the Pacific Coast. The Velero III was pressed into military service, and not returned to USC following the war when collecting trips by field parties from the university and other institutions resumed after 1945.

Deep-water decapods received new attention in the 1950's. The California Academy of Sciences and the California Department of Fish and Game conducted dredging along the coast of California (Goodwin 1952a). The newly-acquired Velero IV of USC was used for traditional collecting methods such as dredging, but was also employed in diving and midwater trawling. From this work, Ebeling et al. (1970) published a distributional analysis of deep-sea decapods from southern California. Two new species of shrimp, *Pasiphaea chacei* Yaldwyn and *Plesionika sanctaecatalinae* Wicksten were described from these midwater collections.

Dredging and trawling provided new records and descriptions of decapods taken off Oregon and the mouth of the Columbia River. The volume by Pruter & Alverson (1972) contains environmental data and records of the offshore decapods. Studies by Oregon State University provided accounts of midwater shrimp (Pearcy & Forss 1966, 1969; Wasmer 1972a; Krygier & Wasmer 1975) and galatheid crabs (Ambler 1980).

A new set of investigators began studies on the decapods of the Pacific Northwest. Jens Knudsen of Pacific Lutheran University published papers on the life cycles of brachyuran crabs, including works on the Xanthidae of California (1960) and Majidae and various anomurans of Puget Sound (1964). T.H. Butler of the Pacific Biological Station published an extensive acccount of the shrimp of British Columbia (1980), and also provided range extensions and redescriptions of carideans of the area. Patsy McLaughlin of Western Washington University produced a monograph on the northern Pacific hermit crabs (1974), and since then has revised the generic classification, described new species, and reared larval stages of these anomurans. She is an active teacher as well as a researcher and is commemorated by *Tomopagurus maclaughlinae* Haig. Gregory C. Jensen (1983, 1987) of the University of Washington described new species of hippolytid shrimps.

Taxonomic and systematic work at various institutions continues on analysis of previously collected and preserved specimens in museums and university collections, as well as on newly collected material sent in by research ships, private collectors, environmental studies, and fisheries biologists. Since 1945, three species of lobster-like decapods, one penaeid, 13 species of carideans, and 11 species of anomurans have been described from the northeastern Pacific; but only one new brachyuran, *Eurypanopeus hyperconvexus* Garth 1986, has been discovered. There have been extensive generic revisions among carideans and anomurans, and further work may change the nomenclature again.

The development of SCUBA diving equipment and underwater photography has proven a boon to crustacean biologists as well as other marine scientists. Four species of west coast hermit crabs and a hippolytid shrimp have been described from material taken by divers and examined alive. Numerous species that previously were almost impossible to collect in their rocky subtidal habitats can be viewed, photographed, or collected readily by divers. Another important innovation is the development of computerized systems for cataloging and analyzing data, which enable researchers to use numerical sorting strategies for systematic studies (e.g. Garth & Stephenson 1966) and analyzing distributional data (e.g. Ebeling et al. 1970, Wicksten 1989).

By now, the intertidal and shallow subtidal decapods of the west coast of the United States seem to be fairly well known. One can expect introductions by human activity, (such as the Oriental shrimp *Palaemon macrodactylus* into Malibu Lagoon, Long Beach Harbor, and San Francisco Bay) and occasional vagrant species (such as the red crab *Pleuroncodes planipes*) carried northward by currents in especially warm years. However, decapods of midwater and the lower continental slope are still poorly known, especially between Point Conception and Oregon. Species inhabiting deeper rocky cliffs and banks are hard to collect and are known today from only a few specimens. There have been few studies of the decapod fauna of the West Coast by cladistic or phenetic methods, nor have electrophoretic or genetic comparisons been made between these species and others related to them. Data are mostly scattered rather than consolidated into up-to-date guidebooks or monographs. Little is known about the life histories and distributions of many west coast species. Even today, many descriptions are based on museum specimens taken decades ago. The diverse northeastern Pacific decapod fauna, rich in endemic species and characteristic cold-water taxa, offers ample opportunities for even the most basic studies in the future.

ACKNOWLEDGMENTS

We thank Karren Elsbernd, California Academy of Sciences, Alan Baldridge, Stanford University, and Paul Illg, University of Washington, for sending helpful historical information.

REFERENCES

Ambler, J.W. 1980. Species of *Munidopsis* (Crustacea, Galatheidae) occurring off Oregon and in adjacent waters. *US Dept. Comm. Fish. Bull.* 78:13-34.

Baker, C.F. 1912. Notes of the Crustacea of Laguna Beach. *First Ann. Rpt. Laguna Mar. Lab.* 1912:100-117.

Bate, C.S. 1864. Characters of new species of crustaceans discovered by J.K. Lord on the coast of Vancouver Island. *Proc. Zool. Soc. Lond.* 1864:661-668.

Benedict, J.E. 1892. Preliminary descriptions of thirty-seven new species of hermit crabs of the genus *Eupagurus* in the US National Museum. *Proc. US Nat. Mus.* 15:1-26.

Benedict, J.E. 1894. Descriptions of new genera and species of crabs of the family Lithodidae, with notes on the young of *Lithodes camtschaticus* and *Lithodes brevipes*. *Proc. US Nat. Mus.* 17:479-488.

Benedict, J.E. 1902. Descriptions of a new genus and forty-six new species of crustaceans of the family Galatheidae, with a list of the known marine species. *Proc. US Nat. Mus.* 26: 243-334.

Benedict, J.E. 1903. Revision of the Crustacea of the genus *Lepidopa*. *Proc. US Nat. Mus.* 26:889-895.

Benedict, J.E. 1904. A new genus and two new species of crustaceans of the family Albuneidae from the Pacific Ocean, with remarks on the probable use of the antennulae in *Albunea* and *Lepidopa*. *Proc. US Nat. Mus.* 27:621-625.

Brandt, J.F. 1850a. Bericht über die für die Reisebeschreibung des Herrn von Middendorf von J.F. Brandt bearbeiteten Krebsthiere aus den Abtheilungen der Brachyuren (Krabben), Anomuren und Makrouren (Krebse). *Bull. phys.-math. Acad. Sci. St. Petersb.* 8:234-238.

Brandt, J.F. 1850b. Vorlaufige Bemerkungen über eine neue aus zwei noch unbeschriebenen Gattungen und Arten gebildete Unterabtheilung (Hapalogastrica) der Tribus Lithodina, begleitet von einer Charakteristik der eben gennanten Tribus der Anomuren. *Bull. phys.-math. Acad. Sci. St. Ptersb.* 8:266-269.

Brandt, J.F. 1851. Krebse. In A.T. von Middendorf (ed.), *Reise in den aussersten Norden und Osten Sibiriens während der Jahre 1843 und 1844*, pp. 77-148, 2 (Zoologie). St. Petersburg.

Brandt, J.F. 1853. Über eine neue Art der Gattung *Cryptolithodes* (*Cryptolithodes sitchensis*). *Bull. phys.-math. Acad. Sci. St. Petersb.* 11:653-654.

Burkenroad, M.D. 1937. Sergestidae (Crustacea Decapoda) from the lower California region, with descriptions of two new species, and some remarks on the organs of Pesta in *Sergestes*. *Zoologica* 22:315-329.

Burkenroad, M.D. 1938. The Templeton Crocker Expedition. XIII. Penaeidae from the region of Lower California and Clarion Island, with descriptions of four new species. *Zoologica* 23:55-91.

Burkenroad, M.D. 1963. The evolution of the Eucarida (Crustacea Eumalacostraca) in relation to the fossil record. *Tulane Studies Geol.* 2:1-16.

Burkenroad, M.D. 1981. The higher taxonomy and evolution of Decapoda (Crustacea). *Trans. San Diego Soc. Nat. Hist.* 19:251-268.

Butler, T.H. 1980. Shrimps of the Pacific coast of Canada. *Can. Bull. Fish. Aquat. Sci.* 202:1-280.

Chace Jr. F.A. 1937. The Templeton Crocker Expedition. VII. Caridean decapod Crustacea from the Gulf of California and the west coast of Lower California. *Zoologica* 22:109-138.

Chace Jr. F.A. 1951. The grass shrimps of the genus *Hippolyte* from the west coast of North America. *J. Wash. Acad. Sci.* 41:353-9.

Chace Jr. F.A. 1990a. Mary J. Rathbun. *J. Crust. Biol.* 10:165-167.

Chace Jr. F.A. 1990b. Waldo L. Schmitt. *J. Crust. Biol.* 10:168-171.

Cole, A.B. 1947. *Yankee surveyors in the Shogun's seas*. New York: Greenwood Press.

Coutiére, H. 1899. Les 'Alpheidae,' morphologie externe et interne, formes larvaires, bionomic. *Ann. Sci. Nat., Zool.* (8)9:1-559.

Coutiére, H. 1909. The American species of snapping shrimps of the genus *Synalpheus*. *Proc. US Nat. Mus.* 36:1-93.

Dana, J.D. 1851. Conspectus Crustaceorum quae in Orbis Terrarum circumnavigatione, Carolo Wilkes e Classes Reipublicae Foederatae Duce. *Proc. Acad. Nat. Sci. Phila.* 5:247-254.

Dana, J.D. 1852. *United States Exploring Expedition during the years 1838, 1839, 1840, 1841, 1842, under the command of Charles Wilkes, USN* Vol. 13, Crustacea, part 1, (viii). Philadelphia.

Deiss, W.A. & R.B. Manning. 1981. The fate of the invertebrate collections of the North Pacific Exploring Expedition, 1853-1856. In *History in the Service of Systematics*: pp. 79-85. London: Soc. Bibliogr. Nat. Hist.

Dodge, E.S. 1971. *Beyond the capes*. Boston: Little, Brown.

Ebeling, A.W., R.M. Ibara, R.J. Lavenberg, & F.J. Rohlf 1970. Ecological groups of deep-sea animals off southern California. *Bull. Los Angeles County Mus. Nat. Hist. Sci.* 6:1-43.

Faxon, W. 1893. Reports on the dredging operations off the west coast of central America to the Galapagos, to the west coast of Mexico, and in the Gulf of California, in charge of Alexander Agassiz, carried on by the US Fish Commission steamer 'Albatross' during 1891, Lieut. Commander Z.I. Tanner, USN, commanding. VI. Preliminary descriptions of new species of Crustacea. *Bull. Mus. Comp. Zool.* 24:149-220.

Faxon, W. 1895. Reports on an exploration off the west coasts of Mexico, central and South America, and off the Galapagos Islands, in charge of Alexander Agassiz, by the US Fish Commission steamer 'Albatross' during 1891, Lieut. Commander Z.I. Tanner, USN, commanding. 15. The stalk-eyed Crustacea. *Mus. Comp. Zool. Mem.* 18:1-292.

Garth, J.S. 1939. New brachyuran crabs from the Galapagos Islands. *Allan Hancock Pac. Exped.* 5:9-48.

Garth, J.S. 1946. Littoral brachyuran fauna of the Galapagos archipelago. *Allan Hancock Pac. Exped.* 5:341-6-00.

Garth, J.S. 1958. Brachyura of the Pacific coast of America Oxyrhyncha. *Allan Hancock Pac. Exped.* 21:1-499.

Garth, J.S. 1986. New species of xanthid crabs from early Hancock expeditions. *Occ. Pap. Allan Hancock Found.* n.s. 4:1-14.

Garth, J.S. & W. Stephenson 1966. Brachyura of the Pacific coast of America Brachyrhyncha: Portunidae. *Allan Hancock Monogr. Mar. Biol.* 1:1-154.

Glassell, S. 1935. New or little-known crabs from the Pacific coast of northern Mexico. *Trans. San Diego Soc. Nat. Hist.* 8:91-106.

Glassell, S. 1937. Hermit crabs from the Gulf of California and the west coast of Lower California. *Zoologica* 22:241-263.

Glassell, S. 1938. New and obscure decapod Crustacea from the west American coasts. *Trans. San Diego Soc. Nat. Hist.* 8:411-454.

Glassell, S. 1945. Four new species of North American crabs of the genus *Petrolisthes*. *J. Wash. Acad. Sci.* 35:223-229.

Goetzmann, W.H. & K. Sloan. 1982. *Looking far north*. Princeton: Princeton Univ. Press.

Goodwin, D.G. 1952a. Some decapod Crustacea dredged off the coast of central California. *Proc. Calif. Acad. Sci.* 28:393-397.

Goodwin, D.G. 1952b. Crustacea collected during the 1950 bottom-fish investigations of the M.V. *N. B. Scofield*. *Calif. Fish & Game* 38:163-181.

Haig, J. 1960. The Porcellanidae (Crustacea Anomura) of the eastern Pacific. *Allan Hancock Pac. Exped.* 24:1-440.

Hart, J.F.L. 1930. Some decapods from the south-eastern shores of Vancouver Island. *Can. Field-Nat.* 44:101-109.

Hart, J.F.L. 1964. Shrimps of the genus *Betaeus* on the Pacific coast of North America with descriptions of three new species. *Proc. US Nat. Mus.* 115:431-466.

Hart, J.F.L. 1971. New distributional records of reptant decapod Crustacea, including descriptions of three new species of *Pagurus*, from the waters adjacent to British Columbia. *J. Fish. Res. Bd. Can.* 28:1527-1544.

Hart, J.F.L. 1982. *Crabs and their relatives of British Columbia*. Brit. Columb. Prov. Mus. Handbook 40. Victoria.

Holmes, S. 1895. Notes on west American Crustacea. *Proc. Calif. Acad. Sci.* (2)4:563-588.

Holmes, S. 1900. Synopsis of California stalk-eyed Crustacea. *Occ. Papers Calif. Acad. Sci.* 7:1-262.

Holthuis, L.B. 1951. A general revision of the Palaemonidae (Crustacea Decapoda Natantia) of the Americas. I. The subfamilies Euryrhynchinae and Pontoniinae. *Occ. Papers Allan Hancock Found.* 11:1-332.

Holthuis, L.B. 1952. A general revision of the Palaemonidae (Crustacea Decapoda Natantia) of the Americas. II. The subfamily Palaemoninae. *Occ. Papers Allan Hancock Found.* 12:1-396.

Jensen, G.C. 1983. *Heptacarpus pugettensis*, a new hippolytid shrimp from Puget Sound, Washington. *J. Crust. Biol.* 3:314-320.

Jensen, G.C. 1987. A new species of the genus *Lebbeus* (Caridea: Hippolytidae) from the northeastern Pacific. *Bull. So. Calif. Acad. Sci.* 86:89-94.

Kingsley, J.S. 1878. A synopsis of the North American species of the genus *Alpheus*. *Bull. US Geol. Geogr. Surv. Territ., US Dept. Int.* 5:189-199.

Kingsley, J.S. 1899. The Caridea of North America. *Amer. Nat.* 33:709-720.

Knudsen, J.W. 1960. Reproduction, life history, and larval ecology of the California Xanthidae, the pebble crabs. *Pac. Sci.* 14:3-17.

Knudsen, J.W. 1964. Observations of the reproductive cycles and ecology of the common Brachyura and Anomura of Puget Sound, Washington. *Pac. Sci.* 18:3-33.

Krygier, E.E. & R.A. Wasmer 1975. Description and biology of a new species of pelagic penaeid shrimp, *Bentheogennema burkenroadi*, from the northeastern Pacific. *US Dept. Comm. Fish. Bull.* 73:737-746.

Lockington, W.N. 1874. On the Crustacea of California. *Proc. Calif. Acad. Sci.* 5:380-384.

Lockington, W.N. 1877. Remarks on the Crustacea of the Pacific coast, with descriptions of some new species. *Proc. Calif. Acad. Sci.* 7:28-36.

MacGinitie, G.E. 1930. The natural history of the mud shrimp *Upogebia pugettensis* (Dana). *Ann. Mag. Nat. Hist.* (5)2:36-44.

MacGinitie, G.E. & N. MacGinitie 1949. *Natural history of marine animals*. New York: McGraw Hill.

McLaughlin, P.A. 1974. The hermit crabs (Crustacea Decapoda, Paguridea) of northwestern North America. *Zool. Verhandel.* 130:1-396.

Manning, R.B. 1986. Fenner A. Chace Jr.: biographical notes and bibliography. *J. Crust. Biol.* 6(3): ii-vi.

Newcombe, C.F. 1898 A preliminary catalogue of the collections of natural history and ethnology in the Provincial Museum, Victoria, B.C. Victoria: Prov. Mus.

Okladnikova, E.A. 1987. Science and education in Russian America. In S.F. Starr (ed.), *Russia's American colony*: pp. 218-248. Durham: Duke Univ. Press.

Owen, R. 1839. Crustacea. In *The zoology of Captain Beechey's Voyage*: pp. 77-92. London: H.G. Bohn.

Papenfuss, G.F. 1976. Landmarks in Pacific North American marine phycology. In I.A. Abbott & G.J. Hollenberg (eds), *Marine Algae of California*: pp. 21-46. Stanford: Stanford Univ. Press.

Pearcy, W.G. & C.A. Forss. 1966. Depth distribution of oceanic shrimps (Decapoda: Natantia) off Oregon. *J. Fish. Res. Bd. Can.* 23:1135-1143.

Pearcy, W.G. & C.A. Forss. 1969. The oceanic shrimp *Sergestes similis* off the Oregon coast. *Limnol. Oceanogr.* 14:755-765.

Pruter, A.T. & D.L. Alverson (eds) 1972. *The Columbia River estuary and adjacent ocean waters. Bioenvironmental studies*. Seattle: Univ. Wash. Press.

Raeff, M. 1973. The enlightenment in Russia and Russian thought in the enlightenment. In J.G. Garrard (ed.), *The eighteenth century in Russia*: pp. 25-47. Oxford: Clarendon Press.

Randall, J.W. 1840. Catalogue of the Crustacea brought by Thomas Nuttall and J.K. Townsend, from the west coast of North America and the Sandwich Islands, with descriptions of such species as are apparently new among which are included several species of different localities, previously existing in the collection of the Academy. *J. Acad. Nat. Sci. Phila.* 8:106-147.

Rathbun, M.J. 1899. List of Crustacea known to occur on and near the Pribilof Islands. In D.S. Jordan (ed.), *The Fur Seals and Fur Seal Islands of the North Pacific*, part 3. Washington, D.C.: US Treasury Dept.

Rathbun, M.J. 1904. Decapod crustaceans of the northwest coast of North America. In *Harriman Alaska Exped.*, Vol. 10, *Crustaceans*: pp. 3-190, Pls. I-X. New York: Doubleday, Page.

Rathbun, M.J. 1918. The grapsoid crabs of America. *Bull. US Nat. Mus.* 97:1-461.

Rathbun, M.J. 1925. The spider crabs of America. *Bull. US Nat. Mus.* 129:1-613.

Rathbun, M.J. 1930. The cancroid crabs of America of the families Euryalidae, Portunidae, Atelecyclidae, Cancridae, and Xanthidae. *Bull. US Nat. Mus.* 152:1-609.

Rathbun, M.J. 1937. The oxystomatous and allied crabs of America. *Bull. US Nat. Mus.* 166:1-278.

Richardson, H. 1904. Isopod crustaceans of the northwest coast of North America. In *Harriman Alaska Exped.*, Vol. 10, *Crustaceans*: pp. 213-230. New York: Doubleday, Page.

Schmitt, W.L. 1921. The marine decapod Crustacea of California. *Univ. Calif. Publ. Zool.* 23:1-470.

Schram, F. 1986. Martin David Burkenroad. *J. Crust. Biol.* 6:302-307.

Scofield, N.B. 1919. Shrimp fisheries of California. *Calif. Fish & Game* 5:1-12.

Smith, S.I. 1880. Notes on Crustacea collected by Dr. G.M. Dawson at Vancouver and the Queen Charlotte Islands. *Geo. Surv. Can. Rep. Progr.* 1878-79 (Append. D):206-218.

Spencer, L.T. 1986. Naturalists of the Pacific shore: Early explorer-naturalists of California. *Amer. Zool.* 26:321-329.

Stevens, B.A. 1925. Hermit crabs of Friday Harbor, Washington. *Publ. Puget Sound Biol. Sta. Univ. Wash.* 3:273-309.

Stevens, B.A. 1927. *Orthopagurus*, a new genus of Paguridae from the Pacific coast. *Publ. Puget Sound Biol. Sta. Univ. Wash.* 5:245-252.

Stevens, B.A. 1928. Callianassidae from the west coast of North America. *Publ. Puget Sound Biol. Sta. Univ. Wash.* 6:315-369.

Stimpson, W. 1856. On some Californian Crustacea. *Proc. Calif. Acad. Sci.* 1:95-99.

Stimpson, W. 1907. Report on the Crustacea (Brachyura and Anomura) collected by the North Pacific Exploring Expedition, 1853-1856. *Smithson. Misc. Coll.* 49:1-240.

Taylor, G. 1912. Preliminary list of one hundred and twenty-nine species of British Columbia decapod crustaceans. *Contr. Can. Biol.* 11:187-214.

Thomson, G.S. 1966. *Catherine the Great and the expansion of Russia*. London: English Univ. Press.

Tilesius, W.C. 1815. De Cancris Camtschaticis, Oniscis, Entomostracis et Cancellis marinis microscopis noctilucentibus, Cum tabulis IV: Aenaeis et appendice adnexo de Acaris et Ricinis Camtschaticis. Auctore Tilesio. Conventui exhibuit die 3 Februarii 1813. *Mem. Acad. Sci. St. Peters.* 5:331-405.

Viola, H.J. & C. Margolis (eds). 1985. *Magnificent voyagers: the US Exploring Expedition, 1838-1842*. Washington, D.C.: Smithson. Inst. Press.

Walker, A.O. 1898. Crustacea collected by W.A. Herdman, F.R.S., in Puget Sound, Pacific coast of North America. *Trans. Liverpool Biol. Soc.* 12:268-287.

Wasmer, R.A. 1972a. A new species of Hymenodora (Decapoda, Oplophoridae) from the northeastern Pacific. *Crustaceana* 22:87-91.

Wasmer, R.A. 1972b. New records for four deep-sea shrimps from the northeastern Pacific. *Pac. Sci.* 26:259-263.

Way, E. 1917. Brachyura and crab-like Anomura of Friday Harbor, Washington. *Publ. Puget Sound Biol. Sta. Univ. Wash.* 1:349-382.

Wells, W.W. 1928. Pinnotheridae of Puget Sound. *Publ. Puget Sound Biol. Sta. Univ. Wash.* 6:283-314.

Weymouth, F.W. 1910. Synopsis of the true crabs (Brachyura) of Monterey Bay, California. *Publ. Leland Stanford Jr. Univ.* 4:1-64.

Weymouth, F.W. 1914. Contributions to the life-history of the Pacific coast edible crab (*Cancer magister*). *Rept. Brit. Columbia Comm. Fish.* 1914:123-129.

Weymouth, F.W. & C.H. Richardson, Jr. 1912. Observations on the habits of the crustacean *Emerita analoga*. *Smithson. Misc. Coll.* 59:1-13.

Whiteaves, J.F. 1878. On some new marine invertebrata from the west coast of North America. *Can. Nat.* n.s. 8:464-471.

Wicksten, M.K. 1983. A monograph on the shallow-water caridean shrimp from the Gulf of California, Mexico. *Allan Hancock Monogr. Mar. Biol.* 13:1-59.

Wicksten, M.K. 1984. Early twentieth century records of decapod Crustacea from Los Angeles and Orange counties, California. *Bull. So. Calif. Acad. Sci.* 82:138-143.

Wicksten, M.K. 1986. A new species of *Heptacarpus* from southern California, with a redescription of *Heptacarpus palpator* (Owen) (Caridea: Hippolytidae). *Bull. So. Calif. Acad. Sci.* 85:46-55.

Wicksten, M.K. 1989. Ranges of offshore decapod crustaceans in the eastern Pacific Ocean. *Trans. San Diego Soc. Nat. Hist.* 21:291-316.

Wicksten, M.K. 1991. Caridean and stenopodid shrimp of the Galapagos Islands. In M.J. James (ed.), *Galapagos Marine Invertebrates*. New York: Plenum Publ.

Time capsule of carcinology: History and resources in the Academy of Natural Sciences of Philadelphia

Earle E. Spamer & Arthur E. Bogan
Department of Malacology (General Invertebrates Section), Academy of Natural Sciences of Philadelphia, Philadelphia, Pennsylvania, USA

1 INTRODUCTION

It is a paradox that historically and scientifically important specimens can be lost in a museum's collections. The definition of every species is preserved in its type specimens; any loss or misplacement of them is detrimental to studies of taxonomy and systematics. Historically important specimens, documenting early research in a given discipline, are equally valuable to workers of later generations. In time, the attrition of specimens is seen in breakage, decay, destruction, and loss during moves; scientific usefulness is often lost, too, even when the specimens remain. Sometimes information about the location of specimens is lost when they are given or sold to other institutions or individuals. It is disconcerting enough when such specimens are in fact lost, destroyed, or become useless; but it is troubling when knowledge of them is known by only a few workers and not broadcast to the scientific community at large. Such academic forgetfulness impedes biological research when investigators are forced to work around the absence (real or presumed) of type and historically important material.

The scientific importance of old collections is overlooked more often today due to a drifting away from education in pure systematics and taxonomy. Collections are less frequently used despite the fact that they contain the very definitions of the species under study. So it is a natural academic phenomenon that when an effort is made to update the census of a museum's holdings, important rediscoveries are made that are significant to taxonomic research. At the same time, new appreciation is gained for aspects of the history of science.

In 1988 to 1990, the General Invertebrates Collection of the Academy of Natural Sciences of Philadelphia (ANSP) was reexamined by collections management staff and an annotated catalogue of type specimens was prepared (Spamer & Bogan 1992). This type-rich collection (which excludes mollusks and most terrestrial arthropods like insects) is composed primarily of polychaete worms and crustaceans. The opportunity is taken here to reintroduce the Academy's crustacean holdings to the scientific community, and to outline the history of crustacean research here, the birthplace of American carcinology. We disseminate this information anew so as to make systemetists and taxonomists aware of the rich holdings of this collection.

Throughout this paper we cite type specimens of species published by authors of the 18th, 19th, and 20th century. In the text and figure legends, binomial combinations for types are as originally used by the authors of those species. Thus we retain these names in their historical context and avoid the subjectivity of revisions of artificial systematics.

We also take the rare opportunity to provide illustrations of a few of the type and other historical specimens that have never been figured, or at least never photographed. Some of these specimens have been believed to be 'lost,' or at least have been long-forgotten, and we here present but a sample of the available material. Plates of type specimens are marked with a diamond ♦ to bring quick attention to them. All specimens are in the dry collection unless indicated otherwise. Our illustrations are specially selected and presented to document representative historical components of the carcinological collection of the Academy. We point out that this paper is not one either of taxonomy or systematic revision.

2   THE BIRTHPLACE OF AMERICAN CARCINOLOGY

In the late 1700's and into the 1800's, Philadelphia was the cultural capital of the Americas. American and immigrant naturalists at one time or another during their careers gravitated to this city. Scientific and other scholarly institutions were founded and thrived here, perhaps the most famous and influential of them being the still-active American Philosophical Society. In 1812, the Academy of Natural Sciences was founded by a small group of educated men whose common avocation was natural history. By 1817, the fraternity was formalized, their own quarters obtained, and an ambitious expedition undertaken in 1817-1818 to the remote Sea Islands of the state of Georgia and to the east coast of Florida (then still under Spanish rule). The early adventurer-naturalists of the Academy – William Maclure, Titian Ramsay Peale, George Ord, and Thomas Say – were arrested and detained in Florida by suspicious military bureaucrats, but they returned from this inaugural Academy expedition laden with specimens (Plate 1A). Very soon the Academy became the premier natural history institution in America.

Thomas Say (1787-1834) was a founding member of the Academy. Briefly employed in an apothecary that quickly bankrupted, Say devoted the remainder of his life to his avocation. He lived frugally and for a time slept in the Academy's collections. He is renowned as the father of American malacology and entomology, but he is less widely appreciated as the father, also, of American carcinology. His brief foray into the carcinology of America immediately gave him an advantage over European systematists. He had access to a wide variety of previously undescribed species in their natural habitat, and he visited and collected widely scattered locales and in addition studied specimens sent to him by travellers. In 1817 and 1818, Say published his first paper on carcinology in the new *Journal* of the Academy (Say 1817a) followed by the first treatise of North American Crustacea (Say 1817b, 1818a); he supplemented that 12-part work with three additional papers later in the volume (Say 1818b, c, d) and another in the second volume (Say 1822). Perhaps most remarkable about this systematic treatment was that Say did not restrict his studies to any particular group of crustaceans. Although his proficiency seemed to be in the Isopoda, Decapoda, and Stomatopoda (as determined from species recognized today from his descriptions; Holthuis 1969), he recognized 7 new genera and 85 new species extending through nearly every group within the Crustacea. (In addition, he also named two new species of limulid xiphosurans, in Say's day regarded as crustaceans.) Say often noted in his papers that the type and described material had been deposited in the Academy's collections. Even though many of the types were in turn sent by him to the Natural History Museum in London, some remain in the Academy (White 1847, Spamer & Bogan 1992; e.g. Plates 2D, E; 10A, C, D).

Say travelled extensively to the American South and West, Mexico, and Canada. He collected voraciously as specimens donated by him are found throughout the Academy's many collections. His acquisitions were concentrated wherever he lived, too, in Philadelphia and in

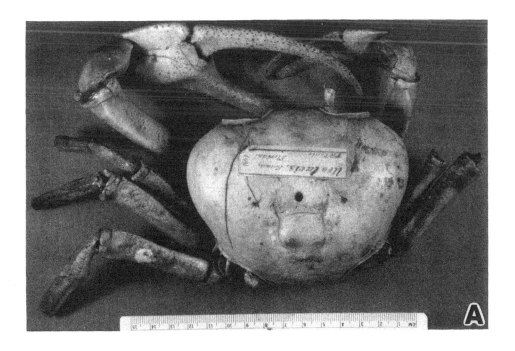

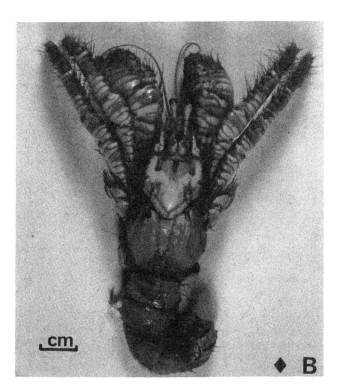

Plate 1. A. *Cardisoma guanhumi* (Marcgraf), (ANSP CA3198, originally identified as *Uca laevis* Milne Edwards, collected by Titian R. Peale on the east coast of Florida, USA, during the first expedition of the Academy of Natural Sciences of Philadelphia, in 1817-1818; loose appendages not photographed). B. *Aniculus longitarsus* Streets, 1871, syntype (ANSP CA807, alcohol-preserved, from Panama).

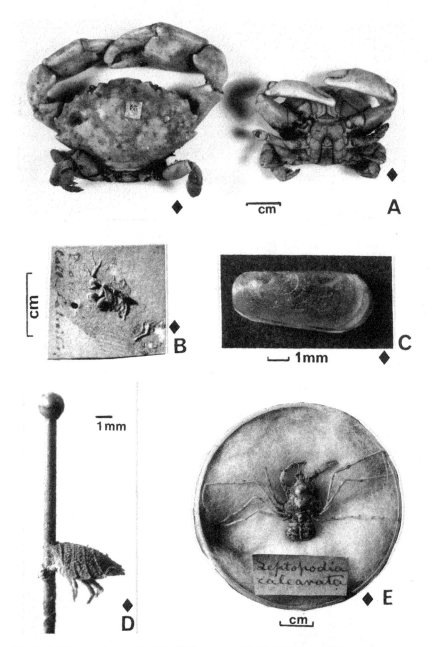

Plate 2. A. *Etisus laevimanus* Randall, 1840, syntypes (♀,♂) (ANSP CA3070, collected by Thomas Nuttall in the Hawaiian Islands). B. *Gammarus pedatus* Abildgaard, 1789, holotype (ANSP CA4199, Guérin-Méneville Collection no. 464, from Denmark; the specimen is glued to a piece of paper (outline of figure), with only the anterior end of the dried body present; the writing reads, 'Proto' and 'Coll. Latreille' (received by Guérin-Méneville from Latreille)). C. *Eulimnadia compleximanus* Packard, 1877, holotype (ANSP CA4258, collected by A. S. Packard 24-29 June 1874 in Ellis, Kansas, USA, during the US Geological and Geographical Survey of the Territories (Hayden Survey)); D. *Porcellio nigra* Say, 1818a, syntype (ANSP CA1607, from Pennsylvania, USA; posterior portion only, still attached to a mounting pin). E *Leptopodia calcarata* Say, 1818c, holotype (♂) (ANSP CA3320, from Charleston Bay, South Carolina, USA).

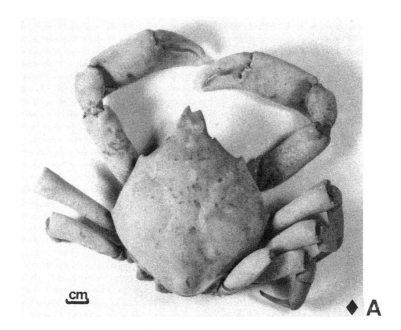

Plate 3. A. *Epialtus nuttallii* Randall, 1840, syntype (ANSP CA4162, alcohol- preserved, collected by Thomas Nuttall in California, USA; loose appendages not photographed); B. *Sesarma recta* Randall, 1840, holotype (♂) (ANSP CA3976, collected by Constantine Hering in Surinam).

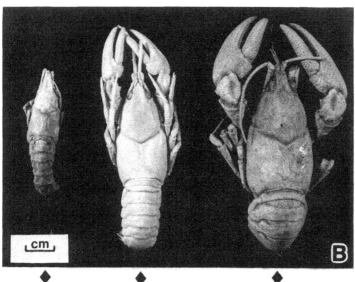

Plate 4. A left. *Astacus fossarum* Le Conte, 1855b, syntype (♀) (ANSP CA314, from Georgia, USA; loose appendages not photographed; A right. *Astacus latimanus* Le Conte, 1855b, syntype (♀) (ANSP CA329, from Georgia, USA; the right first appendage is broken loose from the body). B left. *Cambarus acutissimus* Girard, 1852, probable paratype (♂) (ANSP CA309, from a tributary to the Tombigbee River (fide Hobbs 1989:63), Kemper County, Mississippi, USA (type material listed as not found by Hobbs 1989); loose appendages not photographed. B center. *Cambarus robusta* Girard, 1852, syntype (♂) (ANSP CA328, from the Humber River near Toronto, Ontario, Canada). B right. *Cambarus rusticus* Girard, 1852, probable syntype (♀) (ANSP CA195, from Pittsburgh, Pennsylvania, USA).

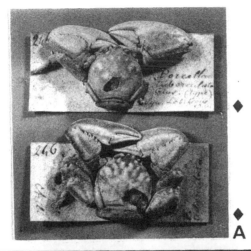

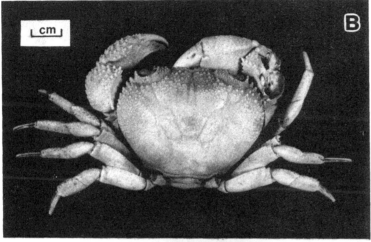

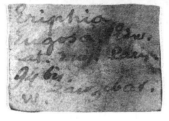

Plate 5. A. *Porcellana tuberculata* Guérin-Méneville, 1836a, syntypes (upper specimen marked as type on original label, lower specimen probably a type) (ANSP CA4116, Guérin-Méneville Collection no. 246, from Valparaiso, Chile; both specimens mounted on cardboard; upper figure reads, 'Porcellana tuberculata Guer. (type) Syn. Lobifrons Edw.'; lower figure reads, 'P. tuberculata' and 'Chili'; B. Specimen ANSP CA3380 (♀) bearing a manuscript(?) name but attributed to Milne Edwards ('*Eriphia rugosa* Edw.') although Guérin-Méneville most often wrote 'MSS.' on his labels to indicate manuscript names; from the Guérin-Méneville Collection (no. 94 bis). C. Label accompanying the specimen in Fig. 5B as an example of the labeling that accompanied the Gurin-Méneville Collection to the Academy of Natural Sciences in 1850-1851; text of label reads, 'Eriphia rugosa. Edw. Cat. Mus. Paris. 94 bis. Zanzibar. W.'; the 'W.' may indicate that this was a lot designated for shipment to T.B. Wilson.

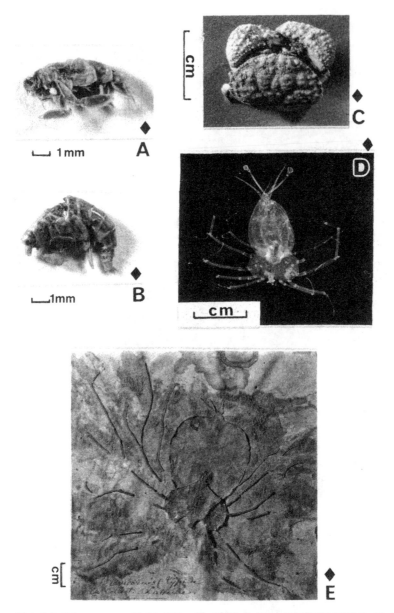

Plate 6. A. *Primno macropa* Guérin-Méneville, 1836b, holotype (ANSP CA 2685, Guérin-Méneville Collection no. 435, from Chile; reported in the literature to be lost (Bowman 1978)); B. *Melita palmata* Leach, 1814, holotype or syntype (ANSP CA 2703, Guérin-Méneville Collection no. 427; originally from the collection of Latreille; Guérin-Méneville's original label reads, 'Melita palmata Leach ... type. Col. Latr.'; Leach's published locality data indicate occurrence as near Plymouth, Devonshire, England). C. *Pilumnoides perlatus* Milne Edwards & Lucas 1844, holotype (ANSP CA 3422, Guérin-Méneville Collection no. 89; data with specimen list locality as Chile, but Guérin-Méneville's published locality is near Lima, Peru). D. *Phyllosoma affinis* Guérin-Méneville, 1833, syntype (ANSP CA 1344, Guérin-Méneville Collection no. 355, alcohol-preserved, specimen photographed in alcohol; type locality only known as 'Atlantic Ocean'). E. *Phyllosoma brevicornes* Leach, 1818, holotype (ANSP CA 2925, Guérin-Méneville Collection no. 361, no locality data; specimen is dried to a piece of paper (outline of figure); writing on paper reads, 'Brevicornes type de la Collect. Latreille').

Plate 7. A. *Gelasimus tangieri* Eydoux, 1835, syntype (ANSP CA3028, Guérin- Méneville Collection no. 151, from Tangier, Morocco; loose appendages not photographed). B. *Pelaeus armatus* Eydoux & Souleyet, 1842, syntype (ANSP CA3709, Guérin-Méneville Collection no. 47, from the Hawaiian Islands; loose appendages not photographed).

Plate 8. A. *Panulirus ricordii* Guérin-Méneville, 1844, holotype (ANSP CA207, Guérin-Méneville Collection no. 276, from 'the Antilles;' additional loose appendages not photographed; Guérin-Méneville's identification is written on specimen as 'P. Ricordi (sic) Antilles type.' B. Specimen of *Astacus wiegmanni* Erichson 1846 (ANSP CA317, from Mexico; this is the female specimen of this enigmatic species cited by Hagan 1871; small fragments not photographed).

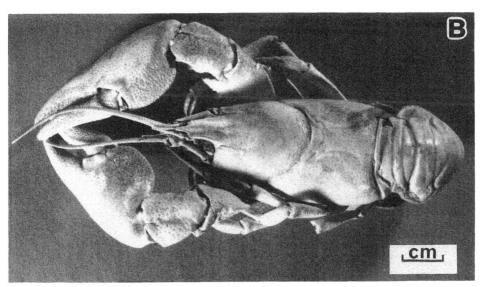

Plate 9. A. Specimen of *Petrolithes cinctipus* Randall, 1840, (ANSP CA4166, from California, USA, received from W. Stimpson and identified by him as *Petrolithes rupicolus* Stimpson). B. Specimen of *Astacus trowbridgii* Stimpson, 1857, (ANSP CA303, from the Columbia River, USA, received from Stimpson). Both of these specimens could constitute type material (see text).

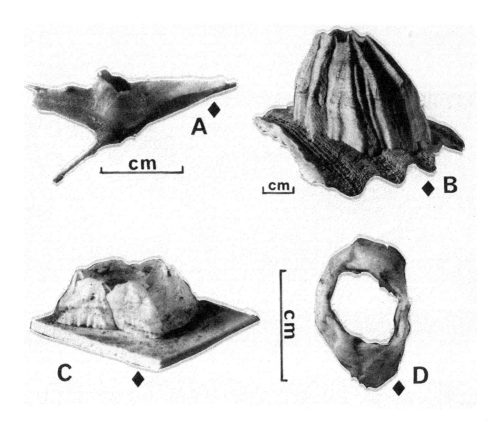

Plate 10. A. *Conopea elongata* Say, 1822, syntype (ANSP CA6889, from Charleston Bay, South Carolina, USA). B. *Balanus proteus* Conrad, 1834, probable holotype (ANSP 18813 (Invertebrate Paleontology), from the Miocene of Suffolk and Yorktown, Virginia, USA; C & D. *Coronula dentulata* Say, 1822, syntypes (ANSP CA6889, from Florida, U.S.A.); E. *Palaemon africanus* Kingsley, 1882, syntype (ANSP CA81, alcohol-preserved, collected by Paul du Chaillu on the west coast of Africa).

New Harmony, Indiana. Many of his specimens were destroyed in a fire in New Harmony, but those that survived at the time of his death were some years later transferred by his widow, Lucy Way Sistaire Say (1800-1886), to the Academy in Philadelphia. The Academy's collection also contains many non-type specimens identified and deposited by Say. So important are Say's early papers to the development of systematic carcinology that they were reprinted in facsimile in 1969, with an introduction by L.B. Holthuis. This was intended 'to bring Say's insufficiently known species again to the attention of carcinologists' (Holthuis 1969:xv). In that introduction, Holthuis summarized the dates of publication of Say's carcinological papers and provided a systematic index, with updated taxonomy, of all of the crustacean species discussed by Say. The present paper reemphasizes and expands on Holthuis' intent by presenting a historical synopsis of carcinological studies and collections of the Academy. Details about all of Say's crustacean types are provided by Spamer & Bogan (1992).

Say of course was the foremost American carcinologist of his time, but carcinology even in the Academy's early years was not Say's monopoly. Timothy Abbott Conrad (1803-1877), another of the Academy's early members (who like Say is most well known for his work in malacology), had some interest in crustaceans, and he described some new species of American cirripedes, including a fossil form (Plate 10B). Indeed, the holdings of fossil crustaceans in the Academy are not to be overlooked; although the numbers are modest, there are many type specimens. These fossils are largely from Neogene deposits of eastern and southern USA, some collected by early members of the Academy; they are held in the Invertebrate Paleontology Collection.

With the departure in 1826 of Say from Philadelphia to the communal social experiment at New Harmony, Indiana, no one in the Academy followed up on his work. However, Say's remarkable first systematic synthesis of North American Crustacea became the groundwork for studies by other workers elsewhere and designated the Academy an important repository of specimens.

3   THE ACADEMY AFTER SAY

As the Academy's second decade ended, it moved to its own building but reeled from financial difficulties that threatened its existence as an institution. Scientifically, its image was untarnished. The Academy's debt was underwritten by William Maclure (1763-1840), a president of the Academy who was a co-founder of the commune of New Harmony. Even after moving away from Philadelphia, he continued his affiliation in waves of enthusiasm, remitting specimens, books, and financial support. This continuing act of generosity bolstered the institution and saved it – and its collections – from what otherwise would have meant dispersal. It allowed the Academy to more surely establish itself as a repository not only for the scientific collections of its members but also from world-venturing expeditions conducted by other agencies.

In 1834-1835, Academy members Thomas Nuttall (1786-1859) and John Kirk Townsend (1809-1851) visited the then-remote west coast of North America and the even more remote Sandwich Islands (Hawaiian Islands). They delivered to the Academy a large number of specimens, including many crustaceans that were described (along with some specimens already in the Academy's collection) in 1840 by John W. Randall (1813-1892) of Boston, who had been elected to Academy membership in 1837. Randall's paper, his only one in the Academy's publications, described 36 new species of decapods (e.g., Plates 2A; 3A, B). This number is not surprising since previous collections from those distant localities were sporadic and delivered to systemetists in Europe whose preoccupations were with European-conducted expeditions to other far-flung places.

Toward the end of the 1830's, the United States of America underwrote a demonstration of its political prowess in the guise of a scientific circumnavigation of the world. This was the United States Exploring Expedition, also known as the Wilkes Expedition because it was under the command of Lieutenant John C. Wilkes. An ambitious undertaking fraught with difficulty ostensibly due to poor command by Wilkes, it extended from 1838-1842. On board were naturalists and collectors, including Academy member Titian Peale, who had been on the Academy's first expedition in 1817. The scientific results of the voyage were issued in ponderous volumes over a period of time, with most of the crustacean research conducted by the renowned naturalist and geologist, James Dwight Dana (1813-1895). Most of the collections were sent to the new Smithsonian Institution, but various specimens were also deposited with other institutions including the Academy of Natural Sciences. Some crustacean specimens survive in the Academy's collection including a paratype of *Alpheus malleator* Dana, 1852.

After the Academy was saved financially by Maclure, one T.B. Wilson was elected to membership in 1832. His relationship with the Academy was destined to make him probably the one most important scientific benefactor the Academy has ever known.

4 THE LEGACY OF THOMAS BELLERBY WILSON

A physician practicing in Wilmington, Delaware, T.B. Wilson (1807-1865) was an accomplished amateur naturalist who never published on his avocation. He used his personal fortune to further his interests, and he deposited his acquisitions with the Academy in Philadelphia. Every collection in the Academy contains hundreds if not thousands of specimens that were donated by Wilson; the crustacean collection contains hundreds of lots from him, and the fossil crustacean collection also contains many of his specimens. The Academy's library holds hundreds of books that Wilson acquired, many of them critical early reference works in all areas of natural history. Wilson also impressed his brother, Edward (born 1808, alive 1863), of Tenby in Wales, Great Britain, with the importance of the Academy's collections. Edward acted as agent for his brother and also was himself a generous donor of specimens and support (Torrens & Taylor 1990, and here). The most remarkable aspect of T.B. Wilson's donations was that they were not made by bequest, but were given to the Academy during his life-long affiliation with the institution. He insisted on anonymity with his gifts, but that request was certainly not heeded. Throughout the collections, specimen labels attribute 'T.B. Wilson' or 'Dr. Wilson' as donor; throughout the library are bookplates that acknowledge his gift. In 1864, he was, reluctantly, elected president of the Academy for a year, but due to poor health he never presided over a meeting.

The current holdings of the Academy contain one of Wilson's most scientifically important acquisitions for the Academy: crustacean specimens from the collection of Félix-Douard Guérin-Méneville (1799-1874), the French naturalist who was editor of the *Magasin de Zoologie* and who published systematic descriptions in most every zoological group. Guérin-Méneville's collection also contained some specimens that he had obtained from other European naturalists, including Pièrre André Latreille (1762-1833), the Baron Cuvier (1768-1832), and William Elford Leach (1790-1836); some of the specimens he received were types of their species (e.g., Plates 2B; 6B, E). This wonderful collection also contains the oldest described type specimen in all of the Academy's collections, an amphipod named as *Gammarus pedatus* Abildgaard, 1789 (Plate 2B).

The Academy's *Proceedings* document several transactions of the Guérin-Méneville collection deposited by Wilson in 1850 and 1851, summing an undetermined number of genera, species, and specimens (ANSP 1850:159, 357; 1851: xxxvi). We accounted for 549 lots, 40 of them alcohol-preserved, from the Guérin-Méneville collection, including the types of 47 species named by him (e.g. Plates 5A; 6A, D; 8A) and several that are the types of species named by other authors: the certain or probable types of species named by Fortuné Eydoux (1803-1841) (Plate 7A), Eydoux and François Souleyet (1811-1852) (Plate 7B), William Elford Leach (Plates 6B, E), H. Lucas (1814-1899), H. Milne Edwards (1800-1885), H. Milne Edwards and Jules Haime (1824-1856), and H. Milne Edwards and Lucas (Plate 6C). In addition, 60 more lots are the 'types' of manuscript names never published by Guérin-Méneville (e.g. Plate 5B). The Guérin-Méneville Collection is significant not only for the number of types it contains, but in that the large number of manuscript names gives insight into the breadth of Guérin-Méneville's ongoing studies of the Crustacea. Many of the smaller crustaceans, once fluid-preserved but long since dried, remain in Guérin-Méneville's original containers, interesting examples of European museum glassware of the 19th century.

Admittedly, many of the available taxa for which the Academy holds types are synonymous with other taxa, and the collection preserves some specimens that can be involved in problems of an antiquated systematics that did not or could not distinguish between larval, juvenile, and adult stages of various organisms. Their recovery and reintroduction to the scientific community may allow interested researchers a better understanding of 19th century systematics. For example, the recovery of the type specimen of *Primno macropa* Guérin-Méneville, 1836 (Plate 6A), will be of value to amphipod systematists if only to verify taxonomic discussions that were made in the absence of the type specimen and the absence of known specimens from the type locality (the Pacific Ocean off Chile; see Bowman 1978).

5 THE DECLINE OF PHILADELPHIA CARCINOLOGY

During the time that the Academy was the recipient of T.B. Wilson's generosity, some significant research on crustaceans was published in the *Proceedings*. Charles Frédéric Girard (1822-1895) published 'A Revision of the North American Astaci' in 1852, from which the types of five species remain in the collection (e.g. Plate 4B). In 1855, John Le Conte (1818-1891) presented a paper on new species of *Astacus* from the American state of Georgia (e.g. Plate 4A) and in another paper the same year erected a new species of *Gelasimus*. However, it is evident in browsing the collection that no large acquisitions of crustaceans were received by the Academy in the decades after the Wilson donation of the Guérin-Méneville Collection. Thus the stage was set that would affect crustacean studies in Philadelphia, the birthplace of American carcinology.

Little work on crustaceans was continued by Academy researchers despite the presence of some remarkable collections; academic affiliates of the Academy also largely disregarded the crustaceans. It was a time of rapid change and stress in the Academy. Its renown, then, was more in vertebrate paleontology; overtures were made to advance the Academy through merger with the University of Pennsylvania; discontent was voiced by some members in areas of management; and the Academy suffered the pangs of physical growth. After enlarging its quarters in a building (no longer standing) at Broad and Sansom Streets in Philadelphia, the Academy finally constructed a new building on its present site on Logan Square in 1876. No doubt during all of these transitions many parts of its collections fell victim to abuse and loss. The carcinological collection, however, was one small part of the whole and academically was overshadowed by the bones of dinosaurs and other creatures that could inspire the public.

Around 1865, an organizational restructuring of the Academy formalized various departments of study, each attending to its own collection, and for the first time established a small paid curatorial staff. As in the early days of the Academy, many of the Academy's workers involved themselves in collecting specimens that were unrelated to their own areas of greatest proficiency. Consequently, the Academy's collections received many specimens unrelated to research at hand.

By the process of sporadic acquisition and miscellaneous collecting during formal expeditions, the Academy managed to accession specimens of crustaceans that proved to be new and interesting species to later workers (e.g. Plates 1B; 10E). In this manner the Academy gained specimens that may be of greater importance now than they were when received. While it is historical fact that the collections of William Stimpson (1832-1872) were destroyed in the great fire in Chicago, Illinois, in 1871 (see Manning, this volume), a few specimens were received by the Academy from Stimpson that may have formed part of his type series when describing some species (e.g. Plates 9A, B). It is conjecture whether these specimens constitute type material, and the matter is left to other professionals to determine; but these few lots remain in the Academy's collection for researchers who need to refer to them.

Various published papers, some in the Academy's *Proceedings*, just before and after the turn of the 20th century periodically bolstered the Academy's position in carcinological research, but never to the status it enjoyed during the period from 1817 to about 1850. The late 1800's and early 1900's were a time of growing speciality in scientific studies, and few younger workers could claim to have proficient knowledge of the whole of their subject. Thomas Hale Streets (born 1847), John Sterling Kingsley (1854-1929), Samuel N. Rhoads (1862-1952), and Henry W. Fowler (1878-1965), much better known in other areas of taxonomic study, made some minor contributions toward carcinology that staved off complete stagnation of the Academy's collections (e.g. Plates 1B, 10E). Still other researchers of the time published on a few specimens that are held by the Academy.

Fowler, known mostly for his work on fishes, in 1912 produced *Crustacea of New Jersey*; it remains the definitive treatment of that subject, based largely on specimens held by the Academy. This geographically restricted study actually was the first research project of its kind issued by an Academy researcher, a precursor of the kinds of census studies that would be the Academy's mainstay later in the century. However, it also had the distinctive character of 19th-century pansystemic treatises, i.e., describing and revising the whole of a taxonomic class as occurring in one region.

Throughout this period, infusions of material from expeditions, even if left unstudied after initial identification, kept the carcinological collection in a state of growth, although few publications were derived from these resources (e.g. Plate 2C, the type of a branchiopod collected in a reconnaissance peripheral to the great Hayden Survey of the geography and geology of the American West in the 1870's). Some crustacean specimens were deposited from the many cruises of the US Fish Commission Steamer Albatross during the late 1800's and early 1900's.

Most of the other significant Academy collections of that time were deposited by Academy members who were very active in collecting specimens outside of their immediate areas of research. Notable among them were Joseph Leidy (1823-1891), inveterate collector and scholar who was probably the best-known natural scientist of his day (and president of the Academy 1883-1891); Angelo Heilprin (1853-1907), the explorer-naturalist who is best known for his Arctic and Caribbean exploits; Charles Willison Johnson (1863-1932), who is remembered for his entomological work in the Boston Society of Natural History; and Joseph Willcox (1829-1918), who maintained a low academic profile yet whose interest in and knowledge of natural history was keen. The enthusiasm of these people and the broad view of natural history they held attracted young, aspiring researchers to the scholarly environments centered around them. It is this enthusiasm that indirectly seems to have positioned the Academy for a resurgence in carcinological research in the 20th century.

6 ACADEMIC TRANSITION IN THE ACADEMY AND THE FIRST REVIVAL OF PHILADELPHIA CARCINOLOGY

For Leidy, Heilprin, Johnson, and Willcox, the close of the 1880's was a vigorous time. Leidy's world reputation was one both of academic and public renown, and the careers of the other men were ascending rapidly. At the Academy, the business of operating a growing research museum was forcing out some of the familiar academic comforts of pure science. Administrative decisions, sometimes unpopular, certainly led to disillusionment among the older workers who had created and nurtured the traditional, less businesslike, academic environment of the late 1800's. Growth of the Academy again created a need for more space, and the collections required a complete physical and systematic reorganization (which began in the mid-1890's after a new wing was added to the building).

Leidy, Heilprin, Johnson, and Willcox also were affiliated with the Wagner Free Institute of Science, in Philadelphia. The Institute is a teaching museum no longer on the cutting edge of natural history research; it is today a National Historic Landmark, in America the sole unchanged survivor of Victorian-era natural history museums. Willcox was a long-time trustee of the Institute (1878-1918), and after the death of the founder, William Wagner (1796-1885), Heilprin was made a curator of the museum until his departure from Philadelphia in 1899; Johnson was a curator from 1888 to 1903. Leidy was the director of the Institute's museum from 1885 until his own death in 1891.

It seems that the Institute served as a refuge from the politics and bureaucracies of the larger institutions with which these prominent men were affiliated. In 1889, the trustees of the Institute sponsored an expedition to the largely unexplored west coast of Florida, an ambitious undertaking that harkened back to earlier times of exploration and research. From that expedition large numbers of Recent and fossil specimens of all kinds were sent back to the collections of both the Institute and the Academy. The time and effort applied to the Wagner Free Institute by Leidy, Heilprin, Johnson, and Willcox were taken from what they might have spent at the Academy. They left more of the daily routines of curating and research to the full-time staff and coadjutant researchers from other institutions such as the University of Pennsylvania.

Still, the Academy at that time was not completely out of the business of exploration. In 1890, an expedition to the Yucatan area of Mexico yielded much material for the Academy's collections, but it was not solely an Academy-funded project. This was a harbinger of the 20th-century style of scientific expeditions that the Academy periodically conducted, funded partly if not wholly by patrons of scientific exploration.

Into the very midst of this dynamic, changing period at the Academy came Henry Augustus Pilsbry (1862-1957). He entered the Academy's employ as a young man in 1885, and remained there for 70 years to become a paramount malacologist and cirripedologist. He was responsive to the growing needs of the Academy and began to expand and organize the collections in his charge. He also pursued his research interests with enthusiasm and precision. Even though he was first a malacologist, his interests in natural history also inspired him to study the cirripedes often found with mollusks. He published occasionally on this subject, and by the second decade of the 20th century his reputation as a natural scientist was world class.

Up to the time of Pilsbry, the authoritative work on cirripedes was that of Charles Darwin. Pilsbry worked on the collections of the National Museum of Natural History and the Academy and produced his first monograph on cirripedes (Pilsbry 1907) after publishing a few separate papers on various Pacific cirripedes. He became the preeminent systematist of cirripedes of the world, revising the systematics of the group and authoring dozens of new species (Clench & Turner 1962). Despite his amazing proficiency and productivity in malacology, Pilsbry continued to publish on the cirripedes throughout his life; his definitive monograph on them (Pilsbry 1916) remains an important reference on the subject, particularly of the American species.

Although Pilsbry's work on cirripedes was largely based on the extensive collections in the National Museum, he assembled a collection for his own use in the Academy. For awhile, the disposition of this material was uncertain, but in the mid-1960's Zullo (1965, 1968) researched and resolved the matter. Most of the specimens were found and returned to the National Museum. The Academy retains part of Pilsbry's working collection, including many types, which remains much as he left it.

7   THE ACADEMY AND THE SECOND REVIVAL OF PHILADELPHIA CARCINOLOGY

While H.A. Pilsbry revitalized the Academy's position as a center for the study of cirripedes, the Academy's collections of other crustaceans lay largely unused. Specimens were sporadically accessioned into the General Invertebrates Collection, but research was stagnant. It was during this period of the first half of the 20th century that the Academy's holdings began to be neglected as a resource by carcinologists. Researchers who depended upon the early literature knew of the early accessions of the Academy, such as the Say material and the specimens described by Randall, Girard, and Le Conte among others; but the generous donations of collections, such as the Guérin-Méneville Collection in 1850-1851, which had gone mostly unstudied, faded into obscurity in carcinological systematics. Although a few workers travelled to Philadelphia to use the Academy's collection and were aware of the presence of some of these important specimens, the older collections fell into disuse and general disrepair.

The absence of carcinological curators at the Academy in the early 20th century, the financial ravages of the Great Depression in the 1930's, the suspension of travel and research during World War II, and the attendant fall in numbers of students in systematic zoology served to diminish the practical usefulness of the Academy's collection of crustaceans. However, in the 1950's a new awareness of the importance of America's freshwater resources emerged. The Academy, beginning in the 1930's, had established itself as a resource in freshwater ecology, a position which steadily grew to the point that the former Department of Limnology is today a separate division of the Academy (the Division of Environmental Research). At this time the Academy entered the area of contractual research, surveying the freshwater resources of North America for government agencies and private industry. The outfall of these surveys required the support of carcinologists and the need to preserve voucher specimens. Specimens collected during field surveys in the 1950's through 1970's are now part of the General Invertebrates Collection and well represent the crustacean fauna of the streams of the eastern and southern USA. During this time some of the principal workers, either at the Academy or through research exchanges with other institutions, included C.W. Hart Jr., Horton H. Hobbs Jr., Horton H. Hobbs III, Samuel L.H. Fuller, and Robert H. Gore. The collection includes type material studied by these and other researchers. In the early 1980's, Gore reorganized the entire General Invertebrates Collection, physically and systematically, concentrating on the decapod crustaceans.

In the area of fossil crustaceans, the Academy has maintained a footing in the 20th-century literature through the work of (again) Henry A. Pilsbry, Axel A. Olsson (1889-1977), Henry B. Roberts (1910-1979, affiliated for a time with the Wagner Free Institute of Science), Mary Jane Rathbun (1860-1943), Wilhelm Bock (1897?-1972, also known for his work in paleobotany and vertebrate paleontology), and long-time curator of paleontology Horace Gardiner Richards (1906-1984). Unfortunately, some of the work by Bock is plagued with technical and clerical errors (Spamer 1988) and his systematic syntheses are ongoing subjects of discussion in the literature. However, Bock's only contributions to fossil carcinology pertain to the branchiopod genus *Cyzicus* Audouin, 1837; but indicative of Bock's problems, his new genus and species of phyllocarid crustacean, *Gwyneddocaris parabolica* Bock, 1946, known only from the type specimen (ANSP 16850, Invertebrate Paleontology), has been identified by D. Baird and H.F. Roellig as the midline of a ganoid fish (Rolfe 1969:R330).

8 THE HOLDINGS OF CRUSTACEA IN THE ACADEMY

There are many type specimens of crustaceans held in the Academy's General Invertebrates Collection. The taxonomic groups and numbers of available species level taxa for which types are held are as follows: Branchiopoda (9 taxa, 7 dry lots, 2 alcohol-preserved lots); Ostracoda (13 taxa, 23 dry lots); Copepoda (4 taxa, 2 dry lots, 4 alcohol-preserved lots); Cirripedia (71 taxa, 65 dry lots, 26 alcohol-preserved lots); Stomatopoda (11 taxa, 12 dry lots, 1 alcohol-preserved lot); Cumacea (1 taxon, 1 dry lot); Isopoda (35 taxa, 27 dry lots, 11 alcohol-preserved lots); Amphipoda (27 taxa, 26 dry lots); and Decapoda (154 taxa, 111 dry lots, 44 alcohol-preserved lots). In addition, the collection holds the 'types' of 71 manuscript names (mostly those of Guérin-Méneville) and two of unavailable quadrinominal taxa (forms of Pilsbry cirripedes). An annotated catalogue of these types and manuscript names is available, along with a general group census of the entire collection (Spamer & Bogan 1992), which identifies many forgotten types including some that are reported in the literature as missing, e.g. *Primno macropa* Guérin-Méneville, 1836b, (holotype ANSP CA2685, Fig. 6A) reported missing by Bowman (1978); and the unique specimen of *Astacus angustatus* Le Conte, 1856, (ANSP CA444) described as *Procambarus (Ortmannicus) angustatus* by Hobbs (1981:384-388, Fig. 148) and noted by him (three times) as lost.

The physical state of the crustacean collection varies, ranging from perfect to scientifically useless. The effects of Philadelphia's climate over 140 years, transatlantic travel in sailing vessels, cellar storage (involving a major flood in 1899), two major moves within the city and unknown numbers of moves within buildings, infestation by dermestid larvae in years long passed, cramped storage conditions, and periods of little curatorial care – all have contributed to breakage, loss of components, and outright destruction of specimens. Generally speaking, the alcohol-preserved collection, including all of the environmental survey material of the mid-1900's, is in good condition with good data; the dry collection is in less than satisfactory condition, containing some incomplete and broken specimens. The collecting data for this latter material are incomplete, as is typical of very old collections. Many of the smaller crustacean specimens of the Guérin-Méneville Collection were once fluid-preserved, but the specimens dried out long ago. Sometimes, antiquated preparation techniques – drying on paper (Plate 6E), gluing of dried bodies to paper (Plate 2B), pinning specimens (Plate 2D) or abrasion and desiccation (Plate 6A) – have contributed to the loss of scientific usefulness of specimens. In some cases, all that is known of a species is from fragmentary remains of non-type material and scant collecting data, such as with Hagan's (1870) cited specimen of *Astacus wiegmanni* Erichson, 1846 (Plate 8B; and refer to Hobbs 1989:5). Other times, good fortune preserved specimens in remarkable condition (Plate 3B), even to the point of long-term survival of delicate alcohol-preserved specimens without drying (Plate 6D).

The fossil crustacean collection is modest in size, composed primarily of Neogene cirripedes. Type holdings total 10 species of Branchiopoda (all of Triassic age), 24 Cirripedia, and 23 Decapoda; the only available published guide to these types is the catalogue by Richards (1968), but more specific and updated information may be obtained by inquiry.

9   THE FUTURE OF CARCINOLOGY AT THE ACADEMY

At the time of this writing, the Academy's carcinological collection, as part of the General Invertebrates Collection, is supervised by the Department of Malacology; the fossils are maintained by the Invertebrate Paleontology Section of the same department. There is a curator-level paleontologist in the paleontology section, but there is no full-time staffing for the Recent collection. Thus carcinological studies at the Academy, except for survey work performed by the Division of Environmental Research, is once again in a period of remission. Nevertheless, the rich historical and type holdings of the collection are accessible and available to researchers. Hopefully, this paper will make systemetists aware of the scope and scientific value of these holdings, much of it long forgotten or assumed lost. We hope this serves as a stimulus to research and renewed use of the collection.

ACKNOWLEDGMENTS

We wish to thank Robert McCracken Peck (Academy of Natural Sciences) for his remarks and careful review of the historical elements of this paper. Gary Rosenberg, Assistant Curator of Invertebrate Paleontology, provided us with summaries of various data about the collection in his charge.

REFERENCES

Abildgaard, P.C. 1789. *Zoologiae Danica seu animalium Daniae et Norvegiae rariorum ac minus notorum descriptiones et historia. Volumen Tertium. Explicatiioni iconum fasciculi tertii eivsdem operis inserviens. Auctore Othone Friderico Müller.* N. Mölleri et Filii, Havniae.
ANSP (Academy of Natural Sciences of Philadelphia) 1850-1851. (Donations to the museum.) *Proc. Acad. Nat. Sci. Phila.* 5: 133, 159, 357; 6: xxxvi.

Audouin, J.V. 1837. (Communication de M. Audouin.) *Ann. Soc. Ent. France, Bull. Ent.* 6:ix-xi.

Bock, W. 1946. New crustaceans from the Lockatong of the Newark series. *Notulae Naturae Acad. Nat. Sci. Phila.* 183: 1-16.

Bowman, T.E. 1978. Revision of the pelagic amphipod genus *Primno* (Hyperiidea: Phrosinidae). *Smithson. Contr. Zool.* 275: 1-23.

Clench, W.J. & R.D. Turner 1962. New names introduced by H. A. Pilsbry in the Mollusca and Crustacea. *Acad. Nat. Sci. Phila. Spec. Pub.* 4: 1-218.

Conrad, T.A. 1834. Descriptions of new Tertiary fossils from the southern states. *J. Acad. Nat. Sci. Phila.* 7: 139-157.

Dana, J.D. 1852. *United States Exploring Expedition. Vol. XIII. Crustacea. Part I.* Philadelphia: Sherman.

Erichson, W.F. 1846. Uebersicht der Arten der Gattung Astacus. *Arch. Naturgesch.* 12: 86-103.

Eydoux, F. 1835. Gélasime, *Gelasimus* Latr. G. de Tanger: *G. Tangeri*. F. Eydoux. *Mag. Zool.* 5 (Classe 7): (29-32).

Eydoux, F. & F.L.A. Souleyet 1842. Zoologie. In A.N. Vaillant, *Voyage autour du monde, exécuté pendant les années 1836 et1837 sur la corvette 'La Bonite,'* Vol. 1:219-250.

Fowler, H.W. 1912. The Crustacea of New Jersey. *Ann. Rept. New Jers. State Mus.* 1911: 29-650, pls. 1-150.

Girard, C. 1852. A revision of the North American Astaci, with observations on their habits and geographical distribution. *Proc. Acad. Nat. Sci. Phila.* 6: 87-91.

Guérin-Méneville, F.E. 1833. Mémoir su l'organisation extérieure des Phyllosomes, et monographie de ce genre de Crustacés. *Mag. Zool.* 2(3) (Classe 7): (1-30).

Guérin-Méneville, F.E. 1836a. (Observations sur les *Porcellanes*.) *Bull. Soc. Sci. Nat. France* (1835): 115-116.

Guérin-Méneville, F.E. 1836b. (Without title.) *Mag. Zool.*, Classe 7 (Crustacés).

Guérin-Méneville, F.E. 1844. In *Iconographie du règne animal de G. Cuvier.... Crustacés. Tome III, Texte explicatif.* Paris: J.B. Baillire.

Hagan, H.A. 1870. Illustrated catalogue of the Museum of Comparative Zoology, at Harvard College. No. III. Monograph of the North American Astacidae. *Mem. Mus. Comp. Zool.* 3: 1-109, pls. 1-11.

Hobbs Jr. H.H. 1981. The crayfishes of Georgia. *Smithson. Contrib. Zool.* 318: 1-549.

Hobbs Jr. H.H. 1989. An illustrated checklist of the American crayfishes (Decapoda: Astacidae, Cambaridae, and Parastacidae). *Smithson. Contrib. Zool.* 443: 1-50.

Holthuis, L.B. 1969. Thomas Say as a carcinologist. In (facsimile reprinting of Say's) *An account of the Crustacea of the United States*, pp. v-xv. Historiae Naturalis Classica, Vol. 73. Lehre: J. Cramer.

Kingsley, J.S. 1882. Carcinological notes; number V. *Bull. Essex Inst.* 14:103-132.

Le Conte, J. 1855a. Descriptions of new species of *Astacus* from Georgia. *Proc. Acad. Nat. Sci. Phila.* 7:400-402.

Le Conte, J. 1855b. On a new species of *Gelasimus*. *Proc. Acad. Nat. Sci. Phila.* 7: 402-403.

Leach, W.E. 1814. Crustaceology. In D. Brewster, *The Edinburgh Encyclopaedia* 7(2).

Leach, W.E. 1818. A general notice of the animals taken by Mr. John Cranch, during the expedition to explore the source of the River Zaire. In J. Tuckey, *Narrative of an expedition to explore the River Zaire, usually called the Congo, in South Africa, in 1816*: pp. 407-419. London: John Murray.

Milne Edwards, H. & H. Lucas 1843 (1844). Crustacés. In A. d'Orbigny, *Voyage dans l' Amérique méridionale* 6(1).

Packard, A.S. 1877. Descriptions of new phyllopod Crustacea from the West. *Bull. US Geol. Geogr. Survey of the Territories* 3: 171-179.

Pilsbry, H.A. 1907. The barnacles (Cirripedia) contained in the collections of the U.S. National Museum. *Bull. US Nat. Mus.* 60: 1-122, pls. 1-11.

Pilsbry, H.A. 1916. The sessile barnacles (Cirripedia) contained in the collections of the US National Museum; including a monograph of the American species. *Bull. US Nat. Mus.* 93: 1-66, pls. 1-76.

Randall, J.W. 1840. Catalogue of the Crustacea brought by Thomas Nuttall and J.K. Townsend, from the West Coast of North America and the Sandwich Islands, with descriptions of such species as are apparently new, among which are included several species of different localities, previously existing in the collection of the Academy. *J. Acad. Nat. Sci. Phila.* 8: 106-147.

Richards, H.G. 1968. Catalogue of invertebrate fossil types at the Academy of Natural Sciences of Phildelphia. *Acad. Nat. Sci. Phila. Spec. Pub.* 8: 1-222.

Rolfe, W.D.I. 1969. Phyllocarida. In R.C. Moore, (ed.), *Treatise on Invertebrate Paleontology, Part R, Arthropoda 4(1)*: pp. R296-R331. Lawrence: Geol. Soc. Amer. & Univ. Kansas.

Say, T. 1817a. On a new genus of the Crustacea, and the species on which it is established. *J. Acad. Nat. Sci. Phila.* 1: 49-52.

Say, T. 1817b-1818a. An account of the Crustacea of the United States. *J. Acad. Nat. Sci. Phila.* 1: 57-80, 155-169, 235-253, 313-319, 374-401, 423-441.

Say, T. 1818b. Observations on some of the animals described in the account of the Crustacea of the United States. *J. Acad. Nat. Sci. Phila.* 1:442-444.

Say, T. 1818c. Appendix to the account of the Crustacea of the United States. *J. Acad. Nat. Sci. Phila.* 1: 445-458.

Say, T. 1818d. Description of three new species of the genus *Naesa. J. Acad. Nat. Sci. Phila.* 1: 482-485.

Say, T. 1822. An account of some of the marine shells of the United States. *J. Acad. Nat. Sci. Phila.* 302-325.

Spamer, E.E. 1988. Notes on six real and supposed type fossils from the Newark Supergroup (Triassic) of Pennsylvania. *Mosasaur* 4: 49-52.

Spamer, E.E., & A.E. Bogan 1992. General Invertebrates Collection of the Academy of Natural Sciences of Philadelphia. Part 1: Guide to the General Invertebrates Collection. Part 2: Annotated catalogue of Recent type specimens. *Tryonia* 26: 1-305. (Available from Department of Malacology, ANSP.)

Stimpson, W. 1857. Notices of new species of Crustacea of Western North America, being an abstract from a paper to be published in the journal of the society. *Proc. Boston Soc. Nat. Hist.* 6: 84-89.

Streets, T.H. 1871. Catalogue of Crustacea from the Isthmus of Panama, collected by J.A. McNeil. *Proc. Acad. Nat. Sci. Phila.* 23: 238-243.

Torrens, H.S. & M.A. Taylor 1990. Collections, collectors and museums of note. No. 55, Geological collections and museums in Cheltenham 1810-1888: A case history and its lessons. *Geol. Curator* 5(5): 175-213.

White, A. 1847. *List of the specimens of Crustacea in the collections of the British Museum.* London.

Zullo, V.A. 1965. The type collection of Henry Pilsbry. *Cirripedologists' Newsl.* 1(3): 1-2.

Zullo, V.A. 1968. Catalog of the Cirripedia named by Henry A. Pilsbry. *Proc. Acad. Nat. Sci. Phila.* 120: 209-235.

Brooks, stomatopods, and decapods: Crustaceans in research and teaching, 1878-1886

Keith R. Benson
Department of History and Ethics, University of Washington, Seattle, Washington, USA

1 INTRODUCTION

When I was a graduate student at Oregon State University in 1976, Professor Charlie Miller suggested that I write a paper to investigate who 'Brooks' was of 'Brooks' Law' fame. Consulting a standard sourcebook on crustaceans and the major biographical source for the history of science, I quickly obtained cursory answers to the law and the man; the law states that the rate of growth from stage to stage in crustacean larval development is 5/4 the previous stage, and William Keith Brooks (1848-1908, Fig. 1) was one of the major American biologists at the end of the 19th century.[1,2] In this paper, I will provide a more detailed investigation of Brooks in an attempt to place his carcinological interests, including the law of larval development, within an appropriate historical context.

2 BIOGRAPHICAL SKETCH

W.K. Brooks, as he preferred to be addressed, was born and raised near Cleveland, Ohio, where he developed an intense early interest in natural history, a not uncommon avocation for young boys living during antebellum America. He soon collected books on natural history, attended several of the ever-present summer courses in geology and field natural history, and formed his own natural history museum in the family barn. Studies of the natural world continued to dominate Brooks' collegiate experiences, first at Hobart College (1866) and finally at Williams (1868), where he studied botany and zoology with Sanborn Tenney. However, because there were few opportunities to pursue a career in natural history when he completed his education in 1870, Brooks returned home to work in the family business.

Although Brooks maintained his avocational interest in natural history after he returned to Cleveland, he was not content. Hearing of a rare teaching job in natural history and finally succeeding to persuade his reluctant father that his skills were not in the retail business, Brooks accepted a job at DeVaux College in 1871. While in this new position, where he was a very successful and popular teacher, Brooks propitiously received a circular announcing Louis Agassiz's plans for a summer school located at the seaside 'for Teachers who propose to introduce the Study (of natural history) into their Schools and for Students preparing to become Teachers' (Wilder 1898). Brooks was one of the 50 fortunate applicants who was accepted for

Figure 1. William Keith Brooks.

the summertime experiment, and thus Penikese Island in 1873 marked the beginning of his life-long relationship with research by the sea.

The summer experience was so positive for Brooks that he immediately applied to and was accepted at Harvard's Lawrence Scientific School, directed by Agassiz, in the fall of 1873 to pursue more instruction in natural history. Unfortunately, Agassiz's death in December of that same year ended the brief exposure Brooks had to 'the father of American zoology,' but he continued his education at Harvard, eventually being awarded the Ph.D. in zoology from Agassiz's son, Alexander, in 1875. After receiving the doctoral degree, only the fourth awarded by Harvard, he obtained his first professional job as curator of molluscs at the Boston Society of Natural History, working under the tutelage of Alpheus Hyatt. During the first year in Boston, he heard about the new graduate university in Baltimore, Johns Hopkins University, and the plans of Daniel Coit Gilman to build the new university upon the research of young postgraduates by offering twenty graduate fellowships to recent graduates who desired postgraduate education according to the new scientific model, designed liberally after European graduate education. With recommendations from Alexander Agassiz, Hyatt, John McCrady, A.S. Packard, and E.S. Morse, all pillars of the newly formed American biology community, Brooks was appointed not as a graduate student, but as the colleague to H. Newell Martin, the Michael Foster-trained physiologist hired by Gilman to build the new Biology Department at Johns Hopkins, the first such department in the United States. Hence, Brooks began his academic career not as a postgraduate student, but as the second member of the new biology faculty at Johns Hopkins University.

Brooks's charge in Baltimore was to develop a graduate program in morphology, the study of animal form, to complement Martin's half of biology, physiology. Beginning in 1876, he set himself to the task and, over his career, succeeded marvelously. Brooks established the first graduate program in morphology; he established several research journals to publish the results of the laboratories at Johns Hopkins; he began the Chesapeake Zoological Laboratory (CZL) in 1878 as the first graduate-level research laboratory in the United States; and he trained over 40 of America's best-known biologists of the early 20th century, including E.B. Wilson, Ross G. Harrison, E.G. Conklin, Thomas Hunt Morgan, J. Playfair McMurrich, and F.H. Herrick. Additionally, he published well over 160 papers, articles, and books on a variety of biological subjects, including several notable investigations of crustaceans.

While there is little question that Brooks was a productive morphologist and scholar who trained and influenced many colleagues and students, he did not have a flamboyant style or outgoing personality that attracted attention to himself. Furthermore, despite an extensive correspondence with many of his colleagues, he did not become actively involved in the formation of professional scientific societies, such as the American Society of Zoologists. In fact, Brooks lived a retiring and almost reclusive life. Outside of his personal research pursuits and his laboratory and teaching activities with his students, his life revolved around his home in Baltimore. Much of his temperament may be attributed to a congenital heart defect; regardless of the causal explanation, Brooks used his physical limitations to his advantage. One of the most characteristic features of the CZL when it was located in Beaufort, North Carolina, was a centrally located hammock, where Brooks rested and conducted his daily chats with his students. In addition, and perhaps because of his restricted interests, he was fiercely dedicated to his students, helping them not only in learning the new research techniques in morphology that were developed in the 1880's and 1890's, but often unselfishly providing them with research projects and ideas he had initiated himself. These admirable traits contributed to his popularity among young American biologists as they flocked to the Biological Laboratory in Baltimore and served to distinguish his academic career until his death of heart failure in November 1908.

3 BROOKS' CRUSTACEAN WORK

Brooks' involvement with crustacean research represented one of his earliest research projects and may, except for some major problems associated with historical accuracy, argue for his position as one of the founders of modern carcinology. The historical factors, of course, refer to the fact that when Brooks began his biological career, the field of biology consisted of two research areas, morphology and physiology, both of which were usually considered to be the sophisticated successors to natural history, the 19th-century foundation of all American biology. It is fair to note, however, that during his career, biological research diverged into many distinctive areas, some of which eventually emerged as today's specialty disciplines (see Rainger et al. 1988). Nevertheless, it is only with the benefit of hindsight, a perspective usually eschewed by historians of science, that we can credit Brooks with a formative influence upon the 20th-century field of carcinology. In fact, Brooks' initial exposure to crustaceans can not be traced to his personal interests with this taxonomic group. Instead, having been schooled within the late 19th-century tradition of morphology, Brooks dedicated his career to obtaining information relating to the life histories of organisms that might have a bearing on questions of ancestral or evolutionary relationships. Following the publication of Charles Darwin's *On the Origin of Species* in 1859, morphologists sought to develop genealogical schemes for individual organisms, groups of organisms, and, ultimately, entire taxa. After Ernst Haeckel elaborated his 'gastraea-theorie' (1874), in which all metazoan life was traced to a universal gastrula-like ancestor, other morphologists began to seek embryological evidence to either corroborate or contradict the 'Darwin of Germany.' Brooks was an early skeptic of Haeckel's comprehensive theory; as a result of preliminary observations involving gastropods, he cautioned his colleagues against developing any comprehensive ancestral plan because he found evidence that embryonic forms, while less complex than adult organisms, were also the products of evolutionary changes through adaptive selection. Therefore, it could be hazardous to hypothesize that a selected embryo retained only the primitive characteristics of the group, an assumption adopted by Haeckel and other German morphologists in the 1860's and 1870's.

3.1   *Early stomatopod work*

Brooks' hesitance to embrace Haeckel's ideas did not dissuade him from adopting the late 19th-century preference for embryological research. In fact, the one common denominator of all morphologists was their belief in the necessity of examining embryos and larval forms because, regardless of the evolutionary changes within these forms, they at least represented the more primitive condition of the adult forms and, therefore, could illuminate the ancestral condition of the entire group. Not surprisingly, therefore, when Brooks observed the larval form of *Squilla empusa*, one of the only common stomatopods of the Atlantic coast, during the summer of 1878, he seized the moment to investigate its morphological significance. In this sense, his interest in *Squilla* was not as a carcinologist but as a 19th-century morphologist.

After reading the literature, Brooks noted that the zoeal stage of *Squilla* bore a more intimate relation to the adult form than was common in the other stalk-eyed Crustacea. Previous investigations and subsequent morphological descriptions of the group had been confused, at best. Brooks continued the research until he was able to describe completely the life history of *Squilla empusa*, thus elucidating the confused embryological history of squilloid development. Brooks demonstrated that the group represented an exception to the generally accepted rule that the larval stages are less diverse than the adult. In an important generalization, and one that is gaining increased attention among larval ecologists at the present time, Brooks then emphasized that evolutionary selection mechanisms acted on larval and other embryonic forms just as these same pressures acted on adult structures. Given his growing belief in the importance that marine life played toward an understanding of the world's fauna and flora, Brooks began to suggest that diversity in the natural world resulted from changes in all phases of life, including the more primitive developmental stages.

3.2   *Decapod work*

During the summer of 1880, Brooks examined another fascinating crustacean, the natantian decapod *Lucifer ancestra*. Because the adult form of this organism was believed to retain its ancestral condition, the investigation promised to yield information of significance to questions of evolution. In perhaps one of the best descriptions of his own research program and interests, Brooks wrote a long letter to his president, Daniel C. Gilman.

I suppose you know that, while the ontogeny of the individual tends to recapitulate the phylogeny of the group, yet the life history of most of the higher animals has been so much modified that most of the stages which would indicate the phylogeny, are lost, or left out of the embryonic history. Once in a while, though, an animal is found, which has retained the ancestral manner of development, without modification, and these forms are of the greatest scientific interest; for one life history of this kind contributes more to the science of morphology than any number of modified life-histories. There is no way of telling whether a given animal will be found, when it is studied, to have an ancestral, or a modified manner of development, but occasionally some observer has the good fortune to find a form of development, which is of the unmodified kind, and which puts the facts which have been observed upon allied forms in their true light.

Leucifer [sic] happens to be an animal of this kind. Its life history is perfectly unmodified, or ancestral, and stages which have been dropped from the history of all other Crustacea are preserved by the Leucifer embryo with beautiful definitions and simplicity.

In the flood of light which it throws on the history of the Crustacea in general the Leucifer embryology is like the discovery of a key to an unknown litterture [sic].[3]

Not only did Brooks consider the adult to be a 'living relic,' the fact that *Lucifer* had a nauplius stage in its larval history was viewed by him as the piece of evidence that linked the low and high Crustacea through the same remote ancestor. This is the kind of morphological evidence from the larval history of an individual that permitted ancestral speculations.

Brooks shared his observations and ideas with his colleagues in several publications and communications. To the Harvard zoologist, Walter Faxon, he wrote that the presence of a nauplius stage in *Lucifer* was an argument against the metameric ancestry for crustaceans, a position championed by several German morphologists.

The various Decapods are not the descendants of an ancestral crustacean, with twenty similar unspecialized somites, but of an ancestral form with a small number of specialized somites.[1]

The implication of his remarks to Faxon was that Brooks believed morphologists should not search to discover ancestral forms that united diverse groups, but to look at existing larval forms for information concerning shared characters of closely related forms. Only in this manner could morphologists gather the necessary information together eventually to build more comprehensive and accurate genealogical schemes.

The *Lucifer* research also attracted the attention of Thomas H. Huxley, who helped to ensure its inclusion in at least two publications of the Royal Society. The major English publication, in *Philosophical Transactions*, provides an example of Brooks' meticulous care and thorough attention to detail in his research. The paper described a molt-by-molt account of larval development, with an additional discussion of the technical difficulties in conducting the work. He concluded the research with the following observation.

As a result of my four months' efforts I can now state that I have seen the eggs of *Lucifer* pass out of the oviduct. I have seen the Nauplius embryo escape from the same egg which I had seen laid, and I have traced every moult from the Nauplius to the adult in isolated specimens. There is therefore no Crustacean with the metamorphosis of which we are more thoroughly acquainted than we now are with that of this extremely interesting genus (Brooks 1882).

However, exhibiting his typical caution, Brooks was not completely confident that all the problems had been obviated and, therefore, he argued for more work to follow his own.

I am therefore unable to give Claus' interpretation of the significance of these larvae unqualified acceptance at present, and feel that our groundwork in this department of knowledge can be made sure only by new observations. Every naturalist who can trace the whole life-history of a single species of any of the genera of lower Malacostraca by actual moults, will not only help us to a sound and thorough appreciation of the significance of Crustacean embryology, but will also contribute to a better knowledge of the relation between ontogeny and phylogeny in the whole province of biology. (Brooks 1882).

3.3   *Later stomatopod work*

After completing the work on *Lucifer*, Brooks had another opportunity to investigate crustaceans, again with stomatopods. Perhaps due to his exposure to British scientists through the
publication in *Philosophical Transactions*, he was asked to analyze the Stomatopoda from the
Challenger expedition and to write the report on the group for the Challenger *Report*. Brooks
received the material, primarily consisting of larval stages, in 1883 and spent the next 3 years
examining it. Upon completion of the research he published *'Report on Stomatopoda collected
by HMS Challenger during the years 1873-1876'* in 1886.

In several regards, the project on the Challenger specimens represented an extension of the
earlier *Squilla* project, his first exposure to stomatopods. Consequently, he encountered many
of the same technical and pragmatic difficulties; however, this time he did not benefit from
having fresh samples taken from a replenishable source. In addition, what diagnostic aids were
available in the literature did not prove to be helpful, particularly because such aids did not deal
with larval forms in enough detail.

... as I soon found that the characteristics which have been selected for diagnosis are by no means the ones which
are most significant and of most scientific importance. In most of the published descriptions little attention is
given to points which are not regarded as diagnostic, and while most of the general are natural ones the points
which are of greatest value in tracing the relation between the larvae and the adult are entirely ignored in most of
the published descriptions (Brooks 1886).

To surmount these difficulties, Brooks utilized three approaches. First, each larval type or genus
was identified by selecting or comparing larvae that were sufficiently alike to indicate they
belonged to the same series. This method quickly illustrated the diagnostic relations and descriptions for at least three larval types. The second approach was to measure the larvae in each
series (in thousandths of an inch), thereby providing a sized series to reveal the molt stage of
each larval form (this led to the formulation of the famous Brooks' Law). That is, each molt
represented a discrete size increase (5/4) over its preceding stage. In this manner, Brooks was
able to extract distinct species from a series. Finally, he identified the genus through a reference
of the larval type to the adult genus. The end result was that Brooks was able to provide a new
interpretation for a large number of larval forms, several of which had been previously described as distinct planktonic species. It also proved to be an excellent method to link adult
forms to their appropriate larval stages.

To complete the stomatopod analysis, Brooks then developed a diagnostic key that unified
the key features of the larval forms and the main adult features. Armed with the key, he was then
in a position to correct the work of others.

Although Claus decided that they are young Lysiosquillae they show their relationship to the genus Squilla as
distinguished from Lysiosquillae by the following characteristics, all of which are shared by all fully-grown
Alima larvae. The dactylus of the raptorial claw has on its inner edge a small number of marginal spines, usually
about five or six; the hind body is wide and flat, and the postero-lateral angles of the abdominal somites end in
acute spines. The outer edge of the proximal joint of the uropod is bordered by a small number of spines usually
less than eight, and the inner one of the two spines on the ventral process from the posterior edge of the basal joint
of the uropod is longer than the outer, and it has a tooth or lobe on its outer edge; and the telson has six marginal
spines with minute secondary spines between the submedians, and four or more large secondary spines between
the submedian and the second or intermediate, and usually a single one internal to the base of the third or lateral
marginal spine. While it is true that all of these characteristics are not exhibited by every adult Squilla, there are
no Stomatopods except those of this genus in which they are all united, and they are all of them present in most
Squilla and in all the Alimae. (Brooks 1886).

The same type of notable attention to detail was presented by Brooks in his description of the stomatopod genera and species. Generic considerations included a general diagnosis of the group, remarks upon the key characteristics, and ontogenetic details. For the species, the diagnosis of the group was followed by a general description, emphasizing precise measurements of all body parameters, habitat preferences of the species, and general remarks on the species. By providing such a detailed appraisal of the Stomatopoda, Brooks gave a complete and accurate description of the Challenger samples, often shedding new light on the life histories of several species.

4   OTHER CONTRIBUTIONS

Although Brooks' major contributions to carcinology from his own research were primarily restricted to his work on stomatopods (*Squilla*) and natantian decapods (*Lucifer*), he also made important contributions to several other areas of crustacean work. First, all of Brooks' publications in both the United States and Europe included an encouragement for biologists to examine other aspects of the problems he investigated, including different crustacean species. Brooks believed that Crustacea was a particularly important group to study, especially because many species had what appeared to be transitional forms in their larval history and these forms could have a significant bearing on the developmental history of the group.

This fact, joined to the definite character of the changes which make up the life history of a marine crustacean, renders these animals of exceptional value for study of the laws of larval development, and for the analysis of the effect of secondary adaptations, as distinguished from the influence of ancestry; for while Claus has already clearly proved that adaptive larval forms are much more common among Decapods than had been supposed, his writings and those of Fritz Müller show that no other group of the animal kingdom presents an equal diversity of orders, families, genera, and species in which the relation between ontogeny and phylogeny is so well displayed, but, while proving this so clearly, Claus' well known monograph also shows with equal clearness that this ancestral history is by no means unmodified, and that the true significance of the larval history of the higher Crustacea can be understood only after careful and minute and exhaustive comparison and analysis. (Brooks & Herrick 1892).

Parenthetically, Brooks' emphasis on the adaptive nature of changes in the larval history of crustaceans and the need for additional investigations of these stages, appears strikingly modern, underscoring the value of his work to modern carcinology. As another indication of Brooks' encouragement to others, he not only appended exhaustive literature surveys to his crustacean papers, but he also introduced students and colleagues to the less available European work. On one occasion (Brooks 1883), he published a review paper in *Science* that included all the recent German studies on questions of crustacean phylogeny, much of which had not been translated into English. Always the teacher, Brooks used publications as a means to expose others to research he regarded as essential.

As a second major contribution to crustacean work, Brooks exerted an influence upon several of his students to pursue research topics on the group. Starting with his first student, E.B. Wilson, he conducted a research project that culminated in the publication (1883) '*The first zoea of Porcellana.*' The study was a joint effort to study the poorly understood early developmental stages of *Porcellana ocellata*, conducted during the same time that Brooks worked on the stomatopods. In the summer of 1885, Brooks collaborated with another student, A.T. Bruce, to study the king crab, *Limulus polyphemus*. Although Brooks did not consider *Limulus* to be a crustacean (it is a xiphosauran arthropod), its comparative anatomy had important evolutionary implications for the entire arthropodan phylum. This interesting project prompted Brooks to write President Gilman the following description.

Mr. Bruce and I are giving most of our time to the King Crab, Limulus. This is one of the most interesting of our American animals, and European naturalists have often reproached us that so little is known about it, but there are very peculiar technical difficulties.

About three fifths of all known animals are Arthropods, and Limulus is the oldest type of the Arthropods which is represented by living forms, so that a complete knowledge of its life history may possibly give the key to the origin and relationship of considerably more than half of the animal kingdom.[5]

A sketch of their research, *Abstract of researches in the embryology of Limulus polyphemus*, was published in 1885. Bruce continued to pursue similar interests, eventually completing his dissertation on the embryological relationship between insects and spiders. Brooks' own un-published work on the Macrura, which he worked on periodically from 1879-1885, was unself-ishly given to F.H. Herrick, upon which Herrick worked as a dissertation topic (*The development of Alpheus*). And while only Herrick completed his doctoral work on a crustacean, many others conducted their graduate research on other arthropods. Additionally, a number of Brooks' students made important contributions later in their careers to carcinology, perhaps as a result of Brooks' influence during the summer sessions of the CZL.

5   CONCLUSIONS

As much as I would like to conclude with an argument for Brooks' paternal position vis-à-vis modern carcinology, my deference to historical accuracy prevents me from staking this claim. Nevertheless, his contributions to the beginning of a research tradition specializing in questions about crustaceans were impressive indeed. Through his own research projects, through the encouragement of others to pursue similar investigations to his own, and through the training of many advanced students, Brooks illustrated how many morphological problems could be addressed in a productive manner by selecting crustaceans as research specimens. Admittedly, he never defined a specialty area of crustacean research and he never used the word carcinology, but he provided an impressive legacy upon which 20th-century biologists could build.

NOTES

1   For information of 'Brooks' Law,' I consulted T.H. Waterman, *The Physiology of the Crustacea* (New York: Academic Press, 1960).
2   The biographical sources on Brooks are limited. The major references include M.V. Edds, Jr., 'Brooks, William Keith,' *Dictionary of Scientific Biography*, ed. by Charles C. Gillispie (New York: Charles Scrib-ner's Sons, 1970), vol. II, pp. 501-501; Dennis M. McCullough, 'W. K. Brooks' Role in the History of American Biology,' *Journal of the History of Biology*, 1969, 2:411-438; Keith R. Benson, 'William Keith Brooks (1848-1908): A Case Study in Morphology and the Development of American Biology,' (unpub-lished Ph.D. dissertation, Oregon State University, 1979); and H. Newell Martin, W.K. Brooks, and the Reformation of American Biology,' *American Zoologist*, 1987, 27:759-771.
3   Letter, W.K. Brooks to D. C. Gilman, 8 September 1880, Johns Hopkins University, Eisenhower Library Manuscript Room, Gilman collection.

4 Letter, W.K. Brooks to Walter Faxon, (n.d.) February (1880), Harvard University, Museum of Comparative Zoology Archives, Miscellaneous papers.

5 Letter, W.K. Brooks to D.C. Gilman, 15 July 1885, Johns Hopkins University, Eisenhower Library Manuscript Room, Gilman collection.

REFERENCES

Brooks, W.K. 1882. Lucifer: A study in morphology. *Phil. Trans. Roy. Soc. Lond.* 173: 57-137.

Brooks, W.K. 1883. The phylogeny of the higher Crustacea. *Science* 2: 790-793.

Brooks, W.K. 1886. Report on the Stomatopoda collected by H.M.S. Challenger during the years 1873-1876. *The Voyage of the Challenger, Zoology,* 16: 1-116.

Brooks, W.K. & A. Bruce 1885. Abstract of researches on embryology of *Limulus polyphemus. Johns Hopkins Univ. Circ.* 5(43): 2-5.

Brooks, W.K. & F.H. Herrick 1892. The embryology and metamorphosis of the Macroura. *Mem. Nat. Acad. Sci.* 5: 1-135.

Brooks, W.K. & E.B. Wilson 1881. The first zoea of *Porcellana. Stud. Biol. Lab. Johns Hopkins Univ.* 3: 58-64.

Haeckel, E. 1874. Die gastraea-theorie die phylogenetische klassification des tierreiches und homologie der keimblatter. *Jena Z. Naturwiss.* 8: 1-55.

Rainger, R., K.R. Benson & J. Maienschein (eds.) 1988. *The American Development of Biology.* Philadelphia: Univ. Penn. Press.

Wilder, B.G. 1898. Agassiz at Penikese. *Amer. Nat.* 33: 189-196.

The scientific contributions of William Stimpson, an early American naturalist and taxonomist

Raymond B. Manning
Department of Invertebrate Zoology, National Museum of Natural History, Smithsonian Institution, Washington, DC, USA

1 INTRODUCTION

Studies on the systematics of decapod Crustacea have long been of importance at the Smithsonian Institution and its National Museum of Natural History. For anyone working on American decapods the names James Benedict, Mary Jane Rathbun, Waldo L. Schmitt, Horton H. Hobbs Jr., and Fenner A. Chace Jr. evoke images of the museum, its enormous collections, and hundreds of published papers on decapod systematics, many of them monographic. It is impossible to work on almost any group of American decapods without encountering a species first recognized by one of these researchers or without using one of their papers. Curiously, another worker whose name is not often associated with the Smithsonian was the first curator of crustaceans there. He began his studies at the Smithsonian some 2 decades before the establishment of the United States National Museum, the forerunner of the present National Museum of Natural History. That worker was William Stimpson (Fig. 1), who began his tenure at the Smithsonian in 1856 and was associated with it as curator of invertebrates until 1865 when he became Curator and Secretary of the Chicago Academy of Sciences.

It is surprising that so little is known about Stimpson, for he was a remarkable man and a productive and gifted scientist. He was also a tragic figure in that the results of his lifetime of work – in drafts and in unpublished manuscripts – were destroyed in the Great Chicago Fire of 1871. Like another American naturalist, Thomas Say, Stimpson's life came to an early and tragic end.

One of the remarkable things about Stimpson was that he seems to have received more recognition from malacologists for his scientific contributions on mollusks than he has from carcinologists for his work on Crustacea. In the Presidential Address delivered at the eighth meeting of the Biological Society of Washington in 1888, the malacologist William H. Dall (1845-1927) divided the early studies on mollusks in the United States into three periods. The first was named for Thomas Say (1787-1834), known to many as a carcinologist. The second period was named for Augustus A. Gould (1805-1866), who founded the Boston Society of Natural History and introduced Stimpson to Louis Agassiz (1807-1873). The third period was named for William Stimpson, who like Say is known to students of Crustacea as a carcinologist.

Like the American James D. Dana (1813-1895), who studied the crustacean collections of the US Exploring Expedition of 1838-1842, Stimpson proved to be a world-class scientist in two different fields. These in Stimpson's case were malacology and carcinology. Dana, known

Figure 1. William Stimpson (The National Portrait Gallery, Smithsonian Institution).

to most crustacean systematists as a carcinologist, also was a world-class geologist and is considered to be the father of geology in the United States.

All of these men were brilliant, accomplished scientists, and Stimpson easily ranks at the top with them. He had to be a genius to accomplish what he did before he was 40, having spent the better part of 4 years at sea as the scientific officer of the North Pacific Exploring Expedition of 1853-1856.

Between 1848, at the age of 16, and 1871, the year before his death, he published more than 60 papers. Only about half of his papers, 28 in all, were on Crustacea; those dealing with Crustacea are cited in the references.

Stimpson was born on February 14, 1832, in Roxbury, Massachusetts. According to one of his biographers, William H. Dall (1888: 129), 'His parents were Herbert H. Stimpson, who, I am informed, was of Virginian origin, and Mary Ann Brewer, of a good New England family.' Johnson (1976), in his biography of Stimpson, reported that the Brewers were a Virginian family, the Stimpson's having settled in Massachusetts in the 17th century.

Stimpson's father was a successful businessman who apparently left an estate large enough to preclude financial worries for Stimpson. His parents wanted him to go into business, but he was only interested in collecting. He entered Cambridge High School in 1847 and was graduated with the school's top academic honors in 1848 at the age of 16. He then spent a year at the Cambridge Latin School. In 1850, he became a special student of Louis Agassiz and in the same year, at age 18, he was appointed curator of mollusks at the Boston Society of Natural History.

In 1852, he was appointed naturalist for the North Pacific Exploring Expedition, largely through the efforts of Dana, Agassiz, and Spencer Baird, then Assistant Secretary of the Smithsonian. When the expedition sailed in June 1853, Stimpson was 21 years old.

Upon the return of the expedition in 1856, Stimpson took up his appointment at the Smithsonian and began making arrangements to work up the vast collections from the expedition. There were extensive collections of all groups, totaling 12000 specimens representing 5300 species, including almost 2000 species of mollusks and almost 1000 of crustaceans (Stimpson 1857a).

Stimpson himself took on the study of most of the marine invertebrates of the expedition, but some groups were entrusted to other workers. A.E. Verrill studied the corals and polyps (Verrill 1869) and the mollusks were to be studied by A.A. Gould, who made a preliminary report on the collection but never completed its study (Johnson 1964).

Stimpson completed 10 preliminary papers on the marine invertebrates, two in 1855 (Stimpson 1855a, b), the remainder in eight parts in his *Prodromus*, of which only the six parts dealing with crustaceans are cited in the references.

In all, Stimpson described 573 species, including 488 decapods, in his *Prodromus*, in only 116 published pages, all, as one biographer put it, in elegant Latin. The *Prodromus* included preliminary accounts of a wide variety of new invertebrates – actinians, tunicates, mollusks, planarians, annelids, turbellarians, sipunculans, amphipods, isopods, and decapods.

His accounts, though very short, are so good that his species can be easily identified from them. He was a master at characterizing taxa in a few short phrases. Among the genera he recognized are two callianassids, *Callichirus* and *Glypturus* (both in Stimpson 1866), that are so distinctive that they are readily distinguished today (Manning & Felder 1986; Manning 1987).

Apparently Stimpson spent his summers, beginning in 1849 at Grand Manan, New Brunswick, Canada, dredging along the Atlantic coast, and he made at least one visit to England, where he also collected; no field notes remain from these excursions. Elliott (1979) noted that Stimpson was credited as a pioneer in methodical dredging along the Atlantic coast.

The results of one of his trips were summarized in a brief report on a trip to Beaufort, North Carolina (Stimpson 1860b). In this paper, he described a species of the brachiopod *Lingula* that he found there in abundance. Apparently Agassiz had one specimen of this species in his holdings that he guarded jealously, allowing no one to study it, and Stimpson, alluding to the possible existence of this species in other collections, named it anyway. This caused a serious rift between Stimpson and his teacher and mentor, Agassiz.

Stimpson's collecting companions included Addison E. Verrill (1839-1926), E.S. Morse (1838-1925), Alpheus Hyatt (1838-1902), and Alpheus S. Packard Jr. (1839-1905), and other prominent biologists of that time. Several of his colleagues participated in the Salem Secession in the 1860's, leaving Agassiz's laboratory to found the Peabody Academy of Science in Salem, Massachusetts.

While in Washington, Stimpson founded two social clubs, including the Megatherium Club, whose members included Cope, Verrill, Ordway, Meek, Gill, and others, and the Polymythian Society of Monosyllabies, members of which contributed 9% of the contents of the *Proceedings of the Academy of Natural Sciences of Philadelphia* in 1861. He was also a founder of the Potomac-side Naturalists' Club, the forerunner of the Washington Academy of Sciences. At the age of 36 he was elected to the National Academy of Sciences.

While curator of invertebrates at the Smithsonian, Stimpson instituted an extensive exchange program with many museums, and records of some of these exchanges are extant in the Smithsonian Archives. One such collection, containing at least some of both Dana's and

Stimpson's types, was discovered recently at the Natural History Museum, London (Evans 1967), and a second was discovered by the author at the Zoological Museum, Copenhagen (Deiss and Manning 1981). Such collections sent out by Stimpson comprise the remnants of the extensive collections of the US Exploring Expedition and the North Pacific Exploring Expedition still in existence.

In 1860, Columbia University awarded Stimpson an honorary Doctor of Medicine, his only college degree. Also in 1860, he called for the formation of an international commission to draft rules for zoological and botanical nomenclature.

In 1864, he married Annie Gordon of Ilchester, Maryland.

In 1865, Stimpson went to the Chicago Academy of Sciences, where he became Director in 1866. He took with him some 10000 lots of Crustacea from the Smithsonian collection, then the largest and most important collection in the world. The collection comprised his type specimens, Dana's types, and all of the collections of the US Exploring Expedition of 1838-1844 and the North Pacific Exploring Expedition of 1853-1856. All of these collections and all of Stimpson's unpublished manuscripts and some 3000 figures, many in color, were lost in the great Chicago fire of 1871. As he noted in a footnote in one of his last published papers (1871c: 119): 'Since these pages were placed in the hands of the printer, the remainder of these materials were involved in the disaster of the great fire of Chicago. The manuscript descriptions of the North American Schizopods, Stomapods, and Tetradecapods, intended to form a part of the present paper, with numerous drawings and the specimens upon which they were based, were all burnt in this third and finally complete destruction of the author's scientific property.'

Characteristically Stimpson spent his last winter dredging with the Coast Survey until his poor health forced him to return to Maryland, where he died in May 1872 at the age of 40.

Had Stimpson published all of the work that he had completed, including a review of the marine invertebrates of the Atlantic coast of the United States, he certainly would be considered the father of American carcinology. The loss, however, of all of the writings of the synthetic phase of his career contributed to the general lack of recognition for him by subsequent generations.

Biographies of Stimpson have been published by Dall (1888), Higley (1902), Mayer (1918), Baker (1936), Johnson (1976), and Elliott (1979). A biographical sketch also was published by Debus (1968).

2 THE NORTH PACIFIC EXPLORING EXPEDITION

The North Pacific Exploring Expedition (Fig. 2) left Hampton Roads, Virginia, in June 1853, returning to New York in July 1856. Initially C. Ringgold was in command, but he was later replaced by John Rodgers. The squadron included the sloop Vincennes, the flagship in which Stimpson sailed, the brig Porpoise, the steamer John Hancock, the schooner James Fenimore Cooper, and the storeship John Pendleton Kennedy; the Vincennes and Porpoise had participated in the earlier US Exploring Expedition. During the expedition, the ships took different routes. It is assumed that most of the collections were made by Stimpson on board the Vincennes, the route of which is shown in Figure 2. Stimpson maintained a journal, excerpts of which were published by Johnson (1964).

The vessels made their first landfall at Madeira in July 1853, then sailed for South Africa via the Cape Verde Islands. In November, they crossed the southern Indian Ocean, arriving at Sydney, Australia, in December 1853. They then proceeded to Hong Kong via the Santa Cruz

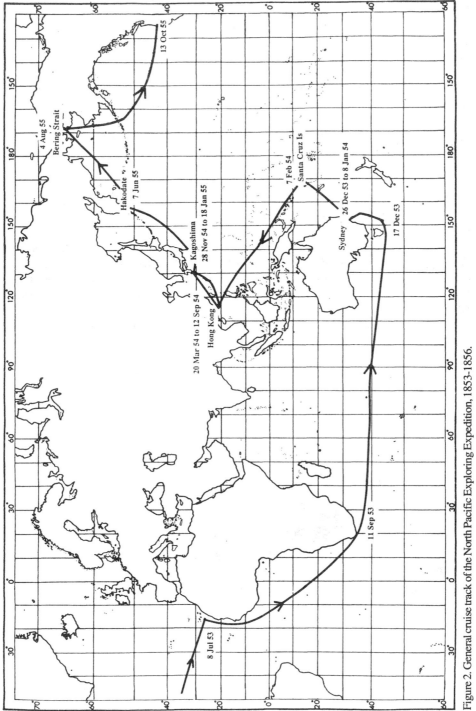

Figure 2. General cruise track of the North Pacific Exploring Expedition, 1853–1856.

Islands and Macao, arriving in Hong Kong in March 1854. In October, they went northward to Japan, stopping in the Bonin and Ryukyu Islands, returning to Hong Kong in January 1855. They departed Hong Kong for good in April 1855 stopping in the Ryukyus and Hakodate, Japan, and Kamchatka, en route to Glassenappe Harbor, Siberia. They sailed again in mid-September, arriving in San Francisco on 13 October, and sailed for New York in March 1856, going via the Hawaiian and Society Islands. They arrived in New York in July 1856. Stimpson maintained a journal for most of the voyage, but daily coordinates are the only entries for the return voyage.

Wherever he could, and whenever the ship's route and its commander's inflexibility permitted it, Stimpson made collections, either by hand or by dredging. He tried to limit collecting to what could be processed in one day. Upon returning to the ship from collecting he would make color notes and sketches of as many species as he could, especially the more fragile forms. New taxa were described on the spot. He held contracted specimens in dishes until they relaxed so he could sketch and describe them. He was considered a pioneer in describing molluscan soft anatomy.

One of his discoveries on the expedition was an anemone associated with the dorippid crab *Dorippoides facchino* (Herbst), a species which he named *Cancrisocia expansa* (see Holthuis & Manning 1990; Fig. 3). The upper figure (Fig. 3a) was published by Verrill (1869); the lower (Fig. 3b) was found in the files of the Division of Crustacea, National Museum of Natural History.

Stimpson's scientific activities were actually severely limited by his ship's commanders. Apparently the Navy Department laid claim to his work and the specimens taken on the expedition, perhaps explaining why his major manuscript on the crabs of the expedition was found in that department. His accounts of new taxa had to be prepared on the stationery used by the ship's captain and submitted for publication through the ship's captain. He apparently side-stepped this restriction by including species descriptions and accounts of his activities in letters to colleagues at home.

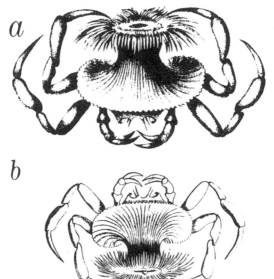

Figure 3. The crab *Dorippoides facchino* (Herbst) and its commensal anemone *Cancrisocia expansa* Stimpson (A) published by Verrill 1869 (B) found in files of the Division of Crustacea, National Museum of Natural History (from Holthuis & Manning 1990).

3   STIMPSON'S CONTRIBUTIONS TO KNOWLEDGE OF CRUSTACEA

Many of Stimpson's earlier papers were on mollusks and some were on other groups of marine invertebrates, including echinoderms and tunicates. In all, he named about 950 species, mostly marine invertebrates, in a career that spanned a mere 24 years. His last paper, published in 1907, was edited by Mary Jane Rathbun based on an unpublished manuscript discovered in the Navy Department after his death. That report was based on the Anomura and Brachyura of the North Pacific Exploring Expedition and included accounts of 353 species (numbered 1-358 to correspond to species numbers used by Stimpson in his *Prodromus*), all of which were named in his *ProdromusD; that report omitted the family Rhizopidae and its five species, apparently removed by Stimpson himself (Stimpson 1907: 95, footnote by M J. Rathbun).*

The numbers of crustacean taxa reported on by Stimpson are summarized by publication date, as follows:

1852 - 1 species (new): decapod.

1854 - 62 species (38 new; 5 new genera): 2 barnacles, 2 parasitic copepods, 1 ostracode, 1 cumacean (new), 10 isopods (9 new species, 1 new genus), 1 tanaid (new), 34 amphipods (27 new species, 4 new genera), 1 mysid, 9 decapods (1 new).

1855a - 14 species (new): amphipods. Also *Cancrisocia expansa*, an anemone associate of a dorippid crab.

1855b - 10 species (new): 4 isopods, 6 amphipods.

1856 - 23 species (15 new): 14 decapods (6 new), 2 isopods (new), 7 amphipods (new).

1857b - 16 species (15 new; 2 new genera): 12 decapods (11 new; 2 new genera), 4 isopods (new).

1857c - 137 species (14 new; 1 new genus): 95 decapods (2 new), 1 stomatopod (new), 20 isopods (7 new; 1 new genus), 20 amphipods (4 new), 1 parasitic copepod.

1857d - 40 species (18 new; 2 new genera): decapods.

1858a - 96 species (34 new; 3 new genera): decapods.

1858b - 108 species (51 new; 16 new genera; 5 new family-group taxa): decapods. Also synopsis of grapsid generic characters.

1858c - 31 species (13 new; 4 new genera): decapods.

1858d - 83 species (51 new; 17 new genera; 2 new family-group taxa): decapods. Also revision of Anomura.

1859a - 1 species (new; 1 new genus): decapod.

1859b - 85 species (39 new; 4 new genera): decapods.

1860a - 111 species (71 new; 14 new genera): decapods.

1860b - 0 species (2 new genera): decapods; revision of Mithracidae.

1860c - 38 species (1 new; 1 new genus): decapods.

1860d - 130 species (79 new; 13 new genera): decapods.

1861 - 9 species (2 new): decapods.

1862 - 1 species: isopod.

1863b - 22 species (1 new): 9 decapods, 1 mysid, 4 amphipods (1 new), 1 isopod, 2 branchiopods, 2 parasitic copepods, 3 barnacles.

1863c - 1 species (1 new; 1 new genus): decapod (fossil).

1864 - 17 species (17 new): 5 decapods, 5 isopods, 7 amphipods.

1866 - 4 species (4 new; 2 new genera): decapods. The publication ends in mid-sentence on p. 48. Fire at the Academy destroyed the remaining text and plates (R. Vasile, in litt.).

1871a - 81 species (52 new; 19 new genera; 10 family-group taxa): decapods.

1871b - 4 species: 1 mysid, 2 amphipods.

1871c - 71 species (42 new; 6 new genera): decapods.

1907 - 358 species: decapods.

In all, Stimpson named 483 decapods, 1 cumacean, 31 isopods, 1 tanaid, 66 amphipods, and 1 stomatopod. He also named 113 genera, 107 of them for decapods, and 17 family-group taxa (as families or subfamilies). Most of the names he introduced are in use today.

REFERENCES

Baker, F.C. 1936. Stimpson, William. In D. Malone (ed.), *Dictionary of American Biography* 18:31-32. New York: Scribner.

Dall, W.H. 1888. Some American conchologists. *Proc. Biol. Soc. Wash.* 4:95-134.

Debus, A.G. (ed.) 1968. Stimpson, William. In *World Who's Who in Science*: p. 1613. Chicago: Marquis-Who's Who.

Deiss, W.A. & R.B. Manning. 1981. The fate of the invertebrate collections of the North Pacific Exploring Expedition, 1853-1856. In A. Wheeler & J.H. Price (eds), *History in the Service of Systematics*: pp. 79-85. London: Society for the Bibliography of Natural History.

Elliott, C.A. 1979. Stimpson, William. In *Biographical dictionary of American science. The seventeenth through the nineteenth centuries*: p. 242. Westport: Greenwood Press.

Evans, A.C. 1967. Syntypes of Decapoda described by William Stimpson and James Dana in the collections of the British Museum (Natural History). *J. Nat. Hist.* 1:399-411.

Higley, W.K. 1902. Historical sketch of the Academy. *Chicago Acad. Sci., Spec. Pub.* 1:1-48.

Holthuis, L.B. & R.B. Manning 1990. Crabs of the subfamily Dorippinae MacLeay, 1838, from the Indo-West Pacific region (Crustacea: Decapoda: Dorippidae). *Res. Crustacea, Carcinological Soc. Japan* Spec. No. 3:1-151.

Johnson, R.I. 1964. The Recent Mollusca of Augustus Addison Gould. *US Nat. Mus., Bull.* 239:1-182.

Johnson, R.I. 1976. Stimpson, William. In C.C. Gillespie, (ed.) *Dictionary of Scientific Biography* 13:pp. 65-66. New York: Scribner's.

Manning, R.B. 1987. Notes on western Atlantic Callianassidae (Crustacea: Decapoda: Thalassinidea). *Proc. Biol. Soc. Wash.* 100:386-401.

Manning, R.B. & D.L. Felder 1986. The status of the callianassid genus *Callichirus* Stimpson, 1866 (Crustacea: Decapoda: Thalassinidea). *Proc. Biol. Soc. Wash.* 99:437-443.

Mayer, A.G. 1918. Biographical memoir of William Stimpson 1832-1872. *Biogr. Mem. Natn. Acad. Sci.* 8:419-433.

Stimpson, W. 1852. 'Description of a new crustacean belonging to the genus *Axius*, of Leach.' *Proc. Boston Soc. Nat. Hist.* 4:222-223. (Untitled.)

Stimpson, W. 1854. Synopsis of the marine Invertebrata of Grand Manan: or the region about the mouth of the Bay of Fundy, New Brunswick. *Smith. Contrib. Knowl.* 6(5):1-67, 3 pls.

Stimpson, W. 1855a. Descriptions of some of the new marine Invertebrata from the Chinese and Japanese seas. *Proc. Acad. Nat. Sci. Phila.* 7:375-384. (Separate without pagination.)

Stimpson, W. 1855b. Descriptions of some new marine Invertebrata. *Proc. Acad. Nat. Sci. Phila.* 7:385-394. (Separate without pagination.)

Stimpson, W. 1856. On some Californian Crustacea. *Proc. Calif. Acad. Sci.* 1:87-90.

Stimpson, W. 1857a. Notice of the scientific results of the expedition to the North Pacific Ocean, under the command of Com. John Rodgers. *Amer. J. Sci.* 23:136-138.

Stimpson, W. 1857b. Notices of new species of Crustacea of western North America; being an abstract from a paper to be published in the journal of the society. *Proc. Boston Soc. Nat. Hist.* 6:84-89.

Stimpson, W. 1857c. Crustacea and Echinodermata of the Pacific shores of North America. *J. Boston Soc. Nat. Hist.* 6:444-532, pls. 18-23. (Pages 1-92 on separate).

Stimpson, W. 1857d. Crustacea Maioidea. Maiidae. Prodromus descriptionis animalium evertebratorum, quae in Expeditione ad Oceanum Pacificum Septentrionalem a Republica Federata missa, Cadwaladaro Ringgold et Johanne Rodgers ducibus, observavit et descripsit, Pars 3. *Proc. Acad. Nat. Sci. Phila.* 1857:216-221. (Pages 23-28 on separate).

Stimpson, W. 1858a. Crustacea Cancroidea et Corystoidea. Prodromus descriptionis animalium evertebratorum, quae in Expeditione ad Oceanum Pacificum Septentrionalem a Republica Federata missa, Cadwaladaro Ringgold et Johanne Rodgers ducibus, observavit et descripsit, Pars 4. *Proc. Acad. Nat. Sci. Phila.* 1858:31-40. (Pages 29-37 on separate).

Stimpson, W. 1858b. Crustacea Ocypodoidea. Prodromus descriptionis animalium evertebratorum, quae in Expeditione ad Oceanum Pacificum Septentrionalem a Republica Federata missa, Cadwaladaro Ringgold et Johanne Rodgers ducibus, observavit et descripsit, Pars 5. *Proc. Acad. Nat. Sci. Phila.* 1858:93-110. (Pages 39-56 on separate).

Stimpson, W. 1858c. Crustacea Oxystomata. Prodromus descriptionis animalium evertebratorum, quae in Expeditione ad Oceanum Pacificum Septentrionalem a Republica Federata missa, Cadwaladaro Ringgold et Johanne Rodgers ducibus, observavit et descripsit, Pars 6. *Proc. Acad. Nat. Sci. Phila.* 1858:159-163. (Pages 57-61 on separate).

Stimpson, W. 1858d. Crustacea Anomoura. Prodromus descriptionis animalium evertebratorum, quae in Expeditione ad Oceanum Pacificum Septentrionalem a Republica Federata missa, Cadwaladaro Ringgold et Johanne Rodgers ducibus, observavit et descripsit, Pars 7. *Proc. Acad. Nat. Sci. Phila.* 1858:225-252. (Pages 63-90 on separate).

Stimpson, W. 1859a. '(*Hapalocarcinus marsupialis*), a remarkable new form of brachyurous Crustacean on the coral reefs at Hawaii.' *Proc. Boston Soc. Nat. Hist.* 6:412-413. (Untitled).

Stimpson, W. 1859b. Notes on North American Crustacea, No. 1. *Ann. Lyceum Nat. Hist. N.Y.* 7:49-93, pl. 1. (Pages 1-47 on separate dated 1859. Stimpson 1871c cited this as having been published in 1860. Paper is dated March 1859 on p. 49, 1860 on cover).

Stimpson, W. 1860a. Notes on North American Crustacea, in the museum of the Smithsonian Institution, No. 2. *Ann. Lyceum Nat. Hist. N.Y.* 7:176-246, pls. 2, 5.

Stimpson, W. 1860b. Sketch of a revision of the genera of the Mithracidae. *Amer. J. Sci.* (2)29:132-133.

Stimpson, W. 1860b. A trip to Beaufort, North Carolina. *Amer. J. Sci.* (2)29:442-445.

Stimpson, W. 1860d. Crustacea Macrura. Prodromus descriptionis animalium evertebratorum, quae in Expeditione ad Oceanum Pacificum Septentrionalem a Republica Federata missa, Cadwaladaro Ringgold et Johanne Rodgers ducibus, observavit et descripsit, Pars 8. *Proc. Acad. Nat. Sci. Phila.* 1860:22-47. (Pages 91-116 on separate).

Stimpson, W. 1861. Notes on certain decapod Crustacea. *Proc. Acad. Nat. Sci. Phila.* 1861:372-373.

Stimpson, W. 1862. On an oceanic isopod found near the south-eastern shores of Massachusetts. *Proc. Acad. Nat. Sci. Phila.* 1862:133-134.

Stimpson, W. 1863a. On the classification of the Brachyura, and on the homologies of the antennary joints in decapod Crustacea. *Amer. J. Sci.* 1863:139-142.

Stimpson, W. 1863b. Synopsis of the marine Invertebrata collected by the late Arctic Expedition, under Dr. I.I. Hayes. *Proc. Acad. Nat. Sci. Phila.* 1863:138-142.

Stimpson, W. 1863c. On the fossil crab of Gay Head. *Boston J. Nat. Hist.* 7:583-589, pl. 12.

Stimpson, W. 1864. Descriptions of new species of marine Invertebrata from Puget Sound, collected by the naturalists of the North-west Boundary Commission, A.H. Campbell Esq., Commissioner. *Proc. Acad. Nat. Sci. Phila.* 1864:153-161.

Stimpson, W. 1866. Descriptions of new genera and species of macrourous Crustacea from the coasts of North America. *Proc. Chicago Acad. Sci.* 1:46-48.

Stimpson, W. 1871a. Brachyura. Preliminary report on the Crustacea dredged in the Gulf Stream in the Straits of Florida by L.F. de Pourtales, Assist. US Coast Survey, Part 1. *Bull. Mus. Comp. Zool.* 2:109-160.

Stimpson, W. 1871b. On the deep-water fauna of Lake Michigan. *Amer. Nat.* 4:403-405.

Stimpson, W. 1871c. Notes on North American Crustacea, in the Museum of the Smithsonian Institution, No., 3. *Ann. Lyceum Nat. Hist. N.Y.* 10:92-136. (Pages 119-163 on separate). @BIBL = Stimpson, W. 1907. Report on the Crustacea (Brachyura and Anomura) collected by the North Pacific Exploring Expedition, 1853-1856. *Smith. Misc. Collns.* 49:1-240, 26 pls.

Verrill, A.E. 1869. Actiniaria, with supplement and geographical lists. Synopses of the polyps and corals of the North Pacific Exploring Expedition, under Commodore C. Ringgold and Captain John Rodgers, USN, from 1853 to 1856, collected by Dr. William Stimpson, Naturalist to the Expedition. *Proc. Essex Inst.* 6:51-104, pls. 1, 2.

A correspondence between Martin Burkenroad and Libbie Hyman: Or, whatever did happen to Libbie Hyman's lingerie

Frederick R. Schram

Instituut voor Systematiek en Populatiebiologie, Amsterdam , Netherlands

1 INTRODUCTION

In sorting through the papers of Martin Burkenroad while incorporating them into the archives of the San Diego Society of Natural History, I discovered an exchange of correspondence between Burkenroad and Libbie Hyman extending from about 1934 to 1963. Neither Burkenroad nor Hyman suffered fools lightly, yet the letters between them clearly reveal a warm human side to both their personalities. Hyman acted as a mentor to Burkenroad, encouraging him to pursue a doctoral degree and take a 'high road' in his dealings with other scientists. Burkenroad had some influence on Hyman's focus on flatworm research and provided encouragement and advice in the initial stages of writing *The Invertebrates*.

The documents in the San Diego archives consist of Hyman's letters to Burkenroad, copies of either some of his letters to her or edited hand written drafts of letters he later sent, and his notes on miscellaneous issues dealt with in the letters. Most of the archive consists of Hyman letters and we lack much of Burkenroad's half of the exchange. Attempts to locate an archive of Hyman letters at the American Museum revealed that she did not keep her correspondence.

I encountered some trouble in placing the letters in their proper sequence. Hyman often did not date her correspondence, and the chronology of her documents had to be determined from internal evidence. Sometimes I could only place them in relative order. Burkenroad's habits differed in that he carefully dated even his hand written drafts. I have injected explanatory and editorial notes in brackets in the transcriptions below where appropriate. After leaving service with the Louisiana Department of Conservation in 1931, Burkenroad began a two year period during which he moved through several institutions in an independent capacity (Schram 1986). These included the Peabody Museum at Yale, the American Museum in New York, and the National Museum in Washington. Burkenroad first met Hyman while at the American Museum studying a collection of Gulf Coast penaeids.

Libbie Hyman had left her position at the University of Chicago as assistant to Prof. C.M. Child in 1931 and after travel in Europe relocated to New York City to establish an affiliation with G.K. Noble's Department of Experimental Biology at the Museum. Hyman's move to New York arose from a combination of factors. She anticipated Child's eventual retirement from the university, and she also wanted to escape the negative effects she then experienced from relatives in Chicago (Stunkard 1970, 1974). However, her relocation also coincided with the move of her friend Dorothea Rudnick, who completed her Ph.D. degree at Chicago in 1931 and who went to Yale to take up a series of fellowships with the Osborn Zoological Laboratory.

(Hyman did not join the staff of the American Museum formally until 1936, and then still only in an honorary capacity, i.e., she had an office and enjoyed use of the library but received no salary.)

Burkenroad and Hyman had a complex relationship. Perhaps it grew from the meeting of two free spirits, or their shared Jewish heritage, or their mutually held low tolerance of scientific pomposity. None the less, they had a positive influence on each other. After his two year 'sabbatical,' Burkenroad took a position as an assistant at the Yale Bingham Oceanographic Collection in early 1934. There he fell under the guidance of Albert E. Parr, who eventually became director of the Peabody Museum. Tulane asked Burkenroad to leave in 1929 before he completed the last year of his bachelor's degree. Notwithstanding, his settling at Yale involved plans to possibly pursue a Ph.D. there, plans that Hyman strongly urged him to make. Burkenroad soon opened up a correspondence with Hyman that, while focused around business issues, quickly came to contain matters of a personal nature.

2   LETTERS IN 1934 – THE CORRESPONDENCE BEGINS

An undated letter from Hyman in response to something from Burkenroad begins the archived exchange. She sent this letter apparently early in 1934 shortly after Burkenroad took the position at Peabody. At the time Burkenroad was 24 and Hyman 46, and she clearly started to act as a mentor and advisor to her young protege.

Sunday, 15
Dear Martin:
It is very nice you are so well satisfied with your work at the Peabody and I trust the position will materialize into something quite satisfactory. It seems like just the sort of thing you want. Let me urge you by all means to go after a doctor's degree as quickly as possible. It doesn't mean anything in itself but with the American worship of education so rampant everywhere the degree is an asset not to be underestimated. It's like having your pants pressed and your shoes shined. In themselves they have no value at all but they are a necessary as a concession to the usages of society. I quite agree with Parr's advice about university red-tape. The thing for you to do is to go ahead and take whatever courses you want and whatever ones the instructors will let you into. At the same time, I'm not so optimistic that you will be accepted into the graduate school but they will probably allow you to take courses and give you credit for them. They may eventually make some sort of concession [concerning Burkenroad's lack of a bachelors degree] but colleges are usually pretty insistent in the matter of proper credits and etc. Here's luck, though. By all means get acquainted with Harrison; he's one of the very finest, absolutely sincere and honest. [Ross Granville Harrison, a student of W. K. Brooks of Johns Hopkins, founded the Osborn Laboratory and was the former chairman of Zoology at Yale.] I recommend for your acquaintance my friend Dorothea Rudnick who is holding a national research fellowship in the department this year. She's a girl of the finest breeding, culture, and intelligence. She has been told about you so that you need not hesitate to make yourself known to her. But I shall be around next week to make the introduction if necessary. My young man seems to think I owe him a visit. [Hyman's second cousin, Jack Greenberg, the only member of her family to whom she was deeply attached.] He is terribly busy interning at the hospital, which as you may have found out is on the other side of town from the Peabody; and consequently I shall be around Osborn a good deal in the interludes between Jack's off hours. I arrive on Wednesday and if I do not happen to meet you in Osborn, I will look you up in Peabody. At Osborn, I'll be most apt to be in Dorothea's room in the basement or in the library. One of my reasons for coming is to hear Spemann [Hans Spemann, experimental embryologist and Nobel laureate]; but the lectures extend over a couple of weeks and I can't stay that long.

Coe's paper came out in the Biological Bulletin the other day but I haven't read it. [Apparently this was Wesley R. Coe's paper in the June 1934 issue, vol. 66: 304-315, on regeneration in nemertines.] From all I have heard about him he is a harmless old duck.

A couple of weeks ago I found myself stared at in the library of the museum by a stoutish, middle-aged woman. The girls in the library told me she was the famous or infamous Miss Boone [Lee Boone, who had just been forced out of the National Museum in Washington see McLaughlin, this volume]. I presume she had learned my name and thought she would get an eyeful while the opportunity offered....

Saturday I was out in Pelham Bay Park with a field class from C.C.N.Y. under one Dr. Spieth [Herman T. Spieth, instructor in biology at CCNY and an authority on Ephemeroptera]. That young woman you met a couple of times in my place, Mrs. Etkin, is taking the course and wanted me to go along. The park is really quite nice and I'm sorry now I didn't go with you when you suggested it in the Spring. There is quite a lot of marine stuff available and I also saw under a bridge a number of those prawns you had at the museum. We ran into a sort of nudist colony who did not seem at all alarmed at or embarrassed by our intrusion. Most of them had on an extremely brief garment around the essential region but one was entirely naked. I wonder if they inhabit the place regularly. There seem to be what are usually called tramps around the region too. The park is so big and so much of it apparently left in a natural state that I suppose men could live there in summer eating clams and fish. But Saturday was rather a cold day to be going around with nothing on but one's skin.

The girls in the library gave me *Ulysses* that you had thoughtfully left for me. I had expected to see you again before you went away but other things turned up and I didn't get to the museum. After handing me the book one of the girls very diffidently asked to borrow it, so I left it there anyway. Surprisingly enough it was that one who doesn't seem very bright – the pretty one you know, who has sort of a brown look about her. I don't know any of their names. I warned her she probably would not like it but she seemed to think she wanted to read it. I shall see you in a few days, then.

Regards, Libbie

The rather cordial and gossipy nature of this first surviving letter continued for the rest of the correspondence. Unfortunately we lack most of Burkenroad's letters, but obviously we can see from what survives of the exchange that they established a friendly and familiar status with each other.

Part of Burkenroad's duties at Yale included participation in the Bingham Oceanographic Expeditions in 1934, 1935, and 1937. Apparently he had bragged about his sailing skills while looking forward to going to sea and collecting, among other things, his beloved prawns. Two undated letters by Hyman posted apparently in the spring of 1934 replied to epistles from Burkenroad that seem to have contained some more boasting on his part. The first relatively short missive hints at difficulties Hyman had at the museum. Both letters reveal Burkenroad's and Hyman's love of music – Burkenroad in his characteristic manner picked a fight, but Hyman clearly had the upper hand.

[no date]

I'll be writing to Dorothea one of these days but I am sure she is worth far more than anyone to whom she could have introduced you. I could guess who both the people are and both of them together do not equal Dorothea. So there.

I received not only one but two copies of your museum publication and shall return one of them to you some time. But why do you persistently address me at the museum. I asked you not to do so, you know, and the house address is much surer. I probably shan't stay there much longer for Noble after pressing me to come now seems to think I'm not spending enough time on research to justify my presence there although when he asked me to come he knew I could not put in much time on a problem. It is unpleasant to be in a laboratory where you are constantly urged to grind out research as if it were a pound of coffee and to publish large quantities of print. He's got infected with ambition, wants to make a big splash with his laboratory. I'm afraid he's picked on the wrong person when he inveigled me there because I balk at any suggestion that I must publish something. Kindly do not pass any of this on.

I'm very sorry you have been seasick – it's about the most miserable experience in the world, far worse than unrequited love. I thought you were immune to it for you always expressed so much fondness for sailboats and such, the mere sight of which makes me seasick....

The question about Stravinsky struck me as slightly delirious but I seem to remember that he really was here and was given an ovation by the Friends and Enemies of Modern Music or some such organization. I presumed that the occasion was for members only and did not attempt to push myself in. There was also some Stravinsky at the symphony, the Psalms I believe, but I was reading in the library when the concert, to which I had a ticket, came off and I forgot to go until it was too late. This, I believe, was the first time in some twenty years or more of attendance on symphonies that I've forgotten to go. Probably my cortex is coagulating.

I hope you'll find the shrimp where you expect them to be and that they'll come up all properly labelled with their scientific names. Thanks for writing – I enjoyed the letter.

Best, Libbie

[P.S.] I meant to tell you above that I thought your remarks about the shrimp investigation people in your paper were offensive and that it was obvious that they were intended to be offensive.

Hyman's reference was to Burkenroad's paper (*Bull. Amer. Mus. Nat. Hist.* 68:62-143) on the Penaeidea of Louisiana in which, in a discussion of *Penaeus setiferous*, he dealt very severely (pp. 81-84) with the work of F.W. Weymouth, M.J. Lidner, and W.W. Anderson, Federal Shrimp Investigation biologists. (The disagreement occurred over the federal scientists' interpretation of the life history of *P. setiferous*, an argument that apparently began when Burkenroad worked in the Louisiana Dept. of Conservation after leaving Tulane. As harsh as his printed attack was, Burkenroad's personal copy of the paper in the possession of this author has many annotations with additional vitreol.) After she posted the above letter and Burkenroad returned from sea, Hyman sent the following undated letter. Her troubles with Noble at the museum began to affect her health and she records prostration with an attack of colitis. Her unhappiness with the situation in Noble's lab led her to consider leaving New York. This letter also illustrates how forcefully Hyman counseled Burkenroad.

Dear Martin:

You are quite wrong in thinking I was annoyed by anything you said – I knew you were merely being playful. I warned you in advance that I already had more correspondents than I could take care of but you would insist on writing to me. I have taken on the job of abstracting in toto three very long-winded German journals for Biological Abstracts; that would not be so bad but as soon as they found out I was a weak person who can never refuse my services, they immediately wished on me immense numbers of back articles, most of them 100-200 pages long, which they had not been able to find voluntary abstractors for. Of course I am paid for this, at a poor rate, naturally; but the occasional checks are a welcome addition to a rapidly dwindling income. As soon as I get the back articles out of the way, it won't be so much work keeping up with the current issues; but at present I have to spend two or three evenings a week on the job. What with that and the necessity of seeing a play or a picture once in a while and at least one evening a week for correspondence and occasional visits to the three or so friends I have in N.Y., the evenings always seem to vanish with nothing much to show for them. Besides I have been sick again, or is it yet? My digestion went on one of the worst rampages it has had since I began taking the HCl. I spent, very reluctantly, fifty dollars in less than a week, in an attempt to get a diagnosis. I was terribly afraid by the severity and location of the pain that I had gall-bladder trouble; but the diagnosis, right or wrong, came out an irritation of the colon. I was much better for a while on the diet recommended but these last few days I am worse again and compelled to lie down most of the time. I can't eat anything much but proteins and potatoes and I am damn sick of potatoes which I have eaten every day for a couple of months now. It pains me terribly to pass up strawberries, asparagus, and other dainties I see now in the markets; but apparently I have to suffer for the least bit of food I eat which isn't completely digestible.

No, I really don't want to go to Rochester. [Apparently Hyman must have spoken of this with Burkenroad. Dorothea Rudnick moved to Rochester in the Fall of 1934 to take a position with the university there.] I have no use whatever for the head of zoology there, the honorable professor Benjamin Harrison Willier [an embryologist whom Hyman knew from her Chicago days, where he received his Ph.D. in 1920 and then stayed to rise to full professor before going on to Rochester in 1933]. I have a long memory and a stiff neck like the rest of my tribe. And I haven't any passion for Dorothea either. I admire her as about the most perfect specimen of a human being I have ever met and am fond of her, but she probably makes me too conscious of my own innumerable imperfections. Yes, I know you did not fall in love with Dorothea but it would have been far more sensible of you than your other innumerable infatuations and you had a good start, because she really liked you. But then the male sex is that way, never show any sense about the women they select.

Any institution could have me for nothing if they would offer me a nice office lined with shelves on which I could dispose of this mountainous pile of reprints and journals which grows daily more Everestian, to coin a non-existent word. It is one of the major problems of my life what to do with these stacks of journals. My conscience won't let me give up the subscriptions because I know how badly our journals need support. So there it is. You get me a nice office in Osborn or the Peabody and I'll move to Yale where I could have the one dream of my life, a nice little house in the country, with a garden. I'm afraid I'd rather have a garden than the most exalted scientific reputation. The trouble is, however, that I ought to move to Arizona or New Mexico.... I have been in a dilemma for some time now and haven't been able to arrive at a decision. If I had the means, I could have a summer home in New England and a winter home in New Mexico but I'm not likely to amass the necessary coin again. I did at one time have enough for at least a shack in both places but I lost it all. [Hyman suffered considerable loses in the Great Depression (Stunkard 1974).]

You really are wrong about Stravinsky, so why not give up the argument. There never was any idea of his appearing with the Philharmonic, if that is what you meant. The season is over now and if he should come, he'd find the hall empty. He was here as I told you, months ago.... I admire some of Stravinsky's music, notably Le Sacre du Printemps, but none of his works since that one seem to be of much moment. I consider Sibelius the greatest modern composer and a great composer of all time. The appreciation of Beethoven is one of the fruits of maturity; he is not for the young, and some of his music is still too profound for any of us. ...The trouble with you and Beethoven, besides your emotional immaturity is, I think, that you confuse classical harmonic form with crudity and banality because naturally classical harmonic form is employed by every noise-maker in Tin Pan Alley. I say, naturally, since such vermin have no trace of originality and therefore can only copy what has gone before.... I am genuinely sorry that the expedition was a failure. I did sincerely hope you would solve the mystery of the shrimps. What I was razzing you about was that business of rushing into print and of getting revenge on the bureau of fisheries. That kind of attitude is really too childish. The business of a scientist is to find out the truth thoroughly and conscientiously and not to go about insulting anybody who happens to disagree with him. You were wrong to take the tone you did in your paper and you'll find you don't gain anything with that sort of behavior. If you think the bureau of fisheries is going to come crawling to your feet, you had better reflect that human beings do not behave in that way. They are far more apt to retaliate on you somehow or other. One of the rules of human conduct is that if you wish people to acknowledge your superiority you have to treat them as equals.

I am really ashamed that I haven't been to visit your mother but I could not seem to find an evening when I was neither sick nor occupied by necessary duties. Just now I am taking a drawing lesson once a week in addition to the other occupations but I fear I'm not capable of improvement. It would be such a tremendous help for me if I could draw decently but I just can't and that's all. I don't see how I can pay a good artist for all the drawings necessary for my book and I'll simply have to take an amateur.

I shall expect to see you soon and think you really ought to eat with me especially in view of the fact that I'm in no state to undertake a French dinner, with its various courses. A couple of eggs is more my style at present.

Sincerely, Libbie

Hyman obviously became an important mentor and advisor to Burkenroad. However, he never did quite completely take her advice that could have lead to a kinder and gentler approach to his colleagues. He in turn came to act as her confidant in regards her personal problems. An exchange of visits occurred with Burkenroad visiting the museum that spring. Hyman then headed for New Haven.

June 12, 1934 [postcard]
Dear Martin:
Expect to be in New Haven this week-end and hope to see you – will stay with Dorothea of course. I'm sorry, I completely forgot to retrieve your hat. I've been so busy recently in an attempt to make up for time lost during the winter because of illness that I don't go anywhere. Haven't seen any of the friends you recommended to me but will really try some day.

Regards, Libbie

And again the following week a followup missive hints at the legendary Burkenroad capacity for argument-for-argument's-sake and his early addiction to cigarettes.

June 20, 1934 [postcard]
Dear Martin:
Safely back on dear old Broadway. Sorry I did not get to say goodby to you; I was so badly sunburned that the climbing of the stairs to your perch would have been painful. The book we talked about is [J.H.] Woodger (not Woodridge), Biological Principles [1929]. I think you'll find it interesting. Sorry, your plant is not the tobacco used in making cigars, etc. That has small, rose-purple flowers. I thought it did but wasn't sure at the time of the argument. Your plant is the common garden nicotine – comes from Brazil, and so far as I know is never used for making tobacco. Sorry to disillusion you if you were counting on a carload of cigarettes from yours.
    Regards, Libbie

Burkenroad's curatorial duties in the Bingham Collection covered the entire spectrum of invertebrates. However, he also had at that time rather catholic research interests and did not come to focus his work routine on penaeid shrimps for a few more years. Until going to Yale, he exclusively worked in the realm of marine fisheries, first for the Carnegie Marine Biological Laboratory in the Tortugas and then for the Louisiana Department of Conservation. Hyman offered considerable help to Burkenroad in orienting him to the general invertebrate literature and indicating recognized authorities on various invertebrate groups, and she offered him guidance as to which authorities served best to contact for identifications of material. Nevertheless, even their purely business letters often contain very personal items. For example, a letter of Hyman's dated simply Nov. 10 (no year but apparently 1934) after an extensive litany of authorities and references concludes with the following.

I am trying to get used to new and very strong glasses for reading. I was getting to the point where I could not read the telephone book and often mistook the lettering on drawings. I suppose you will be running into New York now and then. I'll always be glad to see you if you haven't anything better to do.

3   LETTERS IN 1935 – BURKENROAD AND HYMAN'S INVERTEBRATE WORK

The start of 1935 found Burkenroad preparing for another cruise and Hyman thinking about going away to a marine biological station for the summer. She had already conceived the idea of a treatise on invertebrates. Although library work had the central role in that project, she also realized that she needed to study living animals as well. Where to do that best concerned her greatly. Burkenroad advised her to consider the Bermuda Biological Laboratory. Hyman wanted a place where she could have access to a wide variety of lower invertebrates as she was beginning to plan the first volumes of *The Invertebrates*.

March 5, 1935
Dear Martin:
I was very interested in the report about Bermuda. I had already understood that the fee for bed and board was fifteen a week but I did not know this also included space in the laboratory. If you would give me the address... I would like to write the director about spending July and August there. I am pretty certain to go, I think, as I am giving up the apartment anyway and so would have to move, and I might as well take the opportunity of going off when there's no rent to pay in N.Y. I'll have to find cheaper quarters in the fall. I'm rather fed up with N.Y., which I do not love as you do, and would appreciate a smaller place for a while. All islands are naturally damp

with so much water round about, you know, but I've heard that Bermuda is blistering hot in the summer and that would be splendid. Anyway, I don't know where else to go for invertebrates. I hate the idea of going to Woods Hole the Maine coast is cold and foggy, and at Beaufort the fauna is rather limited. Florida, perhaps – can you tell me anything about that? Why flaunt Tortugas in my face when you know I can't go there? [No woman was ever invited to the Carnegie Tortugas Laboratory (Colin 1980).] Any suggestions would be welcome. Of course there is a very rich fauna on the Pacific coast but it would be too expensive going there from here. I used to ride a bicycle, thirty or forty or fifty years ago, and I suppose I still can if necessary.

[Apparently, Burkenroad was considering paying page charges to speed up the publication of a note on a fresh-water medusa from Louisiana eventually published in 1936, *Science* 84:155-156.] Such an uproar was raised about the sale of space in Science for money that I doubt if Science will ever make that mistake again.... The affair was all the more notorious because the space was bought [page charges] by Princeton to enable Swingle [Wilbur W. Swingle, a professor there and an authority on endocrine physiology] to advance some ideas about adrenal function which so it turned out were not original ideas at all but were either swiped from predecessors or deliberately presented without any acknowledgement of the work of others.... Anyway aren't you counting your chickens even before the eggs are laid?

I have been much better this winter than last, thanks, but I can't say I am getting any more done in consequence. In fact, my progress is appallingly slow, chiefly I suppose because I put in too much time on research, if my small efforts are entitled to this name.

Your mother wrote me a card with her address and I expect to pay her a call whenever I can find an evening when I'm not too tired or too busy. I suppose she found New Haven too tame in your absence.

[not signed]

Another an undated letter of late spring or early summer followed.

Dear Martin:

I have not been able to make up my mind about Bermuda because I have had no reply to my letter of enquiry sent to Dr. Wheeler [J. F. G. Wheeler, director of the Bermuda Biological Station and a nemertine expert] about three weeks ago.... I think you must have been mistaken in telling me that the expenses at Bermuda were only fifteen a week. Others who have been there confirm my original idea that this covers only food and room and there is an additional charge of fifty dollars a month for the laboratory. This makes the place too expensive and I never really would have had an idea of going there if you hadn't told me otherwise.... I still don't fancy the New Haven idea simply because I haven't much faith in the fauna of the Sound. Anyway, Dorothea will be far nicer to have around than me.

I have a habit lately of clearing out all day Sunday to walk in the woods so if you're coming in on a Sunday it would be well to warn me in advance (so that I can be here, not so I can get away!) My digestion is better and I even took a chance with some strawberries which were strictly forbidden. However, I don't want to gamble with it very much – it's too painful.

I do wish you would drop the Dr. Hyman business – it's not consistent, anyway, with the general tone of your communications!

Libbie

Matters finally sorted themselves out and the following postcard posted on July 3, 1935 teased Burkenroad with his previous formality.

Dear Assistant Curator Burkenroad:

Believe it or not, I am leaving for Bermuda this afternoon on a boat. They decided not to charge me any laboratory fee.... I hope you have a nice summer in your cottage with Dorothea and Hortense [another Burkenroad friend]. Goodby [sic]; shall expect to see you in the Fall.

Doctor Hyman

Hyman's stay on Bermuda lasted about a month. Before departing she penned this hand written letter to New Haven. While in Bermuda, among other things, she agreed to work up the turbellarian flatworms of the island for the laboratory but expressed disappointment with the fauna as a whole. Her letter interests us for its insight into scientific life in the 1930's on what was a very isolated ocean island.

Aug. 8 [1935]
Dear Martin:
...It has been very pleasant here but the fauna is certainly disappointing. There is a good variety of sponges but very little else among the lower phyla....

I was in a diving helmet and found it an exciting experience but one of my ears suffered so much that I'd hesitate to go again. I'm afraid sea water was driven up my Eustachian tubes.

Wheeler is very nice but his wife, in plain american, is a 'pain in the neck.' She's an insufferable snob. ...

There is hardly anyone here. The scientific visitors consist of Buchsbaum and his wife [Ralph and Mildred, co-authors of Animals Without Backbones] from Chicago, Fries from C.C.N.Y. [Eric F.B., then asst. Prof., with interests in physiology of fish and distribution of chaetognaths], and Barnes [Thomas Cunliffe, an Asst. Professor at Yale from 1935-1940 and at that time an authority on insect behavior and crustacean nerve physiology] from Yale.... The Buchsbaums are very pleasant company and I tag along with them on various excursions. Barnes is a alcoholic as I suppose is well known at Yale and is always full of rum. He usually doesn't show any obvious effects but one evening he got thoroughly tanked up and went around to the rooms of each of the workers and put a big swastika on the wall. I suppose he thought it was funny but as three of the four are Jews we did not appreciate the humor. However, he also put one on his own wall.

After I leave here I'm expecting to go to Woods Hole for two or three weeks. On the whole the fauna there is certainly superior to that of Bermuda....

I hope you've had a pleasant summer in your cottage surrounded by two females. I'm sure I would only have disrupted the harmonies by my numerous peculiarities.

Best to the others and to you, Libbie

Hyman's visit to Bermuda began a research interest with turbellarians, particularly polyclads, found in the Sargassum. It thus turned out that Burkenroad's flippant suggestion to Hyman to go to Bermuda for the cheap rates fortuitously started Hyman along a major line of investigation, as this letter reveals.

Sept. 1st
Dear Martin:
Woods Hole always reduces me to a state of gloom so do not expect any bright remarks. Everyone seems to be here and if all of them really worked the journals would not be large enough to contain the results. Fortunately, most of them regard the place as a summer resort and I doubt if science will get much of a push forward as a consequence.

I hang out in the old main building where I fiddle a bit with such invertebrates as I can wheedle the supply department into getting for me. I must say they have not been overly energetic in the matter but then I probably have not been very energetic in prodding them into activity. What with dozens of people demanding sea-urchins, or Crepidula, or Chaetopterus, or some other unfortunate creature from which eggs can be extracted, one lone person who wishes unaccountably to look at a sponge or a hydroid or a polyclad receives scant attention. In fact, I doubt if the supply department understands the word polyclad.

Speaking of polyclads, you really must be reasonable. When I declined to undertake the Sargassum polyclads I honestly did not at the time entertain the idea of working on that group. Since then, however, I have decided I might as well take on all the Turbellaria, the more the merrier. It's a large order I fear, but no one else is doing anything with them.... The polyclads of Bermuda were so pretty that I could not resist collecting them and figuring out Verrill's mistakes ...[A. E. Verrill (1839-1926) of Yale and a pioneer invertebrate zoologist and oceanographer]. Wheeler is very anxious to have the fauna worked up taxonomically and undoubtedly a thorough knowledge of the fauna is highly desirable.... I bet Nigrelli [Ross F. Nigrelli, then a recent Ph.D. in 1936 from NYU, but eventually better known as a fish physiologist] has a job with those Sargassum polyclads you gave him if he wants to give them up I'll undertake them now but of course that is up to him....

I see the Atlantis is in dock here she looks queer with those excessively tall masts. Every time I see her I think of your remarks about the way she pitches and I hope I'll never set foot on her. Bigelow [Henry B. Bigelow, first director of the Woods Hole Oceanographic Institution] seems to object to females so that I'm probably in no danger.

I'm staying here until the 15th and as I suppose you are giving up your cottage about that time, I'm afraid I'll not be able to visit your domestic surroundings. But I should like to come to New Haven soon to see what I can wrangle out of Coe in the way of specimens and slides of nemerteans ™ consequently better be polite and invite me as convincingly as possible. When I get back to N.Y. I'll have to spend a couple of weeks looking around for an apartment – horrible prospect. The Etkins will probably pick one out for me in the end.

...I expect that I am cured for life of going down in diving helmets. My sufferings with that ear have been incalculable and I know it will cost me fifty dollars with a specialist to get the damage repaired. I should have been warned not to go swimming before making a descent.

I shall hope to see you soon, either in N.Y. or New Haven. Best to Dorothea if she is still on hand. You can reach me after I leave here c/o W. Etkin, 175 East 151st St.

Sincerely, Libbie

Then, six weeks later Hyman sent this postcard.

Oct. 19, 1935 [postmark]
Dear Martin:
Just to let you know I am settled (more or less) at 85 West 166th St., where I hope you will come to see me if it's not too much trouble. I am at the Museum during the day, of course mostly in the Laboratory of Exp. Biology. My place is best reached by the 6th-9th Ave. L to Jerome Anderson Ave. You walk one block west to Woodycrest Ave. and along Woodycrest to 166th St. I am very well pleased with my new place and the rent is much lower than I paid before. I am coming to N.H. some time or other and shall be glad to accept your hospitality.

Libbie

The order of correspondence becomes obscure at this point. Some disagreement developed over who could work up the Bermuda polyclads, Hyman or Nigrelli. Burkenroad drafted notes to A. E. Parr and C. M. Breder at Yale dated Aug. 27, 1936, in which he indicated he let Nigrelli know of Hyman's interest and left the matter up to him. The latter had a research associateship at the Bingham Collection and later became a noted authority on fish and curator and chairman of the Dept. of Fish and Aquatic Biology at the American Museum. Either before or after this, Hyman posted Burkenroad some undated letters with this and other scientific matters on which Burkenroad requested some information. One of these reveals Hyman scolding Burkenroad for his personal habits.

[undated]
Martin:
But I positively decline to take the polyclads away from Nigrelli; That would be bad manners.... [She went on to reiterate her feeling and tell Burkenroad she would only go ahead with the Sargassum polyclads she would be getting from Wheeler in Bermuda]...

I don't see how I am to invite you for a definite day since I haven't the least idea when you will be in N.Y. If you will write me the day before, I'll expect you to come home to eat with me when you do put in an appearance.

I decline to have anything further to do with your philanderings. You really offend me in your attitude and remarks about women. Miss Morton [a close personal friend of Hyman's] is sick at home with an infected foot.

Sincerely, Libbie

An exchange occurred over some use of the term *gonosomes* by Burkenroad in his paper in *Science* on fresh-water medusae. All through his life, Burkenroad loved to argue the meaning and derivation of words. In this he would delight in splitting hairs. He often did this quite spontaneously, i.e., without any reference to dictionaries or other sources relying only on his own background knowledge and innate aggressiveness. Burkenroad finally made an effort to document his position.

September 30, 1936
Dr. L. H. Hyman
85 West 166th Street
New York, N. Y.
Dear Libbie:
Gonosome seems widely used, not in the limited sense of gonad-bearing bodies attached to trophosomes, as I understood you to say, but in the sense of gonad-bearing bodies attached or free (Pratt [H. S. Pratt, 1935, A Manual of Common Invertebrate Animals Exclusive of Insects], p. 99: 'The *medusoid stage* or *gonosome* is either a free-swimming medusa or a sessile medusoid individual or *gonophore*...' Nutting [C. C. Nutting who authored a 3 part treatise on American Hydroids in 1900, 1904, & 1915], III, p. 93: 'Gonosome. Medusae with 4 radial canals...'). *Gonosome*, like *trophosome*, can certainly be used in the plural when referring to the gonad-bearing bodies of different species.... The question of usage in my note therefore seems whether the plural form *gonosomes* is properly applied to an aggregation of specifically identical medusae. Hincks [T. Hincks, 1868, A History of the British Hydroid Zoophytes], p. iv, says, 'Trophosome. The whole company of alimentary zooids associated in a hydroid community.' Therefore, the plural *trophosomes* can evidently be used for a group of distinct colonies of a single species; but in the earliest stages of these colonies, trophosomes would refer to individual zooids. Clearly, then, all the medusa *buds* of a single colony would before liberation be called collectively, *gonosome*; but when individualized by liberation, *edch* is a gonosome, and in the aggregate they are *gonosomes*.

I wanted much to see you but had to get back to New Haven Monday night; and had a great quantity of Lab business to attend. I hope to see you soon, however.

Best regards,

Hyman, however, would have none of what she considered Burkenroad's nonsense, and her level of exasperation with his behavior appeared clearly evident in her reply.

Oct. 4
Dear Martin:
I knew you'd go scurrying around to try to find some justification of your misuse of the word gonosome. I'm afraid your quotations are not very convincing; in fact they plainly indicate that the word is not used in the sense in which you employed it. Gonosome is not synonymous with medusa and is not used when speaking of medusae as such.... It wasn't very politic to pick Nutting for a quotation since Nutting very decidedly does not employ gonosome as equivalent to medusae. Pratt is about the worst 'authority' you could pick as far as I am concerned.

...I protest that the word can not be used as you employed it. ...just what is wrong with the word medusa that you have to hunt around for some other and more highfalutin word? Everybody else has talked about fresh-water medusae – why can't you? Imagine writing about the fresh-water gonosome! ...I confess that the word gonosome in your paper was so annoying to me and so distracting that I did not pay much attention to what else was said and I rather think other people would react in the same way. As we read, we have thought – why on earth does he call them gonosomes – why can't he say medusae?

...consult some specialist in the Hydroida.

You've been promising so many times to come to see me that I don't take much stock any longer in your assurances. Anyway, I'm in the Museum practically every day from 9.30 to 5.30 including Saturday. I'd like to have you eat with me in my apartment but in that case I wish you'd let me know the day before so that I can secure a fatted calf.

Sincerely, Libbie

4   LETTERS IN 1937 – BURKENROAD DEPRESSION AND PREPARATIONS FOR EUROPE

The correspondence seems to have stopped for awhile at that point. The following summer of 1937 found Hyman at the Mt. Desert Island Biological Laboratory at Salisbury Cove, Maine. She continued with her ongoing general invertebrate studies and spent most of her time concentrating on studying rotifers from the local ponds and lakes, the marine faunas seemed not particularly interesting to her – 'mostly nemertines and snails.' Socializing between the two of them still seemed to have had its ups and downs. However, this noteworthy letter documents that even at this early date Burkenroad suffered from the fits of depression that afflicted him so severely later in life. In this instance, despondency with his job and over lack of progress towards a degree may have triggered the attack.

Salisbury Cove
July 25th
Dear Martin:
You were a little late. Your letter was forwarded to me here. I passed through New Haven last Tuesday, the 20th, around 7 P.M. standard time. I looked around and didn't see you, so I continued on my way. Seriously, I had decided to wait until the return journey, assuming that I am going to visit you at all. If you really want me to stop off for a visit, I will do so on the way back to N.Y. That will be about the last week in August. I was very sorry I did not see you that day you were in the museum. Miss Morton and I had been talking about you that very day and wondering what had become of you. I was tied up with three out-of-town guests.... They were the Buchsbaums from Chicago.... It was quite an occasion but everything went well. They stayed for five days.

This place is really perfectly lovely, one of the most beautiful I've ever seen....

So I'm having a very pleasant time and hope you're not too bored at New Haven. You certainly sounded forlorn and neglected over the phone, for which reason I will see you in a few weeks.

Libbie

Aug. 20th
Dear Martin:
Just a line to let you know that I'm leaving Salisbury Cove on Monday the 23rd, according to present plans. The Etkins have come up here for a visit and I am being driven back with them, as far as Boston anyway. They are going back to Woods Hole but I don't know if I shall accompany them, to Woods Hole for a day or two. In either case I should be in New Haven in the middle of the week. I shall wire you when to expect me.

Dr. Dahlgren says he would be very glad to have the Sergestes. [Ulric Dahlgren of Princeton, an authority on bioluminescence whom she met at Mt. Desert Island.] He says he has always wanted to examine the light organs of this form but never was able to get any material before....

Well, I'll see you soon.

Best regards, Libbie

The visit acted as a tonic to Burkenroad. When he had his attacks, he retreated from social contacts, took to bed, and read fiction endlessly. Hyman swept in and appears to have reorganized his life. Her prescription directed him to get up and about and quit reading the trash novels he read so fondly under those circumstances.

IX/3/37
Dear Libbie,
Enclosed, what remarks I can find on Sargassum Turbellaria.... I won't vouch for the notes as they were made in haste and under bad conditions....

Your visit did me good; I made a resolve to exercise a least once a week and read science instead of other sorts of fiction. I even caught you some polyclads; note is enclosed.

Dr. Parr can't tell yet whether we have money for more outside jobs in the Bulletin, but he'd like very much to have you let us know when the Bermuda thing is done. ...

With my best,

Sept. 25th

Dear Martin:

I am very pleased to learn that I had a stimulating effect on you. I hope the stimulation doesn't die away with the distance like an axial gradient. I guess women in general have more conscience than men, for with the mountainous load of journals always on hand to be inspected, I don't see how anyone wanting to be a scientist can kill time with cheap fiction.

I'm very grateful for your bothering to look for polychaetes, I mean polyclads, must have annelids on the brain. .....But I don't have time for taxonomic work with this eternal book on my mind [she was in the midst of writing *The Invertebrates*, vol. 1]. However, I am at present writing up the Turbellaria sent me for identification by the US National Museum.... Every damn one except one land planarian from Java was a new species. ...The next thing I am going to tackle is that Bermuda collection. ...There are some new species but most are Verril's species that needed to be re-examined and placed in the correct genera. Needless to say he hit the wrong genus almost every time.

I was rather put out by Grace's [Grace Pickford, an authority on annelidans] refusal to identify that oligo-chaete from Mt. Desert that's being extensively used there for experiments and just had to be identified. I thought she should do it far better than I could since oligochaetes are her line. However, I took it on since she wouldn't and after a day's labor in the library, I ran it to its lair. It was sort of interesting.

...I should be very pleased to have the Sargassum polyclads. ...I was sent some 25 vials of land planarians by the Bishop Museum in Honolulu for identification. They came from those god-damn Pacific Islands each one of which seems to have its own little darling land planarians. The fact is, I don't like land planarians. They are a terrible mess, with some of the genera...having three hundred described species and God knows how many undescribed ones. And the identification depends mostly on color pattern – and see what you can do with that with some worm that's been pickled five to ten years in alcohol.

Excuse this Jeremiad. I enjoyed seeing you and staying in your attic. Do come to see me in N.Y.

Libbie

Burkenroad promptly replied and it reflects new found vigor. He started preparation for a long leave of absence to Europe, first to work in Copenhagen with a grant from the Danish govern-ment to study the crustaceans of the Schmidt Collection at the laboratory in Charottenlund and then to visit several European museums to examine penaeid types.

September 29, 1937

Dear Libbie:

Action at last on that material. .....Both lots will arrive at the A.M.N.H. in the same package, together with an item you left behind on your visit.

I expect to depart for Europe in December. University is taking care of me, so I expect to return here in a year. Ante-departure matters overwhelm; but I do hope to see you before leaving, here or in N. Y.

My very best,

Oct. 19

Dear Martin:

You did such a beautiful job of packing that I am sure it is superfluous for me to report that the box arrived in perfect condition....

Thanks for returning the piece of lingerie. I knew I had left it somewhere between Maine and N.Y. but could not recall where.

I thought I would write up the Bermuda polyclads and the Sargassum polyclads all in one article. Do you think the Bingham would be interested in publishing such an article? It would probably be of some length and with many line drawings. However, I feel I have no philosophical conclusions or applications of the data. In fact, there seems to be nothing of any value about the distribution of polyclads. What I mean to say, I guess, is whether you want me to offer it first to the Bingham; because I am sure the American Museum will publish anything for me.... I am limiting myself to Saturdays for taxonomic work; but even so I shall probably get the Bermuda stuff written up by spring, since it is all sectioned. It is the making of sections that takes the time and polyclads cannot be worked out taxonomically without serial sections.

I am glad you are staying on at the Bingham. I think you would be foolish to give up a position so suited to your tastes and abilities. I can't think of anything more empty in this world than running after money. I suppose we've all got to have a minimum of money but putting the making of money as the end and aim of experience is dreadful. I suppose by saying that the University is taking care of you, you mean they are increasing your salary. I'm very glad; I do think you ought to stick by the job.

I hope I'll see you before you go away. I'll probably be returning the collection soon.

Libbie

Hyman received on Dec. 9, 1936, (Judith Winston, personal communication) an official appointment at the American Museum in Noble's Laboratory of Experimental Biology. This occurred despite her continued dissatisfaction with his treatment of her. Her appointment may have occurred because of her increasing interest in experimental work with flatworms and in the extensive laboratory work necessary for preparing specimens for taxonomic identification. At any rate, Hyman wrote the following letter on AMNH stationary, the first in the archive on which she did so.

[?November, 1937]

Dear Martin:

Anticipating that you will depart for Denmark without kissing me goodby, I thought I'd better wind up the Sargassum business as far as I can. ...I am now returning three of the four jars you sent me.

I hope to see you before you leave but if I don't, bon voyage. I suppose you're sailing from N.Y. – if you tell me the name of the boat and the sailing time, I'll try to see you off. Miss Morton hopes you will come to see her also before you leave. There are rumors in the library that you have been sick – I'm very sorry, I hope you have recovered.

Sincerely, Libbie

The following handwritten draft by Martin replied to the above.

XI/26/37

Dear Libbie,

Haven't been sick, just rushed getting ready to go; library [in A.M.N.H.] should have known better since I have been borrowing many books not here [in New Haven] from them. I expect to leave as soon after the end of December as I can get a boat; probably you will be at AAAS meetings when I come to N.Y., but otherwise will surely see you; also Miss Morton to whom my best.

Thank you much for letting me have this information on the polyclads....

Remember that Parr would be glad to have your Bermuda paper submitted, though he can't yet say if money will be available.

Hope you are well. See you presently, ['Best regards' – crossed out] Affectionately,

The reply from Hyman contains a great deal of comment on polyclad matters, but it closes with another story concerning another incident in a long series of Hyman's ongoing problems with male supervisors. Burkenroad had apparently asked after the specimens of sergestids he had sent to Dahlgren.

Dec. 2, 1937
Dear Martin:

I thought you were leaving early in December, rather than after the end of December, hence my haste. I had no intentions of going to AAAS meetings at Indianapolis but now my boy Jack has got a research appointment in Detroit and there's a possibility that I will go and visit him at Christmas. However, I still hope to see you before you go and if you don't leave until after New Years I am sure I shall be back by then. But I may not go at all – it's quite a journey.

...Before I knew Dahlgren I had heard that he never answered letters but he has always been prompt with me. However, I am ready to believe it's a matter of sex. The old gink is very amorous and I have a hard time to stop him from kissing me when he sees me. Not that I lay that to my personal attractions for he is always kissing the other women of his acquaintance, married or not. He's not the first old duck I've met who thinks his age puts him above suspicions. I think he's really interested in your sergestids but it may take him ten years to do anything with them. He doesn't get much done I suspect. At present he is very busy with the fisheries exhibit at the New York Fair, of which he is in charge. He's going to get as much publicity as he can. He's got an inflated and impenetrable ego; nothing makes a dent in it. He's always telling me how wonderful he is in a variety of ways.... I feel like staying away from Mt. Desert permanently – he's too much of a dose for me.

I'm trying to make as much an impression on this museum as I can in order eventually to wangle a decent office out of them. So I think I'll give them the polyclad paper if they want it.

Hope to see you soon. Libbie

5 LETTERS IN 1938 – BURKENROAD IN EUROPE

Martin sailed for Europe in January – without a send off. He worked for a year in Denmark and then went on a tour of European museums, meeting in the process several notable figures (Burkenroad 1985, Burkenroad & Schram 1986) including among others Stephensen in Copenhagen, Buitendijk in Leiden, Calman and Gordon in London, Balss in Munich (Fig. 1). The visit resulted in one major paper (Burkenroad 1940) in which he gave 21 new species of penaeids 'preliminary descriptions.' This meant that he provided the bare minimum to establish the names. He intended to publish a more complete and detailed study later. The war intervened and afterward Burkenroad went on to shrimp fishery research; he never published the detailed study. Burkenroad always felt bad about this, and indeed the stimulus to his returning to the San Diego Museum in 1978 involved a try to relocate the manuscript, notes, and detailed drawings upon which he based the 1940 paper and publish them, but in this he failed.

April 1st, 1938
Dear Martin:

I honestly did intend to write to you soon after your departure but you know how it is. There's always so much work to do and I'm always so tired at the end of every day. Although I apparently give the impression of being in good health, I am never free of pain and I usually can't summon the extra energy needed for letter writing.

It did not surprise me that you took French leave without any farewell ceremonies. Miss Morton, however, seems very much put out and says that I should scold you for neglecting your friends like that. So consider yourself scolded....

I still have to put up with old Dahlgren at frequent intervals. The last time he insisted on taking me out to lunch, saying he had many important things to tell me. They boiled down to a tale about how his one remaining son (much addicted to the bottle, and not milk, either) had gotten married again (his first wife having died) without telling his papa first; and how important he is in connection with the fisheries exhibit of the N.Y. World Fair. The old duck is a trial to me but I don't know how to get rid of him without hurting his feelings. Now he wants me to go collecting with him in the wilds of N.J. By the way, I did remember to ask him about your deep-sea shrimps; he said he had done some work on them. But he does not think there are any light organs in the digestive tract. But he probably won't get much done until after this world fair excitement. [Burkenroad had an ongoing interest in bioluminescence and published a short paper on this in 1943.]

Figure 1. Upper left, Alida Buitendijk at her desk in the Rijksmuseum in Leiden. Upper right, Heinrich Balss, taking a coffee break in Munich. Lower left, Martin Burkenroad (center) with unidentified staff members in Charlottenlund, Denmark. Lower right, Knud Stephensen at his desk in the Zoological Museum in Copenhagen. All photographs taken by or for Martin Burkenroad in 1938-1939.

I hope it will make you violently jealous to learn that my boy Jack Greenberg is back in this vicinity again, practicing medicine with an older brother in a small nearby New Jersey town. He wasn't able to get any kind of appointment that suited him and therefore he was driven as a last resort into private practice. He seems to like it fairly well, although he doesn't get on too amicably with his brother. I see him quite often now although usually only for brief periods as he hasn't much leisure.

If you do any browsing through second hand book shops, I wish you'd look out for a copy of Lang's Polycladen Fauna und Flora des Golfes von Neapel, 1884. A catalogue from a firm in the Hague wants $30.00. It's worth all of that but naturally I'd like to pay less. If you run into a copy at less than this, grab it for me. Another book I'd love is Hanstrom's Vergleichende Anatomie des Nervensystems der Wirbellosen Tiere. It costs around $20 at current German prices (25% discount from the list price) but compared to American books, it is worth about ten and that's what I'd be willing to pay.

[unsigned]

And later in the spring Hyman wrote again. She was still in quest of the perfect invertebrate fauna.

[no date]

I am preparing, chiefly keeping an eye on my bank account, to spend a couple of months this summer at the Puget Sound Biological Station. The expense of getting back and forth is considerable but the costs at the station are low compared to other American marine laboratories. I was there several times fifteen to twenty years ago and liked the place very much but somehow haven't gone there in a number of years. The fauna is by far the best of any I've seen at American marine stations and after viewing the scanty supply seen at Mt. Desert Island (terribly exaggerated and lied about by Dahlgren) I wished I had gone there last summer. They have lots of shrimps there, by the way, but I don't think they are penaeids, not that I'd know the difference. ...I also hope to visit the Canadian station on Vancouver Island nearby. I have been told that the fauna is even richer there than at Friday Harbor. If faunas get progressively richer as the water gets colder, why is the fauna of the Gulf of Maine so poor? heaven knows the water is cold enough.

I'm expecting to visit Dorothea [then at the Storrs Agricultural Experimental Station] at her new location sometime this spring. There is a standing invitation as soon as the birdies begin to sing and the flowerets to peep forth in the woodlands (poetic again, eh? must be the season). ...I suppose they have spring in Denmark also. Are you going to get to see the tulips in Holland? That has been one of my dreams for years but maybe you don't care for flowers....

I'm glad you are having such a pleasant time – I rather thought you would. Denmark, Holland, and the Scandanavian countries remain about the only peaceful, pleasant lands in Europe. The goings on of the Nazis have made me ill and I hardly dare look at the newspaper any more. I feel as if each of us ought to do something rather than hide in a laboratory with a bottle of specimens but I can't think what I could do.

I hope this deserves a real letter in reply. Warmly, Libbie

However, Martin apparently posted only postcards and short notes from Europe. Hyman, nevertheless, continued sending him her news filled letters.

Aug. 2nd

Dear Martin:

I was pleased even to receive the scribble. I expect you are lonely in Europe. There is no wall quite like that of being encircled by a foreign language. I remember that when I came back to New York after more than a year abroad I was surprised to find everyone speaking English.

I arrived at Friday Harbor on the fourth of July after a pleasant journey across the continent broken by stops in Cleveland, Chicago, and Glacier Park. Since you have never been at the University of Chicago there would be little use in retailing my encounters there or elsewhere in Chicago. I was able to avoid encountering the remains of my family and was only the more thankful that I no longer live among them or have to see them. They are harmless, well-meaning folk but I have nothing in common with them nor do they have the least conception of me. My only pleasure in coming to Chicago is the contact with my beloved cousin Jack. It is really his mother who is my cousin... He is a doctor, an opthamologist, and we cling to each other as two persons on a wrecked rock, surrounded by an ocean of philistinic non-understanding relatives. We are so much of a kind and understanding each other so well, and all the others, his family and mine, fail so completely in sensitivity and understanding. It was a pain to me to abandon Jack so to speak when I left Chicago and a pain to him to have me go but I had to save myself from that life of continual vexation. He has built up a successful practice in two or three years and I think in the course of time will rank among the foremost oculists of the country.

Friday Harbor...consists of a group of buildings near the shore, three zoology buildings, and one each for botany, chemistry, and bacteriology. All aspects of oceanography are taught here.... We live in tents in the woods above the lab buildings. The tents have wooden floors and sides and canvas tops and are equipped with a cot and mattress; but one has to bring ones own bedding and linens. There is a general mess, where the food is well selected and cheap (board $5.75 per week). I think on the whole I like this place best of any marine station on our coasts. Unfortunately, the seashore climate has had the usual effect on me and I have been ill most of the time I have been here. For this reason I am leaving sooner than I expected and hope to recuperate in the dry climate of Wyoming before returning to N.Y. It is surely a curse when a zoologist cannot remain at the seashore. I have been so badly off all year with my nose and throat that I am gradually coming around to the hard decision that I must leave N.Y. and settle in a dry climate.

...I am glad you are getting so much material in Europe; soon you will be the ultimate authority on the penaeids. I doubt that anything could induce me to go to Germany. I have always disliked the Germans since I spent a summer in Germany in 1912. Everything I then thought of them has turned out to be only too true. Though I was only a young girl at the time I made a correct diagnosis of their character and nothing they have done since has surprised me in the least. Surely you will be glad to return to the peace and quite of the Danish coast, not to say the sanity of the Danish people.

You will perhaps be interested to know that Parr accepted without question or probably without even reading it my manuscript on the polyclads of Bermuda and the Sargassum. ...I have now begun to work up polyclad material I have been collecting from the New England coast. There is never any end to this taxonomic business; I have been accumulating polyclad material from the Pacific coast which I will have to work up some day and so it goes....

[unsigned]

6   LETTERS IN 1939 – BURKENROAD TRIES TO BRING HYMAN TO YALE

Burkenroad returned to America in 1939. Hyman's problems at the American Museum continued even with her formal appointment in 1936. Noble pressed her about the kind of research she did. She continued to consider the possibility of moving to some other institution and discussed the issue with Burkenroad. He believed that her unhappiness in New York presented a prime opportunity for Yale to lure her to New Haven. Hyman had some concerns, however, about getting any technical help to assist her.

II/8/39 [handwritten draft]

Dear Libbie,

Parr is quite taken with the idea of acquiring you. We have room to spare; and an incubator, and microtome, and a rather decrepit B&L monocular that you could furnish it with, and other equipment would I feel sure be forthcoming. But as for assistance of technical grade, I don't know, I don't think likely (I can't get it myself). A bursary (assisted) student, however, might be procured if you could train him for your needs. A sort of dollar-a-year job instead of an associateship might also be arranged I think. Parr's suggestion is, that if you think seriously of joining us, you can come up for a few months on trial (for you, not us: You might find the library and collections inadequate, etc.)

Hyman at first liked the idea of a move to Yale, but her answer betrayed the indecision she exhibited throughout this period about leaving New York. The letter details, however, the slights that women often had to put up with then in a clearly a male-dominated field. However, she may have imagined her problems with G.K. Noble. He obviously was not easy to get along with, and she believed it better to transfer to the Department of Invertebrates headed by R.W. Miner. However, Horace Stunkard, Hyman's obituarist, recalled Miner as refusing to allow either women or Jews in his department (Judith Winston, personal communication). Noble's behavior toward Hyman, which she outlines below, may reflect merely Noble's clumsy attempt to protect Hyman's feelings.

[undated]

Dear Martin:

You're really very sweet to me and I am profoundly grateful to you and Dr. Parr for your kind invitation. It is very pleasant to know that one would be welcome at New Haven. I fear, however, I am as bad as you about coming to a definite decision on any important matter. The fact is I like the American Museum very much but I decidedly do not like being in Dr. Noble's department. I doubt that I can put up with it much longer. I do hate a liar and a hypocrite. On the other hand, I can see his point of view, that he is giving me space in an experimental laboratory and I do not produce experimental work in return. What I really want is to change to Dr. Miner's department but I have never had the courage to tackle Dr. Miner. For one thing it is almost impossible to ever find him free. Then Dr. Noble told me that Miner did not want me, fearing I would outshine him. Recently, however, Dr. Noble changed his story saying Dr. Miner was offended because I had not come to his department instead of to Noble's. Now I do not know what to believe, probably neither story is the truth. I have never really tried to get what I want at the American Museum. Whatever place and position I have there was given to me by Dr. Noble on his own initiative. He invited me there in the first place with the idea that I would produce research in experimental biology and he had me appointed to a research associateship without me saying anything. I myself am hopelessly and impossibly diffident about approaching anybody for anything and asking for anything directly is beyond me. Not long ago, Noble had quite a talk with me on the question whether I wished to be in his department or Dr. Miner's. I said that I preferred morphological work and that I would like to transfer to Dr. Miner's department and that I hoped he would break the ice for me with Dr. Miner. But nothing further has been heard. It is as I said before impossible to know what Dr. Noble means by what he says. My idea was that I'd try to get one volume of my book published this year and then I'd go to Dr. Miner and talk to him on the basis that if he considered the project valuable and worth continuing, would he not give me some adequate space in his department and some help from his staff with drawing and photography. If he proves indifferent, then I'll really move to New Haven. I'll certainly make up my mind during this year and sometime this spring I'll visit you and inspect the possibilities. One big advantage removal to New Haven would bring is that at last I could have a house in the country. All my life I've wanted to live in the country and I only succeed in being shut up in cities. I really never was cut out by nature for a scientific career. I'm held up as an example to my sex of a successful woman and I'm the last person on earth to relish any such reputation. For years I saved my money to buy a place in the country and then I lost of it in the depression in so-called safe investments. However, I still have some, maybe enough and I realize that if I don't make a move out to the grass soon, all the grass I'll have will be a mound in the cemetery. If I had ever guessed in my youth just what a slave a scientist is, I'd probably taken up cookery or farming as a career. With the mountain load of journals and books always waiting to be read, one has hardly a moment to himself. For several years I've felt as if I weren't really alive. I often wish I had the courage to chuck the whole business. But of course I have no one to look after me and my living depends on my ability to produce books that sell. I feel as if this invertebrate treatise was a really important project of great value to teachers of zoology and that it is more worth while in the long run than any research I might produce in the same time. However, I shall always expect to carry along taxonomic and morphologic work on invertebrates. I don't seem to feel the same enthusiasm I once did for experiment. For years I have been sitting uneasily on the horns of several dilemmas and never able to decide what course to follow. There is that business of longing for the country. Yet how can I carry on my line of work in the country? Without a library I could not possible write the invertebrate treatise or continue taxonomic studies. The library at the American Museum is the best I've ever used and I hate to move myself away from it. I'm fairly well pleased in N.Y. and an ache that has gone on for as many years as mine for the country gets dulled in the course of time, so that you don't mind so much after ten or twenty or forty years. Then there's music and so on – the lack of that would be a real deprivation. Another dilemma that remains insoluble is that for the sake of my health I ought to have removed myself long ago to Arizona or New Mexico. But that brings the same problem again – how to carry on one's chosen work in such places, how to face the giving up of metropolitan culture and one's circle of friends. I don't have many friends but those few are very dear to me, especially since I have nothing else to take their place. The one solution for all my difficulties would seem to be to live in the country near some large city. The cities that have good enough libraries are very few – probably N.Y., Boston, Washington, possibly Philadelphia, Chicago. None of them has a climate suitable for me. I'd have to keep a car and whether my earnings could stand the expense of the constant journeys back and forth is a question. The only city that would afford a dry climate within driving distance is San Francisco. I haven't inspected the library at Berkeley but believe it is quite good. The climate of San Francisco Bay would probably kill me in six weeks but there is good dry country not too far from the city. Last summer when I was in such misery in Friday Harbor I had nearly made up my mind to move to the Sacramento Valley near San Francisco but since then I have been vacillating again and so it goes.

I guess this is quite enough of my affairs. I guess you have plenty of troubles of your own. Have you moved back to your attic? It was an awfully attractive place. I hope you succeed in keeping your mamma away – perhaps you should move into a tiny place where there would be no possible room for her! You've got too much inviting space in that loft.

Thanks again for everything. Next time you are in town, you must come home with me to supper.

With warmest thanks, Libbie

Hyman's indecision dragged on with no immediate resolution. The problem eventually disappeared through what can only be described as 'divine intervention.' Noble suddenly took ill with a severe streptococal infection the following year and died in December, 1940, at age 47 (for details of his career, see Mitman & Burkhardt 1991). A few years after that, Miner retired in 1943, and Hyman quietly shifted from Experimental Biology to the Department of Invertebrates. Meantime, Burkenroad apparently withdrew from social interactions again.

March 31, 1939
Dear Martin:
I can't say I've had any response to my recent numerous communications but this time I really need a reply. The expedition to Boston to the anatomy meetings is leaving N.Y. on the 5th and probably returning so as to bring me to New Haven late on the 8th or sometime on the 9th. I am wondering if you are going to be on hand as I recently had a visitation from [Waldo] Schmitt who told me you were due at the US National Museum in April. If you are planning to go soon don't stay in N.H. on my account as I can come visit you some other time. I'm afraid I can't stay but a day or two this time as there is always so much work on hand and I shall have been away several days as it is. I have birds and plants that need attention and I dislike to bother other people with them.

Please let me know if you're to be in N.H. and where to reach you with a telegram and where to find you on arrival.

I had a letter from Dorothea that she has a job at Wellesley, beginning next fall I suppose. Also wants me to visit her. And I have two other invitations besides hers. Apparently I should spend most of my time touring around from one friend to the next!

Affectionately, Libbie

Hyman visited Burkenroad in New Haven for two days that April.

[April 1939]
Dear Martin:
When I arrived home, the weather wasn't much of an improvement on what I had left behind in N.H. I found my furniture coated with dust and spent the rest of the day cleaning the apartment.

I hope you escaped the cold that seemed to be coming on when we parted. I felt very guilty about taking you out in that weather although really your own chivalry is responsible since I am quite accustomed to seeing myself off on trains without masculine assistance. In fact I have very seldom been treated in my life with the solicitude which you display. But I hope you are all right and got off to Washington as planned.

I am up to my ears in work as usual and feeling rather guilty over the long holiday. I must keep my nose to the grindstone or maybe the microtome for some time now and I think I'll have to cancel a projected visit to Dorothy in May. Since I just saw her at the meetings, there's no great reason for a visit, except to be in the country. ...

I was very pleased with the prospects at N.H. but it is not fair of you to go about saying that I'm coming. Be it distinctly understood that I have not even begun to make up my mind. I visited yesterday the place near Yonkers where I am contemplating buying a lot if I stay in N.Y. It's a lovely spot near the arboretum of the Boyce Thompson Plant Institute – this means there will never be any houses across the road for that is the permanent preserve of native woodland... I do wish I could make up my mind to something definite but you know well enough how difficult that is.

...I hope you have a pleasant and profitable time in Washington. Please stop in N.Y. on the way back and dine with me at my place. Thanks a lot for everything but you mustn't do it again, paying my way like that. [Burkenroad paid Hyman's bill at the Bishop Hotel in N.H. for a two day stay – $5.] Remember me to your mother whom I really will visit soon.

Love, Libbie

The rest of the year saw only an exchange of purely business letters. Concern for the events happening then in Europe started to occupy everyone's minds. In a letter to Hyman on September 20, just after the start of World War II, Burkenroad closed with: 'It sounds as if research on the other side would be badly hit. I feel awfully sorry for all those good people.'

Letters exchanged over Hyman's paper on Bermuda polyclads ('Acoel and polyclad Turbellaria from Bermuda and the Sargassum,' *Bull. Bingham Oceanogr. Coll.* 7:1-36) that had so long occupied her attentions. Burkenroad's health remained unstable, but by late 1939 he married for the first time as indicated in a letter from Hyman towards the close of the year, 'I am sorry you have been sick again; you seem to have a variety of afflictions....Thanks for various favors and remember me to your wife. Libbie'

7 LETTERS IN THE 1940's – BURKENROAD CONSIDERS A Ph.D. PROGRAM AT COLUMBIA

The issue of Hyman's moving from New York continued to drag on, and while she delayed a final resolution slowly things began to improve for her around the American Museum due undoubtedly in part to the publication of the first volume of *The Invertebrates* as well as some of the staff changes alluded to above.

January 30, 1940
Dear Libbie:
Many thanks for the fine separates. It always astonished me to see how much you get done...

...you might come up and pay us a visit... We are now living in our house in the country; very incompletely furnished but satisfactory heating and plumbing and lots of rooms, chairs, and beds. The weather has been unusually dry and very beautiful; if it lasts you might like a winter week-end.

With all the best,

Feb. 22, 1940
Dear Martin:
Am I to understand that you have bought or maybe rented a house in the country? That certainly must be very pleasant and satisfying and no doubt you spend most of your time fixing it up. Thanks for the invitation but I don't think I will do any travelling at present.

...I suppose you saw that the first volume of my work on invertebrates has come out. It came off the press on Feb. 9th. Everyone seems to like it and I believe it will be useful to teachers of zoology. I should like to give you a copy but I received very few and felt I should present them to friends who were actually teaching invertebrate zoology. If you wish to buy a copy, I would be glad to get one for you at 25% discount.

I at last made a genuine effort to transfer from Noble's to Miner's department but met with a cold shoulder. Dr. Miner said he did not have one inch of spare space, which of course I do not believe. It also came out that Dr. Noble had spoken to him about my transferring to his department a year ago and received the same reply. Since that time, however, Miner took somebody into his department and gave him the best room in the row on that side where you used to be. I had had my eye on that room. The man's name is Armstrong [J. C. Armstrong who worked on carideans of Bermuda and the Caribbean] and you must know him as he works on Crustacea. I talked to him and learned that he considers himself permanently established there. I haven't made up my mind to anything yet but don't believe I care to stay at the museum much longer under such circumstances. I'm getting pretty tired of the lack of space and of privacy....

  With best wishes, Libbie

The publication of volume one of *The Invertebrates* really perked up Hyman's spirits. Although her innate humility and self effacement still prevailed despite her success. The death of Noble at the end of 1940 relieved some of the pressure in Hyman's mind about staying at the Museum.

April 9 [probably sent in 1941]
Dear Martin:
I was sorry to hear that you have one of your respiratory infections again. You seem to be a frequent victim.

  My book has been received with the utmost enthusiasm everywhere; or perhaps it is that the dissenters have failed to write me. I was surprised that you found misspelled words as I have been through the book looking for errors and so far have failed to find any misspellings. As I said in the preface the work is a compilation; hence the opinions therein are mostly not my own, but reflect those of the leading students of the group. This answers the question about the priapuloids, sipunculoids, etc; they are separated by all current students of these groups and the old Gyphyrea has been eliminated. Aschelminthes is also widely accepted. All students of ctenophores are unanimous that platyctenids have no phylogenetic significance, etc. Ruebush [T. K. Ruebush, a hydrozoan authority associated with the Osborn Laboratory] did not like Aschelminthes either. But really I expected a great deal more criticism of my innovations (or what appear to be innovations to those who never read German) than I have so far received. It certainly was a pleasure to me to see the thing in print and even if I never get any more of it done that one volume should prove useful.

  ...Just now I am working on a revision of my vertebrate manual. I'm sorry I can't go ahead with the next part of the invertebrate treatise but that vertebrate manual is my bread and butter, not to say cake also, and it has been in need of a new edition for a very long time. I can't put off the unpleasant job any longer and so am deep in gill slits, etc. already. Most of our college texts on this subject certainly do stink, but I'm just not going to sit down and write a text on vertebrates. I never did like them anyway, and feel very annoyed that I have to spend time on them. I also have several taxonomic pieces going and some experiments on planarians that are bringing results.

  I am well enough as my health goes.

  Dorothea was here over the week-end, said she had seen you and you were thinking of abandoning your house and buying a neighborhood one that you liked better. It was very nice to see her and I had a gathering of old Chicagoans in my place in her honor. I suppose you know she is looking for a job; unfortunately I haven't heard of a single opening.

  Remember me to your wife.

  Affectionately, Libbie

At the end of 1941, the director of the American Museum, Roy Chapman Andrews, announced his retirement rather suddenly. Albert Parr of the Yale Peabody Museum replaced him, the same who with Burkenroad wanted Hyman to move to New Haven. Parr's departure from Yale, however, while good for Hyman, removed for Burkenroad his advisor and principal protector from the Yale staff. Burkenroad hoped to secure a full curatorship in the Peabody but still lacked the necessary Ph.D. degree. Informed that his promotion under those circumstances was out of the question, he once again decided to pursue a degree.

  However, instead of taking his degree at Yale, he chose for some reason to try and enter Columbia. Obviously, one of the major attractions to doing so involved working at the American Museum along side Hyman. The following handwritten draft exhibits heavy editing, and the parts Burkenroad wanted deleted from his final draft I place here in parentheses and italics as they perhaps more adequately reflect what he felt at the time. Oddly, when I asked Burkenroad about this incident he claimed that he remembered absolutely nothing of it and denied he ever considered going to Columbia.

IV/7/42

Dear Libbie,

You wouldn't come to Parr, so he came to you.

As you may have guessed, (*I don't get the*) after last year the curatorship is out of my reach (*because of last year. Instead of my feeling sad, however, I find myself terrifically bucked up. I can stay here*) I can stay of course, but (*I want to make a new start on what the past has taught and retrieve my errors. To slip Yale from my shoulders is like dropping a load*) but have the idea of trying to retrieve my errors instead; and in a way feel terrifically pleased to have the chance thrust at me.

(*What I want to do is to take a Ph.D. at Columbia*) Present line of enquiry is into feasibility of taking a degree at Columbia. Parr and [G. Evelynn] Hutchinson heartily approve. Can probably get a leave of absence from here; a hope – in the war-induced dearth of students and what with my many advantages if it is accepted that I am (*really*) finally ready to do my best to appear (*look glittering enough to get a good break*) as a promising student.

As reason for not taking the degree at Yale, I want to point out the value to me of facilities of A.M.N.H. Parr approves but would prefer at this time for me to arrange such facilities myself. He suggested application to you; (*though when*) I pointed out that you were not (*really*) in control...My demands for space would probably be (*more or less*) nominal only, since adequate facilities would no doubt be available at Columbia; but for several reasons, including the provision of formal explanation why I do not wish to take the degree at Yale, I would like to be able to say that I can make use of facilities at AMNH. As I mentioned, Parr allows me to say that he has no objection....

With the above letter to Hyman, Burkenroad posted the next day a short note to Prof. Dobzhansky at Columbia asking for an interview. Burkenroad made detailed notes of just what issues he wanted to bring up with Dobzhansky such as his desire for a Ph.D. to take of 'irregularities' in his academic background and correct deficiencies in his training, that the time to do this is now with the departure of Parr from Yale and the lack of a 'Ph.D. begins to become (a) serious block to advancement,' and that Columbia is the school of choice to do this because a natural historian such as himself could really profit from training in the 'new evolutionary views.' Burkenroad also hoped to launch himself into a serious career in museum systematics and teaching invertebrate zoology and hoped to get from Columbia permission to complete his work within two years if possible and accordingly sought a teaching fellowship.

Burkenroad did not just focus on Columbia, however, since he made notes as to other options he might pursue, such as attending the California Institute of Technology and the University of California at Berkeley, but rejected both these as not particularly appropriate for systematic training. He also had some hopes that he might yet stay at Yale and work under Hutchinson.

Hyman posted a reply to Burkenroad's note almost immediately. She repeated essentially the advice she had given him eight years before when he went to Yale.

[no date]

Dear Martin:

The decision to take a Ph.D. (providing you stick to it) is a sound move on your part. Because of the artificial value that has been placed on collegiate degrees it is very difficult to get any advanced position in academic circles without a Ph.D. degree. However, why Columbia? I do not know anyone there interested in invertebrate zoology. At our other university, N.Y. University, there is [Horace] Stunkard, who is a competent invertebratist but much disliked by everyone. I wonder if you are not looking around for some excuse to absent yourself from New Haven.... With Parr gone, do you think your position there is really secure? It might be well to bear that in mind.

The situation here is very unsettling at present. As you know Miner repeatedly refused to take me into the invertebrates department as a guest on the grounds that he has no space. However, no one believed this reason. ...I understand Miner is...slated to retire next year, having reached an age when according to a new ruling here, men will be forced to retire as an economy measure. I have heard that Miner proposes to fight his retirement but one of the best things Parr can do would be to get him out. He has no proper training as an invertebratist and has a general narrow and uncooperative attitude...

In the department of experimental biology (now called animal behavior), there has been a great improvement in every way since Noble died and Beach took over [Frank A. Beach, an authority on the effects of hormones on behavior]. ...At present time, there is really no good space available here for an additional person. ...Although Beach is always generous and obliging, in general he has wished to limit the laboratory to those working on experimental problems, mostly vertebrates. I am an exception of course; he has kept me on and given me every facility because I suppose he thinks I add to the prestige of the place.

...Now that gardening is feasible [Hyman finally fulfilled her long time dream and purchased some property in the country], I stay at home on Tuesdays and Thursdays, as well as Sundays. This is to say that I am at the museum only four days a week. But I am there on Fridays and Saturdays which would be the days you would be most likely to put in an appearance. Dorothea attended the anatomy meetings but unfortunately I did not see her. I did not go...as I was too tired after the final struggle with the vertebrate manual. I finished it somehow and sent it away to U. of C. Press about two weeks ago. I then fell into a state of being almost incapable of anything but am recovering now.

I was tremendously pleased to learn of Parr's appointment here. I wish you would tell him so for me. This museum certainly needs men like him of comprehensive and exact scientific training. It has had too much of men who were mostly self-taught naturalists and explorers [R.C. Andrews] I do not wish to belittle men of that type who have done such admirable work but they are often inadequate in present situations.

I shall look forward to seeing you soon in N.Y. or perhaps you would like to come to Millwood which is beginning to get pretty.

Affectionately, Libbie

IV/9/41
Dear Libbie,
Very many thanks for the information, which helps.

Does it seem foolish of me not to want further training as an invertebratist? ...Naturally, I do not think my future here secure with Parr gone; which is why I aim to take a degree. In this circumstance, N.Y.U, and Stunkard are of no value whatsoever to me. I want a prominent institution and a great man who can do well by me if I do well under him....

Burkenroad had an interview with Dobzhansky and Professor William King Gregory in regards his application and admission to Columbia. The issue of his lack of a B.S. degree came up but Gregory believed that entry to Columbia still could be granted since Burkenroad certainly had the 'equivalent education' to a bachelor's degree. Nevertheless, despite the general support of people at the American Museum, Gregory's feelings on the matter, and Dobzhansky's willingness to take Burkenroad on as a student, the Columbia admissions office had other ideas. Professor L.C. Dunn, head of the zoology department at Columbia finally wrote Burkenroad a letter on May 11, 1942 which summarized the decision of the admissions office. '...questions have been raised and I think are serious enough so that you might find it more economical to complete your work at Yale.' Among other things, Columbia noted his lack of any courses in Physics, only one course in Chemistry, no proof that he could use French or German, as well as lack of formal course work in histology, embryology, and genetics. They admitted that his published work indicated a superior ability for any graduate work, still they held that rules were rules and could not be broken – at least not in the numbers that would admit Burkenroad.

People at Yale had supported Burkenroad and his desire to work towards a Ph.D. The issue remained unclear as to just why he never did finish a doctorate from Yale. One of the problems always remained a lack of the right kind of course work. Burkenroad didn't want to suffer what he felt were pointless courses. After the Columbia debacle, several people at Yale suggested they confer at the very least a M.S. degree on him if he just would finish one complete course.

He apparently did so and he received the M.S. degree in 1943, and at that time, so he told me years later, the university gave him a permanent waiver of the time limit on writing and submitting a dissertation for a Ph.D., a waiver Burkenroad could have exercised any time up until the time he died.

After the affair with Columbia, Burkenroad's correspondence with Hyman broke off. His first marriage apparently ended sometime during this period as well, and after remarrying he entered another itinerant period. In 1945 he decided to leave Yale. He moved on to Morehead City as Chief Biologist with the North Carolina Survey of Marine Fisheries. In a few years, he and his new family relocated to the marine laboratory of the University of Texas in Port Aransas, spent six months in 1953 with Scripps Institution of Oceanography, and finally ended up in Panama working for the United Nations FAO (Schram 1986). The old friendship with Hyman faded away.

A last purely business exchange occurred in 1963. Burkenroad apparently enquired about where to send a manuscript for consideration in *Systematic Zoology*. Hyman's brief reply to the point had none of the gossipy background of earlier days. More tellingly, the signature that once flowed with a flourish had by then become stilted and shaky.

Aug. 4, 1963
Dear Martin:
The September issue of Systematic Zoology, now in press, will be the last under my editorship. The new editor is Dr. George Byers, Dept. of Entomology, University of Kansas, Lawrence, Kansas.
    Libbie

ACKNOWLEDGMENTS

Drs. John Harris, Joel Martin, and Edward Wilson of the Los Angeles Museum read the manuscript and offered comments. Dr. Judith Winston of the American Museum provided some background information concerning some of Hyman's history; and Dr. Frank Truesdale helped fill in some details about people mentioned in the letters. Ms. Carol Barsi of the San Diego Society of Natural History provided access to the Burkenroad archive and assisted in copying materials.

REFERENCES

Burkenroad, M.D. 1940. Preliminary description of twenty-one new species of pelagic Penaeidae from the Danish Oceanographical Expedition. *Ann. Mag. Nat. Hist.* (11)6:35-54.
Burkenroad, M.D. 1985. W.T. Calman. *J. Crust. Biol.* 5:362-363.
Burkenroad, M.D. & F.R. Schram 1986. Heinrich Balss (1886-1957). *J. Crust. Biol.* 6:300-301.
Colin, P.S. 1980. A brief history of the Tortugas Marine Laboratory and the Department of Marine Biology, Carnegie Institution of Washington. In *Oceanography, the Past* (M. Sears & D. Merriman, eds) pp. 1318-147. New York: Springer Verlag.
Mitman, G. & R.W. Burkhardt. 1991. Struggling for identity: The study of animal behavior in America, 1930-1945. In *The Expansion of American Biology* (K.R. Benson, J. Maienschein, & R. Rainger, eds), pp. 164-194. New Brunswick: Rutgers Univ. Press.
Schram, F.R. 1986. Martin David Burkenroad (20 March 1910-12 January 1986). *J. Crust. Biol.* 6:302-307.
Stunkard, H.W. 1970. Dr Libbie Henrietta Hyman. *Nature* 225:393-394.
Stunkard, H.W. 1974. In Memoriam, Libbie Henrietta Hyman, 1888-1969. In *Biology of the Turbellaria* (N. W. Riser & M.P. Morse, eds), pp. iv-xiii. New York: McGraw-Hill.

Georg Ossian Sars (1837-1927), the great carcinologist of Norway

Marit E. Christiansen
Zoological Museum, University of Oslo, Oslo, Norway

1 EARLY LIFE

Georg Ossian Sars was born 20 April 1837 in Kinn parish (now in Flora municipality) on the west coast of Norway where his father, Michael Sars, was the vicar. His mother, Maren Cathrine Sars, was a sister of the great Norwegian poet Johan S. Welhaven. The boy was usually called Ossian, and when he was 2 years old, the family moved to Manger parish on the coast not far from Bergen.

In addition to his work as vicar, Michael Sars was keenly interested in animal life in the sea and published over the years a number of seminal papers in marine zoology. In 1854, he was appointed extraordinary professor at what was at the time called Det Kongelige Norske Frederiks Universitet (the Norwegian Royal Frederiks University – now the University of Oslo), and the family moved to Christiania (now Oslo).

Ossian transferred from the Bergen Cathedral School, where he had been taking classes for 2 years, to Christiania Cathedral School. He completed the matriculation examination in 1857, and the second university examination in 1858. Even as a schoolboy, Ossian Sars was interested in natural history. Nordgaard (1918) gives a picture of him as a shy, solitary schoolboy, fond of wandering alone and filling his pockets with animals and plants. Ossian's greatest interest was zoology, but to secure his future he started to study medicine.

His first zoological studies were devoted to birds. With a gun lent to him by the zoologist Professor Halvor Rasch, he roamed the suburbs of Christiania, identifying and making colored drawings of the birds he shot. Ossian was a very gifted artist and his uncle, J.S. Welhaven, was of the opinion that Ossian ought to be sent to Düsseldorf, in Germany, to be trained as a painter. Ossian was also very interested in music, and he remarked on one occasion that if he had chosen to use his artistic talents, it would have been as a musician, rather than as a painter.

Ossian Sars' zoological interests gradually shifted from birds to whales, and then, in the middle to late 1850's, he started to study microscopic organisms. The event which led to this shift was his reading of the Swedish Professor W. Lilljeborg's book (1853) on *Cladocera, Ostracoda, and Copepoda of Skåne*. Sars had obtained a copy of this book at about the same time that he acquired his first microscope. Armed with a fine-meshed dipnet, bottles, jars, and vials for preservation of material, he collected in lakes and ponds around Christiania. He found a series of undescribed species that he proceeded to name. His first scientific publication: 'Om de i Omegnen af Christiania forekommende Cladocerer' (On the Cladocera occurring in the neighborhood of Christiania) was issued in *Forhandlinger i Videnskabs-Selskabet i Christiania Aar 1861*.

In 1862, he received the Royal Gold Medal as winner of a prize for young scientists, administered by the University, for a paper on freshwater crustaceans. He made drawings of the species and painted them to approximate their live appearance. The original handwritten manuscript of this paper and many of the original drawings are deposited at the Department of Manuscripts, the University of Oslo Library. The manuscript is about 340 pages long, of which the first 280 pages are concerned with Cladocera, 55 with copepods and 5 with ostracodes.

2 SARS' CRUSTACEAN INTERESTS MATURE

After Ossian Sars became a student, he accompanied his father on zoological excursions around the country, and his growing devotion to zoology made it clear that he wanted to be a professional zoologist. In the summer of 1862, he travelled with his father north through the inland valleys to Trøndelag County. Father and son collected freshwater and marine representatives of many animal groups during the trip. Ossian was, however, mostly interested in studying the crustaceans of fresh and shallow marine waters. On this trip, he also discovered that marine benthic copepods live as deep as 100 fathoms and even deeper. Up to then it was believed that these animals did not live deeper than algae do. In the largest lake in Norway, Mjøsen (Mjøsa), he found glacial relicts similar to those Sven Lovén had found in the two large Swedish lakes Vättern and Vänern a few years earlier. During the summer of 1863, Ossian Sars again went collecting, and he studied freshwater and marine crustaceans, this time in the lowlands and mountains of south and southeastern Norway including Christianiafjord (now Oslofjord).

Between 1863 and 1866 Sars published papers on copepods, isopods, cumaceans, and ostracodes in *Forhandlinger i Videnskabs-Selskabet i Christiania*. These papers contain the descriptions of many new species, even though much had already been published on these animals in Nordic oceans (e.g. by O.F. Müller, Kröyer, and Lilljeborg).

3 SARS AND FISHERIES BIOLOGY

Ossian Sars started his work in fisheries biology in 1864 when he became a fellow in fisheries studies. During the first winter in Lofoten, northern Norway, Sars made the seminal discovery that the cod has pelagic eggs. Later, he discovered that nearly all the important food fishes except herring have pelagic eggs. The commonly held opinion at that time was that fishes spawned on the bottom. It has been claimed that the discovery of the pelagic eggs of cod represents the start of modern fisheries biology. Sars' investigations in Lofoten continued through the winters until 1870. During this period he described the larval development of cod and followed the juveniles in their migrations. Sars continued his fisheries investigations until 1893, and expanded them to include all different species of commercially important fishes in Norway.

The fisheries investigations, however, made up only a small part of Sars' scientific work. Most of his travels around the country were made during vacations from the University, to which he had become attached as a fellow in 1870. During these travels, he collected crustaceans and other invertebrates in addition to information about fisheries.

During the fisheries investigations in Lofoten, Sars made a series of interesting discoveries. At that time, current zoological opinion held that animal life could not exist much deeper than 300 m because of the high pressure in deep waters. Michael and Ossian Sars demonstrated that animals were present in Norwegian fjords far deeper than that. Using a simple triangular dredge with a fine mesh bag, Ossian Sars collected masses of deep-sea crustaceans and other organisms. Among his publications of this time are descriptions of three new species of shrimps in *Forhandlinger i Videnskabs-Selskabet i Christiania Aar 1869*. On his first trip to Lofoten, Ossian discovered the remarkable stalked crinoid *Rhizocrinus lofotensis*, and he drew the illustrations for the memoir (M. Sars 1868) in which his father described it. The year after, he investigated the carcass of a huge rorqual which had been brought ashore. He described this in his first paper on whales published in 1866.

4 SARS' MIDDLE PERIOD

Michael Sars died in 1869, and over the next several years Ossian spent a lot of time completing and publishing several papers started by his father. At this time, Ossian Sars' fame as an outstanding carcinologist already was known outside Norway. He received cumaceans from the Swedish Arctic Expeditions 1861 and 1868 and from the expedition of the Swedish corvette Josephine to the Atlantic Ocean in 1869.

On 10 July 1874 Ossian Sars was appointed professor in zoology, replacing Professor Halvor Rasch, who had retired. Sars was at this time travelling in Finnmark, northern Norway, studying whales, part of his characteristic pattern of summer collecting trips. Sars rarely travelled abroad, but in 1875 to 1876 he went on a trip along the coast of the Mediterranean. He visited many places in Italy and also went to Malta and Tunisia. He collected various animals, especially molluscs and crustaceans. After returning to Norway, he and his fellow professors Sophus Lie and Jakob Worm Müller started a new scientific journal, *Archiv for Mathematik og Naturvidenskab*. Between 1877 and 1887 Sars published on new species of mysids, cumaceans, isopods, and ostracodes from the Mediterranean trip, in several articles in this journal. Sars was also interested in larval development of decapods, and described larval stages of species belonging to ten different genera in two papers (1884 and 1889) in the same journal.

In 1876 Sars participated in the planning and execution of another major scientific undertaking. With the meteorology professor H. Mohn he planned the Norwegian North Atlantic Expedition to the Norwegian Sea and Arctic Ocean. The field work was done in 1876 to 1878 with the naval vessel Vøringen. As a participant in the expedition Sars collected a large quantity of animals. He studied not only benthic fauna, but also planktonic animals that had not been studied before in a systematic way in these areas of the Atlantic. In his publications on crustaceans from the North Atlantic Expedition, Sars treated 337 species belonging to various classes and orders. Of these, 64 were described as new species, and 7 new genera were erected. (It is amazing that Sars at about the same time, 1878, published a 466-page book, with 52 plates, on Norwegian molluscs.) Several other scientists participated in the expedition, doing both hydrographic and biological studies. As a result of these cruises, the boundaries between southern warm-temperate fauna and northern Arctic fauna largely were identified.

Except for one publication on Norwegian freshwater crustaceans that was published in French in 1867, all Sars' papers up to 1872 were in Norwegian with diagnoses in Latin. From the late 1880s, however, most of his papers were published in English. This was probably connected with his admiration of Darwin's revolutionary contribution to biology and of English zoological science on the whole. In the preface of his first article in English from 1872, he wrote,

'I have preferred the English language, as well because it has most affinity with our own, and consequently affords greater facility for rendering the Norwegian expressions, as in acknowledgement of the great progress which zoological science has made in recent time, through the medium of English language.'

Sars, through his lectures at the University, was the first to introduce Norwegian students to Darwinian evolution. For the students, these lectures were fascinating, even if they provoked nasty comments in the community at large. At the Department of Manuscripts, The University of Oslo Library, there is a letter from Charles Darwin to Sars dated 29 April 1877 in which Darwin thanks Sars for the memoir on *Brisinga* (1875). In this paper, Sars gives a description of this asteroid genus completely in accordance with Darwinian principles. Darwin writes,

'Allow me to thank you much for your kindness in having sent me your beautiful memoir on Brisinga. It contains discussions on several subjects about which I feel much interest. I congratulate you on your discovery of the new proofs of Autography which promises to be of much service to those who like yourself are good draftsmen. With the most sincere admiration for your varied works in *Science*, I remain,....'

Between his fathers death in 1869 and until the book on Norwegian molluscs was published in 1878, Sars gained an exceptional knowledge of Norwegian marine fauna, partly through the publication of several papers based on his father's manuscripts, partly through his collection trips to many of the fjords along the coast. He published articles on marine hydroids, annelids, molluscs, pycnogonids, echinoderms, bryozoans, fishes and fisheries, and whales, and also on a few faunal surveys that included a number of animal groups.

However, Sars did not neglect the crustaceans during the decade after his fathers death. He published several papers on Mysidacea and Cumacea from Norwegian waters and the Mediterranean, and on Cumacea from the Arctic, the West Indies, the South Atlantic, and the Mediterranean. In 1877, preliminary descriptions of Crustacea and Pycnogonida from the first year (1876) of the Norwegian North-Atlantic Expedition appeared.

After Sars had finished his big mollusc book in 1878, he gave more and more of his attention to crustaceans. During his travels throughout Norway he had studied both freshwater and marine crustaceans, but except for the trip to the Mediterranean in 1875 to 1876 and the Norwegian North Atlantic Expedition in 1876 to 1878, he did not travel outside Norway as a scientist. How could he then become such a well-known, worldwide specialist on crustaceans? Nordgård's explanation (1927) was that Ossian Sars, together with his father, described peculiar forms of Norwegian deep-sea fauna, and thus attracted the attention of the scientific world. The publications on deep-sea species were also a source of inspiration to the many great scientific ocean expeditions. G.O. Sars published many papers on Crustacea from some of these expeditions, including the Challenger Expedition and the expeditions of Prince Albert I of Monaco. His last expedition publication (1924, 1925) was a 408-paged work, with 127 plates, on bathypelagic copepods from Prince Albert's expeditions.

Sars did not only work with marine crustaceans, but also treated prodigious materials from several continental lakes and from fresh and saline waters in Central Asia. Between 1893 and 1914 he published 12 papers on crustaceans from the Caspian Sea, based on material collected by several scientists. In his first paper on Amphipoda (1894) from the Caspian Sea, Sars wrote: 'By the investigations of Dr. Grimm and Mr. Warpachowsky, a rather extensive material has now been brought together, the examination of which shows indeed the Amphipodous Fauna of that isolated basin to be both rich and diversified.' In the 1907 publication on Mysidae from the Caspian Sea, Sars wrote:

In the year 1904 the Russian government has instituted a very careful and extensive exploration of the Kaspian Sea in order to acquire a fuller knowledge of the fauna of that isolated basin and the physical and biological conditions relating to the fisheries there occurring... The conductor of the expedition was the distinguished Russian naturalist, Prof. N. Knipovitsch... The expedition was in work from March 12 [February 28] to June 19 [June 6], and during that time traversed the whole of the Kaspian Sea... Of the rich zoological collections thereby acquired, those relating to the Crustacea have been placed in my hands for examination. It is indeed quite a colossal material, embracing, as it does, several hundreds of bottles of different size, some of them litterally filled up with specimens.

In the *Festschrift für Prof. N. M. Knipowitsch* (1927), Sars made remarks about the general character of the crustacean fauna of the Caspian Sea and the origin of the fauna.

Sars also got the opportunity to examine crustacean material from freshwater basins on a completely different continent. In 1904-1905 Dr. W.A. Cunnington collected plankton and benthos in the African lakes Victoria Nyanza, Nyasa, and Tanganyika, and Sars published articles on copepods, ostracodes, and some larval stages of prawns from this expedition in *Proceedings of the Zoological Society of London* between 1909 and 1912. Of special interest is

Figure 1. Georg Ossian Sars in 1871. (Copy of a photograph by F. Klem; deposited in the University of Oslo Library).

Figure 2. Georg Ossian Sars. (Copy of a photograph by L. Forbech; deposited in the University of Oslo Library).

Figure 3. (Left) Georg Ossian Sars in his office at the University. (Copy of a photograph donated to the Zoological Museum, Copenhagen, in 1925 by Professor Hjalmar Broch, University of Oslo). (Right) Georg Ossian Sars on his way from the University, with the Royal Palace in the background. (Copy of a photograph taken by Professor Carl Størmer between 1893 and 1897; deposited at the Biological Station, Drøbak).

Figure 4. Georg Ossian Sars in his office at the University. (Copy of a photograph donated to the Zoological Museum, Copenhagen, in 1925 by Professor Hjalmar Broch, University of Oslo).

the description of the fauna of Lake Tanganyika, which like Lake Baikal is an isolated fresh-water basin. No cladocerans were found in Tanganyika, but of the copepods Sars described, several new species were regarded as endemic. He also described two copepod genera that he assumed to be of marine origin.

5 SARS' WORK ON CLADOCERA AND OTHER FORMS

His worldwide studies on crustaceans were expanded further when Sars, in the first part of the 1880's, got the idea to place dried mud from Australia, South Africa, and South America in freshwater aquaria in his laboratory at the University. In this way, Sars was given the opportunity to study living crustaceans from other continents without having to travel himself. His first paper (1885) on Crustacea raised from dried mud was on Australian Cladocera. In the introduction Sars wrote:

Last winter I received, thanks to the kindness of the Norwegian traveller, Mr. Lumholtz, a considerable quantity of dried mud from a fresh-water lake in the tropical part of Australia. When requesting Mr. Lumholtz to forward me such material, it had been my intention to institute, during the spring and summer of the following year, a series of experiments, with a view to obtain the ova of Entomostraca, probably enclosed within the mud, artificially hatched, and thus become enabled to examine some of the entomostracous forms occurring in that remote tract of the globe... According to the statements of Mr. Lumholtz, the mud was taken from a rather large lake, called Gracemere Lagoon, situated at a distance of about 7 miles west of Rockhampton in North Queensland.

Sars described five new species in this first paper on material from Australia, and he named one *Daphnia lumholtzii* after the collector of the mud. Later, he published 14 articles on Branchipoda, Copepoda, and Ostracoda from several localities in Australia. Sometimes he also received preserved material, and names like A. Archer, A.M. Lea, O.A. Sayce, I. Searle, and J. and Th. Whitelegge are mentioned as collectors of dried mud or animals. Sars also got dried material and preserved specimens from the Norwegian biologist Knut Dahl, who had been travelling for 2 years in the northern and western parts of Australia. Dried mud from New Zealand was sent to him by G.M. Thomson, and resulted in a publication (1894) on Cladocera, Copepoda, and Ostracoda.

In 1895, Sars published his first paper on South African Cladocera, Ostracoda, and Copepoda raised from dried mud collected from a swamp at Knysna, some distance from the Cape of Good Hope. Throughout the years, he published ten more papers from South Africa, the last one on Copepoda in the year before he died. Dried mud and preserved specimens were sent to him by Mr. Theson, Mr. Hodgson, and Dr. F. Purcell, and he also received preserved specimens from the South African Museum.

From H. von Ihering, Director of the Museu Paulista in Sao Paulo, Brazil, Sars received both dried mud and preserved samples, and dried mud was also sent to him from Argentina by a Norwegian gentleman, a Mr. Schiander. These materials resulted in four publications between 1900 and 1902 on Cladocera, Copepoda, and Ostracoda. In the publication from 1901 on Cladocera from South America, Sars wrote:

Among the several species derived from these two regions of South America, there is a considerable number which have proved to be common to both the new and old world, some even being identical with species found as far north as Norway... Indeed, as will be shown in this paper, the geographical distribution of some of the species appears quite perplexing, and can scarcely be accounted for without the assumption of a rather different relation between the great continents and oceans in ancient times.

In part two of the paper on freshwater Entomostraca of South America (1901), Sars wrote concerning Copepoda:

... as usual, only Calanoid forms were obtained, not a single Cyclopoid or Harpactoid, though both these groups are undoubtedly well represented in South America, as in all other parts of the world. Indeed, though I have made these experiments for a long series of years, I have never succeeded, in obtaining any form of these two groups by artificial hatching, apparently owing to the circumstance that, as far as its yet known, no resting ova are produced by these forms, whereas probably all the fresh-water Calanoids at times produce such ova.

Sars also treated species from material (dried or preserved) collected in Algeria (1896), India (1900), China and Sumatra (1903), South Georgia (1909), and Malaya (1929).

6   THE CRUSTACEA OF NORWAY

The publication Sars himself considered as his magnum opus, *An Account of the Crustacea of Norway*, is still one of the high points in Norwegian zoology. The first of the nine volumes is on Amphipoda. It was published between 1890 and 1895 and consists of 32 parts.

After the volume on Amphipoda was finished, Sars intended to publish an account of Norwegian Cladocera. In a letter of the 24 August 1895 (deposited in the Department of Manuscripts, the University of Oslo Library and addressed to 'Direktionen for Christiania Videnskabs-Selskab' – now the Norwegian Academy of Science and Letters of which Sars had been a member since 1865), he asked whether the Academy could pay for the publication of a work on Norwegian Cladocera that he was working on at the time. He mentioned the importance of publishing such a work in connection with the biological research on freshwater fishes, which had just begun in accordance with a recent parliamentary decision, because cladocerans constitute the main component of plankton in lakes and ponds and are food for a large part of the freshwater fishes. Sars intended to publish the work in Norwegian with text and figures of all species (also the varieties), and he planned to publish it in 10 parts over approximately 2 years. The work was, however, never published. In the book *Det Norske Videnskaps-Akademi i Oslo 1857-1957* (Amundsen 1957), one can read about he Academy's serious financial problems in 1895 to 1896, and further that the Academy had to refuse the continuation of G.O. Sars' important work on Norwegian crustaceans. Since most of the text and drawings on Cladocera in Sars' Royal Gold Medal work from 1862 have never been published (Frey 1982), with the help of the late Professor David G. Frey, at Indiana University, we will now publish the illustrations and manuscript pages of cladocerans (with the Norwegian text translated into English).

The second volume of *An Account of the Crustacea of Norway* is on Isopoda, the third on Cumacea; then, come four volumes on Copepoda. The last volume, which is on Ostracoda, appeared between 1922 and 1928. The title page, index, and the two last parts were issued the year after Sars' death. Some crustacean groups on which he had already published, were omitted from consideration, but even so, it is nearly impossible to understand how a single person could master the systematics of all the different groups and prepare the text and brilliant illustrations of all the various species. His intensity and hard work is attested by his own remark, overheard once by his colleague Professor Kristine Bonnevie, and cited by her at Sars' 50-year anniversary as a professor in 1924, 'When one gets to be as old as I am, one does not have time to rest any longer' (Bonnevie 1924).

A good deal of the material described by G.O. Sars is deposited in the collections of the

Zoological Museum, University of Oslo. There are, however, problems associated with the material. Sars was not a museum zoologist; he did not seem to have been interested in the technical side of museum collections. Data associated with the specimens is sparse; often only a rough location identifier accompanies his specimens, and even this may be missing. Thus, working with Sars' collections can be very frustrating, both for scientists studying the collections and for the managers of the collections. This is of particular concern because of the enormous number of new species that he described. Sars himself did not designate type specimens. Although he sometimes described new species based on a single specimen, we have occasionally found several specimens in samples from type localities. Since no date is given on the labels, it is difficult to tell whether the holotype is put together with specimens that he collected later at the type locality or whether the holotype is missing. For some species, there are several samples from the type locality without any dates. For other species, there are samples (especially slides) without any more information than the name of the species. Sars' diaries and many of the notebooks in which he made notes about his collecting trips are deposited in the University Library in Oslo. Thus, it might be possible to solve some of the problems related to his material by studying these books.

Sometimes Sars sent material, in exchange, to colleagues in other countries. Type material of species described by Sars are, for example, deposited in the British Museum (Natural History); and the South African Museum has material described by Sars from South Africa. In the G.S. Brady ostracode collection at the Hancock Museum, Newcastle-upon-Tyne, there are probably many species sent by Sars to Brady of which a number are not represented in the Zoological Museum in Oslo. In the Zoological Museum in Oslo, there is material which A.M. Norman collected in the Hardangerfjord, Norway, and which he later sent to Sars.

7 SARS' LATER YEARS

Ossian Sars never married. His mother, Maren, gave birth to 14 children, but 5 died young (between birth and 13 years). Ossian was fourth. His elder brother, Johan Ernst, became a famous history professor, and his youngest sister, Eva Helene, married to Fridtjof Nansen.

The home of the Sars family in Christiania was very special. The family was extremely hospitable, and as long as Maren Sars lived, the home was a meeting place of artistic, literary, and musical people. Ossian's mother was an excellent storyteller and his brother Ernst, a brilliant causeur. Ossian himself was not very active in the entertainment, but contributed with his lovable nature to heighten the atmosphere. The family also had a distinct musical talent, and two of Ossians' sisters, Maria Cathrine (Mally) Lammers and Eva Nansen, became nationally known singers. Ossian played flute in his youth. Later he got a hardanger fiddle, playing it entirely by ear and taking it on many of his zoological travels. In his diaries, one can read that he often went to concerts in the evenings.

Both Ernst (also unmarried) and Ossian lived with their mother until she died in 1898. Later they lived with their sister Mally and her husband, the musician Thorvald Lammers.

Sars retired from the University in 1918, but because of unusual services to his country and science, he was permitted to retain both his salary and laboratory at the University. His handwriting became increasingly illegible in his later years, but his illustrations demonstrate that his hand remained as steady as ever.

In a newspaper article from 1924 in *Aftenposten* in connection with Sars' 50-year anniversary as professor, the journalist who interviewed Sars in his office wrote:

For the real citizens of Kristiania, Ossian Sars is one of the figures which belongs to the city's physiognomy. The citizens know the upright, thin man who every morning at about nine o'clock for many years passed from the railway station to the University to work with his microscope and his manuscript until dinner time, and then just as regular returned to his home at Bestum to sit at his desk and work with another microscope and write another monograph (Anon. 1924).

Since Sars never travelled outside Scandinavia, except for the trip to the Mediterranean, he was personally unknown to most of his foreign colleagues. His publications, however, and many letters deposited at the Department of Manuscripts, the University of Oslo Library, show that he constantly was in contact with colleagues all over the world right up to a short time before he died on 9 April 1927, after a brief illness and less than 2 weeks away from his 90th birthday.

Georg Ossian Sars was a member or honorary member of learned societies and institutions in many countries. He was honored by the King of Sweden as 'Knight of the Order of St. Olav' in 1892 for his scientific contributions; and in 1911 he was advanced to 'Knight-Commander' of the same order.

A characterization by Nordgaard (1918) of Michael and Ossian Sars is as valid today as it was then: 'Michael and Ossian Sars are the two highest alpine peaks (Glitretind and Gald-høpiggen) of Norwegian zoological science.'

REFERENCES

Most of the information on G.O. Sars is from sources listed here; only some of these are cited in the text. Complete citations to papers by G.O. Sars mentioned in the text are in a complete bibliography of Sars that follows this paper.

Amundsen, L. 1957. *Det Norske Videnskaps-Akademi i Oslo 1857-1957*. Oslo: H. Aschehoug & Co.
Anonymous 1897. The great carcinologist of Norway. *Nat. Sci. Lond.* 10: 156-157.
Anonymous 1924. Ossian Sars – professor i 50 aar. Et jubilaums-interview. *Aftenposten* no. 391. 10 July 1924.
Anonymous 1928. Georg Ossian Sars. *Proc. R. Soc. Edinb.* 47: 366-368.
Bonnevie, K. 1924. Professor Georg Ossian Sars 1874-1924. *Aftenposten* no. 391, 10 July 1924.
Brinkmann, A. 1927. Georg Ossian Sars. In memoriam. *Naturen* 1927: 129-132.
Broch, H. 1954. *Zoologiens historie til annen verdenskrig*. Oslo: Akademisk forlag.
Bull, F. 1963. *Tradisjoner og minner*. Oslo: Gyldendal.
Christiansen, M.E. 1989. Georg Ossian Sars (1837-1927). Biografi på høytidsdagen 26. februar 1988. *K. norske Vidensk. Selsk. Forh.* 1988: 29-33.
Frey, D.G. 1982. G.O. Sars and the Norwegian Cladocera: a continuing frustration. *Hydrobiologia* 96: 267-293.
Liljeborg, W. 1853. *De crustaceis ex ordinibus tribus: Cladocera, Ostracoda et Copepoda, in Scania Ocurrentibus*. Lund: Berlingska.
Nordgaard, O. 1917. Georg Ossian Sars. 80 aar. *Naturen* 1917: 132-134.
Nordgaard, O. 1918. *Michael og Ossian Sars*. Kristiania: Steenske forlag.
Nordgård, O. 1927. Georg Ossian Sars. 1837-20/4-1927. *Nyt Mag. Naturvid.* 65:1-18.
Nordgård, O. 1928. Minnetale over professor G.O. Sars. *Årbok norske VidenskAkad.* 1927: 78-84.
Odhner, T. 1917. 80 år. G.O. Sars. *Aftenposten* no. 195. 20 April 1917.
Økland, F. 1955. *Michael Sars*. Et minneskrift. Oslo. Det Norske Vidensks-Akadem: Oslo.
Sars, M. 1868. *Mémoires pour servir à la connaissance des crinoïdes vivantsD. Programme de l'Université royale de Norvège. Christiania: Brøgger & Christie*.
Stephensen, K. 1927. Georg Ossian Sars. Født 20. april 1837, død 9. april 1927. *Naturens Verd.* 11: 289-292.

APPENDIX: A BIBLIOGRAPHY OF GEORG OSSIAN SARS

A complete list of G.O. Sars' publications has never been published. Based on a handwritten list by G.O.Sars (deposited in the Department of Manuscripts, the University, of Oslo Library), the Librarian Sverre Løkken at the Biology Library, University of Oslo has compiled this bibliography. The greatest problem in compiling Sars' publications is to cite the correct year for many of his papers. Sometimes, a separate was published and distributed 1 or 2 years before the article was published in a journal. The title of the separate and the title in the journal are also sometimes slightly different. This occurred, for instance, with his very first publication, cited below. None of the other separates are included in this bibliography. It cannot be guaranteed that all the publication years given below are correct.

1861   Om de i omegnen af Christiania forekommende cladocerer. *Forh. VidenskSelsk. Krist.* 1861 (special separate): 1-25.

1862   Hr. studios. medic. G.O. Sars meddeelte en af talrige afbildninger ledsaget oversigt af de af ham i omegnen af Christiania iagttagne Crustacea cladocera. *Forh. VidenskSelsk. Krist.* 1861: 144-167. (See 1861)

Hr. studios. medic. G.O. Sars fortsatte sit foredrag over de af ham i omegnen af Christiania iagttagne Crustacea cladocera. *Forh. VidenskSelsk. Krist.* 1861: 250-302.

1863   Stud. med. G.O. Sars gav en af talrige afbildninger ledsaget oversigt af de indenlandske ferskvands-copepoder. *Forh. VidenskSelsk. Krist.* 1862: 212-262.

Beretning om en i sommeren 1862 foretagen zoologisk reise i Christianias og Trondhjems stifter. *Nyt Mag. Naturvid.* 12: 193-252.

1864   Hr. studios. medic. G.O. Sars holdt et foredrag om en anomal gruppe af isopoder. *Forh. VidenskSelsk. Krist.* 1863: 205-221.

Beretning om en i sommeren 1863 foretagen zoologisk reise i Christiania stift. *Nyt Mag. Naturvid.* 13: 225-260.

1865   Hr. cand. phil. G.O. Sars foredrog bemarkninger om den aberrante krebsdyrgruppe Cumacea og dens nordiske arter. *Forh. VidenskSelsk. Krist.* 1864: 128-208.

Norges ferskvandskrebsdyr. Første afsnit. Branchiopoda. I. Cladocera Ctenopoda (fam. Sididae & Holopedidae). Med Hs. Majt. Kongens guldmedaille prisbelønnet afhandling. Udgivet som Universitetsprogram for 1ste halvaar 1863 ved dr. M. Sars. Brøgger & Christie's bogtrykkeri, Christiania (Oslo). VIII + 71 pp., pls. 1-4.

1866   Hr. cand. phil. G.O. Sars meddeelte følgende oversigt af Norges marine ostracoder. *Forh. VidenskSelsk. Krist.* 1865: 1-130.

Hr. cand. phil. G.O. Sars holdt følgende foredrag: Om vintertorskens (*Gadus morrhua*) forplantning og udvikling. *Forh. VidenskSelsk. Krist.* 1865: 237-249.

Hr. cand. phil. G.O. Sars holdt følgende foredrag: Beskrivelse af en ved Lofoten indbjerget rørhval (*Balaenoptera musculus* Companyo). *Forh. VidenskSelsk. Krist.* 1865: 266-295, pls. 1-3.

Untersuchungen der norwegischen Hochlandsseen und Küsten auf Crustaceen. *KorrespBl. zool. - min. Ver. Regensburg* 20: 147-160, 167-172. (Translation by Dr. Haupt, of Sars, G.O., 1863)

1867   *Histoire naturelle des crustacés d'eau douce de Norvège. 1e livraison. Les Malacostracés.* Chr. Johnsen, Christiania (Oslo). IV + 146 pp., pls. 1-10.

1868   *Indberetning (til Indredepartementet) om de af ham i aarene 1866 og 1867 anstillede undersøgelser over skreiens eller vintertorskens yngel.* Christiania (Oslo). 8 pp. (Also printed in *Morgenbladet* 1867 no. 342 and in *Aftenbladet* 1867 no. 289 and 291)

Beretning om en i sommeren 1865 foretagen zoologisk reise ved kysterne af Christianias og Christiansands stifter. *Nyt Mag. Naturvid.* 15: 84-128. (special separate 1866)

1869   Om individuelle variationer hos rørhvalerne og de deraf betingede uligheder i den ydre og indre bygning. *Forh. VidenskSelsk. Krist.* 1868: 31-74.

Undersøgelser over Christianiafjordens dybvandsfauna, anstillede paa en i sommeren 1868 foretagen zoologisk reise. *Nyt Mag. Naturvid.* 16: 305-362.

Indberetninger til Departementet for det Indre fra cand. G.O. Sars om de af ham i aarene 1864-69 anstillede praktisk-videnskabelige undersøgelser angaaende torskefiskeriet i Lofoten. Det Steenske bogtrykkeri. Christiania (Oslo). 61 pp.

Om de nyeste undersøgelser af dyrelivets udbredning i havets dybder. *For Idé og Virkelighed* 1869: 397-402.

1870   Nye dybvandscrustaceer fra Lofoten. *Forh. VidenskSelsk. Krist.* 1869: 147-174.

Carcinologiske bidrag til Norges fauna. I. Monographi over de ved Norges kyster forekommende mysider. Første hefte. Brøgger & Christie's bogtrykkeri, Christiania (Oslo). 64 pp., pls. 1-5.

1871   *Indberetning til Departementet for det Indre fra cand. G.O. Sars om de af ham i sommeren 1870 anstillede fortsatte undersøgelser over torskefiskeriet ved Lofoten.* Ringvolds bogtrykkeri, Christiania (Oslo). 27 pp. (Also printed in *Morgenbladet* no. 59 1870)

Nya arter af Cumacea, samlade under K. svenska korvetten Josephines expedition i Atlantiska Oceanen år 1869 af F.A. Smitt och A. Ljungman. Utdrag ur en större afhandling. *Övers. K. VetenskAkad. Förh.* 28: 71-81.

Beskrivelse af de paa fregatten Josephines expedition fundne cumaceer. *K. svenska VetenskAkad. Handl.* 9(13): 1-57, pls. 1-20.

1872   Nye echinodermer fra den norske kyst. *Forh. VidenskSelsk. Krist.* 1871: 1-31.

Undersøgelser over Hardangerfjordens fauna. *Forh. VidenskSelsk. Krist.* 1871: 246-286.

Diagnoser af nye annelider fra Christianiafjorden, efter professor M. Sars' efterladte manuskripter. *Forh. VidenskSelsk. Krist.* 1871: 406-417.

Cumaceer fra de store dybder i Nordishavet, insamlede ved de svenske arktiske expeditioner aarene 1861 og 1868. *Övers. K. VetenskAkad. Förh.* 28: 797-802.

Beskrivelse af fire vestindiske cumaceer opdagede af dr. A. Goës. *Övers. K. VetenskAkad. Förh.* 28: 803-811.

On *some remarkable forms of animal life from the great deeps off the Norwegian coast. I. Partly from posthumous manuscripts of the late professor dr. Michael Sars.* University-program for the 1st half-year 1869. Printed by Brøgger & Christie, Christiania (Oslo). VIII + 82 pp., pls. 1-6.

Carcinologiske bidrag til Norges fauna. I. Monographi over de ved Norges kyster forekommende mysider. Andet hefte. Brøgger & Christie's bogtrykkeri, Christiania (Oslo). 34 pp., pls. 6-8.

1873   Bidrag til kundskaben om dyrelivet paa vore havbanker. *Forh. VidenskSelsk. Krist.* 1872: 73-119.

Bidrag til kundskab om Christianiafjordens fauna. III. Annelida. Vasentlig efter prof. M. Sars' efterladte manuskripter. *Nyt Mag. Naturvid.* 19: 201-281, pls. 14-18.

Beskrivelse af syv nye cumaceer fra Vestindien og det Syd-Atlantiske Ocean. *K. svenska VetenskAkad. Handl.* 11(5): 1-30, pls. 1-6.

Om cumaceer fra de store dybder i Nordishafvet. *K. svenska VetenskAkad. Handl.* 11(6): 1-12, pls. 1-4.

1874   Om en dimorph udvikling samt generationsvexel hos *Leptodora. Forh. VidenskSelsk. Krist.* 1873: 1-15, pl. 1.

Bemarkninger om de til Norges fauna hørende phyllopoder. *Forh. VidenskSelsk. Krist.* 1873: 86-90.

Bidrag til kundskaben om Norges hydroider. *Forh. VidenskSelsk. Krist.* 1873: 91-150, pls. 2-5.

Om en hidtil lidet kjendt markelig slagtstype af polyzoer. *Forh. VidenskSelsk. Krist.* 1873: 386-400, pls. 8-9.

Indberetning til Departementet for det Indre fra cand. G.O. Sars om de af ham i aarene 1870-73 anstillede praktisk-videnskabelige undersøgelser angaaende torskefiskeriet i Lofoten. Det Steenske bogtrykkeri, Christiania (Oslo). 61 pp., 2 maps.

Mohn, H. og G.O. Sars. (*Om vigtigheden af en undersøgelsesexpedition til havet udenfor Norges vestkyst*). Til den kongelige norske regjerings Departement for det Indre. Werners bogtrykkeri, Christiania (Oslo). 9 pp.

1875   Om hummerens postembryonale udvikling. *Forh. VidenskSelsk. Krist.* 1874: 1-27, pls. 1-2.

Om 'blaahvalen' (*Balaenoptera sibbaldii*, Gray) med bemarkninger om nogle andre ved Finmarkens kyster forekommende havdyr. *Forh. VidenskSelsk. Krist.* 1874: 227-241, 1 pl.

Om autografiens praktiske anvendelse i zoologien, samt om en ny autografisk methode. *Morgenbladet* (special separate), Christiania (Oslo). 14 pp.

On the practical application of autography in zoology, and on a new autographic method. Christiania (Oslo). 8 pp., pls. 13-14.

On some remarkable forms of animal life from the great deeps off the Norwegian coast. II. Researches on the structure and affinity of the genus Brisinga, based on the study of a new species: Brisinga coronata. University-program for the last half-year 1875. Aschehoug, Christiania (Oslo). IV + 112 pp., pls. 1-7.

1876   Vorläufige Berichte über die von der (norwegischen atlantischen) Expedition gewonnene Resultate. *Circ. dt. Fisch.-Verein* 1876(4): 147-152.

1877   Nye bidrag til kundskaben om Middelhavets invertebratfauna. I. Middelhavets mysider. *Arch. Math. Naturv.* 2: 10-119, pls. 1-36.

Prodromus descriptionis crustaceorum et pycnogonidarum, quae in expeditione Norvegica anno 1876, observavit. *Arch. Math. Naturv.* 2: 237-271.

Fischerei-Beobachtungen in der Nord-See 1876. *Circ. dt. Fisch.-Verein* 1877(2): 25-30.

Den norske Nordhavsexpedition i 1876. *Tidsskrift for populare Fremstillinger af Naturvidenskaben* R. 5, 4: 227-235.

1878   *Indberetninger til Departementet for det Indre fra professor, dr. G.O. Sars om de af ham i aarene 1874-1877 anstillede undersøgelser vedkommende saltvandsfiskerierne.* Bergh & Ellefsens bogtrykkeri, Christiania (Oslo). 59 pp.

Bidrag til kundskaben om Norges arktiske fauna. I. Mollusca regionis arcticae Norvegiae. Oversigt over de i Norges arktiske region forekommende bløddyr. Universitetsprogram for første halvaar 1878. Aschehoug, Christiania (Oslo). XV + 466 pp., 52 pls., 1 map.

Nye bidrag til kundskaben om Middelhavets invertebratfauna. II. Middelhavets cumaceer. *Arch. Math. Naturv.* 3: 461-512 b, pls. 1-18.

1879   Bidrag til en nøiere characteristik af vore bardehvaler. *Forh. VidenskSelsk. Krist.* 1878(15): 1-20, pls. 1-4.

Indberetninger til Departementet for det Indre fra professor, dr. G.O. Sars om de af ham i aarene 1864-1878 anstillede undersøgelser angaaende saltvandsfiskerierne. Bergh & Ellefsens bogtrykkeri, Christiania (Oslo). 221 pp.

Indberetning til Departementet for det Indre fra professor, dr. G.O. Sars om de af ham i vaaren 1879 anstillede praktisk-videnskabelige undersøgelser over loddefisket ved Finmarken. Bergh & Ellefsens bogtrykkeri, Christiania (Oslo). 32 pp. (Also printed in *Morgenbladet* no. 275 and 282 1879).

Reports made to the Department of the Interior of investigations of the salt-water fisheries of Norway during the years 1874-1877. *Rep. US Comm. Fish.* 1877, appendix A: 663-705. (Translation of Sars, G.O., 1878.)

Carcinologiske bidrag til Norges fauna. I. Monographi over de ved Norges kyster forekommende mysider. Tredie hefte. Universitetets-program for første halvaar 1880. Aschehoug, Christiania (Oslo). 131 pp., pls. 9-42.

Nogle bemarkninger om den marine faunas character ved Norges nordlige kyster. *Tromsø Mus. Aarsh.* 2: 58-64, 2 maps.

Nye bidrag til kundskaben om Middelhavets invertebratfauna. II. Middelhavets cumaceer. *Arch. Math. Naturv.* 4: 1-144, pls. 19-60.

1880   Indberetning om Selskabets zoologiske samlings tilstand. *K. norske Vidensk. Selsk. Skr.* 1879: 140-143.

Crustacea et Pycnogonida nova in itinere 2do et 3tio expeditionis Norvegicae anno 1877 & 78 collecta. (Prodromus descriptionis). *Arch. Math. Naturv.* 4: 427-476.

1881   Fortsatte bidrag til kundskaben om vore bardehvaler. 'Finhvalen' og 'knølhvalen'. *Forh. VidenskSelsk. Krist.* 1880(12): 1-20, pls. 1-3.

1882   Revision af gruppen: Isopoda chelifera med characteristik af nye herhen hørende arter slagter og. *Arch. Math. Naturv.* 7: 1-54. (special separates 1880 and 1881)

1883   Oversigt af Norges crustaceer med foreløbige bemarkninger over de nye eller mindre bekjendte arter. I. (Podophtalmata – Cumacea - Isopoda - Amphipoda). *Forh. VidenskSelsk. Krist.* 1882(18): 1-124, pls. 1-6.

1884   Preliminary notices on the Schizopoda of H.M.S. 'Challenger' expedition. *Forh. VidenskSelsk. Krist.* 1883(7): 1-43.

Bidrag til kundskaben om decapodernes forvandlinger. I. *Nephrops - Calocaris - Gebia. Arch. Math. Naturv.* 9: 155-204, pls. 1-7.

1885   On some Australian Cladocera, raised from dried mud. *Forh. VidenskSelsk. Krist.* 1885(8): 1-46, pls. 1-8.

Zoology. Crustacea, I. *The Norwegian North-Atlantic Expedition 1876-1878* 6(14): 1-280, pls. 1-21, 1 map. Printed by Grøndahl & søn, Christiania (Oslo). Also with Norwegian text: Zoologi. Crustacea, I. *Den Norske Norhavs-Expedition 1876-1878* 6(14): 1-280, pls. 1-21, 1 kart.

Note sur les crustacés schizopodes de l'estuaire de la Seine, suivie de la description d'une espèce nouvelle de *Mysis, Mysis kervillei*, G.O. Sars. *Bull. Soc. Amis Sci. nat. Rouen* 1885: 87-99, pl. 5.

Report on the Schizopoda. *Report on the scientific results of the voyage of H.M.S. Challenger during the years 1873-76.* Zoology. 13. Part 37: 1-228, pls. 1-38. Longmans & Co., London.

1886   Nye bidrag til kundskaben om Middelhavets invertebratfauna. III. Middelhavets saxisopoder (Isopoda chelifera). *Arch. Math. Naturv.* 11: 263-368, pls. 1-15.

Zoology. Crustacea, II. *The Norwegian North-Atlantic Expedition 1876-1878* 6(15): 1-96, 1 map. Printed by Grøndahl & søn, Christiania (Oslo). Also with Norwegian text: Zoologi. Crustacea, II. *Den Norske Nordhavs-Expedition 1876-1878* 6(15): 1-96, 1 kart.

1887   Report on the Cumacea. *Report on the scientific results of the voyage of H.M.S. Challenger during the years 1873-76.* Zoology. 19. Part 55: 1-78, pls. 1-11. Eyre & Spottiswoode, London.

Report on the Phyllocarida. *Report on the scientific results of the voyage of H.M.S. Challenger during the years 1873-76.* Zoology. 19. Part 56: 1-38, pls. 1-3. Eyre & Spottiswoode, London.

On *Cyclestheria hislopi* (Baird), a new generic type of bivalve Phyllopoda; raised from dried Australian mud. *Forh. VidenskSelsk. Krist.* 1887: 1-65, pls. 1-8.

Nye bidrag til kundskaben om Middelhavets invertebratfauna. IV. Ostracoda mediterranea. (Sydeuropaiske ostracoder). *Arch. Math. Naturv.* 12: 173-324, pls. 1-20.

1888   Pycnogonidea borealia & arctica. (Prodromus descriptionis). *Arch. Math. Naturv.* 12: 339-356.

Additional notes on Australian Cladocera, raised from dried mud. *Forh. VidenskSelsk. Krist.* 1888(7): 1-74, pls. 1-6.

Indberetning til Departementet for det Indre fra professor dr. G.O. Sars om de af ham i sommeren 1888 anstillede praktisk-videnskabelige undersøgelser vedkommende saltvands- fiskerierne samt angaaende hvalfredningen i Finmarken. Tønsbergs bogtrykkeri, Christiania (Oslo). 19 pp.

1889   On some freshwater Ostracoda and Copepoda, raised from dried Australian mud. *Forh. VidenskSelsk. Krist.* 1889(8): 1-79, pls. 1-8.

Bidrag til kundskaben om decapodernes forvandlinger. II. *Lithodes - Eupagurus - Spiropagurus - Galathodes - Galathea - Munida - Porcellana - (Nephrops).* *Arch. Math. Naturv.* 13: 133-201, pls. 1-7.

On a small collection of freshwater Entomostraca from Sydney. *Forh. VidenskSelsk. Krist.* 1889(9): 1-9.

1890   Oversigt af Norges crustaceer, med foreløbige bemarkninger over de nye eller mindre bekjendte arter. II. (Branchiopoda - Ostracoda - Cirripedia). *Forh. VidenskSelsk. Krist.* 1890(1): 1-80.

Amphipoda. Part I. Hyperiidea. *An Account of the Crustacea of Norway, with short descriptions and figures of all the species* 1: 1-20, pls. 1-8. Cammermeyer, Christiania [Oslo].

Amphipoda. Part II. Orchestiidae and Lysianassidae (part). *An Account of the Crustacea of Norway, with short descriptions and figures of all the species* 1: 21-44, pls. 9-16. Cammermeyer, Christiania [Oslo].

Amphipoda. Part III. Lysianassidae (continued). *An Account of the Crustacea of Norway, with short descriptions and figures of all the species* 1: 45-68, pls. 17-24. Cammermeyer, Christiania [Oslo].

Bidrag til kundskaben om decapodernes forvandlinger. III. Fam. Crangonidae. *Arch. Math. Naturv.* 14.: 132-195, pls. 1-6.

1891   Zoology. Pycnogonidea. *The Norwegian North-Atlantic Expedition 1876-1878* 6(20): 1-163, pls. 1-15, 1 map. Sampson, Low, Marston, Searle & Rivington, London. Also with Norwegian text: Zoologi. Pycnogonidea. *Den Norske Nordhavs-Expedition 1876-1878* 6(20): 1-163, pls. 1-15, 1 kart. Aschehoug, Christiania [Oslo].

Amphipoda. Part IV. Lysianassidae (continued). *An Account of the Crustacea of Norway, with short descriptions and figures of all the species* 1: 69-92, pls. 25-32. Cammermeyer, Christiania [Oslo].

Praktisk-videnskabelige undersøgelser af Trondhjemsfjorden. (Indberetning til Departementet for det Indre om en i sommeren 1890 foretagen reise). Tønsbergs bogtrykkeri, Christiania [Oslo]. 15 pp.

Amphipoda. Part V. Lysianassidae (concluded). *An Account of the Crustacea of Norway, with short descriptions and figures of all the species* 1: 93-120, pls. 33-40. Cammermeyer, Christiania [Oslo].

Amphipoda. Part VI. Pontoporeiidae. *An Account of the Crustacea of Norway, with short descriptions and figures of all the species* 1: 121-140, pls. 41-48. Cammermeyer, Christiania [Oslo].

Amphipoda. Part VII. Phoxocephalidae. *An Account of the Crustacea of Norway, with short descriptions and figures of all the species* 1: 141-160, pls. 49-56. Cammermeyer, Christiania [Oslo].

Amphipoda. Part VIII. Ampeliscidae (part). *An Account of the Crustacea of Norway, with short descriptions and figures of all the species* 1: 161-184, pls. 57-64. Cammermeyer, Christiania [Oslo].
Amphipoda. Part IX. Ampeliscidae (concluded), Stegocephalidae. *An Account of the Crustacea of Norway, with short descriptions and figures of all the species* 1: 185-212, pls. 65-72. Cammermeyer, Christiania [Oslo].

1892   Amphipoda. Part X. Amphilochidae, Stenothoidae (part). *An Account of the Crustacea of Norway, with short descriptions and figures of all the species* 1: 213-236, pls. 73-80. Cammermeyer, Christiania [Oslo].
Fortsatte praktisk-videnskabelige undersøgelser af Trondhjemsfjorden. (Indberetning til Departementet for det Indre om en i sommeren 1891 foretagen reise). Tønsbergs bogtrykkeri, Christiania [Oslo]. 12 pp.
Amphipoda. Part XI. Stenothoidae (continued). *An Account of the Crustacea of Norway, with short descriptions and figures of all the species* 1: 237-252, pls. 81-88. Cammermeyer, Christiania [Oslo].
Amphipoda. Part XII. Stenothoidae (continued). *An Account of the Crustacea of Norway, with short descriptions and figures of all the species* 1: 253-272, pls. 89-96. Cammermeyer, Christiania [Oslo].
Amphipoda. Part XIII. Stenothoidae (concluded), Leucothoidae, Oediceridae (part). *An Account of the Crustacea of Norway, with short descriptions and figures of all the species* 1: 273-296, pls. 97-104. Cammermeyer, Christiania [Oslo].
Amphipoda. Part XIV. Oediceridae (continued). *An Account of the Crustacea of Norway, with short descriptions and figures of all the species* 1: 297-320, pls. 105-112. Cammermeyer, Christiania [Oslo].
Amphipoda. Part XV. Oediceridae (concluded). *An Account of the Crustacea of Norway, with short descriptions and figures of all the species* 1: 321-340, pls. 113-120. Cammermeyer, Christiania [Oslo].

1893   Amphipoda. Part XVI. Paramphitoidae, Epimeridae (part). *An Account of the Crustacea of Norway, with short descriptions and figures of all the species* 1: 341-364, pls. 121-128. Cammermeyer, Christiania [Oslo].
Amphipoda. Part XVII. Epimeridae (concluded), Syrrhoidae (part). *An Account of the Crustacea of Norway, with short descriptions and figures of all the species* 1: 365-388, pls. 129-136. Cammermeyer, Christiania [Oslo].
Amphipoda. Part XVIII. Syrrhoidae (concluded), Pardaliscidae (part). *An Account of the Crustacea of Norway, with short descriptions and figures of all the species* 1: 389-412, pls. 137-144. Cammermeyer, Christiania [Oslo].
Amphipoda. Part XIX. Pardaliscidae (concluded), Eusiridae. *An Account of the Crustacea of Norway, with short descriptions and figures of all the species* 1: 413-432, pls. 145-152. Cammermeyer, Christiania [Oslo].
Amphipoda. Part XX. Calliopiidae (part). *An Account of the Crustacea of Norway, with short descriptions and figures of all the species* 1: 433-452, pls. 153-160. Cammermeyer, Christiania [Oslo].
Amphipoda. Part XXI. Calliopiidae (concluded), Atylidae. *An Account of the Crustacea of Norway, with short descriptions and figures of all the species* 1: 453-472, pls. 161-168. Cammermeyer, Christiania [Oslo].
Om brislingen og brislingfisket. Indberetning til Departementet for det Indre fra professor dr. G.O. Sars. Tønsbergs bogtrykkeri, Christiania [Oslo]. 15 pp.
Crustacea Caspia. Contributions to the knowledge of the carcinological fauna of the Caspian Sea. Part I. Mysidae. *Izv. imp. Akad. Nauk* 13: 399-422, pls. 1-8.

1894   Amphipoda. Part XXII. Gammaridae (part). *An Account of the Crustacea of Norway, with short descriptions and figures of all the species* 1: 473-500, pls. 169-176. Cammermeyer, Christiania [Oslo].
Amphipoda. Part XXIII. Gammaridae (continued). *An Account of the Crustacea of Norway, with short descriptions and figures of all the species* 1: 501-516, pls. 177-181. Cammermeyer, Christiania [Oslo].
Amphipoda. Part XXIV. Gammaridae (concluded), Photidae (part). *An Account of the Crustacea of Norway, with short descriptions and figures of all the species* 1: 517-540, pls. 182-192. Cammermeyer, Christiania [Oslo].

Amphipoda. Parts XXV & XXVI. Photidae (concluded), Podoceridae (part). *An Account of the Crustacea of Norway, with short descriptions and figures of all the species* 1: 541-588, pls. 193-208. Cammermeyer, Christiania [Oslo].

Crustacea Caspia. Contributions to the knowledge of the carcinological fauna of the Caspian Sea. Part II. Cumacea. *Izv. imp. Akad. Nauk* 13: 461-502, pls. 1-12.

Amphipoda. Parts XXVII & XXVIII. Podoceridae (concluded), Corophiidae, Cheluridae. *An Account of the Crustacea of Norway, with short descriptions and figures of all the species* 1: 589-628, pls. 209-224. Cammermeyer, Christiania [Oslo].

Contributions to the knowledge of the fresh-water Entomostraca of New Zealand as shown by artificial hatching from dried mud. *Skr. VidenskSelsk. Christiania, Mat.-naturv. Kl.* 1894(5): 1-62, pls. 1-8.

Amphipoda. Parts XXIX & XXX. Dulichiidae, Caprellidae, Cyamidae. *An Account of the Crustacea of Norway, with short descriptions and figures of all the species* 1: 629-672, pls. 225-240. Cammermeyer, Christiania [Oslo].

Crustacea Caspia. Contributions to the knowledge of the carcinological fauna of the Caspian Sea. Part III. Amphipoda. First article: Gammaridae (part). *Izv. imp. Akad. Nauk* 1894: 179-223, pls. 1-8.

Crustacea Caspia. Contributions to the knowledge of the carcinological fauna of the Caspian Sea. Part III. Amphipoda. Second article: Gammaridae (continued). *Izv. imp. Akad. Nauk* 1894: 343-378, pls. 9-16.

1895   Descriptions of some Australian Phyllopoda. *Arch. Math. Naturv.* 17(2): 1-51, pls. 1-8.

Amphipoda. Part XXXI & XXXII. Appendix. *An Account of the Crustacea of Norway, with short descriptions and figures of all the species* 1: I-VIII, 673-711, suppl. pls. 1-8. Cammermeyer, Christiania [Oslo].

On some South-African Entomostraca raised from dried mud. *Skr. VidenskSelsk. Christiania, Mat.-naturv. Kl.* 1895(8): 1-56, pls. 1-8.

Crustacea Caspia. Contributions to the knowledge of the carcinological fauna of the Caspian Sea. Part III. Amphipoda. Third article: Gammaridae (concluded), Corophiidae. *Izv. imp. Akad. Nauk* 3: 275-314, pls. 17-24.

Crustacea Caspia. Account of the Mysidae in the collection of dr. O. Grimm. *Izv. imp. Akad. Nauk* 3: 433-458, pls. 1-8.

1896   Development of *Estheria packardi*, Brady, as shown by artificial hatching from dried mud. *Arch. Math. Naturv.* 18(2): 1-27, pls. 1-4.

On fresh-water Entomostraca from the neighbourhood of Sydney, partly raised from dried mud. *Arch. Math. Naturv.* 18(3): 1-81, pls. 1-8.

Crustacea Caspia. Contributions to the knowledge of the carcinological fauna of the Caspian Sea. Amphipoda. Supplement. *Izv. imp. Akad. Nauk* 4: 421-489, pls. 1-12.

Fauna Norvegiae. I. Descriptions of the Norwegian species at present known belonging to the suborders Phyllocarida and Phyllopoda. Aschehoug, Christiania [Oslo]. VI + 140 pp., pls. 1-20. Also with Norwegian text: *Fauna Norvegiae. I. Beskrivelse af de hidtil kjendte norske arter af underordnerne Phyllocarida og Phyllopoda.*

On a new fresh-water ostracod, *Stenocypris chevreuxi*, G.O. Sars, with notes on some other Entomostraca raised from dried mud from Algeria. *Arch. Math. Naturv.* 18(7): 1-27, pls. 1-2.

Description of two new Phyllopoda from North Australia. *Arch. Math. Naturv.* 18(8): 1-34, pls. 1-6.

Isopoda. Parts I & II. Apseudidae, Tanaidae. *An Account of the Crustacea of Norway, with short descriptions and figures of all the species* 2: 1-40, pls. 1-16. Publ. by the Bergen Museum, Bergen. Cammermeyer, Christiania [Oslo].

On some West-Australian Entomostraca raised from dried sand. *Arch. Math. Naturv.* 19(1): 1-35, pls. 1-4.

1897   Pelagic Entomostraca of the Caspian Sea. *Ezheg. zool. Muz.* 2: 1-74, pls. 1-8.

Isopoda. Parts III & IV. Anthuridae, Gnathiidae, Aegidae, Cirolanidae, Limnoriidae. *An Account of the Crustacea of Norway, with short descriptions and figures of all the species* 2: 41-80, pls. 17-32. Publ. by the Bergen Museum, Bergen. Cammermeyer, Christiania [Oslo].

Isopoda. Parts V & VI. Idotheidae, Arcturidae, Asellidae, Ianiridae, Munnidae. *An Account of the Crustacea of Norway, with short descriptions and figures of all the species* 2: 81-116, pls. 33-48. Publ. by the Bergen Museum, Bergen. Cammermeyer, Christiania [Oslo].

Isopoda. Parts VII & VIII. Desmosomidae, Munnopsidae (part). *An Account of the Crustacea of Norway, with short descriptions and figures of all the species* 2: 117-144, pls. 49-64. Publ. by the Bergen Museum, Bergen. Cammermeyer, Christiania [Oslo].

On some additional Crustacea from the Caspian Sea. *Ezheg. zool. Muz.* 2: 273-306, pls. 13-16.

The Phyllopoda of the Jana-expedition. *Ezheg. zool. Muz.* 2: 463-494, pls. 23-30.

1898   On some South-African Phyllopoda raised from dried mud. *Arch. Math. Naturv.* 20(4): 1-43, pls. 1-4.

Description of two additional South African Phyllopoda. *Arch. Math. Naturv.* 20(6): 1-23, pls. 1-3.

Isopoda. Parts IX & X. Munnopsidae (concluded), Ligiidae, Trichoniscidae, Oniscidae (part). *An Account of the Crustacea of Norway, with short descriptions and figures of all the species* 2: 145-184, pls. 65-88. Publ. by the Bergen Museum, Bergen. Cammermeyer, Christiania [Oslo].

On *Megalocypris princeps*, a gigantic fresh-water ostracod from South Africa. *Arch. Math. Naturv.* 20(8): 1-18, 1 pl.

On the propagation and early developement of Euphausiidae. *Arch. Math. Naturv.* 20(11): 1-41, pls. 1-4.

Isopoda. Parts XI & XII. Oniscidae (concluded), Bopyridae, Dajidae. *An Account of the Crustacea of Norway, with short descriptions and figures of all the species* 2: 185-232, pls. 89-96. Publ. by the Bergen Museum, Bergen. Cammermeyer, Christiania [Oslo].

1899   The Cladocera, Copepoda and Ostracoda of the Jana expedition. *Ezheg. zool. Muz.* 3: 324-359, pls. 6-11.

On the genus *Broteas* of Lovén, with description of the type species: *Broteas falcifer*, Lov. *Arch. Math. Naturv.* 21(2): 1-27, pl. 4.

Additional notes on South African Phyllopoda. *Arch. Math. Naturv.* 21(4): 1-29, pls. 1-3.

Isopoda. Parts XIII & XIV. Cryptoniscidae, appendix. *An Account of the Crustacea of Norway, with short descriptions and figures of all the species* 2: I-X, 233-270, pls. 97-100, suppl. pls. 1-4. Publ. by the Bergen Museum, Bergen. Cammermeyer, Christiania [Oslo].

Account of the postembryonal development of *Pandalus borealis*, Krøyer, with remarks on the development of other Pandali, and description of the adult *Pandalus borealis*. *Rep. Norw. Fishery mar. Invest.* 1(3): 1-45, pls. 1-10.

Cumacea. Parts I & II. Cumidae, Lampropidae (part). *An Account of the Crustacea of Norway, with short descriptions and figures of all the species* 3: 1-24, pls. 1-16. Publ. by the Bergen Museum, Bergen. Cammermeyer, Christiania [Oslo].

1900   Cumacea. Parts III & IV. Lampropidae (concluded), Platyospidae, Leuconidae. *An Account of the Crustacea of Norway, with short descriptions and figures of all the species* 3: 25-40, pls. 17-32. Publ. by the Bergen Museum, Bergen. Cammermeyer, Christiania [Oslo].

Crustacea. *Scient. Results Norw. N. polar Exped. 1893-1896* 1(5): 1-141, pls. 1-36. Longmans, Green, and Co., London.

Description of *Jheringula paulensis*, G.O. Sars, a new generic type of Macrothricidae from Brazil. *Arch. Math. Naturv.* 22(6): 1-27, pls. 1-2.

Cumacea. Parts V & VI. Diastylidae. *An Account of the Crustacea of Norway, with short descriptions and figures of all the species* 3: 41-68, pls. 33-48. Publ. by the Bergen Museum, Bergen. Cammermeyer, Christiania [Oslo].

Cumacea. Parts VII & VIII. Pseudocumidae, Nannastacidae, Campylaspidae. *An Account of the Crustacea of Norway, with short descriptions and figures of all the species* 3: 69-92, pls. 49-64. Publ. by the Bergen Museum, Bergen. Cammermeyer, Christiania [Oslo].

On some Indian Phyllopoda. *Arch. Math. Naturv.* 22(9): 1-30, pls. 1-4.

On *Epischura baikalensis*, a new calanoid from Baikal Lake. *Ezheg. zool. Muz.* 5: 226-239, pl. 6.

Cumacea. Parts IX & X. Anatomy, development, supplement. *An Account of the Crustacea of Norway, with short descriptions and figures of all the species* 3: I-X, 93-115, pls. 65-72. Publ. by the Bergen Museum, Bergen. Cammermeyer, Christiania [Oslo].

1901   Contributions to the knowledge of the fresh-water Entomostraca of South America, as shown by artificial hatching from dried material. I. Cladocera. *Arch. Math. Naturv.* 23(3): 1-102, pls. 1-12.

On the crustacean fauna of Central Asia. Part I. Amphipoda and Phyllopoda. *Ezheg. zool. Muz.* 6: 130-164, pls. 1-8.

Copepoda Calanoida. Parts I & II. Calanidae, Eucalanidae, Paracalanidae, Pseudocalanidae, Aetideidae (part). *An Account of the Crustacea of Norway, with short descriptions and figures of all the species* 4: 1-28, pls. 1-16. Publ. by the Bergen Museum, Bergen. Cammermeyer, Christiania [Oslo].

Contributions to the knowledge of the fresh-water Entomostraca of South America, as shown by artificial hatching from dried material. Part II. Copepoda-Ostracoda. *Arch. Math. Naturv.* 24(1): 1-52, pls. 1- 8.

1902   Copepoda Calanoida. Parts III & IV. Aetideidae (concluded), Euchaetidae, Phaennidae. *An Account of the Crustacea of Norway, with short descriptions and figures of all the species* 4: 29-48, pls. 17-32. Publ. by the Bergen Museum, Bergen. Cammermeyer, Christiania [Oslo].

Copepoda Calanoida. Parts V & VI. Scolecithricidae, Diaixidae, Stephidae, Tharybidae, Pseudocyclopiidae. *An Account of the Crustacea of Norway, with short descriptions and figures of all the species* 4: 49-72, pls. 33-48. Publ. by the Bergen Museum, Bergen. Cammermeyer, Christiania [Oslo].

On the Polyphemidae of the Caspian Sea. *Esheg. zool. Muz.* 7: 31-54, pls. 1-4.

Copepoda Calanoida. Parts VII & VIII. Centropagidae, Diaptomidae. *An Account of the Crustacea of Norway, with short descriptions and figures of all the species* 4: 73-96, pls. 49-64. Publ. by the Bergen Museum, Bergen. Cammermeyer, Christiania [Oslo].

Copepoda Calanoida. Parts IX & X. Temoridae, Metridiidae, Heterorhabdidae (part). *An Account of the Crustacea of Norway, with short descriptions and figures of all the species* 4: 97-120, pls. 65-80. Publ. by the Bergen Museum, Bergen. Cammermeyer, Christiania [Oslo].

On a new South American phyllopod, *Eulimnadia brasiliensis*, G.O. Sars, raised from dried mud. *Arch. Math. Naturv.* 24(6): 1-12, pl. 1.

Copepoda Calanoida. Parts XI & XII. Heterorhabdidae (continued), Arietellidae, Pseudocyclopidae, Candaciidae, Pontellidae. *An Account of the Crustacea of Norway, with short descriptions and figures of all the species* 4: 121-144, pls. 81-96. Publ. by the Bergen Museum, Bergen. Cammermeyer, Christiania [Oslo].

1903   Copepoda Calanoida. Parts XIII & XIV. Parapontellidae, Acartiidae, supplement. *An Account of the Crustacea of Norway, with short descriptions and figures of all the species* 4: I-XIII, 145-171, pls. 97-102, suppl. pls. 1-6. Publ. by the Bergen Museum, Bergen. Cammermeyer, Christiania [Oslo].

Fresh-water Entomostraca from China and Sumatra. *Arch. Math. Naturv.* 25(8): 1-44, pls. 1-4.

Copepoda Harpacticoida. Parts I & II. Misophriidae, Longipediidae, Cerviniidae, Ectinosomidae (part). *An Account of the Crustacea of Norway, with short descriptions and figures of all the species* 5: 1-28, pls. 1-16. Publ. by the Bergen Museum, Bergen. Cammermeyer, Christiania [Oslo].

On the crustacean fauna of Central Asia. Part II. Cladocera. *Ezheg. zool. Muz.* 8: 157-194, pls. 1-8.

On the crustacean fauna of Central Asia. Part III. Copepoda and Ostracoda. *Ezheg. zool. Muz.* 8: 195-232, pls. 9-16.

On the crustacean fauna of Central Asia. Appendix. Local faunae of Central Asia. *Ezheg. zool. Muz.* 8: 233-264.

1904   Pacifische Plankton-Crustaceen. (Ergebnisse einer Reise nach dem Pacific. Schauinsland 1896-1897). I. Plankton aus Salzseen und Süsswasserteichen. *Zool. Jb., Abt. Syst., Ökol. Geogr. Tiere* 19: 629-646, pls. 33-38. (special separate 1903)

On a new (planktonic) species of the genus *Apherusa*. *Publs Cironst. Cons. perm. int. Explor. Mer.* 10: 1-4, pl. 1.

Copepoda Harpacticoida. Parts III & IV. Ectinosomidae, Harpacticidae (part). *An Account of the Crustacea of Norway, with short descriptions and figures of all the species* 5: 29-56, pls. 17-32. Publ. by the Bergen Museum, Bergen. Cammermeyer, Christiania [Oslo].

Description of *Paracartia grani*, G.O. Sars, a peculiar calanoid ocurring in some of the oyster-beds of Western Norway. *Bergens Mus. Aarb.* 1904(4): 1-16, pls. 1-4.

Copepoda Harpacticoida. Parts V & VI. Harpacticidae (concluded), Peltidiidae, Tegastidae, Porcellidiidae, Idyidae (part). *An Account of the Crustacea of Norway, with short descriptions and figures of all the species* 5: 57-80, pls. 33-48. Publ. by the Bergen Museum, Bergen. Cammermeyer, Christiania [Oslo].

On a remarkable new chydorid, *Saycia orbicularis* G.O. Sars, from Victoria, South Australia. *Arch. Math. Naturv.* 26(8): 1-15, 1 pl.

1905   Pacifische Plankton-Crustaceen. (Ergebnisse einer Reise nach dem Pacific. Schauinsland 1896-

1897). II. Brackwasser-Crustaceen von dem Chatham-Inseln. *Zool. Jb., Abt. Syst., Ökol. Geogr. Tiere* 21: 371-414, pls. 14-20.

Copepoda Harpacticoida. Parts VII & VIII. Idyidae (continued), Thalestridae (part). *An Account of the Crustacea of Norway, with short descriptions and figures of all the species* 5: 81-108, pls. 49-64. Publ. by the Bergen Museum, Bergen. Cammermeyer, Christiania [Oslo].

Liste préliminaire des Calanoïdés recueillis pendant les campagnes de S.A.S. le prince Albert de Monaco, avec diagnoses des genres et des espèces nouvelles. (1re partie). *Bull. Mus. océanogr. Monaco* 26: 1-22.

Liste préliminaire des Calanoïdés recueillis pendant les campagnes de S.A.S. le prince Albert de Monaco, avec diagnoses des genres et des espèces nouvelles. (2e partie). *Bull. Mus. océanogr. Monaco* 40: 1-24.

On two apparently new Phyllopoda from South Africa. *Arch. Math. Naturv.* 27(4): 1-16, pls. 1-2.

Copepoda Harpacticoida. Parts IX & X. Thalestridae (continued). *An Account of the Crustacea of Norway, with short descriptions and figures of all the species* 5: 109-132, pls. 65-80. Publ. by the Bergen Museum, Bergen. Cammermeyer, Christiania [Oslo].

1906   Copepoda Harpacticoida. Parts XI & XII. Thalestridae (concluded), Diosaccidae (part). *An Account of the Crustacea of Norway, with short descriptions and figures of all the species* 5: 133-156, pls. 81-96. Publ. by the Bergen Museum, Bergen. Cammermeyer, Christiania [Oslo].

Postembryonal development of *Athanas nitescens*, Leach. *Arch. Math. Naturv.* 27(10): 1-29, pls. 1-4.

Copepoda Harpacticoida. Parts XIII & XIV. Diosaccidae (continued). *An Account of the Crustacea of Norway, with short descriptions and figures of all the species* 5: 157-172, pls. 97-112. Publ. by the Bergen Museum, Bergen. Cammermeyer, Christiania [Oslo].

Copepoda Harpacticoida. Parts XV & XVI. Diosaccidae (concluded), Canthocamptidae (part). *An Account of the Crustacea of Norway, with short descriptions and figures of all the species* 5: 173-196, pls. 113-128. Publ. by the Bergen Museum, Bergen. Cammermeyer, Christiania [Oslo].

1907   Copepoda Harpacticoida. Parts XVII & XVIII. Canthocamptidae (continued). An Account of the Crustacea of Norway, with short descriptions and figures of all the species 5: 197-220, pls. 129-144. Publ. by the Bergen Museum, Bergen. Cammermeyer, Christiania [Oslo].

Notes supplémentaires sur les Calanoïdés de la Princesse-Alice. (Corrections et additions). *Bull. Inst. océanogr. Monaco* 101: 1-27.

On two new species of the genus *Diaptomus* from South Africa. *Arch. Math. Naturv.* 28(8): 1-17, pls. 1-2.

Copepoda Harpacticoida. Parts XIX & XX. Canthocamptidae (concluded), Laophontidae (part). *An Account of the Crustacea of Norway, with short descriptions and figures of all the species* 5: 221-240, pls. 145-160. Publ. by the Bergen Museum, Bergen. Cammermeyer, Christiania [Oslo].

Mysidae. *Trudov' Kaspijskoj ekspedicii 1904 goda.* 1: 1-71, pls. 1-12. Petrograd [St. Petersburg]. (With Russian and English text).

1908   Copepoda Harpacticoida. Parts XXI & XXII. Laophontidae (continued). *An Account of the Crustacea of Norway, with short descriptions and figures of all the species* 5: 241-256, pls. 161-176. Publ. by the Bergen Museum, Bergen. Cammermeyer, Christiania [Oslo].

On the occurrence of a genuine harpacticid in the Lake Baikal. *Arch. Math. Naturv.* 29(4): 1-13, 1 pl.

Fresh-water Copepoda from Victoria, Southern Australia. *Arch. Math. Naturv.* 29(7): 1-24, pls. 1-4.

Copepoda Harpacticoida. Parts XXIII & XXIV. Laophontidae (continued). *An Account of the Crustacea of Norway, with short descriptions and figures of all the species* 5: 257-276, pls. 177-192. Publ. by the Bergen Museum, Bergen. Cammermeyer, Christiania [Oslo].

1909   Note préliminaire sur trois formes remarquables de Copépodes, provenant des campagnes de S.A.S. le Prince Albert de Monaco. *Bull. Inst. océanogr. Monaco* 147: 1-8, figs. 1-3. (Reprinted in: *Résult. Camp. scient. Prince Albert I* 97: 36-40, pl. 1 figs. 5-6, pl. 3 fig. 5)

Report on the Copepoda. Zoological results of the third Tanganyika expedition, conducted by dr. W.A. Cunnington, F.Z.S., 1904-1905. *Proc. zool. Soc. Lond.* 1909: 31-77, pls. 6-23.

Crustacea. *Report of the second Norwegian Arctic expedition in the 'Fram' 1898-1902* 18: 1-47, pls. 1-12. Publ. by Videnskabs-selskabet i Kristiania [Oslo].

Copepoda Harpacticoida. Parts XXV & XXVI. Laophontidae (concluded), Cletodidae (part). *An*

Account of the Crustacea of Norway, with short descriptions and figures of all the species 5: 277-304, pls. 193-208. Publ. by the Bergen Museum, Bergen. Cammermeyer, Christiania [Oslo].

Fresh-water Entomostraca from South Georgia. *Arch. Math. Naturv.* 30(5): 1-35, pls. 1-4.

Copepoda Harpacticoida. Parts XXVII & XXVIII. Cletodidae (concluded), Anchorabolidae, Gylindropsyllidae, Tachidiidae (part). *An Account of the Crustacea of Norway, with short descriptions and figures of all the species* 5: 305-336, pls. 209-224. Publ. by the Bergen Museum, Bergen. Cammermeyer, Christiania [Oslo].

1910   Copepoda Harpacticoida. Parts XXIX & XXX. Tachidiidae (concluded), Metidae, Balaenophilidae, supplement (part). *An Account of the Crustacea of Norway, with short descriptions and figures of all the species* 5: 337-368, pls. 225-230, suppl. pls. 1-10. Publ. by the Bergen Museum, Bergen. Cammermeyer, Christiana [Oslo].

Report on the Ostracoda. Zoological results of the third Tanganyika expedition, conducted by dr. W.A. Cunnington, F.Z.S., 1904-1905. *Proc. zool. Soc. Lond.* 1910: 732-760, pls. 64-73.

1911   Copepoda Harpacticoida. Parts XXXI & XXXII. Supplement (continued). *An Account of the Crustacea of Norway, with short descriptions and figures of all the species* 5: 369-396, suppl. pls. 11-26. Publ. by the Bergen Museum, Bergen. Cammermeyer, Christiania [Oslo].

Platycopia perplexa, n. gen. and sp., a remarkable new type of deep-water Calanoida from the Norwegian coast. *Arch. Math. Naturv.* 31(7): 1-16, pls. 1-2.

Copepoda Harpacticoida. Parts XXXIII & XXXIV. Supplement (continued). *An Account of the Crustacea of Norway, with short descriptions and figures of all the species* 5: 397-420, suppl. pls. 27-42. Publ. by the Bergen Museum, Bergen. Cammermeyer, Christiania [Oslo].

Copopoda Harpacticoida. Parts XXXV & XXXVI. Supplement (concluded). *An Account of the Crustacea of Norway, with short descriptions and figures of all the species* 5: I-XIV, 421-449, I-XII suppl. pls. 43-54. Publ. by the Bergen Museum, Bergen. Cammermeyer, Christiania [Oslo].

1912   On the genera *Cryptocheles* and *Bythocaris* D, G.O. Sars, *with description of the type species of each genus. Arch. Math. Naturv.* 32(5): 1-19, pls. 1-2.

Account of the postembryonal development of *Hippolyte varians*, Leach. *Arch. Math. Naturv.* 32(7): 1-25, pls. 3-5.

Report on some larval and young stages of prawns from Lake Tanganyika. Zoological results of the third Tanganyika expedition, conducted by dr. W.A. Cunnington, 1904-1905. *Proc. zool. Soc. Lond.* 1912: 426-440, pls. 57-60.

Notes on Caridea (sexual differences - mimicry). *Arch. Math. Naturv.* 32(9): 1-12, pl. 6.

Additional notes on fresh-water Calanoida from Victoria, Southern Australia. *Arch. Math. Naturv.* 32(13): 1-20, pls. 7-9.

List of Crustacea from selected stations. In Murray, J. & J. Hjort: *The depths of the ocean*: 654-655. Macmillan, London.

On the problematic form '*Moina lemnae* King' and its true relationship. *Arch. Math. Naturv.* 32(14): 1-13, pl. 10.

1913   Copepoda Cyclopoida. Parts I & II. Oithonidae, Cyclopinidae, Cyclopidae (part). *An Account of the Crustacea of Norway, with short descriptions and figures of all the species* 6: 1-32, pls. 1-16. Publ. by the Bergen Museum, Bergen. Cammermeyer, Christiania [Oslo].

Thaumatopsyllus paradoxus, G.O. Sars, a remarkable copepod from the Norwegian coast, apparently referable to the monstrilloid group. *Arch. Math. Naturv.* 33(6): 1-11, 1 pl.

Copepoda Cyclopoida. Parts III & IV. Cyclopidae (continued). *An Account of the Crustacea of Norway, with short descriptions and figures of all the species* 6: 33-56, pls. 17-32. Publ. by the Bergen Museum, Bergen. Cammermeyer, Christiania [Oslo].

1914   Copepoda Cyclopoida. Parts V & VI. Cyclopidae (continued). *An Account of the Crustacea of Norway, with short descriptions and figures of all the species* 6: 57-80, pls. 33-48. Publ. by the Bergen Museum, Bergen. Cammermeyer, Christiania [Oslo].

Daphnia carinata, King, and its remarkable varieties. *Arch. Math. Naturv.* 34(1): 1-14, pls. 1-2.

Report on the Cumacea of the Caspian expedition 1904. *Trudy Kaspijskoj ékspedicii 1904 goda.* 4: 1-32, pls. 1-12. Petrograd [St. Petersburg]. Also with Russian text: Cumacea Kaspijskoj ékspedicii 1904 goda. *Trudy Kaspijskoj ékspedicii 1904 goda.* 4: 1-34, pls. 1-12.

1915   Copepoda Cyclopoida. Parts VII & VIII. Cyclopidae (concluded), Ascomyzontidae. *An Account of the Crustacea of Norway, with short descriptions and figures of all the species* 6: 81-104, pls. 49-64. Publ. by the Bergen Museum, Bergen. Cammermeyer, Christiania [Oslo].

Copepoda Cyclopoida. Parts IX & X. Ascomyzontidae (concluded), Acontiophoridae, Myzopontiidae, Dyspontiidae, Artotrogidae, Cancerillidae. *An Account of the Crustacea of Norway, with short descriptions and figures of all the species* 6: 105-140, pls. 65-80. Publ. by the Bergen Museum, Bergen. Cammermeyer, Christiania [Oslo].

Entomostraca of Georgian Bay. *Contr. Can. Biol. Fish.* 1911-1914(2) (=*Rep. Dep. mar. Fish. Ottawa* 47 suppl.): 221-222.

1916   Liste systématique des Cyclopoidés, Harpacticoidés et Monstrilloidés recueillis pendant les campagnes de S.A.S. le Prince Albert de Monaco, avec descriptions et figures des espèces nouvelles. *Bull. Inst. océanogr. Monaco* 323: 1-15, pls. 1-8. (Reprinted in: *Résult. Camp. scient. Prince Albert I* 97: 40-51, pls.)

The fresh-water Entomostraca of the Cape Province (Union of South Africa). Part I: Cladocera. *Ann. S. Afr. Mus.* 15: 303-351, pls. 29-41.

On the juvenile state of *Lophogaster typicus* M. Sars. *Arch. Math. Naturv.* 34(13): 1-9, 1 pl.

1917   Copepoda Cyclopoida. Parts XI & XII. Clausidiidae, Lichomolgidae (part). *An Account of the Crustacea of Norway, with short descriptions and figures of all the species* 6: 141-172, pls. 81-96. Publ. by the Bergen Museum, Bergen. Cammermeyer, Christiania [Oslo].

Urocopia singularis G.O. Sars, a peculiar semiparasitic copepod from great deeps of the North Atlantic Ocean. *Bergens Mus. Aarb., Naturv. R.* 1916-17(4): 1-11, 1 pl.

1918   Copepoda Cyclopoida. Parts XIII & XIV. Lichomolgidae (concluded), Oncaeidae, Corycaeidae, Ergasilidae, Clausiidae, Eunicicolidae, supplement. *An Account of the Crustacea of Norway, with short descriptions and figures of all the species* 6: I-XIII, 173-225, pls. 97-118. Publ. by the Bergen Museum, Bergen. Cammermeyer, Christiania [Oslo].

1919   Copepoda Supplement. Parts I & II. Calanoida, Harpacticoida (part). *An Account of the Crustacea of Norway, with short descriptions and figures of all the species* 7: 1-24, pls. 1-16. Publ. by the Bergen Museum, Bergen. Cammermeyer, Christiania [Oslo].

1920   Copepoda Supplement. Parts III & IV. Harpacticoida (continued). *An Account of the Crustacea of Norway, with short descriptions and figures of all the species* 7: 25-52, pls. 17-32. Publ. by the Bergen Museum, Bergen. Cammermeyer, Christiania [Oslo].

Copepoda Supplement. Parts V & VI. Harpacticoida (continued). *An Account of the Crustacea of Norway, with short descriptions and figures of all the species* 7: 53-72, pls. 33-48. Publ. by the Bergen Museum, Bergen. Cammermeyer, Christiania [Oslo].

Copepoda Supplement. Parts VII & VIII. Harpacticoida (continued). *An Account of the Crustacea of Norway, with short descriptions and figures of all the species* 7: 73-92, pls. 49-64. Publ. by the Bergen Museum, Bergen. Cammermeyer, Christiania [Oslo].

Calanoidés recueillis pendant les campagnes de S.A.S. le Prince Albert de Monaco. (Nouveau supplément). *Bull. Inst. océanogr. Monaco* 377: 1-20.

1921   Copepoda Supplement. Parts IX & X. Harpacticoida (concluded), Cyclopoida. *An Account of the Crustacea of Norway, with short descriptions and figures of all the species* 7: I-V, 93-121, pls. 65-76. Publ. by the Bergen Museum, Bergen. Cammermeyer, Christiania [Oslo].

Copepoda Monstrilloida and Notodelphyoida. Parts I & II. Thaumatopsyllidae, Monstrillidae, Notodelphyidae (part). *An Account of the Crustacea of Norway, with short descriptions and figures of all the species* 8: 1-32, pls. 1-16. Publ. by the Bergen Museum, Bergen. Cammermeyer, Christiania [Oslo].

Copepoda Monstrilloida and Notodelphyoida. Parts III & IV. Notodelphyidae (concluded), Doropygidae, Buproridae, Ascidicolidae. *An Account of the Crustacea of Norway, with short descriptions and figures of all the species* 8: 33-68, pls. 17-32. Publ. by the Bergen Museum, Bergen. Cammermeyer, Christiania [Oslo].

Copepoda Monstrilloida and Notodelphyoida. Parts V & VI. Botryllophilidae, Enterocolidae, supplement. *An Account of the Crustacea of Norway, with short descriptions and figures of all the species* 8: I-V, 69-90, pls. 33-37. Publ. by the Bergen Museum, Bergen. (Cammermeyer, Christiania [Oslo]).

1922   Ostracoda. Parts I & II. Cypridinidae, Conchoeciidae, Polycopidae (part). *An Account of the*

Crustacea of Norway, with short descriptions and figures of all the species 9: 1-32, pls. 1-16. Publ. by the Bergen Museum, Bergen. Cammermeyer, Christiania [Oslo].

1923   Ostracoda. Parts III & IV. Polycopidae (concluded), Cytherellidae, Cypridae (part). *An Account of the Crustacea of Norway, with short descriptions and figures of all the species* 9: 33-72, pls. 17-32. Publ. by the Bergen Museum, Bergen. Cammermeyer, Christiania [Oslo].

1924   The fresh-water Entomostraca of the Cape Province (Union of South Africa). Part II: Ostracoda. *Ann. S. Afr. Mus.* 20(2): 105-193, pls. 2-20.
Copépodes particulièrement bathypélagiques provenant des campagnes scientifiques du Prince Albert Ier de Monaco. Planches. *Résult. Camp. scient. Prince Albert I* 69: pls. 1-127.
Contributions to a knowledge of the fauna of South-West Africa. I. Crustacea Entomostraca, Ostracoda. *Ann. S. Afr. Mus.* 20(3): 195-211, pls. 21-25.

1925   Ostracoda. Parts V & VI. Cypridae (continued). *An Account of the Crustacea of Norway, with short descriptions and figures of all the species* 9: 73-104, pls. 33-48. Publ. by the Bergen Museum, Bergen. Cammermeyer, Christiania [Oslo].
Ostracoda. Parts VII & VIII. Cypridae (continued). *An Account of the Crustacea of Norway, with short descriptions and figures of all the species* 9: 105-136, pls. 49-64. Publ. by the Bergen Museum, Bergen. Cammermeyer, Christiania [Oslo].
Ostracoda. Parts IX & X. Cypridae (concluded). Cytheridae (part). *An Account of the Crustacea of Norway, with short descriptions and figures of all the species* 9: 137-176, pls. 65-80. Publ. by the Bergen Museum, Bergen. Cammermeyer, Christiania [Oslo]).
Ostracoda. Parts XI & XII. Cytheridae (continued). *An Account of the Crustacea of Norway, with short descriptions and figures of all the species* 9: 177-208, pls. 81-96. Publ. by the Bergen Museum, Bergen. Cammermeyer, Oslo (Christiania).
Copépodes particulièrement bathypélagiques provenant des campagnes scientifiques du Prince Albert Ier de Monaco. Texte. *Résult. Camp. scient. Prince Albert I* 69: 1-408.

1926   Description of *Scolecimorpha insignis*, G.O. Sars, a remarkable new asidicole parasit. *K. norske Vidensk. Selsk. Skr.* 1925(2): 1-12, pls. 1-2 (= *Meddr Trondhjems biol. Stn* 23).
Freshwater Ostracoda from Canada and Alaska. *Rep. Can. arct. Exped. 1913-18* 7: Crustacea. Part I: Ostracoda: 1-23, pls. 1-5.
The fresh-water Entomostraca of the Cape Province (Union of South Africa). Part III: Copepoda. *Ann. S. Afr. Mus.* 25(1): 85-149, pls. 5-16.
Ostracoda. Parts XIII & XIV. Cytheridae (continued). *An Account of the Crustacea of Norway, with short descriptions and figures of all the species* 9: 209-240, pls. 97-112. Publ. by the Bergen Museum, Bergen. Cammermeyer, Oslo (Christiania).

1927   Notes on the crustacean fauna of the Caspian Sea. *Sbornik v chest professora Nikolaya Mikhailovicha Knipovicha: 1885-1925. Festschrift für Prof. N. M. Knipowitsch 1885-1925*: 315-329. Moskva.

1928   Ostracoda. Parts XV & XVI. Cytheridae (concluded). *An Account of the Crustacea of Norway, with short descriptions and figures of all the species* 9: I-XII, 241-277, pls. 113-119. Publ. by the Bergen Museum, Bergen. Cammermeyer, Oslo (Christiania).

1929   Fauna of the Batu Caves, Selangor. VIII. Description of a remarkable cave-crustacean, *Parabathynella malaya* G.O. Sars, sp. nov. with general remarks on the family Bathynellidae. *J. fed. Malay St. Mus.* 14: 339-351, pls. 7-8.

Women's contributions to carcinology

Patsy A. McLaughlin
Shannon Point Marine Center, Western Washington University, Anacortes, Washington, USA

Sandra Gilchrist
Division of Natural Sciences, New College of the University of South Florida, Sarasota, Florida, USA

1 INTRODUCTION

Frank Truesdale, in his letter asking us to prepare an overview on women's contributions to carcinology, noted that the only woman suggested to the Crustacean Society for biographical coverage in the History of Carcinology Symposium was Mary Rathbun. Unquestionably Mary Rathbun was a major force in carcinology, but certainly not the only substantive female contributor. There have been a great number of women, world-wide, whose pioneering efforts advanced the knowledge of crustaceans on many fronts. We include individuals who not only made major inroads into research subjects, but also those whose contributions included the training of future generations of carcinologists. Time and space do not permit us to present detailed accounts of the lives of each, but we hope to summarize the work of many and emphasize that of a few. As the thrust of this volume is historical, emphasis is placed on those whose contributions in their respective countries was made during the first 100 or so years that women were influential in shaping carcinology. We are sure, even after the diligent search that we made, that there will be some whom we overlooked. It is out of ignorance rather than intentioned neglect that omissions may have occurred.

Others in this volume present historical insight into the origins of carcinological work, and it is quite apparent that women did not figure directly into these early endeavors. This is not surprising in view of the sentiments still prevalent among many men of science in the 1800's. As a case in point, the British naturalist William Buckland, in a letter discussing the June meeting of the British Association for the Advancement of Science in 1832 remarked, 'Everybody whom I spoke to on the subject agreed that, if the meeting is to be of scientific utility, ladies ought not to attend the reading of the papers...as it would at once turn the thing into a sort of Albemarle-dilettanti-meeting, instead of a serious philosophical union of working men' (Ogilvie 1986). As late as 1941 the Royal Society of Edinburgh was exclusively a man's Society with women sometimes barred from its conversations and permitted to use the library only after permission was granted by the Council (Campbell & Smellie 1983).

To get an idea of the inroads women have made in some of the scientific societies today, we reviewed the membership rolls, membership retention, and publication history of the Crustacean Society. We found that although women participate in the Society, when they join as graduate students their retention rate after graduation is much lower than that of their male counterparts. In addition, the percentages of papers published in the *Journal of Crustacean Biology* by women as first or second authors has not increased significantly since the beginning of the journal 10 years ago. The number of papers in the journal coauthored exclusively by women over the years is on average less than 2%.

Although space limitations prevent us from listing individual contributions fully, for each biographical entry we have cited a few major works.

2 WOMEN CARCINOLOGISTS AT THE US NATIONAL MUSEUM: MARY J. RATHBUN, LEE BOONE & HARRIET RICHARDSON

2.1 *Mary Jane Rathbun*

Mary Jane Rathbun (Figs 1A, 5A), whose carcinological career was initiated with her appointment as a 'copyist' in the Division of Marine Invertebrates of the US National Museum in 1886, is one of the earliest, if not the earliest, woman to have a direct impact on carcinology in the United States. Because of her substantial contributions, some 166 published papers that included innumerable descriptions of new Recent and fossil species, and the special fondness felt for her by her museum colleagues, several accounts detailing her life and works have been published (Schmitt 1943, 1973; McCain 1943; Chace 1990). Between 1891 and 1904 Rathbun published 47 papers in which she diagnosed 25 new genera, a new family, and 344 new species. In the subsequent 14 years, Rathbun added another 45 papers to her growing list of contributions, which included an additional new family, new superfamily, 22 new genera, 330 new species, and 12 new subspecies. However, she is best known to modern carcinologists by her four-volume monumental work on the American brachyuran crabs. Three of these volumes, together with an additional 72 papers, were published after she had resigned her formal position with the Museum to make funds available for a male assistant curator, viz., Waldo LaSalle Schmitt. Although her formal education extended no further than high school, she was awarded an honorary master's degree from the University of Pittsburgh and then went on to qualify for a doctorate at George Washington University in 1917. Mary Rathbun died in 1943, at the age of 83, having devoted nearly 45 years to carcinological research.

1918	The Grapsoid Crabs of America. *US Natl Mus. Bull.* 97: i-xii, 1-461, 172 figs, 161 pls.	
1925	The Spider Crabs of America. *US Natl Mus. Bull.* 129: i-xx, 1-598, 153 figs, 283 pls.	
1930	The Cancroid Crabs of America. *US Natl Mus. Bull.* 152: i-xvi, 1-593, 85 figs, 230 pls.	
1937	Oxystomatous and Allied Crabs of America. *US Natl Mus. Bull.* 166: i-vi, 1-278, 47 figs, 86 pls.	

2.2 *Lee Boone*

Lee Boone (who sometimes referred to herself as Pearl Lee Boone, Pearl Boone, or even Mr. Lee Boone) was a contemporary of Mary Rathbun and perhaps one of the most controversial carcinologists. Boone received her A.B. (1919) and her M.S. and B.Ed. degrees (1920) from George Washington University. She became a member of the scientific staff at the US National Museum in 1913 and remained there until 1922. It was during her tenure at the Museum that she gained a great deal of the negative notoriety that followed the rest of her career. Like many other early carcinologists, Boone began her work with other organisms. Her first position was in the Division of Insects, although the majority of her time at the Museum was spent in the Division of Marine Invertebrates. While working under Division curator Waldo Schmitt her competence as a scientist and assistant was questioned. This lead to a barrage of emotional and explicit letters submitted by her to the Director of the Museum claiming harassment, and from Schmitt and

Figure 1. A: Mary J. Rathbun. B: Isabella Gordon. C: Olive Tattersall. D: Sidnie Manton. E: Atie Vorstman. F: Alfreda Berkeley. G: Lina Zelickman. H: Shakuntala Shenoy. I: Dorothy Travis.

others suggesting her incompetence. She was dismissed from the Museum in 1922, where upon Boone gained employment first at the Aquarium and Biological Laboratory in Miami and then with the US Department of Agriculture. She later became a research associate at the Bingham Oceanographic Foundation at Yale where she remained until the end of her carcinological career. She published several works dealing with crustaceans, including the results of the cruises of the Vanderbilt yachts Eagle, Ara and Alva. However, the scientific rigor and veracity of many of her identifications were frequently called into question by other workers.

1927   Crustacea from tropical east American Seas. Scientific results of the first oceanographic expedition of the 'Pawnee.' *Bull. Bingham Oceanogr. Coll.* 1: 1-147, 33 figs.

1930   Crustacea: Anomura, Macrura, Schizopoda, Isopoda, Amphipoda, Mysidacea, Cirripedia, and Copepoda. Scientific results of the cruises of the yachts 'Eagle' and 'Ara,' 1921-1928, William K. Vanderbilt, commanding. *Bull. Vanderbilt Marine Mus.* 3: 1-221, 83 pls.

1931   A collection of anomuran and macruran Crustacea from the Bay of Panama and the fresh waters of the Canal Zone. *Bull. Amer. Mus. Nat. Hist.* 63: 137-189, 22 figs.

1935   Crustacea and Echinodermata. In: Scientific results of the world cruise of the yacht 'Alva', 1931, Wm. K. Vanderbilt commanding. *Bull. Vanderbilt Marine Mus.* 6: 1-264, 13 figs, 96 pls.

2.3   *Harriet Richardson*

As Mary Rathbun's name is synonymous with early brachyuran systematics, the name of another contemporary of hers at the Museum, Harriet Richardson, is equally recognized in the field of isopod systematics. After completing her bachelor's degree at Vassar in 1896, and master's at the same college in 1901, Richardson went on to earn her doctorate from George Washington University in 1903. In the interim she was appointed a collaborator at the Museum in January of 1901. Her first published work, the description of a new species of *Sphaeroma* appeared in the *Proceedings of the Biological Society of Washington* in 1897. In 1913 she became Mrs. William Searle, and apparently ceased formal work at the museum; however, she continued to publish occasional papers with the last, a study of the isopods crustaceans of the Dutch West Indies, appearing in 1919. During her relatively short carcinological career, Harriet Richardson (Searle) published a total of 80 papers on isopods and tanaidaceans. Only 6 years before her death in 1958, her museum title was changed to Research Associate (Cattell 1955, in part).

1901   Key to the isopods of the Atlantic coast of North America with descriptions of new and little known species. *Proc. US Natl Mus.* 23: 493-579, 33 figs.

1903   Contributions to the natural history of the Isopoda. *Proc. US Natl Mus.* 27: 1-89, 92 figs.

1905   A monograph on the isopods of North America. *US Natl Mus. Bull.* 54: i-lii, 1-727, 740 figs.

1909   Isopods collected in the northwest Pacific by the US Bureau of Fisheries steamer 'Albatross' in 1906. *Proc. US Natl Mus.* 37: 75-129, 50 figs.

3   WOMEN CARCINOLOGISTS IN THE BRITISH ISLES

About the same time as Rathbun and her contemporaries were active, across the Atlantic the legendary pillars of British carcinology, W.T. Calman, H. Graham Cannon, and E.R. Lancaster were influencing and later being succeeded by women such as evolutionist Florence Buchanan at University College, London; artist-systematist Elsie Sexton and developmentalist Marie Lebour at the Plymouth Laboratory; systematists Susan Finnegan, Isabella Gordon, and Olive Tattersall at the British Museum; embryologist-biological oceanographer Sheina Marshall at

Millport; and developmentalist-morphologist Sidnie Manton at the University of London and later at the British Museum.

3.1　*Florence Buchanan*

Florence Buchanan appears to have been one of the earliest women, perhaps the earliest, to enter carcinological research in Great Britain. In a paper read before the Biological Society of University College in 1889, she proposed a phylogeny for the Crustacea based upon the structure of respiratory organs. In 1910, she was awarded the prestigious Ellen Richards Research Prize (formerly the Naples Table Association Prize) for 'the best thesis written by a woman on a scientific subject embodying new observations and new conclusions' (Rossiter 1982).

1889　On the ancestral development of the respiratory organs in the decapodus Crustacea. *Quart. J. Micro. Soc. N.S.* 29: 451-569, 11 figs, 1 tabl.

3.2　*Elsie Wilkins Sexton*

Like many early women carcinologists, Elsie Wilkins Sexton had no formal zoological training; however, she had considerable artistic talent that was put to good use by E.J. Allen, then director of the Plymouth Marine Laboratory. Although she illustrated polychaetes for Dr. Allen, Crustacea, particularly amphipods, became her major interest. Between 1908 and 1951 she published more than 30 scientific papers, including the elucidation of the complex taxonomy of European gammarid amphipods. She also made a detailed study of postembryonic development, molting, and intersexes in local amphipod species, and became an expert in rearing animals in the laboratory. Her discovery of a red-eyed mutant in one *Gammarus* species led her to a long series of experiments on the inheritance of eye color. Her work was hampered by the advent of World War II, as it was for many carcinologists, female and male alike. She died in 1959, at the age of 91, having been a Fellow of the Linnean Society for 43 years. (Compiled from Gordon 1960).

1911　On the amphipod genus *Leptocheirus*. *Proc. Zool. Soc. Lond.* 1911: 561-594, 1 fig, 3 pls.
1924　The moulting and growth-stages of *Gammarus*, with descriptions of the normal and intersexes of *G. chevreuxi. J. Mar. Biol. Assc. UK* 13: 340-401, 4 figs, 21 pls.
1940　An account of *Marinogammarus* (Schellenberg) gen. nov. (Amphipoda), with a description of a new species, *M. pirloti. J. Mar. Biol. Assc. UK* 24: 633-682, 11 figs, 1 Fig. (with G.M. Spooner).
1951　The life-history of the multiform species *Jassa falcata* (Montagu) (Crustacea Amphipoda) with a review of the bibliography of the species. *J. Linn. Soc., Zool.* 51: 29-91, 30 pls. (with D.M. Reid).

3.3　*Marie Victoire Lebour*

Although born in 1876, Marie Victoire Lebour (Fig. 2D) did not begin her formal scientific education until the turn of the century. She published her first paper in 1900 on land and freshwater mollusks chiefly from the lower Tweed and Corbridge-on-Tyne where her family lived at the time. It is presumed that she acquired her interest in natural history by accompanying her geologist father on his excursions. She received her bachelor's degree in 1904, master's in 1907, and doctorate in 1917 from Durham University. In 1906, she was appointed junior demonstrator in the zoology department at Leeds University, advancing to assistant lecturer and demonstrator by 1909. She stayed at Leeds until 1915, when she was 'loaned' to the Plymouth

Figure 2. A: Sheina Marshall. B: Josephine Hart. C: Bella Galil. D: Marie Lebour. E: Janet Haig. F: Elizabeth Pope. G: Dora Henry. H: Marit Christiansen. I: Dorothy Skinner, center front, with postdoctoral fellows at 1986 ASZ meeting.

laboratory. That loan became a permanent position in 1917, and she remained on the staff of the laboratory until her retirement in 1946. Despite her retirement, she continued working at the laboratory for another 18 years.

She began her professional career with faunistic work on the mollusks of the northeast coast of England, followed by years of work with parasitic trematodes, dinoflagellates, larval fishes and diatoms. However, Marie Lebour is claimed by the carcinological community for her work with the larval development of euphausiids and decapods, publishing original descriptions of the larval stages of well over 100 species. Her success can be attributed to her acute powers of observation, the intensity and speed with which she worked, and her ability to find suitable food sources for developing larvae long before the culture of algae, rotifers, and brine shrimp took the guesswork out of larval rearing. She too was among the first to recognize the importance of larval development to taxonomy. Her establishment of two new species of shrimps came about after she recognized the larvae as being distinct and then found specific differences in the adults. During her long career she published some 140 papers, the last in 1959 at the age of 82. She was honored by His Majesty King of the Belgians for her help in classifying natural history collections, was a Fellow of the Linnean Society, and a life fellow of the Zoological Society of London. (Compiled from Russell 1972).

1926   A general survey of larval euphausiids, with a scheme for their identification. *J. Mar. Biol. Assc. UK* 14: 519-527, 1 fig.
1928   The larval stages of the Plymouth Brachyura. *Proc. Zool. Soc., Lond.* 1928: 473-560, 16 pls.
1940   The larvae of the Pandalidae. *J. Mar. Biol. Assc. UK* 24: 239-252, 12 figs.
1959   The larval decapod Crustacea of tropical West Africa. *Atlantide Report* No 5: 119-143.

3.4 *Susan Finnegan*

Susan Finnegan's zoological career was short-lived because of her marriage and subsequent retirement. However, she holds the distinction of being the first woman appointed to a post at the British Museum. She became head of the Arachnida and Myriapoda Section in September 1927, a post she held until her resignation in July 1936. During her tenure most of her research was on the Acari (mites and ticks); however, she made contributions to other arthropod groups as well. Work related to her doctoral thesis led to the publication of a major report on the brachyuran crabs of the tropical eastern Pacific.

1931   Report on the Brachyura collected in Central America, the Gorgona and Galapagos Islands by Dr. Crossland on the 'St. George' Expedition to the Pacific, 1924-25. *J. Linn. Soc. (Zoology)* 37: 607-673, 6 figs.

3.5 *Isabella Gordon*

A contemporary of Finnegan's at the British Museum, Isabella Gordon (Figs 1B, 5B), has been referred to by many as the 'Grand Old Lady of Carcinology' and is perhaps the best known of women carcinologists other than Mary Rathbun. In fact, during a visit to the United States in 1927 these two remarkable women had the occasion to meet. Isabella Gordon began her zoological career in 1923 when she was awarded the first Kilgour Research Scholarship and began a study of the taxonomy of the Alcyonaria at Aberdeen University. This was followed by a postgraduate research scholarship at Imperial College of Science (University of London) where she studied embryology and development in echinoderms.

Her entrance into carcinology came in 1928 when she was a successful candidate for the post of Assistant Keeper of Crustacea at the Natural History Museum, vacated by W.T. Calman. She continued to work in the crustacean section of the Museum until her retirement in November 1966. Soon after her initial appointment she established herself as a respected international carcinologist, publishing numerous papers on nearly all groups of malacostracans. She was elected a Fellow of the Linnean Society and the Zoological Society of London. Perhaps her most prestigious honor was the Order of the British Empire, awarded on November 5, 1961, by the Queen Mother, Queen Elizabeth. Prior to this formal acknowledgement of her work by the British Government, Isabella had spent April and May 1961 in Japan as a guest of the Japanese scientific community, where she received an informal, 2-hour audience with Emperor Hirohito. It was at the time of this visit that the Carcinological Society of Japan was founded, and Isabella Gordon was elected as one of three Honorary Founder Members, the only woman so honored. When queried about her popularity with the Japanese scientific community, she quipped to her friends that one reason was because they had no trouble understanding her pure (Scottish) accent. Her accent and her soft-spoken voice were her trademarks. During her nearly 50 years of active scientific research, Gordon produced 133 publications, ranging from reviews, obituaries, and short scientific encyclopedic articles to monographic treatments of brachyuran decapods.

1930   Brachyura from the coasts of China. *J. Linn. Soc., Lond. (Zool.)* 37: 525-558, 36 figs.
1935   On new or imperfectly known species of Crustacea Macrura. *J. Linn. Soc., Lond. (Zool.)* 39: 307-351, 27 figs.
1950   Crustacea: Dromiacea. Part I. Systematic account of the Dromiacea collected by the 'John Murray' Expedition. Part II. The morphology of the spermatheca in certain Dromiacea. *Sci. Repts. 'John Murray' Exped. 1933-34* 9: 201-253, 26 figs, 1 Fig.
1963   On the relationship of Dromiacea, Tymolinae, and Raninidae to the Brachyura. In *Phylogeny and Evolution of Crustacea. Mus. Comp. Zool. Spec. Publ.* 4: 51-57, 5 figs.

3.6   *Sidnie Milana Manton*

A contemporary of Isabella Gordon's and the person with whom she shared an office at the British Museum after retirement was Sidnie Milana Manton (Figs 1D, 3H). Manton received her M.S., Ph.D., and Sc.D. from Cambridge University but was also awarded an honorary doctorate from the University of Lund (Sweden) in acknowledgement of her many contributions. She was both a teacher and a researcher, holding positions as demonstrator and director of natural science studies at Girton College from 1927-1942 and lecturer and reader at the University of London (1943-1960). In research, she is probably best remembered for her work on jaw mechanisms and locomotion of arthropods, which provided the foundations for her evolutionary theories. However, much of her initial efforts were dedicated to embryology and functional morphology of Crustacea and Onychophora. Her work on development in the mysid *Hemimysis* and the phyllopod *Nebalia* represented the first application of developmental study to crustacean phylogeny. Conclusions drawn from her developmental studies were corroborated by her work on crustacean feeding mechanisms and excretory organs. Dr. Manton was one of the first women to be elected to the Royal Society of London, in 1948, and was the recipient of the Gold Medal of the Linnean Society of London in 1963. (For further information, see Schram this volume).

1928   On some points in the anatomy and habits of the lophogastrid Crustacea. *Trans. Roy. Soc. Edinb.* 56: 103-119, 3 pls.
1934   On the embryology of *Nebalia bipes*. *Phil. Trans. Roy. Soc. Lond.* (B)223: 163-238, 17 figs, 7 pls.

Figure 3. A: Maria (and Fredrick) Dahl. B: Mildred Wilson. C: Michèle de Saint Laurent. D: Sheina Marshall. E: Galina Zevina. F: Georgiana Deevey. G: Eupraxie Gurjanova. H: Sidnie Manton. I: Charlotte Holmquist.

1973   Arthropod phylogeny. A modern synthesis. *J. Zool.* 171: 111-130, 8 figs.
1977   *The Arthropoda. Habits, Functional Morphology and Evolution.* xx + 527 pp., 185 figs, 8 pls., 2 tabls. New York, Oxford University Press.

3.7   *Olive Tattersall*

The carcinological career of Olive Tattersall (Fig. 1C) began as an assistant to her husband, Walter Tattersall. From 1922 to 1939 she was his draftsman, typist and 'handywoman.' When her husband died in 1943, Olive Tattersall undertook the task of completing the monograph on the British Mysidacea that he had begun. Following the publication of that volume she moved from South Wales to London and the British Museum, where she continued her work. In addition to the Tattersall and Tattersall monograph, Olive Tattersall published 17 scientific papers on her own. (Compiled from Gordon 1980).

1955   Mysidacea. *Discovery Report* 28: 1-190, 46 figs.
1961   Mysidacea from the coasts of tropical West Africa. *Atlantide Report* 6: 143-159.
1967   A survey of the genus *Heteromysis* (Crustacea: Mysidacea) with descriptions of five new species from tropical coastal waters of the Pacific and Indian Oceans, with a key for identification of the known species of the genus. *Trans. Zool. Soc., Lond.* 31: 157-193, 48 figs.
1969   A synopsis of the genus *Mysidopsis* (Mysidacea, Crustacea) with a key for the identification of its known species and descriptions of two new species from South African waters. *J. Zool.* 158: 62-79, 21 figs.

3.8   *Sheina Macalister Marshall*

The zoological studies of Sheina Macalister Marshall (Figs 2A, 3D) at the University of Glasgow were interrupted by national service during World War I. In 1922, she was appointed assistant naturalist at the laboratory of the Scottish Marine Biological Association at Millport, and during the subsequent 42 years rose to deputy director of the station formally retiring in 1964. During most of those years Marshall worked cooperatively with Andrew Orr on marine food and food chains. In 1928 Marshall and Orr both joined the Great Barrier Reef Expedition led by Sir Maurice Younge. Their work on the chemistry and phytoplanktonic contents of the coral reef waters were crucial to the success of the expedition. Marshall also made important contributions to our knowledge of coral planulae. Following the expedition they returned to Millport and resumed work on the biology of *Calanus finmarchicus* and other copepods. Marshall authored more than 50 scientific publications, including the famous *The Biology of a Marine Copepod*. In the words of Younge, Sheina Marshall 'was a person of high natural dignity and yet with a keen sense of humor; her scientific work was impeccable.' She was one of five women elected Fellows of the Royal Society of Edinburgh in 1949 when a 1941 proposal to admit qualified women to the Society was finally implemented. In 1961, she was elected a Fellow of the Royal Society of London and received the Neill Prize that same year. In 1964, she was honored with the Order of the British Empire. Her death in April 1977 came shortly before she was to receive an honorary degree from the University of Uppsala (Sweden) on the occasion of its 500th anniversary. (Compiled from Younge 1978, Barnes 1966, Campbell & Smellie 1983).

1933   The production of microplankton in the Great Barrier Reef Region. *Sci. Rept. Great Barrier Reef Exped.* 2: 111-158.
1949   On the biology of small copepods in Loch Striven. *J. Mar. Biol. Assc. UK* 28: 45-122, 32 figs.

1955   *The Biology of a Marine Copepod, Calanus finmarchicus (Gunner)*. Oliver and Boyd, Edinburgh
and London, 188 pp (with A.P. Orr). Reprinted with new preface and selection of papers on the
subject published since 1953 by S.M. Marshall in 1972 by Springer-Verlag, New York, Heidelberg
Berlin, i-vii, 1-195, frontpiece, 63 figs.

1973   Respiration and feeding in copepods. *Adv. Mar. Biol.* 11: 57-120, 10 figs, 5 tabls.

4   WOMEN CARCINOLOGISTS ON THE EUROPEAN CONTINENT

Women made inroads into the carcinological world on the European continent as well. In the
Netherlands, these early pioneers included Annie van Dame, Atie Vorstman, Anna de Vos,
Jentena Leene, and Alida Buitendijk; in the Soviet Union, Eupraxie Gurjanova, Natalia Loma-
kina, Sinaida Kobjakova, and M.A. Dolgopol'skaya; and in Germany, Maria Dahl. This van-
guard was soon followed by Charlotte Holmquist in Sweden; Marit E. Christiansen in Norway;
Jeanne Renaud-Mornant, Michele de Saint Laurent, and Daniele Guinot in France; Galina B.
Zevina, Stella V. Vassilenko, Nina L. Tsvetkova, Elena B. Makkaveeva, and Elizabet Pavlova
in the Soviet Union; and Livia Pirocchi in Italy.

4.1   *Annie van Dame*

Other than knowing that she worked at the Zoological Museum in Amsterdam, little information of the professional career of Annie van Dame is available. However, her impact on the
systematics of galatheid anomurans was considerable. Quite recently a new species of galatheid
was named for her in recognition of her contributions (Baba 1988).

1933   Die Decapoden der Siboga-Expedition. VIII (no. 119). Galatheidea: Chirostylidae. *Siboga Expeditie*
39a(7): 1-46, 50 figs.

1938   Die Gattung *Bathymunida* Balss. *Zool. Anz.* 121: 194-202, 4 figs.

1939   Ueber einige *Uroptychus*-Arten des Museums zu Kopenhagen. *Bijd. Dier.* 27: 392-407.

1940   Anomura, gesammelt vom Dampfer 'Gier' in der Java-See. *Zool. Anz.* 129: 95-104.

4.2   *Adriana Vorstman*

Adriana (Atie) Geertruida Vorstman (Fig. 1E) was born on Sumatra and grew up in West Java.
Vorstman completed her early education in the Netherlands, entering the University of Amsterdam in 1913 as a student of biology. She received her B.Sc. in 1917, M.A. in 1920, and her Ph.D.
in biology in 1922, with her thesis topic on fish teeth. She returned to Java in 1922 where she
remained until 1928. During this period she did extensive field work to collect freshwater
organisms, building upon her interest in hydrobiology. After leaving Java she returned to
Amsterdam with an appointment as assistant, later curator, under the supervision of Professor
Ihlie at the Zoological Institute of the University of Amsterdam. She flourished in this environment both as a teacher and as a researcher, becoming at one point the president of the Dutch
Society of Hydrobiology. In September 1949 she retired from the University. Vorstman pub-
lished some 50 papers during her career with several of these specifically dealing with crusta-
ceans. She brought to her studies a thoroughness which serves as a model for current research.
After retirement she suffered ill health for several years and died in 1963. (Condensed from Van
Benthem Jutting 1964).

1939   Biologische Notizen betreffs der Zuiderzeekrabbe *Pilimnopeus tridentatus* (Maitland) syn. -

Heteropanope tridentata (Maitland). Mit einer nathehatischen Bearbeitung des Materiales von A. Witt van Herk. *Bijd. Dierk.* 27: 369-391, 1 fig, 1 Fig.

1944   Over het voorkomen van *Peridinium borgei* Lemm. in de Botscholsche plassen en van *Leander longirostris* M.E. in de kom bij de suikerfabriek te Halfweg. *Handl. Hydrobiol. Club Amsterdam* 6: 23.

1949   Limnologische gegevens. *Bijd. Dierk.* 28: 530-539, 2 figs.

1951   A year's investigations on the life cycle of *Neomysis vulgaris* Thompson. *Proc. Internl. Assc. Theoret. Applied Limnol.* 11: 437-445, 5 figs.

4.3   Anna de Vos

Anna Petronella Cornelia de Vos (Fig. 6A), 'Nel', had one of the more unusual beginning for a carcinologist. She qualified as a primary school teacher in 1913, then taking the diploma for needlework in 1914. The final examination did not qualify her for a university degree; however, she matriculated as a student of biology at the University of Amsterdam that same year. By 1918 she had been certified for secondary school teaching in biology and geology, but her interest in hydrobiology and appointment as an assistant at the Rijksinstituut voor Biologisch Visserijonderzoek precluded any further teaching career. She devoted all her time to the study of hydrobiology, particularly of the Dutch fresh and brackish waters, until economy measures eliminated her post. In 1938, she was appointed an adjunct assistant to the Director of the Amsterdam Zoological Museum and in 1946 named conservatrix of the Museum. By that time she had gained recognition for her work with ostracodes, aquatic insects, copepods, and even annelids. She died in 1958 at the age of 66 after a sudden illness. During her 37-year professional career, she published a total of 26 scientific papers on the Dutch fauna. (Condensed from Engel & Stock 1959.)

1939   Over de oever- en bodemfauna der binnendijksche kolken langs de kust van het IJsselmeer. *Handl. Hydrobiol. Club Amsterdam* 2: 1-9.

1945   Contributions to the copepod fauna of the Netherlands. *Archives Néerlandaises de Zool.* 7: 52-90, 99 figs.

1953   Three new commensal ostracods from *Limnoria lignorum* (Rathke). *Beaufortia* 4: 21-31, 7 figs.

1957   Liste annotee des ostracodes marins des environs de Roscoff. *Arch. Zool. Exp. Gén.* 95: 1-74, 28 pls.

4.4   Jentena Emma Leene

Jentena Leene (Fig. 4A), known to her friends as 'Tine' (pronounced Teena), was born and raised in Amsterdam. In 1931, she obtained her master's in biology and became a biology teacher in the Amsterdam area. During her spare and vacation times she worked on her Ph.D. thesis, 'The Portunidae of the Siboga Expedition I,' publishing several papers in connection with this work. On June 22, 1938, she was awarded the degree with highest honors. Teaching (she worked for at least three schools simultaneously) absorbed so much of her time that she had to neglect her work on portunids after receiving the degree. In 1945, she was offered the curatorship in the section of fiber technology at the Laboratory for Mechanical Technology, Polytecnical University of Delft and in 1949 also became a lecturer in fiber technology. She held these two positions until her retirement in 1971.

1936   Note on *Charybdis erythrodactyla* (Lam.), *Charybdis acutifrons* (DeMan), and *Charybdis obtusifrons* nov. spec. *Zool. Mededel.* 19: 117-127, 12 figs.

1937   Notes on *Charybdis demani* nov. spec., *Charybdis variegata* var. *brevis pinosa* nov. var. and other *Charybdis* species. *Zool. Mededel.* 19: 165-176, 4 figs.

1938   The Portunidae of the Siboga Expedition, I. *Siboga Expeditite* 39(c3) (no. 131): 1-156, 87 figs.

1940   The Portunidae of the Snellius Expedition (Part I.). Biological Results of the Snellius Expedition. VI. *Temminckia* 5: 163-188, 5 pls.

4.5 *Alida Margaretha Buitendijk*

Alida Buitendijk (Fig. 4G) inherited her interest in zoology from her father, a ship's surgeon with one of the principal Dutch steamship companies. She received her early education in Leiden and studied at the University for a teaching degree in biology. However, her secondary education was not the kind that would permit her to obtain a more advanced degree. She began her work at the Rijksmuseum in Leiden through an administrative post in the Arthropod Section, where her scientific abilities were quickly recognized. She started her work with crustaceans by studying the Paguridae of the Snellius Expedition, later studying the brachyurans from the expedition and other tropical crab collections. When Lipke Holthuis began his work on Fauna van Nederland, Buitendijk studied the crabs for the series. In 1937, she was appointed to the position of assistant curator of Crustacea and Chelicerata and in 1944 was appointed to a full curatorship. Miss Buitendijk was only 47 at the time of her death from cancer in 1950. During her short professional career she published 21 scientific papers, two of which appeared after her death. (For further information, See Holthuis this volume).

1937   The Paguridae of the Snellius Expedition. *Temminckia* 2: 252-280, 19 figs.

1939   The Dromiacea, Oxystomata and Oxyrhyncha of the Snellius Expedition. *Temminckia* 4: 223-276, 27 figs, 5 pls.

1950   On a small collection of Decapoda Brachyura, chiefly Dromiidae and Oxyrhyncha, from the neighbourhood of Singapore. *Bull. Raffles Mus., Singapore* 21: 59-82.

1960   Brachyura of the families Atelecyclidae and Xanthidae. (Part I). Biological Results of the Snellius Expedition XXI. *Temminckia* 10: 252-338, 9 figs (posthumous publication).

4.6 *Maria Johanna Dahl*

Maria Dahl (Fig. 3A) was born in Boromlya, near Kharkov, Russia, in 1872. An honor graduate from the Girls' Gymnasium in Kharkov, she moved with her family to Kiel where she expected to become a medical student. However, Germany's medical schools were open only to men, so Maria's academic hopes were quickly ended. She found employment as an assistant to Karl Brandt, sorting and illustrating collections from the 1889-90 Plankton-Expedition, and while working for Brandt she met and later married Friedrich Dahl. When Friedrich became burdened with other responsibilities, he encouraged Maria to pursue the work he had begun on the corycaeid copepods of the Expedition. While caring for their four children, Maria worked at home to complete the first monograph of a planned series. World War I interrupted work and after the war the Dahls' attention turned to arachnid research, culminating in *Die Tierwelt Deutschland* of which Friedrich was the founder, author of the first 3 volumes, coauthor with Maria of volume 5 and editor of 15 volumes. After his death in 1929, Maria continued as editor of the series until 1968. She died in 1972 a few months before her 100th birthday. (Condensed from Damkaer and Mrozek-Dahl 1980).

1912   Die Copepoden der Plankton-Expedition. I. Die Corycaeinen, mit Berücksichtigung aller bekannten Arten. *Ergebnisse der Plankton-Expedition der Humboldt-Stiftung* 2(G)f(1): 1-134, 16 pls.

Figure 4. A: Jentena Leene. B: Georgiana Deevey, in the field. C: Dorothy Bliss, in the field. D: Elizabeth Batham. E: Jeanne Renaud-Mornant. F: Shigeko Ooishi, center. G: Alida Buitendijk. H: Isobel Bennett.

4.7 *Eupraxie Fedorovna Gurjanova*

Although it was difficult for women to break into male dominated science around the world, it was particularly difficult for women in the USSR. Nonetheless, Eupraxie Fedorovna Gurjanova (Fig. 3G) became one of the Soviet Union's most respected carcinologists. Born in 1902, she began her scientific work in the field of hydrobiology in 1920 while a student at Petrogradskago University. In the summer of 1921, she undertook an apprenticeship with Professor K.M. Deryugina, founder of Leningrad's School of Oceanography and Hydrobiology. In 1922, Gurjanova was involved in a large scale and complex expedition to the White Sea. After completing her university work in 1924, she participated in a number of additional expeditions and distinguished herself as both a researcher and manager. From 1939 to 1952 she was a member of the department of hydrobiology and ichthyology at Leningrad State University and, after the sudden death of Professor Deryugina, she took over the department, sucessfully developing its direction toward marine hydrobiology. She continued her interest and participation in exploratory marine biology, working in the White and Bering seas, Kuril and Sakhalin islands areas, and the northwestern Pacific Ocean. Gurjanova was participant in the IX Pacific Science Congress in Thailand in 1957, and in 1963 was part of the organizing committee for the Institue of Oceanology for the Cuban Academy of Sciences. In 1966, she was invited to present a paper on marine hydrobiology at the Royal Society of London, and in 1967 chaired a section at the International Symposium in Norway. Not only was she a respected scientist, she was also an exceptionally talented lecturer and a teacher whose scientific enthusiasm always stimulated her students. She has been described as a woman with inexhaustible energy, enormous ability for hard work, and an exceptional devotion to interests of science. She died in 1972, having published over 170 scientific works; however, several additional papers have been published posthumously (condensed from Baranova and Ushakov 1972).

1935   Zur Zoographie der Crustacea-Malacostraca des arktischen Gebietes. *Zoographica* 2: 555-571.
1951   Bokoplavy Morey SSSR i Sopredel'nykh Vod. (Amphipoda-Gammaridea of the seas of the USSR and adjoining waters.) *Opredeliteli po Faune SSSR, Izv. Zool. Inst. Akad. Nauk SSSR* 41: 1-1029, 705 figs.
1955   Novye vidy bokoplavov (Amphipoda, Gammaridea) iz severnoi chasti Tikhogo Okeana. (Studies on the amphipods (Amphipoda, Gammaridea) of the North Pacific Ocean.) *Trudy Zoo. Inst. Akad. Nauk SSSR* 18: 166-218, 23 figs.
1962   Bokoplavy severnoi chasti Tikhogo okeana (Amphipoda Gammaridea). Chasti I. (Scud shrimps (Amphipoda Gammaridea) of the northern part of the Pacific Ocean. Part 1). *Opredeliteli po Faune SSSR, Izv. Zool. Inst. Akad. Nauk SSSR* 74: 1-440, 143 figs.

4.8 *Natalia Borisovna Lomakina*

Natalia Lomakina was a systematist perhaps best known for her work on Euphausiacea. Her collection in the Zoological Institute of the Academy of Science (Leningrad) represents specimens gathered from all of the world oceans. However, before she began her work on this group she had already established herself as an authority on the morphology, systematics, and distribution of Cumacea in the Soviet Seas. At the time of her death in September 1972, she was working on a monograph of the Euphausiacea of the world that included complete descriptions and figures of 83 species with keys to their identification. This monograph was finally published in 1978.

1955   Kumbye raki (Cumacea) dalnevostochnykh morey. (Cumacea of the Far-East Seas). *Trudy Zool. Inst. Akad. Nauk SSSR* 15: 112-165.

1958   Kumbye raki (Cumacea) morey SSSR. (Cumacea of the Seas of the USSR). *Opredeliteli po Faune SSSR, Izv. Zool. Inst. Akad. Nauk SSSR* 66: 1-301, frontispiece 201 figs.

1964   Fauna evfauziid (Euphausiacea) antarkticheskoy i notalnoy oblastey. (The euphausid fauna (Euphausiacea) of the Antarctic and Notal regions.) V ki *Issledovanpya Fauny Morey* 2: 254-334, 23 figs.

1978   Zufauziidy mirovogo okeana (Euphausiacea). (Euphausiids (Euphausiacea) of the world ocean.) *Opredeliteli po faune SSSR, Izv. Zool. Inst. Akad. Nauk SSSR* 118: 1-222, 133 figs.

4.9   *Sinaida Ivanovna Kobjakova*

We have been able to gather little personal information on Sinaida Kobjakova. Apparently she spent her entire professional career at Leningrad State University. She is best remembered for her systematic and faunistic studies of decapod crustaceans from eastern and Arctic Soviet seas published over a period of 50 years. Her last paper published in 1986, 2 years before her death, represents a diversion from her crustacean work, being an ecological study of agonid fishes, coauthored with Z. V. Krasyukova.

1937   Desyatinogie raki (Decapoda) Okhotskogo i Yaponskogo morei. (Decapod crustaceans (Decapoda) of the Seas of Okhotsk and Japan.) *Uch. Zap. Leningradskii Gosudarstvenni Universitet* 15: 93-228, 10 figs.

1956   Zakonomernosti raspredeleniya desyatinogikh rakov (Decapoda) v raione Yuzhnogo Sakhalina. (Regularities of the distribution of decapod crustaceans (Decapoda) in the south Sakhalin area.) *Trudy Problemnykh i Tematischeskikh Sovenshchanij Zool. Inst. Akad. Nauk SSSR* 6: 47-64, 1 fig, 2 tabls.

1967   Desyatinogie raki (Crustacea, Decapoda) zaliva Poc'et (Yponskoe rope). (Decapod crustaceans (Crustacea, Decapoda) of the Pos'yet Bay (Sea of Japan.) V kn.: *Issledovaniya Fauny Morei* 5(13): 230-247, 1 tabl. Izdavaemye Zoologicheskiy Institut Akademii Nayk SSSR.

1979   Osobennosti raspredeleniya desyatinogikh rakov (Crustacea, Decapoda) na shelfe Kuril'skikh ostopovov. (Peculiarities of distribution of decapod crustaceans (Crustacea, Decapoda) on the Kuril Islands shelf.) V kn.: *Biologiya shelfa Kuril'skikh Ostrovov, Moskva* pp. 95-111.

4.10   *M. A. Dolgopol'skaja*

A contemporary of Kobjakova's and another Russian carcinologist for whom we have little personal information is M. A. Dolgopol'skaja from the Institute of Biology of the Southern Seas of the Academy of Science of the Ukraine in Sebastopol. Although Dolgopol'skaja specialized in the decapods of the Black Sea and their larval development, she also contributed to our knowledge of anostracans, barnacles, cladocerans, copepods, and fouling organisms.

1938   Dopolnenie k faune rakoobrazhnykh Chernogo morya. (Some additions to the crustacean fauna of the Black Sea.) *Trudy Azovo-Chernomorskii Nauchno-issl. Inst. Morskogo Ryb. Khozyaist. Okenaogr.* 11: 134-153, 5 pls.

1959   Cladocera Chernogo morya. (Black Sea Cladocera). *Trudy Sevastopol'skoy Biologicheskoy Stantsii* 10: 27-75.

1969   Otryad Kalanoida-Calanoida G.O. Sars. (Free-living Calanoida.) V kn: F.D. Mordukhai-Boltovskoi, Klass Rakoobraznye-Crustacea. *Opredelitel' fauny Chernogo i Azovskogo Morey* 2: 34-48, 4 pls.

1974   O biologicheskom deistvii postoyannogo magnitnogo polya (PMP) na *Artemia salina*. (Biological effect of a constant magnetic field on *Artemia salina*.) *Gidrobiol. Zh.* 10: 63-69.

4.11   *Charlotte Holmquist*

Like so many other women carcinologists, Charlotte Holmquist (Fig. 3I) began her professional career as a secondary school teacher, but found that rather dull. In 1946, she entered Lund

University to study botany, although her doctoral research was on the 'Problems on marine-glacial relicts on account of investigations on the genus *Mysis.*' In 1960, she joined the staff of the Naturhistoriska Riksmuseet in Stockholm, a position she held until her retirement in 1983. During those years she studied not only mysids, but turbellarians, polychaetes, and other freshwater faunal elements of lakes in Sweden, Norway, Denmark, England, Ireland, and northern Alaska, and similar groups in the marine environments of Greenland, northwestern Canada, southern Alaska, and northwestern Washington. Holmquist published some 57 scientific papers over the course of her career, the last a key to the Mysidacea of the Pacific northwest in 1986 with Kendra Daly. Her accomplishments are particularly notable when it is recognized that she worked in the male dominated environment of the Swedish Natural History Museum. After 23 years she decided to 'shake the dust off her feet' because the 'atmosphere for a female scientist, a woman who had the courage of her convictions was not the best at the department.' She declined the departmental offer of a tiny study in an 'out-of-the-way' spot in the department and moved to her old hometown of Goteborg where she still lives.

1957	Reports of the Lund University Chile Expedition 1948 - 49. 28. Mysidacea of Chile. *Lunds Univ. Årsskr. N.S. 2* 53: 1-53.
1959	*Problems on Marine-Glacial Relicts on Account of Investigations on the Genus, Mysis.* Berlingska Boktryckeriet, Lund. 270 pp., 80 figs.
1973	Taxonomy, distribution and ecology of the three species *Neomysis intermedia*, (Czerniavsky), *N. awatschensis* (Brandt) and *N. mercedis* Holmes (Crustacea, Mysidacea). *Zool. Jahrb. (Syst.)* 100: 197-222, 9 figs.
1982	Mysidacea (Crustacea) secured during investigations along the west coast of North America by the National Museums of Canada, 1955-1966, and some inferences drawn from the results. *Zool. Jahrb. (Syst.)* 109: 469-510, 7 figs.

4.12 *Marit Ellen (Hammerstad) Christiansen*

Marit Christiansen (Fig. 2H) received an Honors Degree in Science in 1951 from the University of Oslo, during which time she worked at the University's biological station at Drobak on the life history of chitons. Following graduation she accepted a part-time position as senior research assistant at the Station. Her work with decapod crustaceans began after her transfer to the Tromso Museum in 1955. Funding from the Norwegian Research Council for Science and the Humanities (NAVF) allowed Christiansen to continue her studies of prosobranch mollusks and brachyuran decapods with sojourns to the Stazione Zoologica in Naples, Swedish Museum in Stockholm, Zoological Museum in Copenhagen, and Rijksmuseum in Leiden. When her husband, also a zoologist, was awarded a fellowship to the University of Miami in 1966, she and their two young sons accompanied him. There she had the opportunity to work with A.J. Provenzano Jr., and learn from him the techniques for rearing decapod larvae. In 1972 Christiansen was awarded a fellowship from the Nordic Council and a Fulbright Fellowship that enabled her to return to the United States and work both at the University of Miami's Rosenstiel School of Marine and Atmospheric Sciences and the Duke University Marine Laboratory.

Following her fellowships, Christiansen returned to the University of Oslo in the position of Curator of the Zoological Museum. In 1974 she was awarded the Ph.D. For the next several years she was able to spend from 6 weeks to 5 months as a research associate at the Duke University Marine Laboratory where she worked cooperatively with John Costlow on the effects of insect growth regulators on development of crab larvae. In addition to her curatorial duties at the Museum and larval rearing studies, Christiansen teaches undergraduate and graduate course at the University, is Chief Editor for *Zoological Scripta* and is a symposium organizer. In the mid-1980's, she was one of three women at the University to be awarded the rank of full professor by the King of Norway. To date Christiansen's scientific publications number 51.

1962   The Crustacea Decapoda of Isfjorden. A comparison with the Swedish Spitsbergen Expedition in 1908. *Acta Borealia A. Scientia* 19: 1-53, 11 figs (with B.O. Christiansen).

1969   Crustacea Decapoda Brachyura. *Marine Invertebrates of Scandinavia* 2: 1-143, 54 figs, 47 maps. Oslo, Universitetsforlaget.

1972   *Crustacea Decapoda Tifokreps.* Zoologiske bestemmelsestabeller, pp. 1-71, 91 figs. Oslo, Universitestsforlaget.

1988   Hormonal processes in decapod crustacean larvae. *Symp. Zool. Soc., Lond.* 50: 47-68.

4.13 *Jeanne Renaud-Mornant*

Jeanne Renaud-Mornant (Fig. 4E) first studied at the University of Bordeaux, then began research on the meiofauna of the southwest Atlantic coast of France. During 1953-1954 Renaud-Mornant studied subtropical marine biology at the University of Miami under the auspices of both Fullbright and Smith-Mundt fellowships. She was awarded her doctorate from the University of Paris in 1961 and remained at the University's Laboratory of Comparative Anatomy until 1967. She then transferred to the Zoological Laboratory of the Musum National d'Historie Naturelle and in 1987 to the institution's Marine Invertebrate Laboratory. Although noted particularly for her work with the Tardigrada, during her nearly 40 years of active research, Renaud-Mornant published 112 scientific publications, covering all aspects of meiofaunal populations. Among these are 14 papers on the systematics, ecology, physiology, and morphology of mystacocarids and 2 on interstitial isopods.

1963   Recherches écologiques sur la faune interstitielle des sables (Bassin d'Arcachon, le de Bimini, Bahamas). *Vie et Milieu* Sup. no. 15: 1-157, 76 figs, 6 pls.

1974   Étude du système nerveux de *Derocheilocaris remanei* Delamare et Chappuis, 1951 (Crustaca, Mystacocarida). *Cah. Biol. Mar.* 15: 589-604, 2 figs, 5 pls. (with S. Baccari).

1976   Un nouveau genre de Crustacé Mystacocaride de la zone néotropicale: *Ctenocheilocaris claudiae* n. g., n. sp. *C. R. Hebd. Séan. Acad. Sci.* 282: 863-866, 2 figs.

1980   Mystacocarides du brésil. Description de deux espèces nouvelles du genre *Ctenocheilocaris* Renaud-Mornant, 1976 (Crustacea). *Vie et Milieu* 28-29, (AB): 393-408, 19 figs.

4.14 *Michèle de Saint Laurent*

Michèle de Saint Laurent (Fig. 3C) has gained international recognition through her systematic and phylogenetic work on anomuran and thalassinoid decapods. She was also among those who first recognized the importance of larval characters in the assessment of phylogenetic relationships. She began her work at the Muséum National d'Histoire Naturelle, Paris, in the 1950's in collaboration with the noted carcinologist Jacques Forest, first publishing under the name Dechancé and later under de Saint Laurent. Her early accomplishments included monographic revisions of several hermit crab genera. She followed with equally comprehensive studies of thalassinids, including a new phylogenetic classification. During a visit in 1975 to the US National Museum she was asked to examine a strange specimen that had been collected some 67 years before, during a cruise of the US Fisheries steamer Albatross. Her preliminary observations, confirmed by Forest, resulted in a description of the 'living fossil' *Neoglyaphea inopinata*. The following year, while participants in MUSORSTOM expedition to the Philippine Islands, she and Forest collected nine additional specimens and were able to document the external morphology of this extremely important genus of a decapod group previously thought to be extinct.

1963   Développement direct chez un Paguride, *Paguristes abbreviatus* Dechancé, et remarques sur le développement des *Paguristes. Bull. Mus. Natl. Hist. Natur.* (2)35: 488-495, 13 figs.

1968   Révision des genres *Catapaguroides* et *Cestopagurus* de quatre genres nouveaux. I. *Catapaguroides* A. Milne Edwards et Bouvier et *Decaphyllus* nov. gen. (Crustacés Décapodes Paguridae). *Bull. Mus. Natl. Hist. Natur.* (2) 39: 923-954, 1100-1119, 57 figs.

1973   Sur la systématique et la phylogénie des Thalassinidea: definition des families des Callianassidae et des Upogebiidae et diagnoses de cinq genres nouveaux (Crustacea Decapoda). *C. R. Hebd. Sean. Acad. Sci.* 277: 513-516.

1981   La morphologie externe de *Neoglyphea inopinata*, espèce actuelle de Crustacé Décapode Glypheide. In: Résultats des Campagnes MUSORSTOM. I Philippines (18 -28 mars 1976). *Mémoires d'ORSTOM* 91: 51-84, 28 figs (with J. Forest).

4.15 *Daniele Guinot*

Daniele Guinot has excelled in the field of brachyuran systematics and phylogeny. Like de St. Laurent, Guinot began her work in the 1950's at the Muséum National d'Histoire Naturelle in collaboration with Jacques Forest . She published extensively on various brachyuran groups before completing her monumental doctoral thesis in 1977 that laid the foundation for her revolutionary classification of brachyuran crabs.

1964   Sur une collection de Crustacés Décapodés Brachyoures de Mer Rouge et de Somalie. *Boll. Mus. Stor. Nat. Venezia* 15: 7-63, 39 figs, 4 pls.

1977   Donnes nouvelles sur la morphologie, la phylogénie et la taxonomie des Crustacés Décapodés Brachyoures, *Thèse de Doctorat d'Etat es Sciences*, soutenue le 21 juin 1977 a l'Université Pierre-et-Marie Curie, 2 volumes in folio p. I-XV, 1-486, XVI-XXIV, 56 f. n. n., 78 figs, 31 pls, 2 fig. n.n., 14 tabl. (Published in 1979, *Mém. Mus. Natl. Hist. Natur. N.S. A, Zool.* 112: 1-354, 70 figs, 27 pls).

1978   Principes d'une classification volutive des Crustacés Décapodés Brachyoures. *Bull. Biol. France et Belg. N.S.* 112: 211-292, 3 figs, 1 tabl.

1989   Le genre *Carcinoplax* H. Milne Edwards, 1852 (Crustacea, Brachyura: Goneplacidae). In J. Forest (ed.) Résults des Campagne MUSORSTOM, 5. *Mém. Mus. Natl. Hist. Natur. (A)* 144: 265-345, 46 figs, 13 pls.

4.16 *Galina Benizianovna Zevina*

Galina Zevina (Figs 3E, 5F) received her biological training at the University of Moscow under the tutelage of Prof. L.A. Zenkevitch. She is best known for her extensive work on the systematics of pedunculate and sessile barnacles; however, she also has published extensively on marine fouling. In addition, Zevina is a teacher, holding the Chair of Invertebrates at the University of Moscow, and is on the staff of the Institute of Oceanology.

1957   Usonogie raki (Cirripedia Thoracica) morey SSSR. (Barnacles of the seas of the USSR.) V. kn. *Fauna SSSR, Rakoobraznye* 6: 1-267, frontispiece, 106 figs, 3 pls. Moskva, Izdatel'stvo Akademii Nauk SSSR (with N.I. Tarasov).

1972   *Obrastaniya v Moryakh SSSR.* (Fouling in the seas of the USSR.) 214 pp., 56 figs, 47 tabls. Moskva, Izdatel'stvo Moskovckogo Universiteta.

1981   Usonogie raki podotryada Lepadomorpha (Cirripedia, Thoracica) mirovogo okeana. Chast' I. Semeystvo Scalpellidae. (Barnacles of the suborder Lepadomorpha (Cirripedia, Thoracica) of the world ocean. Part I: Family Scalpellidae.) *Opredeliteli po faune SSSR, Izv. Zool. Inst. Akad. Nauk SSSR* 127: 1-406, 300 figs.

Figure 5. A: Mary J. Rathbun. B: Isabella Gordon. C: Isabel Perez Farfante. D: Dora Henry, 2nd from left, christening the *Thomas G. Thompson*. E: Dorothy Bliss. F: Galina Zevina, 2nd from left, during the expedition to Maldiven Is. G: Pat Dudley. H: Mary Alice McWhinnie, aboard the USNS Eltanin.

1982   Usonogie raki podotryada Lepadomorpha (Cirripedia, Thoracica) mirovogo okeana. Chasti II. (Barnacles of the suborder Lepadomorpha (Cirripedia, Thoracica) of the world ocean. Part II.) *Opredeliteli po faune SSSR, Izv. Zool. Inst. Akad. Nauk SSSR* 133: 1-222, 162 figs.

4.17 *Stella Vladinirovna Vassilenko*

Originally a student of Gurjanova, Stella Vassilenko, (also spelled in Russian with a single 's') has forged a distinctive place in Russian carcinology with her work at the Zoological Institute in Leningrad. Vassilenko initially followed Gurjanova with systematic work on amphipods, particularly caprellid amphipods, and has now diversified to include systematic studies of decapods.

1967   Fauna Kaprellid (Amphipoda, Caprellidae) zaliva Pos'et izuch. (Caprellid fauna (Amphipoda, Caprellidae) of the Pos'yet Bay.) *Belogo Morya* 1: 391-410.
1972   Novoe semeystvo, novye rody i vidy kaprellid (Amphipoda, Caprellidae) iz Severnoy Chasti. (A new family, new genus and species of caprellids (Amphipoda, Caprellidea) from the North Pacific.) *Trudy Zool. Inst.* 52: 237-250, 9 figs.
1974   Kaprellidy (Morskie Kozochki) Morey SSSR i Sopredel'nych Vod. (Caprellids from the seas of the USSR and adjoining waters.) *Opredeliteli po Faune SSSR, Izv. Zool. Inst. Akad. Nauk SSSR* 1-287 pp, 185 figs.
1982   Paracercopidae (Amphipoda) iz severo-zapadnoy chasti Tikhogo Okeana. (Paracercopidae (Amphipoda) from the northwestern Pacific Ocean.) V. kn. *Issledovaniya Fauny Morey* no. 29: 95-101.

4.18 *Nina Leverevna Tsvetkova*

Another former student of Gurjanova, Nina Tsvetkova, also is a professional in her own right. In addition to her systematic, ecological, and zoogeographic work on gammaridean amphipods, she is the curator of the crustacean collections at the famous Leningrad Zoological Institute.

1965   Novye rod gammarid (Amphipoda, Gammaridae) iz pribrezhnykh uehastkov Japonskogo Morya. (New genus of gammarid (Amphipoda, Gammaridae) from the intertidal zone of the Japan Sea.) *Zool. Zh.* 44: 1631-1636, 2 figs.
1967   O faune i ekologii bokoplavov (Amphipoda, Gammaridea) zaliva Poc'et (Yaponskoe More). (On the fauna and ecology of amphipods (Amphipoda, Gammaridea) of Poc'yet Bay (Japan Sea).) V ki *Issledovaniya Fauny Morey* 5: 160-195.
1975   *Pribrezhnye gammaridy severnykh i dalnovostochnkh morey SSSR i sopredelnykh vod.* (Coastal gammarids of the northern and far eastern seas of the USSR and surrounding waters.) Leningrad, Akademiya Nauk Zoologischeskiy Institut, 257 pp.
1980   Novye vid roda Photis (Amphipoda, Corophioidea) iz Beringova Morei. (Studies on the genus Photis (Amphipoda, Corophioidea) from the Bering Sea.) V ki *Issledovaniya Fauny Morey* 24: 101-104.

4.19 *Elena Borisovna Makkaveeva*

Elena Makkaveeva, of the Institute of Biology of the Southern Seas, is another carcinologist with wider interests. Makkaveeva received her doctorate in Biological Science in 1954 and subsequently joined the staff of the institute in Sevastopol. She has published on the biology and morphology of caridean shrimp, tanaidaceans, and isopods as well as the dynamics of biocenoses.

1963   Zaroslevye biotsenozy Sredizemnogo Morya. (Epigrowth biocenoses of the Mediterranean Sea.) *Trudy Sevastopol. Biol. Sta.* 14.

1969   Opredelitel kleshnenosnykh oslikov Chrenogo mora. (Identification keys of the fauna of the Black and Asov Seas. Freeliving Invertebrates.) *Opredelitel fauny Chernogo i Azovskogo morey.* 2: 402-408. Kiev.

1979   Bespozvonochnye zarosley makrofitov Chernogo morya. *Akademia Nauk, Institut Biologii Yuzhnykh morei*, Kiev: Nauk Dumka. 228 pp (with A.O. Kovalevskogo).

1983   Zaroslevye biotsenozy. Sistemnyy analiz i modelirovanie protsessov na shelfe Chernogo morya. (Epigrowth biocenoses. Systematic analyses and modeling processes on the Black Sea shelf.) *Sevastopol* 1983: 123-131.

4.20 *Elizabet V. Pavlova*

Another eminently successful woman carcinologist at the Institute of Biology of the Southern Seas is Elizabet Pavlova. Born in 1930, Pavlova began publishing on copepods in 1959 and quickly became recognized as a leading authority on planktonic copepods and other crustaceans of the Southern seas. Pavlova and her colleagues at the Sevastopol Institute have a cumulative record of more than 90 years of crustacean research.

1961   Raspredelenie *Penilia avirostris* Dana (Crustacea, Cladocera) v Chernom More i pogloshchenie kisloroda nekotorymi ilanktoniymi rachkami Sevastopol'skoy bukhty. (Distribution of *Penilia avirostris* Dana (Crustacea, Cladocera) in the Black Sea.) *Trudy Sevastopol. Biol. Sta.* 14: 91-101.

1966   Sostav i raspredelenie zooplanktona v Egeiskom more. Ha: *Issledovaniya planktona Yuzhnykh morei.* (Composition and distribution of zooplankton in the Aegean Sea. In: Plankton investigations of southern seas.) *Trudy Okeanogr. Kom. Akad. Nauk SSSR, Moskva*, 38-61, 6 figs, 10 tabls.

1987   *Dvizhenie i znergeticheskiy obmen morskikh planktonnykh organizmov.* (Movement and energetic exchange in marine planktonic organisms.) 212 pp. Kiev, Nauk Dumka.

1988   Methodological and theoretical analyses of the respiration research of planktonic animals. *Polskie Arch. Hydrobiol.* 35: 45-85.

4.21 *Livia Pirocchi*

Livia Pirocchi (Tonalli) was always interested in limnology, and during the early part of her career she specialized in the study of planktonic crustaceans, particularly copepods. As her recognition as a limnologist grew, she was appointed the director of the Institute of Hydrobiology at Pallanza on Lago Maggiore. Shortly before her death in 1985, she was named a corresponding member of the American Society of Zoologists.

1937   I laghi di Antermoia e di Erdemolo, con particolare reguardo al'Arctodiaptomus bacillifer Koelb. *Mem. Mus. Storia Nat. Venezia trident* 4: 19-31, 3 figs, 2 pls.

1941   Diaptomidi d'alta montagna. II. Il Diaptomide del Lago Azzurro de Cervino. *Arch. Zool. Ital.* 29: 89-112, 14 figs.

1944   Distribuzione nella peninsola Italiana di tre specie di Diaptomidi (Eud. vulgaris Schmeil, Arctod. bacillifer Koelb., Acanthod. denticornis Wierz.). *Atti R. Acad. Ital.* 14: 859-888, 5 figs.

1951   Hochendemische Copepoden- und Cladoceren-Lokalformen im Karst. *Archiv Hydrobiol.* 45: 245-253, 4 figs.

5 WOMEN CARCINOLOGISTS IN ISRAEL

5.1 *Bella Galil*

Across the Mediterranean women have also entered Israel's carcinological community. Although younger than some of our other pioneers, Bella Galil (Fig. 2C) already has an impressive publication record. Galil began her graduate work with the noted Israeli carcinologist Chanan Lewinsohn in 1976, obtaining her M.S. (1978) and Ph.D. (1983) from Tel Aviv University. She has held positions as research and teaching assistant in the zoology department of the Tel Aviv University, curatorial assistant at the American Museum of Natural History (1985-1988), senior researcher and curator of Crustacea at Tel Aviv University. The latter affiliation she continues to maintain while currently serving as Senior Research Associate for the Israel Oceanographic and Limnological Research Company in Haifa. From her first publication in 1977 to her most recent (in press), Galil has produced no fewer than 40 scientific papers, most dealing with aspects of the systematics and morphology of coral-inhabiting crabs.

1983   Two new species of *Trapezia* (Brachyura, Decapoda), coral inhabiting crabs from Taiwan. *Microne-sica* 19: 123-129.
1986   *Quadrella* (Brachyura: Xanthoidea: Trapeziidae) - review and revision. *J. Crust. Biol.* 6: 275-293.
1987   The adaptive functional structure of mucus-gathering setae in trapezid crabs symbiotic with corals. *Symbiosis* 4: 75-86.
1988   Trapeziidae (Decapoda, Brachyura, Xanthoidea) of the Red Sea. *Israel J. Zool.* 34: 159-182.

5.2 *Engelina Zelickman*

Engelina Zelickman (Fig. 1G), formerly of the Institute of Oceanology of the Russian Academy of Sciences, now also calls Israel home. Zelickman was born in 1926 in Moscow. She remained there for her early education, attending Moscow University where she earned her Ph.D. in 1955. From 1953 to 1964 she worked at the Murmansk Marine Biological Institute of the Academy of Sciences (Kola Peninsula, Barents Sea) where she directed the department of planktonology and later became the director of the institute. As a provincial laboratory, the Institute had very harsh living and working conditions, sometimes requiring that the scientists forage locally, fishing, hunting, and collecting berries. In 1965, she returned to Moscow and in 1967 took a position at the Institute of Oceanology where she remained until 1985. During this time she was the leader of 24 marine arctic expeditions to the White, Barents, Kra, Okhotsk, and Japan seas. In 1988 she emigrated with her son and his family to Israel where she is now a research associate in the Department of Biology of the Hebrew University of Jerusalem. Zelickman has had three major research thrusts during her long and productive career: Ecology of marine zooplankton in the Northern Hemisphere (Barents, Norwegian, and Greenland seas), parasitology of marine mollusks and crustaceans, and group swarming behavior and communication of crustaceans and hydromedusae. She has just recently added another dimension the systematic study of hyperiid amphipods. She has written some 75 papers since 1950, with 47 of these published since 1965.

1958   Materialy o raspradelenii i razmnozhenii zvfauziid v pribrezhhoy zone Murmana. (Materials on the distribution and reproduction of the euphausiids in nearshore waters Murman (Barents Sea)). *Trudy Murmansk Morsko go Biol. Inst.* (Akad. Nauka SSSR) 4: 79-117.
1961   O podemakh k prverkhnosti morya Barentsevomorskikh z fauziebykh rachkov i nekotorykh cher-takh ikh povedeniya. (On the rising to the sea surface of the Euphausiacea in Barents Sea, and some features of their behaviour.) Ha: *Sbornik Gidrologicheskiye i Gidrobiologischeskie osobennosti pribrezhnych vod Murmana*, pp. 136-152, 4 tabls. Izdanie Kolskogo filiala Academii Nauk SSSR, Murmansk.
1974   Group orientation in *Neomysis mirabilis* (Mysidacea: Crustacea). *Mar. Biol.* 24: 251-258, 6 figs.
1978   Agregirovannost raspredeleniya *Thysanoessa inermis* (Kroyer) i *T. raschii* (M. Sars) (Euphausi-cacea) v Barentsevom More. (Aggregative distribution of *Thysanoessa inermis* (Kroyer) and *T. raschii* (M. Sars) (Euphausiacea) in the Barents Sea.) *Okeanologiya* 18: 1077-1084, 2 figs. 4 tabls. (with I.P. Lukashevich and S.S. Drobysheva).

6   WOMEN CARCINOLOGISTS ASSOCIATED WITH THE PACIFIC BIOLOGICAL STATION AND THE PUGET SOUND BIOLOGICAL LABORATORY

In the early days of marine stations, the Pacific Biological Station at Departure Bay, Nanaimo, British Columbia, Canada, and the Puget Sound Biological Laboratory, later to become the University of Washington's Friday Harbor Laboratories, were the sites of work by several women carcinologists, including Alfreda Berkeley, Josephine F.L. Hart, Belle A. Stevens, Mildred S. Wilson, and Dora P. Henry. As was so frequently the case for early women scientists, most began first as teachers. Some went on to make science a career in its own right, others practiced their science in spare time and on summer vacations.

6.1   *Alfreda Alice Berkeley*

Alfreda Berkeley (Fig. 1F) was born in 1903 in India where her father had recently become employed in the indigo industry. Two years later she was sent back to England because the Indian climate presumably was not suitable for bringing up English children; however, her parents remained in India until 1913. The family was reunited in the Okanagan Valley of British Columbia, but moved to Vancouver in 1916 when career opportunities became available at the new University of British Columbia. Berkeley's mother, Edith, was employed in the zoology department while her father worked for the bacteriology department. It was during the summer months when they spent their time collecting material for the university along Vancouver Island's coasts that they discovered the potentials for research at the Pacific Biological Station. Edith gave up her paid position at the university to devote her time to polychaete taxonomy, subsequently becoming an acknowledged expert in the field. Such an example encouraged Alfreda to also pursue a career in science. She received her doctorate in marine biology from the University of Toronto in the early 1930's and soon married Alfred W. Needler.

Positions for women during the Depression years were virtually nonexistent; however, her husband's position at a small fisheries station gave Berkeley the opportunity to continue her work in marine biology. Her first research efforts dealt with sex reversal in oysters and sexuality and larval development of pandalid shrimp, which resulted in several prestigious publications. In 1941, the Needlers moved to the St. Andrews Biological Station at New Brunswick. Despite the responsibility of caring for three children, Berkeley conducted 'volunteer' research on red tide organisms responsible for food poisoning. She died in 1951 at the age of 48, having made subtantial contributions to our knowledge of caridean larval development and the sex reversal phenomenon (Condensed from Ainley 1986).

1929   Sex reversal in *Pandalus danae. Amer. Nat.* 63: 1-3.
1930   The post-embryonic development of the common pandalids of British Columbia. *Contributions to Canadian Biology and Fisheries, being Studies from the Biological Stations of Canada N.S.* 6: 79-163.
1933   Larvae of some British Columbia Hippolytidae. *Contributions to Canadian Biology and Fisheries, being Studies from the Biological Stations of Canada N.S.* 8: 239-242 (as A.B. Needler).
1938   The larval development of *Pandalus stenolepis. J. Fish. Res. Bd. Canada* 4: 88-95 (as A.B. Needler).

6.2   *Josephine Francine Lavinia Hart*

Josephine Hart (Fig. 2B), or 'Babs,' as she is known to her friends, is a woman carcinologist who also successfully balanced her professional interests with her responsibilities as a wife and mother. She was born in Victoria, British Columbia, in 1909 and received her early education there. She attended the University of British Columbia, 1925-1931, receiving a B.A. with first-class honors in biology as well as an M.A. in this period. During summers from 1929 to 1937 she worked in an unsalaried position at the Pacific Biological Station in Nanaimo where she did extensive analysis of the taxonomy of Cumacea and began her pioneering work on laboratory rearing of decapod Crustacea. After completing her Ph.D. work at the University of Toronto in 1937 and marrying in 1938, Hart worked mostly out of a make-shift laboratory in her basement where, remarkably, she reared complete series of at least 30 species, publishing sporadically on crustacean development and distribution until 1962. As a laboratory assistant at Victoria College in 1960, she wrote for funding of her developmental research. Upon receipt of a National Science Foundation grant in 1962 and purchase of needed equipment, she began a steady production of distributional records and larval development descriptions. During the summer months she and her husband, Clifford Carl, often traveled the waters of British Columbia and Puget Sound aboard their home-built houseboat, 'Carl's Ark,' and were frequent visitors to the Friday Harbor Laboratories. In January 1967, she was awarded a 1-month contract to serve as temporary Supervisor of Plankton and instruct the staff of the Smithsonian Oceanographic Sorting Center on the recognition and identification of crustacean larvae. After Clifford's death in 1970, domestic responsibilities required even more of her time; however, she continued compiling data for her long term project, a handbook of British Columbian decapods. With its completion in 1982, Hart 'officially' retired. In recognition of her contributions, she was appointed Honorary Curator of Marine Biology by the Provincial Museum and received an honorary Doctor of Science in 1986 from the University of Victoria. Her colleagues and other friends will long enjoy the unique floral bookmarks (autographed 'Lavinia') that she created to fund her book purchases over the course of her career.

1937   Larval and adult stages of British Columbia Anomura. *Can. J. Res.* (D)15: 179-220, 11 figs, 1 Fig.
1965   Life history and larval development of *Cryptolithodes typicus* Brandt from British Columbia. *Crustaceana* 8: 257-276, 11 figs, 1 Fig.
1971   Key to planktonic larvae of families of decapod Crustacea of British Columbia. *Syesis* 4: 227-234, 22 figs.
1982   *Crabs and their relatives of British Columbia.* Handbook No. 40, i-iii, 1-266, 102 figs, 12 pls. Victoria, British Columbia Provincial Museum.

6.3   *Belle Alice Stevens*

Belle Stevens was a teacher first, and researcher when she could afford the time. None the less, she made the first contributions to our knowledge of north Pacific hermit crabs and thalassinids. She earned her B.S. in 1919 and her M.S. in zoology in 1924 from the University of Washington. She worked as a teacher during all of her time at the University, later becoming the head of the science department in a local high school. Her interest in hermit crabs was sparked by attending a summer course at the Puget Sound Biological Station in 1918. Her work on hermit crabs and other decapods was directed by Trevor Kincaid, as well as by Waldo Schmitt and Mary Rathbun. She continued teaching high school during the academic year and spent summers and spare time on her research. In the summer of 1924, she was employed as a laboratory assistant at the Puget Sound Biological Station and again during the summer of 1928. Subsequently she was provided with a small office in the University of Washington's zoology department and later obtained moderately larger quarters in the University's oceanography department. During her somewhat intermittent, professional scientific career she amassed a great deal of information on the local shrimp fauna, but her few publications were restricted to papers on hermit crabs and callinassids. Following her death in September 1960, one paper on a mesopelagic shrimp from the northeastern Pacific was published jointly by Fenner A. Chace Jr. Because of her long-standing friendship with Lipke B. Holthuis, to whom she sent care packages during World War II, much of her collection was deposited in the Leiden Museum.

1925   Hermit crabs of Friday Harbor, Washington. *Publ. Puget Sound Biol. Station* 3: 273-309, 41 figs.

1927   *Orthopagurus*, a new genus of Paguridae from the Pacific coast. *Publ. Puget Sound Biol. Station* 5: 245-252, 4 figs.

1928   Callianassidae from the west coast of North America. *Publ. Puget Sound Biol. Station* 6: 315-369, 71 figs.

1965   The mesopelagic caridean shrimp *Notostomus japonicus* Bate in the North-eastern Pacific. *Crustaceana* 8: 277-284 (with F.A. Chace Jr.).

6.4   *Mildred Stratton Wilson*

Mildred Wilson (Fig. 3B) began as a student teacher at the young age of 18, but after 2 exhausting years, enrolled in summer courses at the Puget Sound Biological Station in 1929. That first summer was sufficient to stimulate an interest in biology, and although she continued her teaching duties for an additional 6 years, her marriage to Charles Wilson in 1934 ended that aspect of her professional life. When her husband received his M.S. degree from the University of California and became a research associate, Mildred began her academic work as well. She received her bachelor's degree in 1938, and became research associate to Prof. S.F. Light for whom she curated his entomostracan collections. She also began her own work with copepods at this time. Her husband's transfer to Washington, D.C., placed her in a position to continue Light's copepod studies at the US National Museum, where her first publication was produced in 1941. During World War II she was the only professional in the Division of Marine Invertebrates, and although she moved with her husband back to the West Coast, she maintained an honorary appointment as Collaborator in Copepod Crustaceans. It was through this appointment that she was able to continue her copepod research, although most was done from her home, first in Corvallis, Oregon, and then in Anchorage, Alaska. In 1957, she obtained her first financial support from the National Science Foundation and this support was continued first through the Museum and subsequently through the University of Alaska until the time of her death in 1973. During her carcinology career, she published at total of 40 scientific papers, despite 20 years of poor health. (Condensed from Damkaer 1988).

1953   New and inadequately known North American species of the copepod genus *Diaptomus. Smith. Misc. Coll.* 122: 1-30, 58 figs.

1958   A review of the copepod genus *Ridgewayia* (Calanoida) with descriptions of new species from the Dry Tortugas, Florida. *Proc. US Natl. Mus.* 108: 137-179, 37 figs.

1959   Calanoida. In W.T. Edmondson (ed.), *Freshwater Biology.* Second edition, pp. 738-794, 95 figs. New York, John Wiley & Sons.

1975   North American harpacticoid copepods. 11. New records and species of *Elaphoidella* (Canthocamptidae) from the United States and Canada. *Crustaceana* 28: 125-138, 43 figs, 1 tabl.

6.5   *Dora Priaulx Henry*

Dora Henry (Figs 2G, 5D), who retired in 1989 at the age of 85, the oldest research professor in the history of the University of Washington's School of Oceanography, spent less of her time at the Biological Station than on the University's Seattle campus. After completing her Bachelor's degree in 1925 at the University of California, Henry remained at Berkeley to do her graduate work, attaining her master's in 1926 and Ph.D. in 1931. Her graduate research dealt with protozoan parasites of birds and mammals, and subsequently she branched into protozoan parasites of invertebrates after she and her husband moved to Seattle where he was appointed to the faculty of the University of Washington's microbiology department. Henry became a research associate in oceanography and zoology at the University of Washington. Her interest in protozoan parasites, particularly gregarines, led her to their barnacle hosts, and ultimately her attention was turned not to additional parasites, but to the barnacles themselves.

Prior to World War II the Navy went to considerable expense to develop some bottom material or paint that would repel barnacles. After hearing of the tests, Henry offered her services, but it was at a time when the Navy was exclusively male. Not only was she a woman, but also an academician. She was thanked, but her services were not accepted, an unfortunate decision by the Navy because their tests failed miserably. By filtering the water for their test tanks they filtered out all the barnacle nauplii, a problem Henry could have prevented quite easily. Henry's work with barnacles was interrupted by World War II; in 1942 she became an assistant oceanographer in the headquarters branch of the Army Air Force. In 1943 she transferred to the Navy Hydrographic Office where she worked with another carcinologist, Fenner A. Chace Jr., whose career similarly had been interrupted by the war.

In 1945, Henry returned to the University of Washington as a research associate in oceanography and zoology and to her study of barnacles. In 1960, she was promoted to research associate professor in the Department of Oceanography. During the next several years, summer vacations for Henry and her husband consisted of travels to various parts of the world, particularly the Pacific coast of Central and South America, where she insatiably collected barnacles. The result of these 'vacations' was the accumulation of one of the most complete collections of Pacific coast, intertidal, and shallow-subtidal barnacles available anywhere. Much of this collection has now been deposited in the collections of the US National Museum and the Nationaal Natuurhistorisch Museum, Leiden. In 1973, Henry was promoted to research professor, a rank she held until her retirement.

Although Henry rarely found time to write even letters herself, she was, and continues to be, a marvelous editor. Not long ago when the editor of *Deep Sea Research* died suddenly, she filled in as interim editor until a replacement could be found. There have been few manuscripts, theses, or dissertations to come out of the School of Oceanography in recent years that have not been substantially improved by Henry's critical review. In recognition of her devotion to the University and the School of Oceanography, she was chosen to christen the University's new oceanographic vessel R/V *Thomas G. Thompson* in July 1990. This time the Navy recognized her with the acknowledgement that she was one of the few individuals chosen to christen ships that deserved the honor.

Henry's contributions to our knowledge of cirripedes are substantial and include the discovery of the first complemental male to be found in a sessile barnacle. This discovery, described by her colleague William Newman as the greatest since Charles Darwin's barnacle work, changed the concepts of cirripede evolutionary pathways and set the stage for more intensive scrutiny of sessile barnacles by others.

1940   The Cirripedia of Puget Sound with a key to the species. *Univ. Wash. Publ. Oceanogr.* 4: 1-48, 5 text figs, 4 pls.

1942   Studies on the sessile Cirripedia of the Pacific coast of North America. *Univ. Wash. Publ. Oceanogr.* 4: 99-131, 5 text figs, 4 pls.

1973   Descriptions of four new species of the *Balanus amphitrite*-complex (Cirripedia, Thoracica). *Bull. Mar. Sci.* 23: 964-1001, 11 figs.

1986   The recent species of *Megabalanus* (Cirripedia: Balanomorpha) with special emphasis on *Balanus tintinnabulum* (Linnaeus) sensu lato. *Zool. Verhand.* 235: 14 figs, 2 pls. (with P.A. McLaughlin).

7   WOMEN CARCINOLOGISTS IN SOUTH AMERICA: DYRCE LACOMBE & ELDA FAGETTI

7.1   *Dyrce Lacombe*

Important histological and systematic studies of barnacles have been made by the Brazilian carcinologist, Dyrce Lacombe, who has just retired from the Instituto Oswaldo Cruz, Rio de Janeiro, after 35 years. She began her work in cytochemistry and microphotography under her German mentor, Dr. Barth, and her work with barnacles was stimulated by Admiral Moreiro, the Brazilian naval leader who founded the institute. Following the Admiral's untimely death, the institute became the equivalent of a public health institute, and Lacombe's efforts necessarily were directed toward the vector of Chagas's disease. However, her true vocation has always been her research on barnacles, which she was permitted to continue as a 'hobby.' This work has included comparative histological and ultrastructure studies of the cement glands in balanomorph and lepadomorph species, balanomorph larval development, and distributional and systematic studies of Brazilian barnacles. In her public health role, she is noted for her work with the life cycle of the parasite *Trypanosoma cruzi.*

1968   Histologia, histoquímica e ultra-estrutura das glândulas de cimento e seus canais em *B. tintin-nabulum. Inst. Pesq. Mar.* 17: 1-22, 29 figs.
1970   A comparative study of the cement glands in some balanid barnacles (Cirripedia, Balanidae). *Biol. Bull.* 139: 164-179, 27 figs.
1973   Desenvolvimento larvário de Balanídeos em laboratrio *Balanus amphitrite* (var. *amphitrite*). *Mem. Inst. Oswaldo Cruz* 70: 175-206, 66 figs (with W. Monteiro).
1977   Anatomia e microanatomia de Balanidae da Baía de Guanabara (Crustacea, Cirripedia). *Revist. Brasil. Biol.* 37: 151-165, 71 figs.

7.2   *Elda Fagetti*

Another notable woman carcinologist from South America is Elda Fagetti, formerly of the Estacion de Biologia Marina, Universidad de Chile, and more recently with the Food and Agriculture Organization of the United Nations, headquartered in Rome. Her contributions to our knowledge of the breeding seasons, fecundity, and larval development of anomuran and brachyuran decapods of Chile were among the first for the South American fauna.

1960   Huevos y el primer estadio larval del langostino (*Cervimunida johni* Porter 1903). *Rev. Chilena Hist. Nat.* 55: 33-42.
1967   The larval development of the crab *Cyclograpsus einereus* Dana, under laboratory conditions. *Pac. Sci.* 21: 166-177 (with J. Costlow).
1969   The larval development of the spider crab *Libidoclaea granaria* H. Milne Edwards and Lucas under laboratory conditions (Decapoda, Brachyura; Majidae, Pisinae). *Crustaceana* 17: 131-140, 5 figs.
1970   Desarrollo larval en el laboratorio de *Homalaspis plana* (Milne-Edwards) (Crustacea Brachyura; Xanthidae). *Rev. Biol. Mar.* 14: 29-49, 11 figs.

8   WOMEN CARCINOLOGISTS IN NEW ZEALAND AND AUSTRALIA: ELIZABETH BATHAM, FREDERIKA BAGE & ISOBEL BENNETT

8.1   *Elizabeth Joan Batham*

Although perhaps better known for her general ecological studies and directorship of the

Portobello Marine Biological Station, Elizabeth Batham (Fig. 4D) did much of her personal research on barnacles and sea anemones. Batham spent most of her life in her birthplace, Dunedin, New Zealand, with the exception of her years of postgraduate study in England (1945-1950). She completed her bachelor's (1938) and master's (1939) degrees at the University of Otago. In 1941, she accepted a position as senior demonstrator in zoology at Victoria University College in Wellington but returned to Otago the following year to undertake research on hydatid protozoans. It took only a few years to convince her that she was not interested in a lifelong career in medical or agricultural parasitology, and she turned her attentions to marine biology. Following World War II, she began a Shirtcliffe Fellowship at Cambridge University and for the next 5 years distinguished herself with her research on the structure, neurophysiology, and behavior of sea anemones. In 1951, she returned to Otago to take up the directorship of the Portobello Marine Biological Station. It took several years for Batham to refurbish the station, and only after the laboratory was on a sound footing was she able to return to her own research program, which dealt primarily with aspects of the biology, systematics, behavior, and ecology of such groups as crayfish, crabs, barnacles, and even octopods and pogonophorans. Because she believed that SCUBA-diving was a necessary tool in her research, she became a qualified diver after the age of 50. She continued as director of the Portobello laboratory until ill health forced her to resign in 1974, and that same year went on leave to Victoria University of Wellington. She disappeared, without a trace, near the shore at Seatoun, Wellington, in early July 1974, presumably the victim of a SCUBA accident (Condensed from Jillett 1978).

1945   Description of female, male and larval forms of a tiny stalked barnacle, *Ibla idiotica*, n.sp. *Trans. Roy. Soc. New Zealand* 75: 347-356, 29 figs, 2 tabls.

1967   The first three larval stages and feeding behaviour of phyllosoma of the New Zealand palinurid cray fish *Jasus edwardsii* (Hutton, 1875). *Trans. Roy. Soc. New Zealand* 9: 53-64.

1969   Benthic ecology of Glory Cove, Stewart Island. *Trans. Roy. Soc. New Zealand* 11: 73-81, 3 figs, 2 tabls.

1970   On behaviour of symmetrical hermit crab, *Mixtopagurus* n. sp. (Decapoda, Paguridae). *Crustaceana* 19: 45-48, 1 fig, 1 Fig.

8.2   *Frederika Anna Bage*

Frederika Bage was the earliest of women carcinologists in Australia, and like other women of her time, her interest in science was kindled by her father, a wholesale chemist, who was an amateur scientist on the side. She graduated with bachelor's (1905) and master's (1907) degrees from the University of Melbourne. After graduation she worked as a junior demonstrator in biology, but won a research scholarship in 1908 from the Victorian government. A research scholarship from King's College, London, allowed her to go to England (1910-1911) to work under A. Dendy, and that work led to a fellowship of the Linnean Society. Upon her return to Melbourne she was appointed lecturer in charge of biology at the University of Queensland in 1913 and became the first principal of the Women's College of the University in 1914. So involved in her professional academic career and multitude of extramural activities, particularly those involving women's organizations and activities, she was able to devote little time to research. None-the-less she did complete a report on the Decapoda of the Australian Antarctic Expedition (Condensed from Bell 1979).

1938   Crustacea Decapoda. *Sci. Rept. Aust. Antarctic Exped., series C (Zool. Bot.)* 2: 1-13, 4 pls.

8.3   *Isobel Ida Bennett*

Isobel Bennett (Fig. 4H; Fig. 6B) joined the University of Sydney as a zoology librarian and secretary to W.J. Dakin in 1933. Soon after her employment her talents in science were noted, and she became research assistant, then lecturer, and finally demonstrator for the University program. She was one of the first women to accompany an Australian expedition to Macquarie Island and she participated in numerous other expeditions, including one in Alaskan waters. She has studied many aspects of the ecology of the Great Barrier Reef and produced several guidebooks and handbooks illustrating the crustacean fauna of the reef and various intertidal and shallow subtidal areas of Australia.

1953   Intertidal zonation of the exposed rocky shores of Victoria, together with a rearrangement of the biogeographical provinces of temperate Australian shores. *Aust. J. Mar. Freshw. Re.* 4: 105-159 (with E.C. Pope).

1960   *Australian Seashores. A Guide for the Beachlover, the Naturalist, the Shore Fisherman, and the Student (revised addition)*: 1-372. Angus and Robertson Pty, Ltd.: Sydney. (with W.J. Dakin and E.C. Pope). Reprinted in 1973, 1980, 1987.

1966   *The Fringe of the Sea.* 261 pp., 179 pls., Adelaide, Rigby Limited.

1971   *The Shores of Macquarie Island.* 69 pp, 63 pls. Sydney, Rigby Limited.

8.4   *Elizabeth Carrington Pope*

Elizabeth Pope (Fig. 2F), born in 1912, received her B.Sc. in Zoology with honors from the University of Sydney and won the Haswell Prize for her honors year. Albeit 6 years of postgraduate experience, she joined the staff of the Australian Museum in 1939 as a scientist second class, the first graduate woman-scientist on the staff. Of her own hiring circumstances, she commented, '...(as) a female, ... got a much lower salary.' She ultimately became the deputy director for the Museum, participating in administration as well as developing her own scientific interests. She published many popular articles as well as those of general ecological and specific scientific interest. By the end of her career in the late 1970's, she had published well over 100 articles. Her carcinological interests were the studies of barnacles and crayfish.

1945   A simplified key to the sessile barnacles found on the rocks, boats, wharf piles and other installations in Port Jackson and adjacent waters. *Rec. Aust. Mus.* 21: 351-372, 5 figs, 3 pls.

1947   The endless house-hunt. *Aust. Mus. Mag.* 9(4): 129-132, 3 figs.

1956   The barnacle, *Xenobalanus globicipitis* Steenstrup, in Australian seas. *Proc. Roy. Soc. New South Wales* 1956/57: 159-161, 1 fig.

1965   A review of Australian and some Indomalayan Chthamalidae, Crustacea, Cirripedia. *Proc. Linn. Soc. New South Wales* 90: 10-77, 2 pls.

9   WOMEN CARCINOLOGISTS IN PAKISTAN AND INDIA: NASIMA TIRMIZI & SHAKUNTALA SHENOY

9.1   *Nasima M. Tirmizi*

In Pakistan, Nasima Tirmizi (Fig. 6G) began her carcinological research with systematic, developmental, and morphological studies on penaeid shrimp, but has not restricted her investigations to this group. Her accomplishments include faunistic studies of Pakistani stomatopods and a variety of decapod groups, developmental morphology, and estuarine ecology and fisheries. In her 37-year professional career, she has published more than 130 papers and several

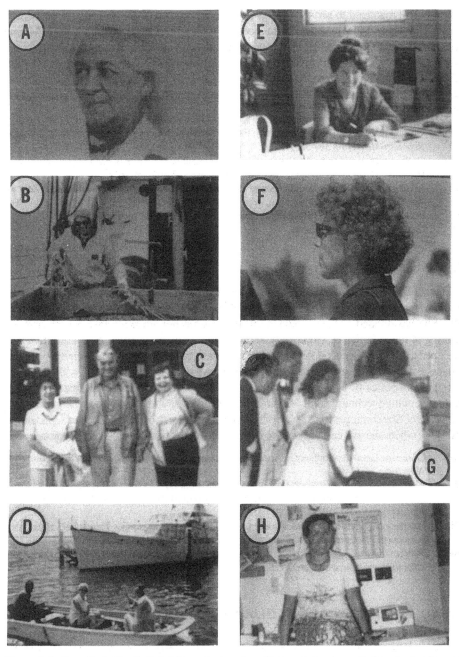

Figure 6. A: Cornelia de Vos. B: Isobel Bennett, aboard an Alaskan shrimp boat. C: Shigeko Ooishi and Pat Dudley, with Paul Illg at the Friday Harbor Laboratories. D: Dorothy Bliss, center. E: Hélène Charniaux-Cotton. F: Mary Alice McWhinnie. G: Nasima Tirmizi, center. H: Geneviève Payen.

books but still managed to serve as professor and chairperson of the zoology department at Karachi University. More recently she has been Director of the Center of Excellence in Marine Biology, a presenter of radio and television programs, and author of a weekly column in the daily *Allah Maaf Kare*. She received the Pakistan Academy of Sciences Gold Medal in 1971 and 1978, and a gold medal from Karachi University in 1987. She is a Fellow of the Zoological Society of London, as well as a Fellow of the Pakistan Academy of Science, and the Zoological Society of Pakistan.

1959   Crustacea: Benthesicymae (Penaeidae), Part II. *The John Murray Expedition 1933-34, Scientific Reports. BM(NH)* 10: 319-383, 96 figs.

1968   On the structure and some developmental stages of genitalia in the prawn *Parapenaeopsis stylifera* H. Milne-Edwards (Decapoda, Penaeidea). *Crustaceana* 15: 193-209, 8 figs.

1972   An illustrated key to the identification of the northern Arabian Sea penaeids. *Pakistan J. Zool.* 4: 185-221, 21 figs.

1983   Four axiids (Decapoda, Thalassinidea) from Indonesia. *Researches on Crustacea* 12: 85-95, 5 figs, Carcinological Society of Japan.

9.2   *Shakuntala Shenoy*

Shakuntala Shenoy (Fig. 1H), who has worked for more than 25 years on aquaculture of decapods, received her M.Sc. (1962) and her Ph.D. (1967) from the University of Bombay, India. She has divided her considerable talents between her aquacultural research and her nurturing of future Indian carcinologists, at times combining these interests. For example she developed a school of decapod larval studies with the Konkan Agricultural University. She is perhaps best known for her team efforts, with her husband Dr. K.N. Sankolli, with the freshwater prawn species of *Macrobrachium*. Shenoy's group was among the first to develop egg-to-egg culture techniques for *M. malcolmsomi* and to hybridize *M. rosenbergii* with *M. malcolmsoni*. She continues her teaching and studies as research officer at Konkan Agricultural University where she will certainly add to her current publication list of 55 contributions.

1967   Studies on larval development in Anomura (Crustacea, Decapoda). *Proc. Symp. Crust., Mar. Biol. Assc. India* 2: 777-804.

1973   Metamorphosis of two species of genus *Polyonyx* Stimpson *P. hendersonii* Southwell and *P. loimicola* Sankolli (Anomura, Porcellanidae). Symposium on the Indian Ocean and Adjacent Seas. *J. Mar. Biol. Assc. India* 15: 710-727.

1978   Adaptational significance of salinity tolerance in some freshwater prawns. Proceedings of the Symposium of Ecology and Animal Populations. Supplement. *Survey of India* Pt. 2: 175-187.

1984   Five new species of the freshwater atyid shrimps of the genus *Caridina* H. Milne Edwards from Dharwar area (Karnataka state) India. *Rec. Zool. Surv. India* Occasional Paper no. 69: 1-40.

10 A WOMAN CARCINOLOGIST IN JAPAN: SHIGEKO OOISHI

At a time in Japanese history when women's roles in carcinology were primarily those of artists (e.g., the two wives of the famous Tune Sakai), Shigeko Ooishi (Figs 4F, 6C) was creating her own unique nitch. Currently a professor in the faculty of biosciences at Mie University, Shigeko has had a diversified career. She was born in 1927 in Kumamoto, Japan, attained her Baccalaureate and later her doctorate at Nara Women's University. From early studies on decapod embryology she went on to participate in expeditionary work to the Ogasawara Volcano and Southwestern Islands (Tokara, Amami and Okinawa groups). During the 1970's and 1980's she was a frequent visitor to the University of Washington's Friday Harbor Laboratories where she worked with Paul Illg on copepod associations and development.

1964   Results of Amami Expedition. No. 3. Invertebrates. *Rept. Fac. Fish. Prefect. Univ. Mie* 5: 187-215, 2 figs.

1970   *Marine Invertebrate Fauna of the Ogasawara and Volcano Islands Collected by S. Ooishi, Y. Tomida, K. Izawa and S. Manabe.* 194 pp. Toba Aquarium, Asahi Shimbun Publishing Co.

1977   Haplostominae (Copepoda, Cyclopoida) associated with compound ascidians from the San Juan Archipelago and vicinity. *Spec. Publ. Seto Mar. Lab.* 5: 1-154, 49 pls., 5 tabls (with P.L. Illg).

1981   The larval development of some copepods of the family Ascidicolidae, subfamily Haplostominae, symbionts of compound ascidians. *Publ. Seto Mar. Lab.* 25: 253-292, 9 figs, 1 tabl.

11 MORE RECENT WOMEN CARCINOLOGISTS FROM THE US AND FRANCE

11.1 *Janet Haig*

What Dora Henry is to barnacles, Janet Haig (Fig. 2E) is to anomuran decapods, although, like Henry, Haig didn't begin her systematic career with anomuran decapods. After completing her bachelor's degree at Whittier College, Haig did her master's work on fishes at Stanford University. Her exposure to Crustacea first came when the opportunity arose for her to work with John Garth at the University of Southern California, performing such tasks as cataloging the *Velero* porcellanids. During the following 35 years, Haig became one of the world's leading authorities on the systematics of anomurans, particularly the porcellanids and the galatheids. She participated in cruises of the Southeast Pacific Biological and Oceanographic Program and examined collections in museums around the world. Although she 'officially' retired from the University of Southern California's Allan Hancock Foundation in 1990, her present publication record of 65 scientific papers promises to be supplemented substantially in her 'leisure' years.

1955   Reports of the Lund University Chile Expedition 1948-49. 20. The Crustacea Anomura of Chile. *Lunds Univer. Årsskr. N.S.* (2) 51: 1-68, 13 figs.

1960   The Porcellanidae (Crustacea Anomura) of the eastern Pacific. *Allan Hancock Pac. Exped.* 24: i-vii, 1-440, frontispiece, 12 figs, 41 pls.

1974   A review of the Australian crabs of family Hippidae (Crustacea, Decapoda, Anomura). *Mem. Qld. Mus.* 17: 175-189, 5 figs, 1 Fig.

1989   Porcellanidae (Decapoda, Anomura) collected during MUSORSTOM 1 and 2. In: J. Forest (ed.), Resultats des campagnes MUSORSTOM, 5. *Mém. Mus. Natl. Hist. Natur.* Paris (A) 144: 93-101.

11.2 *Isabel Canet Perez Farfante*

The mere mention of penaeid shrimp immediately brings to mind the name of another noted carcinologist, Isabel Perez Farfante (Fig. 5C). She was born in Cuba and received her undergraduate training at the University of Havana. She served as professor of biology at the Instituto de La Vibora and assistant professor of zoology at the University of Havana until 1942. Fellowships from the Guggenheim Foundation, Woods Hole Oceanographic Institution, and Harvard University allowed her to continue her graduate education in the United States. She received her Master's degree (1944) and her Ph.D. (1948) from Radcliffe College, and then returned to Havana where she held the rank of professor of zoology at the University of Havana until 1960, with the concurrent position of director of the Center for Fisheries Investigations from 1959-1960. In 1961 Perez Farfante joined the US Fish and Wildlife Service as a researcher in charge of commercial shrimps in US waters. During the period from 1961 to 1969 she was also an associate in invertebrate zoology at the Museum of Comparative Zoology at Harvard, and did independent research under grants from Radcliffe College and the National Science Foundation. In 1966 she was appointed systematic zoologist for the National Marine Fisheries Service, a position she held until her retirement in 1986.

A perusal of her publications shows that she too began her biological and systematic studies in fields other than carcinology. Her early publications, those appearing between 1939 and 1949, dealt with Foraminifera and Mollusca, both fossil and Recent. Her first crustacean paper was published in 1950, and it has been followed by no fewer than 45 additional papers dealing principally with the systematics of penaeid shrimps. For those of us who have been involved in faunistic surveys in the eastern Atlantic and Gulf of Mexico, Isabel Perez Farfante's keys and descriptions of penaeid genera have been indispensable.

1969   Western Atlantic shrimps of the genus *Penaeus*. *Fish. Bull.* 67: i-x, 461-591, 77 figs.
1971   Western Atlantic shrimps of the genus *Metapenaeopsis* (Crustacea, Decapoda, Penaeidae), with descriptions of three new species. *Smith. Contr. Zool.* 79: 1-37, 22 figs.
1977   American solenocerid shrimps of the genera *Hymenopenaeus, Haliporoides, Pleoticus, Hadropenaeus* n. gen., and *Mesopenaeus* n. gen. *Fish. Bull.* 75: 261-346, 63 figs.
1985   The rock shrimp genus *Sicyonia* (Crustacea: Decapoda: Penaeoidea) in the eastern Pacific. *Fish. Bull.* 83: 1-79, 60 figs.

11.3 *Mary Alice McWhinnie*

Mary McWhinnie (Figs 5H, 6F), who contributed significantly to our understanding of krill biology, received her bachelor's and master's degrees from DePaul University. After receiving her Ph.D. from Northwestern University (1952), she returned to DePaul where she became an instructor in biological sciences. She valued her teaching and was an ardent supporter of education throughout her lifetime, being awarded DePaul's highest faculty honor, posthumously, for her dedication to teaching as well as for her scientific accomplishments. As she rose to the rank of professor and chairman of the department, she participated in some 11 cruises to the Antarctic. She was the first female investigator to join the US Antarctic Research Program in 1962 and subsequently became one of the two first women (Sister Mary Odilex from DePaul was the other) to overwinter in Antarctica. On the last biological cruise of the USNS Eltanin (cruise 51), in honor of her scientific excellence and leadership, the Office of Polar Programs of the National Science Foundation appointed her as the chief scientist and US Antarctic Program representative. At one time or another, she served a leadership role on practically every committee dealing with the Antarctic. She was extremely active in promoting the krill industry throughout her career, spending the 3 years before her death in 1980 traveling and amassing an extensive bibliography on krill that contains more than 1800 references. She was the author and the subject of many articles, both professional and popular. (compiled from files of Smithsonian Oceanographic Sorting Center).

1962   Crayfish hepatopancreas metabolism and the intermolt cycle. *Comp. Biochem. Physiol.* 6: 159-170 (with C.J. Mohrher).
1976   Biology of krill (Euphausia superba) and other Antarctic invertebrates. *Antarct. J. U. S.* 11: 55-58, 4 figs (with C. Denys and D. Schenborn).
1978   Biological studies of Antarctic krill. Austral summer, 1977-78. *Antarct. J. U. S.* 13: 133-135, 3 figs (with C.J. Denys).
1981   *Euphausiacea Bibliography. A World Literature Survey.* 731 pp., New York, Pergamon Press (with C.J. Denys and P.V. Angione).

11.4 *Jocelyn Crane*

In one monumental work, Jocelyn Crane brought together practically everything known about fiddler crabs. She knew from early childhood that she wanted to work with small animals and by her teens that she wanted to go to Smith College and study zoology. She graduated Phi Beta Kappa with highest honors and an A.B. in Zoology in 1930. She immediately went to a job with the Tropical Research Department of the New York Zoological Society, a position that she held through her entire career. Despite her academic excellence, Crane did not earn a Ph.D. When the opportunity to return to graduate work arose, she decided that she would forego the advanced degree and instead study the behavior of small animals in their natural habitats. Her work with William Beebe in Bermuda gave her the opportunity to take her interest in the behavior of small animals into the deep recesses of the sea, becoming the first woman to explore the ocean in a bathysphere. Much of the information she used in *Fiddler Crabs of the World* was based on her own work (published and unpublished) during the years of her association with the New York Zoological Society's Department of Tropical Research. She also received funding from the National Science Foundation and cooperation of the Society enabling her to travel to Asia, the South Pacific, and Africa for one third of her time each year over a 5-year period to accumulate data on ocypodid crabs. Her book on fiddler crabs addressed not only the systematics of this fascinating group, but their ecology, behavior, morphology, associations, and biogeography. Because this volume was 10 years in the printing process, the nomenclatural part became somewhat outdated. However, it remains the single most comprehensive coverage of a brachyuran genus ever published (Condensed from Yost 1959).

1941   Eastern Pacific Expeditions of the New York Zoological Society 24. Crabs of the genus Uca from the west coast of Central America. *Zoologica* 26: 145-208, 8 figs, 9 pls.
1957   Basic patterns of display in fiddler crabs (Ocypodidae, genus Uca). *Zoologica* 42: 69-82, 4 figs, 1 Fig.

1967   Combat and its ritualization in fiddler crabs (Ocypodidae) with special reference to Uca. *Zoologica* 52: 49-75, 1 Fig.
1975   *Fiddler Crabs of the World. Ocypodidae: Genus Uca.* i-iii+736 pp., 101 figs, 50 pls, 24 tabls, New Jersey, Princeton University Press.

11.5 *Georgiana Baxter Deevey*

A Floridian carcinologist and specialist in ostracodes and copepods, Georgiana Deevey (Figs 3F, 4B), who until her death in 1982 was an adjunct curator at the University of Florida's State Museum as well as a member of the Bermuda Biological Station, considered herself primarily a biological oceanographer. She received her formal education at Radcliffe (A.B.) and Yale (M.S. and Ph.D.). Like many others, she began her work on animals other than crustaceans. In 1941, she published her first paper on tarantulas. From 1943 to 1946, she worked at Woods Hole Oceanographic Institution where she began her studies on copepods. After returning to Yale in 1946 she extended her work on the coastal copepod fauna of the western Atlantic south to Delaware Bay accumulating data on variations in body length relative to season, temperature, food availability, and the number of generations produced by animals in a year. Shortly after 1965, while she was on staff at the Bermuda Biological Station, she published her first major work on ostracodes, as well as described a new ostracode genus from the Gulf of Mexico. From this point, she divided her research interests between the study of ostracodes and copepods, publishing several descriptions of new species for each of these groups. In 1980, she published her last paper on copepods while her final work on ostracodes was published in 1982. During her 4 decades of crustacean research, Deevey produced 35 papers, leaving behind numerous unfinished projects upon her death (Taken in part from Ferrari 1982, 1983).

1960   Relative effects on temperature and food on seasonal variations in length of marine copepods in some eastern American and western European waters. *Bull. Bingham Oceanogr. Coll.* 17: 54-86, 15 figs.

1978   A taxonomic and distributional study of the planktonic ostracods collected on three cruises of the ELTANIN in the South Pacific and the Antarctic region of the South Pacific. Biology of the Antarctic Seas 8. *Antarct. Res. Ser.* 28(3): 43-70.

1980   The planktonic ostracods of the Sargasso Sea off Bermuda: Species composition and vertical and seasonal distribution between the surface and 2000 m. *Bull. Florida St. Mus. Biol. Sci.* 26: 37-124, 39 figs (with A.L. Brooks).

1982   Planktonic ostracods of the North Atlantic off Barbados. *Bull. Mar. Sci.* 32: 467-488, 11 figs.

11.6 *Patricia L. Dudley*

As Josephine Hart had proposed the use of larval and adult characters in assessing relationships among decapods, Patricia L. Dudley (Figs 5G, 6C) advocated the use of adult features and developmental stages in evaluating ontogenetic and phylogenetic divergences and parallelisms shown in particular copepod groups. She has also been one of the early pioneers in the use of the electron microscope to elucidate fine structures of copepod eyes, 59 sensory receptors, and other organs and tissues. Dudley received her B.A. (1951) and M.A. (1953) degrees from the University of Colorado and her Ph.D. from the University of Washington in 1957. In 1959, she accepted an assistant professorship at Barnard College, Columbia University and has remained there rising to the rank of full professor. In addition to her academic duties, she has continued her systematic and morphological studies with cyclopoid copepods particularly during summers spent at the Friday Harbor Laboratories.

1966   Development and systematics of some Pacific marine symbiotic copepods. A study of the Notodelphyidae, associates of ascidians. *Univ. Wash. Publ. Biol.* 21: i-v, 1-282, 51 figs.

1968   A light and electron microscope study of tissue interactions between a parasite copepod, *Scolecodes huntsmani* (Henderson), and its host ascidian, *Styela gibbsii* (Stimpson). *J. Morph.* 124: 263-282, 4 figs, 5 pls.

1972   The fine structure of a cephalic sensory receptor in the copepod *Doropygus seclusus* Illg (Crustacea: Copepoda: Notodelphyidae). *J. Morph.* 138: 407-416, 2 figs.

1980   The family Ascidicolidae and its subfamilies (Copepoda, Cyclopoida) with descriptions of new species. *Mém. Mus. Natl. Hist. Natur. (Zool.)* 117: 1-192, 62 pls.

11.7 *Dorothy Travis*

Dorothy Travis (Fig. 1I), one of the three contemporary Dorothys (the others being Dorothy Bliss and Dorothy Skinner), began her work in the 1950's with physiological and biochemical studies of the molt cycle in the spiny lobster, while at the University of New Hampshire. From New Hampshire she moved to the Bermuda Biological Station and then to the Harvard University Biological Laboratories. It was while at the Harvard laboratories, that she published her first work on biomineralization in decapod crustaceans. She continued examining histochemical processes of crustacean biomineralization, especially of crayfish, through the early 1960's that resulted in a series of publications. Later she transferred this specialization to the study of bivalve mollusks. In 1973 she began a position at the National Institutes of Health in the Minority Funding Program, which she held for 5 years before being forced to take a disability retirement. She died in 1982, having contributed substantially to our understanding of the physiology of the decapod molt cycle and crustacean biomineralization.

1954   The molting cycle of the spiny lobster *Panulirus argus* Latreille. I. Molting and growth in laboratory-maintained individuals. *Biol. Bull.* 107: 433-450, 11 figs.

1955   The molting cycle of the spiny lobster *Panulirus argus* Latreille. III. Physiological changes which occur in the blood and urine during the normal molting cycle. *Biol. Bull.* 109: 88-112, 34 figs.

1960   Matrix and mineral deposition in skeletal structures of the decapod Crustacea. In *Calcification in Biological Systems*. American Association for the Advancement of Science, Washington D.C. pp. 57-116.

1963   The deposition of skeletal structures in the Crustacea. 4. Microradiographic studies of the gastrolith of the crayfish *Orconectes virilis* Hagen. *J. Ultrastr. Res.* 8: 48-65.

11.8 *Dorothy Bliss*

The name of Dorothy Bliss (Fig. 4c; Fig. 5E; Fig. 6D) is known to all carcinologists through her expert editorship of the treatise *The Biology of Crustacea*, published in 10 volumes over the period 1977-1986. Bliss was a Phi Beta Kappa scholar with bachelor's and master's degrees from Pembroke College and a Ph.D. from Radcliffe College. Her doctoral dissertation on the neuroendocrine structures in the eyestalks of the land crab *Gecarcinus lateralis* was a major contribution to crustacean neuroendocrinology. Much of her professional career was spent on the staff of the American Museum, beginning as Assistant Curator of Living Invertebrates in 1956 and becoming curator emeritus in 1980. However, she was much more than a 'museum carcinologist.' She was an avid field biologist, amassing a wealth of data on growth, locomotion, water balance, and hormones in land crabs. Simultaneously, she was a research professor in the anatomy department of the Albert Einstein College of Medicine and adjunct in the Department of Biology of City College of New York, where she was mentor or advisor for many fledgling carcinologists. In addition to her editorial expertise, Bliss was a writer in her own right, producing more than 40 scientific papers and 1 quasi-popular account of shrimps, lobsters, and crabs. (Condensed from Mantel 1988).

1953   Endocrine control of metabolism in the land crab *Gecarcinus lateralis* (Freminville). I. Differences in the respiratory metabolism of sinusglandless and eyestalkless crabs. *Biol. Bull.* 104: 275-296, 9 figs.

1962   Neuroendocrine control of locomotor activity in the land crab *Gecarcinus lateralis*. In H. Heller and R.B. Clark (eds), Neurosecretion. *Mem. Soc. Endocrin.* 12: 391-410, 12 figs.

1968   Transition from water to land in the Crustacea Decapoda. *Amer. Zool.* 8: 355-392, 21 figs.

1982   *Shrimps, Lobsters and Crabs. Their Fascinating Life Story.* i-xii, 1-242, 104 figs. New Century Publishers, Inc., Piscataway, NJ.

11.9 *Dorothy M. Skinner*

The third of the trio of Dorothys is Dorothy Skinner (Fig. 2I). Born in Newton, Massachusetts, Dorothy received much of her early training in institutions within that state. She was tapped for Phi Beta Kappa in 1951 and obtained her bachelor's degree in 1952 from Jackson College (Tufts University, Medford) where Ken Roeder kindled her early interests in science. Her work as an undergraduate was on the muscle receptor organ of the crayfish. Her early career aspirations were as a neurophysiologist. In the summers, as a graduate student at Harvard she attended the invertebrate biology and embryology courses at the Marine Biological Laboratory (MBL) at Woods Hole where she continued to expand her interests in decapod crustaceans under the guidance of Lew Kleinholz and Ted Bullock. It was during this graduate student period that two major developments occurred that influenced the rest of her career. She met another woman carcinologist, Dorothy Bliss, and through her was introduced to the land crab *Gecarcinus lateralis.* Her work then began to focus on understanding the molting processes. After receiving her Ph.D. in 1958, Skinner held a series of postdoctoral positions where she learned more about ways to study the physiology of molting and its control. Her work was expanded on this subject to include exciting new studies of the DNA structure of *G. lateralis.*

Skinner has influenced many aspiring crustacean biologists with her dynamic personality, strong participation on scientific advisory boards, and her active work in numerous scientific societies including her current positions as a governor of the Crustacean Society and as a member of that society's awards committee. In response to our inquiries about her life in science, Skinner replied, 'It was at MBL that I learned in addition to being a 'way of knowing,' science was (is) a 'way of living' and it is a way that I like very much.' To date, Dorothy has published 77 articles, with no signs of slowing in her productivity.

1967   Satellite DNA's in the crabs *Gecarcinus lateralis* and *Cancer pagurus. Proc. Natl. Acad. Sci.* 58: 103-110, 4 figs.

1970   Molting in land crags: Stimulation by leg removal. *Science* 169: 383-385, 1 fig, 1 tabl. (with D.E. Graham).

1985   Molting and Regeneration. In: D.E. Bliss and L.H. Mantel, eds., *The Biology of Crustacea,* 9: 43-146, 20 figs, 8 tabls. Academic Press, New York.

1990   Atrophy of crustacean somatic muscle and the proteinases that do the job. *J. Crust. Biol.* 10: 577-594 (with D.L. Mykles).

11.10   *Linda Habas Mantel*

One of Dorothy Bliss's most notable protegees and very close associates is Linda Mantel, a moving force among women engaged in crustacean physiological research. A graduate of Swathmore College (1960) with a B.A. in biology, Mantel went on to complete her M.S. (1962) and Ph.D. (1965) degrees at the University of Illinois. Following a National Institute of Health postdoctoral fellowship she joined the Department of Living Invertebrates of the American Museum of Natural History as a research fellow. In 1968, she became an assistant professor in the biology department of City College, City University of New York, where she rose to professor of biology and assistant provost for research and graduate studies (1982-1987), and currently serves as department chairperson. In addition to her academic career, Mantel has conducted research on the comparative physiology of osmoregulation, adaptations, effects of pollutants on physiological processes, and neuroendocrine mechanisms of growth in crustaceans. She has also been at the organizational forefront for symposia, and educational and scientific committees and societies, particularly those involved with women in science. VE

1968   The foregut of *Gecarcinus lateralis* as an organ of salt and water balance. *Amer. Zool.* 8: 433-442, 9 figs, 4 tabls.

1979   Terrestrial invertebrates other than insects. In: G.M.O. Maloiy (ed.), *Comparative Physiology of Osmoregulation in Animals.* 1: 175-219. London, Academic Press.

1983   Osmotic and ionic regulation. In: L.H. Mantel (ed.), *The Biology of Crustacea* 5: 57-161. New York, Academic Press (with L.H. Farmer).

1985   Neurohormonal integration of osmotic and ionic regulation. *Amer. Zool.* 25: 253-263, 3 figs.

11.11   *Gertrude H. Hinsch*

Gertrude Hinsch was trained at Iowa State University, receiving her B.S., M.S., and Ph.D. (1957) from that institution. After a short teaching stint, she enrolled in a course at Woods Hole Oceanographic Institution given by Charlie Metz on the topic of fertilization in Crustacea. She became a research associate with Metz, working both at the University of Miami and at Woods Hole. During this period she concentrated her research efforts on various aspects of spermatogenesis and fertilization in crustaceans. After commencing a position in the biology department at the University of South Florida in 1974, a position which she presently still holds, she continued to publish extensive studies of crustacean sperm ultrastructure and has recently shifted her research focus to include discerning the mechanisms of sperm movement during the process of egg-coat penetration leading to fertilization in brachyurans. She has published in excess of 30 papers or books on the subjects of spermatogenesis and crustacean reproductive ultrastructure, with several additional papers now in review.

1968   Reproductive behavior in the spider crab *Libinia emarginata* (L.). *Biol. Bull.* 135: 273-278.
1971   Penetration of the oocyte envelope by fertilizing spermatozoa in the spider crab. *J. Ultrastr. Res.* 35: 86-97.
1980   Spermiogenesis in *Coenobita clypeatus*. I. Ultrastructure of the sperm. *Internatl. J. Invert. Repro.* 2: 189-198.
1988   The morphology of the reproductive tract and seasonality of reproduction in the golden crab *Geryon fenneri* from the eastern Gulf of Mexico. *J. Crust. Biol.* 8: 278-294, 22 figs.

11.12 *Hélène Charniaux-Cotton*

The name Hélène Charniaux-Cotton (Fig. 6E) will always be linked to crustacean sex determination. She began this work when she entered the Centre National de la Recherche Scientifique (CNRS) to study in the laboratory of Professor Teissier at the Sorbonne. Her studies first focused on relative growth of sexual variants in an amphipod species. Her results led her to suggest hormonal controls of such development during critical periods, these periods associated with modifications of hormonal levels. Her work in crustacean endrocrinology led to the discovery of the androgenic gland and its influence on secondary sexual characters and testicular differentiation. In recognition of the originality of her experimental methods and discoveries she was awarded several prestigious awards, including the Prix Da Gamma Machado in 1957, the Médaille d'argent du CNRX in 1958, the Pelman Biology Prize in 1959, the Grand prix de Sciences chimiques et Naturelles de l'Académie des Sciences in 1977, and the Médaille d'or de la Société d'encouragement au progrès in 1979. Her professional career included Directeur de Recherche at CNRS, French representative to the International Society of Invertebrate Reproduction, and professor in the Faculty of Sciences from 1966 to her retirement in 1986. During this time she published an incredible 319 scientific papers of which many were lengthy contributions (condensed from Payen 1989a, 1989b).

1957   Croissance, régénération et déterminisme endocrinien des caractéres sexuels d'*Orchestia gammarella* Pallas (Crustac Amphipode). *Ann. Sci. Nat.* (B) 19: 411-560.
1962   Détermininisme de l'intersexualiét chez les Crustacés Supérieurs et particulirèment chez les Amphipodes Talitridae. *Bull. Soc. Zool. France* 87: 338-366.
1975   Hermaphroditism and Gynadromorphism in Malacostracan Crustacea. In R. Reinboth (ed.), *Intersexuality in the Animal Kingdom*. pp. 91-105. Berlin, Heidelberg, New York, Springer-Verlag.
1985   Vitellogenesis and its control in malacostracan Crustacea. *Amer. Zool.* 25: 197-206.

11.13 *Geneviève Payen*

Hélèn Charniaux-Cotton's successor as Directeur of Recherche at CNRS is Geneviève Payen (Fig. 6H). She did her early degree at the University of Paris, entering the CNRS in 1967 where she began doctoral work with Charniaux-Cotton. After receiving her doctoral degree, Payen focused her research attentions at the CNRS on androgenic gland organogenesis and the role of the androgenic gland in development of male sexual characteristics. Simultaneously with this work (1978-1985) she headed a project dealing with physiopathology of crustacean reproduction, particularly concerning the effects of parasitic barnacles on reproduction. After the death of Charniaux-Cotton in 1987, Payen turned her research attentions to the study of the sinus glands of lobsters and continued with her responsibility of coordinating the team 'Neuroendocrinologie des Crustacés.' Like her predecessor, Payen has been very prolific, publishing some 77 scientific papers to date.

1974   *Recherches sur la réalisation et le contrôle de la differenciation sexuelle ches les Crustacés Décapodes Reptantia.* Thése d'Etat, AO au CNRS no. 9930, Université Pierre et Marie Curie (Paris VI), pp. 1-224.

1980   Aspects fondamentaux de l'endocrinologie de la reproductions des Crustacés marins. *Oceanis* 6: 309-339.

1985   Sexual differentiation. In L.H. Mantel (ed.), The *Biology of Crustacea.* 9: 217-299. New York, Academic Press (with H. Charniaux-Cotton).

1986   Endocrine regulation of male and female genital activity in crustaceans. A retrospect and perspectives. In M. Porchet, J.-C. Andries & A. Dhainaut (eds), *Advances in Invertebrate Reproduction* 4: 125-134. North-Holland, Elsevier.

12 CONCLUSIONS

It is useful to try to draw some general conclusions about influential women carcinologists. In attempting to construct a common profile, we discovered that most of the early workers blazed their own unique paths in the field. These early pioneers were not necessarily highly educated in a formal sense, although teaching was a common profession for many. Often they were unmarried. Those who were married typically worked in the shadows of their husbands. Others had a significant male figure, such as a father or brother, influencing their entrance into, and prominence in, the field. Women carcinologists often worked with little or no financial compensation. It was not unusual for women to be offered lower salaries or for them to take decreases in salaries to allow their male colleagues more success. Despite personal difficulties, scorn by their male counterparts, and poor working conditions, women began making carcinological contributions before the turn of the century. Most of these early women started out working with other organisms, even vertebrates, finding crustaceans only in their mid- to late-careers. Thus some of these women worked for only a short time on crustaceans. However, they were extremely productive, contributing papers ranging from 1 or 2 monographic works to 316 publications of varying lengths.

ACKNOWLEDGEMENTS

Many individuals have provided information and/or photographs for this compilation of the contributions of women to carcinology, and without their assistance our efforts would have been trivial. We gratefully thank all of the following: Dr. Mary Needler Arai for the information and photo of her mother, Alfreda Berkeley Needler; Dr. P.B. Berents for information and photos of Fredericka Bage, Isobel Bennett, and Elizabeth Pope; Dr. Edward Bousfield for his information on Canadian women carcinologists; Dr. Fenner A. Chace Jr., for information on Lee Boone; Reverend Robert Campbell-Smith and Dr. Keith H. Hyatt for information on Reverend

Campbell-Smith's mother, Susan Finnegan; Dr. David M. Damkaer for information and photographs of Maria Dahl and Mildred Stratton Wilson; Dr. Frank Ferrari for information on Georgiana Deevey; Dr. and Mrs. Arthur Fontaine for information and photograph of Josephine Hart; Christine Hammer for the photograph of Charlotte Holmquist; Dr. L.B. Holthuis for information on Alida Buitendijk, Jentena Leene, and Isabella Gordon; Dr. Paul Illg for information and photographs of Sheina Marshall, Livia Pirocchi, Josephine Hart, Belle Stevens, Shigeko Ooishi, and Patricia Dudley, as well as for his prudent advice; Dr. Ray Ingle for information and photographs of Alida Buitendijk, Jentena Leene, Isabella Gordan, Marie Lebour, and Olive Tattersall; Dr. Boris Ivanov for information and bibliographies for Sinaida Kobjakova and Eupraxie Gurjanova and for his personal contacts with Russian carcinologists on our behalf; Dr. Sophia Jakowska for information on Dyrce Lacombe; Betty Landrum for photos of Mary Alice McWhinnie; Mrs. Ruth Lehmann and Dr. Mark Brenner for information and photos of Mrs. Lehmann's mother, Georgiana Deevey; Dr. Rafael Lemaitre for information on Isabel Perez Farfantae, Mary Alice McWhinnie, Harriet Richardson, and Mary J. Rathbun; Dr. Elena Makkaveeva for information on M.A. Dolgopol'skaya and Elizabet Pavlova; Dr. Raymond Manning for photographs of Marie Lebour, Mary Rathbun, Isabel Perez Farfante, and Dorothy Skinner, as well as for sharing rememberances of Isabella Gordon and for the bibliographies of Alida Buitenijk and Harriet Richardson; Drs. Linda Mantel and Penny Hopkins for information and photographs of Dorothy Bliss; Dr. Goeff Moore for information on Sheina Marshall; Dr. William Newman for information and photographs of Galina Zevina; Dr. Genevieve Payen for information and photograph of Hélène Charniaux-Cotton; Dr. Keith Probert for information and photograph of Elizabeth Batham; Dr. Margaret Rossiter for information on Harriet Richardson; Dr. Fredrick Schram for information and photographs of Sidnie Manton and Isabella Gordon; Dr. Dorothy Skinner for information and photograph of Dorothy Travis; Dr. Jan Stock for information on Cornelia de Vos and Atie Vorstman. We thank William Deiss of the Smithsonian Archives for providing access to the correspondence files of Dr. Waldo L. Schmitt. We are also very grateful to the numerous women carcinologists who provided us with personal photographs and biographies, and to E.J. McGeorge for converting the photographs into prints for this publication.

REFERENCES

Ainley, M. 1986. A family of women scientists. *Le Bulletin/Newsletter Concordia University, Simone de Beauvior Institute* 7: 5-11.

Baba, K. 1988. Chirostylid and galatheid crustaceans (Decapoda: Anomura) of the 'Albatross' Philippine Expedition, 1907-1910. *Res. Crust.* Special No. 2: v, 1-203.

Baranova, Z.I. & P.V. Ushakov, 1972. Evpraksiya Fedorovna Guryanova (k 70-letiyu so dnya rozhdeniya). (Eupraxie Fedorovna Gurvanova (in the 70th year since birth)). *Issled. Fauny morei* 10: 5-7, 1 fig.

Barnes, H. (ed.) 1966. A collection of original scientific papers presented to Dr. S.M. Marshall, O.B.E., F.R.S. in recognition of her contribution with the late Dr. A.P. Orr to marine biological progress. In *Some Contemporary Studies in Marine Science*: pp. 1-13. London: George Allen & Unwin Ltd.

Bell, J. 1979. Frederika Anna Bage. In B.Nairn and G. Serle (eds.), *Australian Dictionary of Biography* 7 (1891-1939): 131-132. Melbourne: Melbourne University Press.

Campbell, N. & R.M.S. Smellie 1983. *The Royal Society of Edinburgh (1783-1983). The first two hundred years.* p. 50. Edinburgh, The Royal Society of Edinburgh.

Cattell, J. 1955. Searle, Dr. Harriet Richardson. In *American Men of Science. A Biographical Directory. II. Biological Sciences*: p. 1007. Lancaster, PA: The Science Press.

Chace, F.A., Jr. 1990. Mary J. Rathbun (1860-1943). *J. Crust. Biol.* 10: 165-167.

Damkaer, D.M. 1988. Mildred Stratton Wilson, copepodologist (1909-1973). *J. Crust. Biol.* 8: 131-146.

Damkaer, D.M. & T. Mrozek-Dahl 1980. The Plankton-Expedition and the copepod studies of Friedrich and Maria Dahl. In M. Sears & D. Merriman (eds), *Oceanography: The Past*: pp.462-473. New York: Springer-Verlag.

Engel, H. & J.H. Stock 1959. Anna Petronella Cornelia de Vos. Amsterdam 14 I 1893 - Loosdrecht 21 III 1958. *Hydrobiol.* 12: 393-395.

Ferrari, F. 1982. Georgiana Baxter Deevey (1914-1982). Her work on copepods. *Monoculus – Copepod Newsletter* No. 5: 9-12.

Ferrari, F. 1983. Publications of Georgiana Baxter Deevey. *Monoculus -- Copepod Newsletter* No. 6: 16-18.

Gordon, I. 1960. Mrs Elsie Wilkins Sexton. *Proc. Linn. Soc. Lond.* 171: 134-135.

Gordon, I. 1980. Walter M. Tattersall and Olive S. Tattersall: 7 decades of peracaridan research. *Crustaceana* 38(3): 511-520.

Jillett, J.B. 1978. Elizabeth Joan Batham (1917-1974) M.Sc., Ph.D., F.R.S.N.Z. *Proc. Roy. Soc. N.Z.* 106: 23-29.

McCain, L. 1943. Mary Jane Rathbun. *Science* 97: 435-346.

Mantel, L.H. 1988. Dorothy E. Bliss (1916-1987). *J. Crust. Biol.* 8: 706-709.

Ogilvie, M.B. 1986. *Women in Science. Antiquity through the Nineteenth Century. A Biographical Dictionary with Annotated Bibliography.* Cambridge, MA: MIT Press.

Payen, G.G. 1989. Hélène Charniaux-Cotton (1918-1986). *Invert. Reprod. Develop.* 16: 1-2.

Payen, G.G. 1989. List of Publications from the laboratories of H. Charniaux-Cotton (until 1986). *Invert. Reprod. Develop.* 16:3-16.

Rossiter, M.W. 1982. *Women Scientists in America, Struggles and Strategies to 1940.* Baltimore: Johns Hopkins University Press.

Russell, F.S. 1972. Dr Marie V. Lebour. *J. Mar. Biol. Assc. UK* 52: 777-788

Schmitt, W.L. 1943. Mary Jane Rathbun. *J. Wash. Acad. Sci.* 351-352.

Schmitt, W.L. 1973. Mary J. Rathbun 1860-1943. *Crustaceana* 24: 283-297.

Van Benthem Jutting, W.S.S. 1964. In memoriam Dr A.G. Vorstman. *Hydrobiol.* 23: 577-581.

Yonge, C.M. 1978. Sheina Macalister Marshall O.B.E., D.Sc. (Glas.), F.R.S. *Roy. Soc. Edinb. Yearbook* 1978: 37-38.

Yost, E., 1959. *Women of Modern Science.* New York: Dodd, Mead.

More than 200 years of crustacean research in Denmark

Torben Wolff

Zoological Museum, University of Copenhagen, Denmark

1 THE 18TH CENTURY

The second half of the 1700's was a golden age in Danish natural history, and four outstanding zoologists contributed to carcinology in that period.

1.1 *Peter Forsskål*

Peter Forsskål (1732-1763; Fig. 1A) was Swedish born and from the age of 10 took up studies of oriental philology. In addition, he also studied botany and zoology under Linnaeus at Uppsala, initially merely as a hobby. These latter studies were a decisive factor for Forsskål being selected as naturalist on an expedition that in 1761 was sent out by the Danish King Frederik V to Arabia Felix, the present North Yemen. Of the six participants only one, the cartographer Carsten Niebuhr survived, returning to Copenhagen in late 1767.

Based on Forsskål's 1800 notes and collections, Niebuhr edited *Flora Aegyptiaco-Arabica* (1775) and *Descriptiones Rerum Naturalium* (1775), which describe 3000 animal species including 33 crustaceans. *Icones Rerum Naturalium* (1776) contains many copperplates based on the excellent drawings by the artist of the expedition. One plate shows the amphipod *Phronima sedentaria* in its salp barrel, the type of which is still in the Copenhagen Zoological Museum, one of the oldest existing type specimens in alcohol. The *Descriptiones* and *Icones* give the first descriptions and illustrations of the surface fauna of the open sea, e.g., the salps, often with biological observations and interpretations. The expedition's zoological collections mainly consist of the famous 'fish herbarium', mollusk shells, corals, and insects.

Forsskål was hard-working and gifted, with a keen eye for details as well as broad outlines. It is a great loss to science that he did not survive to publish his results himself. (Gosch 1873: 439-448, 1878: 210-211; Christensen 1918; Forsskål 1950; Hansen 1964; Wolff 1967: 13-41, 1968).

1.2 *Otto Frederik Müller*

O.F. Müller (1730-1784; Fig. 1B) was the son of a poor court trumpeter. He studied theology and music, and became the teacher of the sons of countess Schulin, the widow of a former prime minister. At Frederiksdal, her estate north of Copenhagen, Müller served as a butler for 17 years, studying and teaching the sons law, languages, art, and music. Probably inspired by the count-

Figure 1. A: Peter Forsskål. B: O.F. Müller. C: J.C. Fabricius. D: Otto Fabricius. E: Henrik Krøyer (drawing by P.S. Krøyer). F: J.C. Schiødte (drawing by P.S. Krøyer). G: Fr. Meinert. H. P.E. Müller. I: G. Budde-Lund.

ess, he became interested in nature as well, and around 1760 he began studies of the insect and freshwater fauna around Frederiksdal. During extensive travels with the countess's eldest son in Europe, Müller became a member of learned societies. When the countess died he managed to marry a wealthy lady and devoted the rest of his life to the study of marine and freshwater invertebrates.

Although dredging implements were used on a few earlier occasions (e.g., by Forsskål), Müller was a pioneer in collecting plants and animals systematically. His marine studies were mainly conducted in coastal waters of southern Norway and in Oslo Fjord close to his summer house, often under considerable dangers and hardships.

Müller's whole research output came in a period of less than 25 years prior to his early death. His *Zoologia Danica* was planned as a companion piece to the famous *Flora Danica*, but he was able to publish only two sections, which included 160 excellent plates by his brother. In a so-called forerunner, the *Prodromus* (1776), he totalled some 3000 species, including many crustaceans. In *Entomostraca seu Insecta Testacea* (1785), he created such well-known genera as *Cyclops*, *Caligus*, *Cypris*, *Daphnia*, *Polyphemus*, and *Argulus*, and introduced classifications still in use.

In his observations and comprehension of what he observed, as well as in his ability to systematize microscopical animals and plants, O.F. Müller was decades ahead of his contemporaries. (Gosch 1873: 354-418, 1878: 170-185; Anker 1943, 1950; Spärck 1932a; Wolff 1967: 41-50).

1.3   *Otto Fabricius*

Otto Fabricius (1744-1822; Fig. 1D) studied theology and Eskimo to qualify as a missionary in Greenland. At the same time he corresponded with O.F. Müller and was probably inspired by him to become interested in natural history as well. From 1768 he stayed for 5 1/2 years in the remote and primitive West Greenland. He did not live in a town like other Danes but in a tiny settlement, sharing the life of the Greenlanders in a tent in summer and an earth hut in winter. Using the silent, low-lying kayak, he made his unique biological observations that ranged from the feeding habits of the huge baleen whales to the life of small animals in shallow water. In the faint glow of an oil lamp, he examined the creatures he had collected and observed their reactions. He had no microscope, only weak handlenses; no glass jars, only a cup and empty shells; and no library, only one book – Linnaeus's *Systema Naturae*.

Based on his own experience and the Greenlanders' detailed knowledge, Fabricius published his *Fauna Groenlandica* (1780) with descriptions of 473 species, including 30 crustaceans and pycnogonids (sea spiders), and documented their habitat, food, breeding habits, and any possible use they were put to by the Greenlanders. This was the first and for a long time only comprehensive work on Arctic animal life. He later became a professor of theology and titular bishop but continued publishing: papers on his beloved Greenland animals (mainly marine) and Eskimo hunting methods; a Greenlandic dictionary; a grammar; and translations from Danish. (Gosch 1873: 418-425, 1878: 212-215; Jensen 1932; Kornerup et al. 1923; Wolff 1967: 50-54).

1.4   *Johan Christian Fabricius*

J.C. Fabricius (1745-1808; Fig. 1C) was a distant relative of Otto Fabricius. For 2 happy years he was a student of Linnaeus at Uppsala and Hammarby where Linnaeus's inspiring teaching and his own receptive mind were equally beneficiary. They became close friends, and Fabricius is the only one of Linnaeus's many pupils who approached the master's stature. Back in Copenhagen Fabricius finished his studies and began the study of several insect collections, including Forsskål's. As he could obtain no permanent occupation in Copenhagen he accepted a position as professor of natural history, economics, and political science at the small Kiel University in Schleswig-Holstein, at the time a Danish duchy. He spent only the teaching semester in Kiel, otherwise working mainly in Copenhagen. Both before and after going to Kiel he travelled extensively throughout Europe, particularly to England. Due to his charm and scientific enthusiasm he was well received everywhere and was given free access to collections.

While Linnaeus classified insects using wing morphology, Fabricius introduced, with Linnaeus's consent, a classification using mouthparts, although the insect orders still carry the previous Linnean names. Based on the study of 1400 insect and crustacean species from all over the world, Fabricius very logically first dealt with their taxonomy in *Systema Entomologiae* (1775), followed by *Genera Insectorum* (1776), *Species Insectorum* (1781) and *Mantissa Insectorum* (1787). These books were all forerunners of his great *Entomologia Systematica Emendata* (1792-94) and its *Supplementum* (1798). Later followed books on separate insect orders. Thousands of type specimens still exist of the almost 10,000 species he dealt with. Two thirds of these are in Copenhagen and include several hundred crustacean types. Ella Zimsen (1964) published an extensive catalogue of all the existing Fabricius types.

Fabricius was a modest and helpful person. He was the leading entomologist of the 18th century and made Copenhagen a world center for the study of the morphology and systematics of arthropods, which was continued in the next century particularly by Krøyer, Schiødte, and Hansen. (Barton 1808; Latreille 1808, Walckenaer 1815: 62-67; Steffens 1841: 186-203, Gosch 1873: 449-508, 1878: 202-210; Schuster 1928; Henriksen 1932a; Jespersen 1946; Tuxen 1959, 1967; Mayer 1967).

2 THE KRØYER-SCHIØDTE SCHOOL

Danish natural history did not prosper in the beginning of the 19th century, mainly for political and economic reasons. However, much was changed with the establishment in 1824 of 'The Society for the Propagation of Natural Sciences' by the physicist H.C. Ørsted, and in 1833 of 'The Danish Natural History Society.' Both are still flourishing.

2.1 *Henrik Nikolaj Krøyer*

Krøyer (1799-1870; Figs 1E, 3-24) studied Latin and Greek and had a lifelong love for classical antiquity. At the age of 22, he wanted to join the Greek revolt against the Turks and vagabondized all the way to Marseille and then sailed to Greece. However, he found that the Greeks of his day were far from his Hellenistic ideals and returned by foot via Rome arriving home, as he said, 'totally stripped of illusions but rich in experience.' His acquaintance with Mediterranean marine life aroused his interest in natural history, which was further supported by university studies in Heidelberg and Göttingen, for 1 year, on his way home from Greece.

Krøyer never took a degree but served as a schoolteacher until he obtained a grant to travel. This allowed him 2 years in an open boat with his young wife, making comprehensive studies of the Danish fish fauna and fisheries that resulted in a large, three-volume treatise on fishes. In 1836, he started the scientific natural history journal *Naturhistorisk Tidsskrift* that he, with great personal sacrifice, edited until 1849. Here during 13 years he published most of his enormous production, collectively distributed as his *Opuscula carcinologia* (most translated into German in L. Oken's *Isis*). These papers were mainly monographic revisions of families and genera, e.g., Greenland's amphipods and the other peracarid groups. Among other things he recognized cumaceans as a separate group and not the young of other crustaceans. He also revised shrimp genera including *Hippolyte, Crangon,* and *Sergestes* and dealt with pycnogonids and free-living copepods, being the first to describe and illustrate the nearly complete development of a marine calanoid copepod. Uncharacteristically for his time, Krøyer deposited his specimens in a museum that was specifically cited in his publications.

Krøyer participated for almost 1 year (1838) in the French *La Recherche* Expedition to Spitsbergen, spent the following winter in northern Norway, and in 1840-41 joined a cruise on a Danish naval vessel to the southern half of eastern and western South America.

Krøyer had a poorly-paid post at the Royal Natural History Museum, one of the forerunners of the present day Zoological Museum. When Professor J. Reinhardt, head of the museum and professor of zoology, died in 1845, the younger Japetus Steenstrup, 14 years Krøyer's junior, was chosen as Reinhardt's successor instead of the more enterprising and productive Krøyer. Krøyer was deeply hurt and full of bitterness for the rest of his life. However, although 'my health is totally broken (and I feel) lethargic and stupid' he, in 1853, once again traveled, this time to western Europe, Newfoundland, and the Gulf of Mexico, bringing back vast collections of marine animals, but he himself was only able to make use of a fraction of the material. His sole major paper after 1849 was the classic piece on parasitic copepods (1863), dealing with 116 species of which 96 were new to science and many collected by Krøyer himself, particularly during his voyage to South America. The 18 plates were excellent and were drawn by Krøyer's 9-year-old stepson, P.S. Krøyer, whose mother was Krøyer's mentally retarded sister-in-law. P.S. Krøyer later became one of the foremost Danish painters.

With his thorough, exact analyses of details Henrik Krøyer became a standard by which to judge later workers. (Gosch 1875: 186-219, 1878: 344-356, 1898-1905, II: 218-224, III: 180-199; Dahl 1941; Wolff 1967: 56-66; Damkaer & Damkaer 1979; Scheele 1982, 1985).

2.2   *Jørgen Matthias Christian Schiødte*

J.C. Schiødte (1815-1884; Figs. 1F & 3-6), who was interested in insects from his early boyhood, left school at the age of 17. Like Krøyer, he never took a degree and also became an assistant at the Royal Natural History Museum, publishing his first large monograph, on Danish beetles, at the age of 26. In the mid-1840's, he travelled in southern Europe where he collected 70000 insects and other tiny soil arthropods, introducing the use of a sieve for this purpose. He was also the first to study – in the Adelsberg, now Postojna, Cave – the subterranean fauna which he systematized according to adaptation, thus being the founder of zoological spele-ology. Two of the cave animals were an amphipod and an isopod for which he established the genera *Niphargus* and *Titanethes.*

The passing over of Krøyer for Professor of Zoology initiated the strife between Schiødte and Steenstrup, and their followers, that was to last for almost 75 years. Both Schiødte and Steenstrup were uncompromising and self-confident persons – Schiødte self-taught and a meticulous student of details, Steenstrup an Aladdin type with powerful tutors, a wealth of scattered interests, and almost no inclination to use a microscope.

In 1854 Schiødte became a titular professor, and in 1861 he revived Krøyer's publication *Naturhistorisk Tidsskrift,* serving as editor until his own death. In Schiødte's systematic-comparative studies, the fundamental rule was the coordination between morphology and mode of life. This is evident in his two major carcinological contributions: one on the sucking mouth of crustaceans (1861) and the other, together with Meinert, the monograph on the parasitic isopods Cymothoidae and Aegidae (1879-84). This 'rule' was also the basis for his enormous entomological production, mainly on beetles and beetle larvae.

J.C. Schiødte was loved by his many students, one of whom claimed that 'he did not give us proficiency but insight.' Although in his attacks on Steenstrup and followers he could be both personal and harsh, Schiødte as a human being as well as a scientist was one of the great minds that mark an epoch. (Gosch 1875: 461-600, 1898-1905; Bourgeois 1889; Henriksen 1932b).

2.3 Frederik Vilhelm August Meinert

Fr. Meinert (1833-1912; Fig. 1G) studied and became a candidate in theology, but he was also a student of Schiødte with whom he published the isopod monograph. In 1870 he became Schiødte's assistant at the new Zoological Museum, but the hot-headed and ambitious Meinert could not adjust to Schiødte's somewhat blunt manner. Either from inclination or from ulterior motives Meinert began to align himself more and more with Steenstrup. The ultimate result of this was that he was chosen to succeed Schiødte as curator after Schiødte's death instead of the more qualified, although much younger H.J. Hansen, who had been Schiødte's choice.

Besides contributing to the isopod monograph, Meinert wrote on malacostracans from Danish waters and on deep-sea pycnogonids from the Danish Ingolf Expedition. He also published on a great variety of other arthropod groups and collected extensively on travels to central and southern Europe and Algeria, 1868-69, and the West Indies and Venezuela, 1891-92. Neither as a scientist nor as a curator did he reach the standard of his contemporary fellow entomologists and carcinologists. (Gosch 1875: 548-561, 1878: 514-518, 1898-1905, I: 147-164, II: 256-287; Dahl 1910: 26-31).

2.4 Jørgen Vilhelm Bergsøe

Vilhelm Bergsøe (1835-1911) graduated in entomology in 1860 but had to give up zoology because of a serious eye disease. His thesis was on an internal parasitic copepod, previously described by Steenstrup. Bergsøe later wrote a large and very popular book on Danish insect life and became a greatly admired novelist.

2.5 Peter Erasmus Müller

P.E. Müller (1840-1926; Fig. 1H) graduated in agriculture and later in forestry but was inspired by Schi179dte to study zoology. A large work on the faunistics and biology of Danish cladocerans won him a university prize (published 1867, with beautiful plates drawn by the author), and was followed the next year by a work on the embryology and propagation of Cladocera.

In 1868 P.E. Müller set out on a 3-year journey through Europe to study freshwater crustaceans. However, being informed on the way that on return he would obtain a position at the Royal Veterinary and Agricultural University, he switched entirely to forestry and later became a leading authority on the ecology of forest soil and forestry in general. (Gosch 1875: 563-567, 1878: 526-527).

2.6 Gustav Henrik Andreas Budde-Lund

G. Budde-Lund (1846-1911; Fig. 1I) was also a student of Schiødte and studied zoology, mainly entomology, but failed to graduate. He then went into business, becoming a wholesale dealer and later the owner of a brush factory. Eventually he accepted political and social posts of honor in the Copenhagen city council and the municipal supreme treasury. At that same time he took up his old love of arthropods and became the leading authority on terrestrial isopods, working on collections sent to him from all over the world. A general, systematic review of the whole group (1885) was followed by several supplements. He created 74 genera and about 500 species, a hundred types of which are in the Copenhagen Museum. (Gosch 1875: 572-573, 1878: 531; Dahl 1910: 35-36).

2.7   Hans Jacob Hansen

H.J. Hansen (1855-1936; Fig. 2A) is perhaps the greatest name in Danish carcinology. He was a student of Schiødte, started working at the Zoological Museum while still a zoology student, and graduated in 1874. His thesis on the mouthparts of flies gave him the nickname 'Flue-(Fly-) Hansen'. In 1885, he failed to become chief curator when Meinert got the position, and he was passed over again 25 years later at Meinert's retirement. The reason was partly an offshoot of the old Schiødte-Steenstrup quarrel, and partly due to Hansen's coarse and uncompromising attitude towards opponents. With his well-established international scientific reputation, he regarded the passing over in 1910 as a personal offense and angrily resigned. His action was followed by an unprecedented open letter in *Nature*, signed by W.T. Calman and 28 other prominent British scientists. This no doubt contributed to the provision of a most unusual civil-list pension enabling him to work productively, free of all administrative burdens, until his death 26 years later.

In his scientific work, Hansen concentrated on studying forms that he regarded of interest from a morphological point of view or of being capable of throwing light on mutual relationships: 'He always labored just where there were real gaps and imperfections in the common stock of zoological knowledge' (Calman et al. 1910). Thus, he was a pioneer in homologizing the limb structure in all groups of arthropods. His technique was simple but his accuracy so great that he brought out details overlooked by previous workers. These studies resulted in his chief work, the small paper *A contribution to the morphology of limbs and mouthparts in crustaceans and insects* (1893) which was followed 35 years later by *Studies in Arthropods II & III* (552 pp., 24 pls.). In the preliminary paper, he set up the new and presently accepted classification within the Malacostraca, and he revised the work by Boas 10 years earlier, introducing Peracarida and separating isopods from tanaids.

Throughout his long life Hansen worked on extensive materials from most major oceanographic expeditions, including Challenger, Ingolf, Albatross, Siboga, and those of Prince Albert of Monaco. Most of these contributions are monographic in nature, e.g., several on sergestid shrimps, and pioneer works on euphausiids that established the systematics of this group.

In 1893, Hansen made extensive collections of mainly land arthropods in southern Italy, and in 1895 he participated in the opening year of the Ingolf Expedition to the deep North Atlantic during which he was the first to secure minute representatives of the deep-sea fauna by means of a special sieve and very careful examination of the residue. Based on one deep-sea species and one littoral species from Siam he established Ingolfiellidea, a fourth suborder of amphipods. Fundamental are his two Ingolf reports on the Tanaidacea (1913) and the Isopoda Asellota (1916). Other isopod monographs are on Cirolanidae (1890) and Sphaeromatidae (1905) [up until then in a chaotic taxonomic state due to sexual differences], and also one on parasitic copepods from the Ingolf (1923). His monograph on the Choniostomatidae (1897), which are tiny ovoid parasites mainly in the brood-pouch of amphipods and closely resemble

Figure 2. A: H.J. Hansen. B: Carl With. C: Jap. Steenstrup. D: R.S. Bergh (drawing by P.S. Krøyer). E: C.F. Lütken. F: J.E.V. Boas. G: Søren Jensen. H: Frits Johansen. I: C. Wesenberg-Lund.

eggs, has been called 'a high point in the literature on copepods' and 'a masterpiece of delicate dissection and exquisite illustration.' Finally, he was also a specialist on millipedes and minor groups of arachnids on which he published five monographs around the turn of the century.

H.J. Hansen was an excellent and accurate draftsman who like Schiødte used copperplates for most illustrations. Besides arthropods, he had a profound knowledge of military politics, agitating in newspaper articles and pamphlets for defense from a clear anti-German viewpoint. He was also an advocate for the use of English as the scientific language by Danish zoologists. He could be unnecessarily aggressive in his fight against real or imagined favoritism, snobbishness, and mediocrity, but his idealism was strong and pure and he never compromised his convictions, scientifically, politically, or ethically. (Dahl 1910; Calman et al. 1910; Calman 1936, 1937; Stephensen 1937).

2.8   Carl Johannes With

Carl With (1877-1923; Fig. 2B) was Hansen's only student. He worked at the Zoological Museum while studying zoology and graduated in 1901. Between 1903 and 1908 he published half a score of splendid arachnoid works quite in the spirit of Schiødte. Since he could not get a position, however, he began to study medicine; after the publication in 1915 of his monograph on the pelagic Copepoda Calanoidea Amphascandria from the Ingolf Expedition he had to abandon zoology completely, eventually becoming an outstanding authority on venereal diseases. A continuation of his copepod studies after an intended early retirement as a physician was prevented by his death at only 46 years.

3   THE STEENSTRUP SCHOOL

While Krøyer-Schiødte and their followers were almost entirely engaged in meticulous comparative morphological and taxonomic studies of arthropods, including crustaceans, the representatives of the Steenstrup line made up a much more heterogeneous group, with arthropods and particularly crustaceans being only a minor part of their scientific activities.

3.1   Johannes Japetus Smith Steenstrup

Japetus Steenstrup (1813-1897; Figs 2C, 3-5) was professor of zoology from 1845 to 1885 and the dominating and inspiring central figure in Danish zoology and natural sciences for half a century. At his death he was an honorary member of at least 35 academies and societies. He wrote extensively on a great variety of subjects, although mainly on cephalopods, fishes, alternation of generations in lower animals, Quaternary fossils, and kitchen middens. His publications include only a few papers on crustaceans: Recent and fossil cirripedes; hermit crab parasites; and, with C.F. Lütken, the monograph on parasitic copepods of the open sea (1861), with excellent copperplates. (Gosch 1875: 221-387; Lütken 1897; Dahl 1913; Spärck 1932b, 1948; Knudsen in Roeleveld & Knudsen 1980: 296-305).

3.2   Christian Frederik Lütken

Chr. Lütken (1827-1901; Fig. 2E) was a student and an ardent admirer of Steenstrup whom he succeeded as professor of zoology. He was a specialist on echinoderms and fishes. In addition to the parasitic copepod monograph with Steenstrup, he published another monograph on whale-lice, and a few minor carcinological papers. (Gosch 1875: 404-424, 1878: 466-483; Bather 1901; Dahl 1910: 9-14; Madsen 1989).

3.3   *Ludwig Sophus Rudolph Bergh*

Rudolph Bergh (1824-1909; Fig. 2D) resembles Carl With in being partial to zoology but choosing medicine for a living. Like With he became a dermatologist and venerologist of very high standing, initiating, for example, a pioneer new hospital in this field.

Probably inspired by Steenstrup, Bergh took up studies of mollusks. He was equally skillful in dissecting and drawing, and his production on the anatomy and taxonomy of particularly the nudibranchs is enormous. Contact with foreign colleagues was considerable, and eventually he became a member of many medical and scientific societies and academies.

Bergh's three major papers in carcinology deal with the embryology of *Mysis* and *Gammarus* and the histology of blood vessels in arthropods. (Gosch 1875: 152-159, 1878: 496-504; Dall 1909; Dahl 1910: 15-21; Vayssière 1910; Schlesch 1946).

3.4   *Johan Erik Vesti Boas*

J.E.V. Boas (1855-1935; Fig. 2F) was also a student of Steenstrup but primarily of Carl Gegenbaur in Heidelberg. Like Schiødte, Steenstrup, and Hansen, he was a comparative morphologist, but contrary to the others he endorsed Darwin's theory of evolution. After graduating in zoology he worked as an assistant at the Zoological Museum, but in 1885, at Lütken's appointment as professor, the Ministry of Education chose G.M.R. Levinsen for Lütken's position as curator. This obvious disregard made Boas so furious that for the next 35 years he was in no way associated with the Museum. Instead he obtained a position at the Royal Veterinary and Agricultural University where in 1893 he became professor of zoology.

Boas's main works deal with both comparative anatomy of mammals and reptiles, and pest animals in forestry and agriculture. He organized teaching and study collections and wrote two excellent textbooks, in forestry zoology and in general zoology; the latter was translated into English and Polish and appeared in no less than 10 editions in German.

His first major carcinological contributions are his famous papers on the relationships of decapods (1880) and of malacostracans (1883), which included a new classification of decapods (Natantia versus Reptantia, the latter including anomurans and crabs) and abolition of the group Schizopoda (euphausiids and mysids). In the 1920's, he returned to carcinology writing on anomurans from the Zoological Museum: including the genera *Lithodes*, and *Paguropsis* and other symmetrical hermit crabs, all beautifully illustrated. (Steenberg 1936).

3.5   *Søren Jensen*

Søren Jensen (1873-1902; Fig. 2G) won the university gold medal for his research on the biology and systematics of Danish freshwater ostracodes and copepods, as a continuation of O.F. Müller's work. A year before graduating in zoology in 1901, he published a paper on two aberrant parasitic copepod species, with an unnecessarily violent attack on Hansen's work in the same field. Hansen's reply was no less savage, and this late offshoot of the Schiødte-Steenstrup quarrel created a great stir in the zoological community. In 1900, Jensen also participated in the Carlsberg Foundation's East Greenland Expedition, and his journal testifies to his skill in life studies of marine crustaceans. His death in 1902 from typhoid fever put a tragic end to a promising carcinological career. The gold medal studies were published posthumously. (Dahl 1910: 154-155; Wolff 1967: 133-135).

3.6 Frits Johansen

Frits Johansen (1882-1957; Fig. 2H) did all his research in the Arctic. During the Danmark Expedition, 1906-1908, he was responsible for collecting all but mammals and birds. He was a formidable field worker, and his unpublished journals are full of biological observations. Due to a lack of concentration he did not manage to pass his degree. After publication of pioneer observations on the ecology and biology of Arctic freshwater entomostracans, he obtained a position at the museum in Ottawa and participated in a number of expeditions in Arctic Canada. In the 1920's, he wrote about 15 papers on the freshwater crustaceans, mainly phyllopods. Other contributions dealt with fishes and insects from northeast Greenland and Arctic Canada. In 1929, he returned to Copenhagen, but in spite of many plans for further publications he was never able to concentrate on their realization. (Wolff 1967: 137-144).

3.7 Carl Jørgen Wesenberg-Lund

C. Wesenberg-Lund (1867-1955; Fig. 2I) was the last of the Steenstrup line. He devoted his whole life to research and field observations of the biology of freshwater animals, being supported most generously by the University and the Carlsberg Foundation. After graduating in 1893, he published a systematic-biological work on Greenland entomostracans, the only paper in his enormous production based on museum material. After participation in the second year (1896) of the Ingolf Expedition, he later took over the ship's deck-house which was set up as a laboratory at Frederiksdal close to where his great idol, O.F. Müller, had worked 150 years before (Wesenberg-Lund 1940). In 1922, he became the first Professor of Limnology at the new freshwater laboratory in North Zealand.

Wesenberg-Lund's first major publication was *Plankton Investigations I, II* (1904, 1908) in which he deals with the seasonal variation in shape, mainly in Cladocera, and finds it to be caused by changing specific gravity rather than viscosity as formerly believed. His ideas are further discussed in *Fundamental features of freshwater plankton* (1910). In the 1920's, he returned to the study of Cladocera with a paper on the morphology of the genus *Daphnia*. With his many monographs on various groups of freshwater animals, Wesenberg-Lund stands as one of the founders of modern limnology. (Berg 1938, 1949, 1957, 1958; Wesenberg-Lund 1940; Thienemann 1949, 1956).

4 CRUSTACEAN RESEARCH IN THIS CENTURY

4.1 Kaj Berg

Kaj Berg (1899-1972; Fig. 4A) was Wesenberg-Lund's first assistant at the laboratory. He also worked on Cladocera, but contrary to Wesenberg-Lund's direct observations in nature, he combined field studies with laboratory experiments. First he worked on cladoceran reproduction in various habitats. Instead of previously acceptable, mainly theoretical conclusions, he found that transition from parthenogenesis to gamogenesis is caused by external conditions.

Figure 3. Portraits from a painting: 'A scene from the Scandinavian Meeting of Natural Scientists in Copenhagen 1847.' 1. Sven Nilsson. 2. Berzelius. 3. Schouw. 4. Sibbern. 5. Steenstrup. 6. Schiedte. 7. Thomander. 8. Stein. 9. Trier. 10. Faye. 11. Grundtvig. 12. Fries. 13. Scharling. 14. Fenger. 15. Firchhammer. 16. Claussen. 17. Johnstrup. 18. Agardh. 19. Conradi. 20. Ørsted. 21. Bang. 22. Hansteen. 23. Eschricht. 24. Krøyer. 25. Retzius.

Figure 4. A: Kaj Berg. B: K. Stephensen. C: Poul Heegaard. D: Thydsen Meinertz. E: P. Jespersen. F: Erik M. Poulsen.

The most significant of several later papers on Cladocera dealt with cyclic reproduction, sex determination and depression (1934), and alternation of generations (1937). In 1939, he succeeded Wesenberg-Lund as professor of limnology. Berg is best known for his comprehensive, quantitative studies of the bottom faunas of Lake Esrum and River Suså and for respiration physiology of freshwater animals (Madsen & Nygaard 1970; Jonasson 1975).

4.2 Knud Hensch Stephensen

K. Stephensen (1882-1947; Fig. 4B) succeeded Hansen as curator at the Zoological Museum 4 years after Hansen's resignation in 1910. Stephensen was a born museum man. He corresponded with all carcinologists of his day, and his activities in exchange of specimens, particularly of amphipods, were considerable. His quiet work among museum specimens was interrupted by only a few collecting travels: to southwest Greenland fjords in 1912 and joining Johannes Schmidt's eel investigation on the Dana to the West Indies in 1921-22.

Stephensen's publications are of a systematic-faunistic nature, with rare attempts to draw general conclusions, but he was one of the very last carcinologists capable of covering almost all groups. He became a world authority on amphipods, writing extensively on pelagic amphipods from the Danish Thor Expedition to the Mediterranean in 1908-10, deep-sea amphipods of the Ingolf, amphipods (and decapods) of the Godthaab Expedition off West Greenland, and amphipods of northern Norway. He had for many years prepared a thorough revision of Stebbing's amphipod catalogue from 1906, but his fairly early death prevented this. In the Zoology of Iceland, of the Faroes, and of eastern Greenland he covered almost all crustacean groups and the pycnogonids. His last major paper was on the crabs of the Iranean Gulf, and in it he drew attention to the systematic significance of the male pleopods.

Stephensen was an amiable man who showed a boundless hospitality in his home to Danish and foreign friends and colleagues (Bruun 1947).

When Stephensen died in 1947, I succeeded him as curator. Thus, from 1842 till today there have been only five persons in charge of the Zoological Museum crustaceans (until the 1950's within the Department of Entomology): Krøyer, Meinert, Hansen, Stephensen, and myself. The combination of many expeditions and 150 years of curating and research is the background for the rich crustacean collections of the Museum.

4.3 Poul Edheru Auker Heegaard

Poul Heegaard (1908-1974; Fig. 4C) was attached to the Zoological Museum and the Fishery Research Institute through grants but failed to obtain a permanent position. From 1952-1959 he taught in Bandung (as the Indonesian state's first Professor of Zoology), and in 1959-1961 in Mosul, Iraq. On his return home an urgent request from 15 colleagues at the university provided him with a senior scholarship till his death. Both before and after his residence in the East he worked as visiting zoologist at many marine stations.

Heegaard's thesis (1947) dealt with the development of copepods (based on excellent laboratory and field studies) and included his theory about the origin of the arthropod limb. During a stay in Texas these ideas were tested on larvae of penaeids, but with disappointing results. Similarly controversial was what his colleagues called 'the Heegaardian heresy' regarding the identity of mouthparts of parasitic copepods on whose taxonomy he wrote about 15 papers. In addition to treating the decapods from East Greenland, he published two fine monographs on the rich Dana Expedition material: One a pioneer study of larvae of three genera of oceanic penaeids, the other on the 16 larval and postlarval stages of the caridean *Amphionides*. (Gunter 1975).

4.4   *Niels Thydsen Meinertz*

Thydsen Meinertz (1894-1984; Fig. 4D) worked at the Institute of Comparative Anatomy in Copenhagen, specializing in facial musculature in mammals and internal organs of mammals and reptiles. However, in addition he collected terrestrial isopods at about 6000 localities throughout Denmark and published several papers on their distribution, propagation, and variation. He also treated the group in *Zoology of Iceland*.

4.5   *Poul Christian Jespersen*

Poul Jespersen (1891-1951; Fig. 4E) worked during his studies at the Carlsberg Laboratory under Johannes Schmidt for whom he collected eel larvae across the Atlantic before the schooner was shipwrecked in the West Indies (the collections were saved). After graduating in 1917, he worked at the Plankton Laboratory at the Fisheries Research Institute till his death. In 1920 and 1921-22, he returned to the West Indies on the Dana, and he also participated in the first half of its circumnavigation, 1928-30, as well as in many cruises in the North Atlantic.

Besides papers on oceanic and commercial fishes and eel larvae, Jespersen concentrated in his plankton studies on copepods, including pioneer studies of their distribution, quantity, seasonal variation, etc., around Greenland-Iceland and in the Mediterranean (including the Godthaab and Thor Expeditions). Much appreciated are the many zooplankton identification sheets he edited for ICES, the International Council for the Exploration of the Sea.

Jespersen's correlation between the quantitative distribution of zooplankton and the frequency of sea birds builds a bridge to his other great interest. He was a keen ornithologist who wrote extensively on Danish birds and was active in international bird preservation (Russell 1952; Wolff 1967).

4.6   *Erik Mellentin Poulsen*

Poulsen (1900-1985; Fig. 4F) was also employed by the Fishery Research Institute. Besides serving for shorter periods as fishery adviser for FAO, the United Nations Food and Agricultural Organization, in Colombia and Chile, he acted for 11 years as Secretary General of ICNAF, the Northwest Atlantic Fishery Commission in Halifax.

Poulsen published many papers on races, fluctuations, biology, etc., of important commercial fishes. His contributions on the significance for fisheries of various invertebrates, including crustaceans, form a transition to his carcinological works, and extensive fishery research collections are the basis for detailed studies of cirripedes, thalassinids, and crabs in Danish waters. Several papers are devoted to Entomostraca from Greenland, Iceland, and the Faroes, and works on Danish daphniids continue a long tradition, dating back to O.F. Müller. Poulsen's pièce de résistance, however, was his study of the huge Dana collections of oceanic Ostracoda Myodocopa, started when he was nearly 60. The four taxonomic sections (1962-73) treat 97 species from 1800 trawls, with the number of individuals of the 7 most common species varying between 21000 and 460000! In 1977, the taxonomic sections were followed by an outline of the biogeography, vertical distribution, and daily migrations (Bagge 1986; Wolff 1986).

5 OTHER CONTRIBUTIONS

For the sake of completeness this survey of Danish crustacean research is rounded off with a brief list of mainly minor contributions by zoologists whose main efforts lay within a great variety of subjects other than carcinology. The list is in strict chronological order.

1764-1800 M.Th. Brünnich: Keys, etc., to Danish insects and crustaceans; P. Ascanius: *Balanus hammeri*; N. Tønder Lund: *Scyllarus*; P.C. Abildgaard: *Monoculus*, *Caligus*; L.Spengler: *Lepas*, *Scyllarus*; J. Rathke: *Nymphon*.

1825-1850 P.W. Lund: Blood vessels; J.Th. Reinhardt: *Phyllamphion*, *Lithotrya*.

1851-1900 L. Lund: Cladocera; G.M.R. Levinsen: parasitic copepods; Th. Mortensen: *Palaemon adspersus*.

1901-1925 J.O. Bøving-Petersen: Lobster; O. Paulsen: *Calanus finmarchicus*; A. Otterstrøm: *Mysis*, *Eriocheir*; C.G.J. Petersen: Prawn fishery; H.F.E. Jungersen: Parasitic copepod; P.L. Kramp: Schizopoda; H. Blegvad: gammarids, mysids; A.M. Hemmingsen: Crayfish blood sugar; A.F. Bruun: *Chiridothea* (*Saduria*), *Thermosbaena*.

1926-1960 A.C. Johansen: Mortality; K.L. Henriksen: Molting; Ad.S. Jensen: *Eriocheir*; H.H. Ussing: Greenland copepods; H.C. Terslin: *Munida*, *Geryon*; E.W. Kaiser: *Gammarus*. - M. Thomsen: hormones; G. Thorson: Decapod larvae; Aa.J.C. Jensen: *Nephrops*.

Recent contributors to the field of carcinology include K. Larsen, E. Smidt, P. Metz, E. Rasmussen, T. Wolff, Aa. M. Christensen, K. G. Wingstrand, U. Røen, J. Bresciani, S.Aa. Horsted, J. P. Jensen, G. H. Petersen, B. Christensen, B. F. Theisen, E. Kanneworf, J. Lützen, J. Just, L. Hagerman, J. Mossin, W. & H. Nicolaisen, K. Jensen, F. Møhlenberg, T. Kiørboe, and J. Høeg.

REFERENCES

Articles in Dansk Biografisk Leksikon (Danish Biographical Dictionary, 3rd edition), and other biographies and obituaries in Danish exist for almost all persons mentioned above. Only biographies and obituaries in English, German, or French and books in Danish are included in the present list.

Anker, J. 1943. Otto Friedrich Müller. *Acta Hist. Sci. Nat. Med.* 2:1-317.

Anker, J. 1950. Otto Friedrich Müller's Zoologia Danica I. *Libr. Res. Monogr.* Univ. Libr. scient. med. Dep., Copenh. 1:1-108.

Bagge, O. 1986. Erik Mellentin Poulsen 14. July 1900 - 12. January 1985. *Dana* 6:53-56.

Barton, B.S. 1808. Obituary (J. Christ. Fabricius). *Philad. Med. Phys. J.*, Part I(3):187-189.

Bather, F.A. 1901. Christian Frederik Lütken. *Science* 13:540-542.

Berg, K. 1938. Prof. Dr.phil. C. Wesenberg-Lund zu seinem 70. Geburtstage am 22. Dezember 1937. *Arch. Hydrobiol.* 32(4):1-6.

Berg, K. 1949. C. Wesenberg-Lund. *Hydrobiol.* 1:322-324.

Berg, K. 1957. Professor Carl Jørgen Wesenberg-Lund. *Proc. Linn. Soc. Lond.* 1955-56(1-2):57-60.

Berg, K. 1958. Professor C. Wesenberg-Lund. *Verh. Int. Verein. Theor. Angew. Limnol.* 13:975-978.

Bourgeois, M.J. 1889. Notice sur la vie et les travaux de Jörgen-Christian Schiödte, membre honoraire. *Ann. Soc. Ent. Fr.* 1889:473-480.

Bruun, A.F. 1947. K.H. Stephensen. *Nature* 160:82.

Calman, W.T. 1936. Dr. H.J. Hansen. *Nature* 138:193.

Calman, W.T. 1937. Obituary notice (H.J. Hansen). *Proc. Linn. Soc. Lond.* 1936-37(4):193-195.

Calman, W.T. et al. 1910. Dr. H.J. Hansen and the Copenhagen Museum of Zoology. *Nature* 83:36.

Christensen, C. 1918. *Naturforskeren Pehr Forsskål* (The Natural Scientist P.F.). Copenhagen: H. Hagerup.

Dahl, S. 1910. *Bibliotheca Zoologica Danica 1876-1906*. Copenhagen: J.L.Lybecker.

Dahl, S. 1913. Bibliographia Steenstrupiana. In H.F.E. Jungersen & E. Warming (eds), *Mindeskrift i Anledning af Hundredaaret for Japetus Steenstrups Fødsel, I, II*: I(8), pp. 1-28. Copenhagen.

Dahl, S. (ed.) 1941. *Til Minde om Zoologen Henrik Krøyer* (In memory of the zoologist Henrik Krøyer: Introduction, Krøyer on his descent, school memories, fragments from his unpublished work on Danish fisheries, and letters). Copenhagen: Hassing.

Dall, W.H. 1909. Ludwig Rudolph Sophus Bergh. *Science* 30:304.

Damkaer, C.C. & D.M. Damkaer 1979. Henrik Krøyer's publications on pelagic, marine Copepoda (1838-1849) (Krøyer's life and work pp. 3-9.). *Trans. Amer. Phil. Soc.* 69(6):1-48.

Forsskål, P. 1950. *Resa till Lyckliga Arabien. Dagbok 1761-63* (Travel to Happy Arabia. Diary 1761-63, with introduction by A.H. Uggla). Uppsala.

Gosch, C.C.A. 1873. *Udsigt over Danmarks zoologiske Literatur* II(1). Copenhagen.

Gosch, C.C.A. 1875. *Udsigt over Danmarks zoologiske Literatur* II(2). Copenhagen.

Gosch, C.C.A. 1878. *Udsigt over Danmarks zoologiske Literatur* III. Copenhagen. (Bibliographies).

Gosch, C.C.A. 1898-1905. *Jorgen Christian Schiødte I-III*. Copenhagen: Gyldendal.

Gunter, G. 1975. Poul E. Heegaard 1904-1974. *Contr. Mar. Sci.* 19:1-2. (1904 should be 1908.)

Hansen, Th. 1964. *Arabia Felix. The Danish Expedition of 1761-67*. London: Collins.

Henriksen, K.L. 1932a. Johann Christian Fabricius (1745-1808). In V. Meisen (ed.), *Prominent Danish Scientists*: pp. 76-80. Copenhagen: Levin & Muuksgaard; London: Oxford Univ. Press.

Henriksen, K.L. 1932b. Jørgen Christian Schiødte (1815-1884). In V. Meisen (ed.), *Prominent Danish Scientists*: pp. 124-127. Copenhagen: Levin & Muuksgaard; London: Oxford Univ. Press.

Jensen, Ad.S. 1932. Otto Fabricius (1744-1822). In V. Meisen (ed.), *Prominent Danish Scientists*: pp. 72-75. Copenhagen: Levin & Muuksgaard; London: Oxford Univ. Press.

Jespersen, P. H. 1946. J.C. Fabricius as an evolutionist. *Svenska Linnésällsk. Årsskr.* 29:35-56.

Jonasson, P.M. 1975. Kaj Berg. *Arch. Hydrobiol.* 76:256-264. (With bibliography.)

Kornerup, B., Schultz-Lorentzen & Ad.S. Jensen 1923. Biskop, Dr. theol. Otto Fabricius. Et Mindeskrift i Hundredeaaret for hans Død (Bishop, Dr.Theol. Otto Fabricius. A Memorial Book at the Centenary of his Death). *Meddr. Grønland* 62:215-400.

Latreille, P.A. 1808. Notice biographique sur Jean Chrétien Fabricius, conseiller d'état du roi de Danemarck, professeur d'histoire naturelle et d'économie rurale Kiell, et membre d'un grand nombre d'académies. *Annls. Mus. Hist. Nat.* 11:393-404.

Lütken, C.F. 1897. Steenstrup. *Nat. Sci.* 11(67):159-165.

Madsen, B. Lauge & G. Nygaard 1970. Professor Kaj Berg – 70 years old. *Hydrobiol.* 35:345-351.

Madsen, F. Jensenius 1989. Christian Frederik Lütken 1827-1901. *Echinoderm Stud.* 3:349-357.

Mayer, K. 1967. Der Mensch und Forscher J.Ch. Fabricius. *Mitt. Dt. Ent. Ges.* 26(2):21-29.

Roeleveld, M. & J. Knudsen 1980. Japetus Steenstrup: On the mermaid (called the Sea Monk) caught in the Øresund in the time of Christian III. A translation into English. With a biography of Japetus Steenstrup. By J. Knudsen. *Steenstrupia* 6:293-332.

Russell, F.S. 1952. Poul Jespersen 1891-1952. *J. Cons. Perm. Int. Explor. Mer* 18:104-106.

Scheele, I. 1982. Die Rostocker Ehrenpromotion des dänischen Zoologen Henrik Krøyer (1799-1870). *Dt. Schiffahrtsarch.* 5:187-202.

Scheele, I. 1985. Der Zoologe Henrik Krøyer (1799-1870) und die deutsch-dänischen Wissenschaftsbeziehungen. *Centaurus* 28(1):17-30.

Schlesch, M. 1946. Rudolph Bergh. *J. Conch., Paris* 22:225-226.

Schuster, J. 1928. Linné und Fabricius zu ihrem Leben und Werk. *Münch. Beitr. Gesch. Lit. Naturw. Med.*, Sonderh. IV: (var.pag.)

Spärck, R. 1932a. Otto Friedrich Müller (1730-1784). In V. Meisen (ed.), *Prominent Danish Scientists*: pp. 60-64. Copenhagen: Levin & Muuksgaard; London: Oxford Univ. Press.

Spärck, R. 1932b. Japetus Steenstrup (1813-1897). In V. Meisen (ed.), *Prominent Danish Scientists*: pp. 115-119. Copenhagen: Levin & Muuksgaard; London: Oxford Univ. Press.

Spärck, R. 1948. *Japetus Steenstrup. Et Bidrag til Vurdering af hans Indsats som Naturforsker* (Japetus Steenstrup A Contribution to an Appraisal of His Effort as a Natural Scientist). Festskr. udg. af Kbh. Univ. Copenhagen.

Steenberg, C.M. 1936. Johan Erik Vesti Boas. Ein Lebensbild. *Morph. Jb.* 78:253-265. (With bibliography.)

Steffens, H. 1841. *Was ich erlebte* III. Breslau. (Danish edition 1841.)

Stephensen, K. 1937. H.J. Hansen as a carcinologist. *Vidensk. Meddr Dansk Naturh. Foren.* 100:409-417.

Thienemann, A. 1949. C. Wesenberg-Lund zu seinem 80. Geburtstag. *Arch. Hydrobiol.* 42:199-200.

Thienemann, A. 1956. C. Wesenberg-Lund. *Arch. Hydrobiol.* 51:578-579.

Tuxen, S.L. 1959. Der Entomologe I.C. Fabricius und die Typen der von ihm beschriebenen Arten. *Zool. Anz.* 163: 343-350.

Tuxen, S.L. 1967. The entomologist J.C. Fabricius. *Ann. Rev. Ent.* 12:1-14.

Vayssière, A. 1910. Rudolph Bergh (1824-1909). *J. Conch.* 68:110-117.

Walckenaer, C.A. 1815. Jean-Chrétien Fabricius. *Biographie Universelle* 14. Paris.

Wesenberg-Lund, C. 1940. *Det ferskvandsbiologiske laboratorium gennem 40 Aar* (The Freshwater Laboratory through 40 Years). Copenhagen: Bianco Luno. (With bibliography.)

Wolff, T. 1967. *Danske Ekspeditioner på Verdenshavene/Danish Expeditions on the Seven Seas*. Copenhagen: Rhodos.

Wolff, T. 1968. The Danish expedition to 'Arabia Felix' (1761-1767). *Bull. Inst. Océanogr. Monaco* No.spec. 2:581-601.

Wolff, T. 1986. Erik M. Poulsen (1900-1985). *J. Crust. Biol.* 6:309-310.

Zimsen, E. 1964. *The Type Material of I.C. Fabricius*. Copenhagen.

History of the carcinological collections of the Rijksmuseum van Natuurlijke Historie, Leiden, Netherlands (1820-1950)

L.B. Holthuis
Emeritus Curator of Crustacea, Rijksmuseum van Natuurlijke Historie, Leiden, Netherlands

1 PRELUDE TO THE FOUNDING OF THE MUSEUM

In the 17th and 18th century, when the Netherlands were at the height of their power and wealth, an amazingly great number of private natural history cabinets existed there. The prosperity of the country made it possible for many persons to give time and money to the gathering of such collections. The fact that the country had colonies and trading posts in many parts of the world, and that its ships could be found on all the seas, made material for such collections relatively easily available. Sailors brought home natural curiosities from distant areas, while merchants and officials of outlying posts, fascinated by the wealth of the nature in their domiciles, made collections of natural products, which they sent or took home. These private collections were very many and of widely diverging interests and sizes; this is well illustrated in the valuable enumeration of the Dutch zoological collections published by Engel (1986). Some collections were purely for aesthetic reasons, but most had a more or less scientific basis, even if it were only to gather rare species. The most popular were the shell collections, but other animals figured likewise in the curiosity cabinets. So in the famous collection of the Amsterdam apothecary Albertus Seba (1665-1736) practically all classes of animals were represented. A well-known collection of mounted birds was owned by the rich Amsterdam merchant Jacob Temminck (1748-1822). Also the Prince of Orange, Willem V (1748-1806), who was stadtholder of the Netherlands from 1751 to 1795, owned an internationally famous cabinet and menagerie.

Notwithstanding this general interest in natural history and natural history collections, the Republic of the Seven United Netherlands could not boast a national museum: The seven provinces evidently too jealously guarded their autonomy to be much interested in such a federal project. It was not before the 19th century that it proved possible in the Netherlands to establish a national museum of natural history. However, much had to happen before that. The end of the 18th century saw the fall of the old republic and the beginning of the so-called 'French era.' In 1795 French revolutionary troops occupied the Netherlands and helped to set up the 'Batavian Republic' with a Dutch puppet government that received its orders from Paris. This Batavian Republic (1795-1806) was followed by the Kingdom of Holland (1806-1810) with Louis Napoleon Bonaparte, brother of Napoleon Bonaparte, as King. Finally the country became part of the French Empire (1810-1813), reaching nationally and economically the lowest point of its existence. The French era meant a period of great poverty for the country. Foreign trade came practically to a standstill due to French measures and English blockade; all the colonies were lost in this time to England. Many once rich persons lost their money, and

225

natural history collections were often considered superfluous luxury. A few of these cabinets, like the Temminck collection, survived the unfavorable times, but many were sold or otherwise disappeared, often leaving hardly any traces. Boeseman's (1970) interesting account of the fate of the once famous Seba collection gives a good example of this. The larger part of the cabinet of the Prince of Orange had been confiscated by the French troops and, together with the menagerie and other possessions of the prince, was sent to Paris in 1795.

After the fall of Napoleon in 1813, Holland and Belgium were united into the Kingdom of the Netherlands, with Willem of Orange (1772-1843), the oldest son of the last stadtholder (Willem V, Prince of Orange) as the first king, Willem I. The new king gave his full attention to the revival, not only of the trade and industry but also of the arts and sciences, and for a great part, thanks to his energy and foresight, the situation of the country improved rapidly. He is also responsible for the foundation of 's Rijks Museum van Natuurlijke Historie, which took place on 9 August 1820 by Royal Decree No. 75.

The collections of the new museum were formed by the fusion of three existing natural history cabinets (one national, one private, and one university collection).

1.1   's Lands Kabinet van Natuurlijke Historie

's Lands Kabinet van Natuurlijke Historie (National Cabinet of Natural History), was originally founded in 1808 during the reign of King Louis Napoleon of Holland as the Cabinet du Roi or the Koninklijke Kabinet van Natuurlijke Historie. On 28 July 1808, the king appointed C.G.C. Reinwardt, who at that time was professor of chemistry, botany, and natural history at the University of Harderwijk, to be the Director of the Jardin du Roi and of 'notre Cabinet d'histoire naturelle, aussitôt qu'ils seront formés.' After the liquidation of the Kingdom of Holland in 1810, and after the downfall of Napoleon and the liberation of the Netherlands in 1813, the cabinet remained in existence and was named 's Lands Kabinet van Natuurlijke Historie. Reinwardt remained director even after he left in 1815 for the Dutch East Indies as Director General of Agriculture. During Reinwardt's absence the supervision of the Museum came into the hands of C.J. Temminck for the Vertebrates and M. van Marum for the Invertebrates, the latter assisted in the Arthropoda by Mr. A.J. d'Ailly an amateur entomologist from Amsterdam. At that time the Museum, which first had been in the Royal Palace at Amsterdam, was housed in the so-called Trippenhuis, also in Amsterdam, which at present is the seat of the Netherlands Academy of Sciences. The collections of the Kabinet were small and unimportant: it had existed too short a time (and very troubled times at that) to have grown to a museum of some standing. A list of the Crustacea that formed part of this collection is kept in the archives of the Museum and gives the following species: *Cancer cursor*, *C. pagurus* L., *C. victor* Fabr., *C. lunarius* Herbst (2), *C. sexdentatus* (4), *C. major*, *C. sanguinolentus*, *C. frascone*, *C. pelagicus* (2), *C. personatus*, *C. calappa*, *C. granulatus*, *C. lophos*, *C. muricatus* (3), *C. ovis*, *C. squinado*, *C. maja*, *C. longipes*, *C. dorsipes*, *C. clibanarius*, *C. arctus*, *C. homarus* (2), *C. mantis digitalis* (3), *C. mantis chiragra*, *C. minor*, *C. longirostris*, *C. septemspinosus* and also 'chela cancri condyliati et sexdentati; thorax cancri homari et sexdentati' and finally *Monoculus polyphemus* (when a name is given more than once in the list, the number of times is given in parentheses behind the name). All this material was evidently considered unworthy to be kept for the new museum, and around 1826 it was turned over to the University of Liége (then, before the independence of Belgium, a Dutch university) for use in its zoology courses; the present whereabouts of the specimens are not known.

1.2 The Temminck collection

Jacob Temminck (1748-1822), treasurer of the Dutch East India Company, owned a remarkable collection of birds in which foreign species were well represented. His son Coenraad Jacob Temminck (1778-1858) shared his father's interest in birds, extended the collection, and became one of Europe's foremost ornithologists. When in 1820's Rijks Museum van Natuurlijke Historie was established, C.J. Temminck was appointed its director on the condition that the Temminck collection would become part of the Museum. This collection contained mostly birds and mammals, and we have no records that there were any Crustacea in it.

1.3 The collection of Leiden University

In the course of its existence, Leiden University (founded 1575) acquired miscellaneous zoological objects, but only in the 18th century were these brought together in a single collection, which was later greatly enlarged and became quite famous. In the French period (1795-1813), however, the collection was badly neglected, partly because of lack of funds, partly because the professor of natural history at that time, S.J. Brugmans, was more interested in botany and in lecturing than in zoological research. In 1815 the university collection greatly increased in size and value when the King donated the zoological collection of his late father to the University. This material, which French troops had carried off in 1795, had been incorporated into the Muséum National d'Histoire Naturelle in Paris, where it stayed until 1815, when professor Brugmans was sent to retrieve it. Brugmans met great resistance from French zoologists, especially Lamarck, who was using a large part of the collection in writing *Histoire Naturelle des Animaux sans Vertèbres*. Brugmans was persistent, but finally gave in when Alexander von Humboldt, who was a Francophile, was sent as a mediator. Part of the original collection was returned, and the specimens that the French zoologists would not part with were exchanged for several thousand duplicates and triplicates from the Paris Museum. The collection of Leiden University, greatly improved by the king's donation, was incorporated in the new 's Rijks Museum van Natuurlijke Historie. A list made in 1834 of the Crustacea of the old university collection enumerated 15 species of Decapoda (in 18 specimens) and 4 of Cirripedia (in 6 specimens; classed with the Mollusca): *Cardiosoma cordata* (2), *Maja squinado*, *Mithrax aculeatus* (*), *M. spinipes* (8), *M. hispidus*, *Libinia muricata* (*), *Portunus spinamanus* (*), *P. diacantha* (*), *Grapsus goniopsis* (2), *Uca una* (2), *Cancer impressus*, *Calappa fornicata* (*), *Lithodes arctica*, *Lupa* n.sp. (*), *Calappa gallus* (*), *Pollicipes scalpellum* (3), *Anatifa villosa*, *A. striata*, *Coronula diadema* (if more than one specimen is present the number is indicated in parentheses behind the name, an asterisk (*) is given if the specimens are still extant). Most if not all of these specimens were duplicates from the Paris Museum. These 19 species must be considered the only crustacean material in the scientific collection of the Museum at its foundation in 1820.

2 THE HISTORY OF THE CRUSTACEAN COLLECTION OF THE MUSEUM

This can be roughly divided into three periods.

2.1 First period. Start and initial explosive growth. The period of De Haan, Herklots, Hoffmann and De Man (1823-1883)

This first period covers the time of the first two directors of the Museum: C.J. Temminck from

1820 to 1858 and H. Schlegel from 1858 to 1884. These two men, though in many respects each other's antipode, both had the knowledge, competence, energy, and devotion to their institution to make it one of the foremost natural history museums of their time. In the following paragraphs, each of the curators responsible for the Crustacea collection is given some attention, either briefly, or, when his person merits, more elaborately. The years indicated before their name is the period of their curatorship.

In 1820 Director Temminck envisaged three curators for the Museum: One for Vertebrata, one for Invertebrata, and one for mineralogy. The first and the last were appointed in 1820 with the establishment of the Museum. The post of Curator of Invertebrata was only filled in the third year. There were two candidates for the Invertebrata position: Wilhem de Haan (1801-1855) and J. van der Hoeven (1801-1868). Temminck thought both very capable, but preferred De Haan; he asked the trustees of the Museum to appoint De Haan as full curator of Radiata and Mollusca and J. van der Hoeven as honorary curator of Arthropoda. The trustees followed his advice, and both gentlemen accepted; their appointments started 1 January 1823. So Van der Hoeven was the first curator in charge of the Crustacea at the Museum; however, within less than a year he resigned from this unpaid job (on 30 September 1823) to take a university position. He became professor of zoology and was one of the foremost Dutch zoologists of his time, but his role as supervisor of the Crustacea was too brief to have had any impact on the collection.

2.1.1   *1823-1845*

Wilhem de Haan (born in Amsterdam, 7 February 1801; died in Haarlem, 15 April 1855), like Temminck, belonged to a patrician Amsterdam family. His father, Pieter de Haan (1757-1833), was a banker, who under King Louis Napoleon had an important post in the Ministry of Naval and Colonial Affairs, and who finally settled in Leiden as the owner of a cloth factory. Wilhem (his first name is rather unusual and looks like intermediate between the usual Dutch name Willem and the German Wilhelm) studied natural sciences at Leiden University (1818-1825). He received his Ph.D. on 7 May 1825 with a thesis entitled 'Monographiae Ammoniteorum et Goniatiteorum Specimen.' In his youth De Haan had shown much interest in cryptogamic botany, but later concentrated entirely on invertebrate zoology, especially on the Arthropoda. He wrote several important entomological papers, e.g., on Lepidoptera and Orthoptera of the East Indies. His only contribution to carcinology is his masterful volume on the Crustacea of Ph.F. von Siebold's *Fauna Japonica*, which he published between 1833 and 1850 in several fascicles. That his interest in Crustacea dated from much earlier is shown by his report of a journey to Germany that he made in 1826. There he visited museums and private collections in Hamburg, Kiel, Berlin, Breslau (now Wroclaw and part of Poland), Dresden, Nurnberg, Frankfurt, Bonn, and Crefeld. In Kiel, he saw the Fabricius collection and even exchanged some Indonesian insects for Fabrician types of *Dorippe* and *Doclea*. In his report, he wrote the following note about his visit to Berlin: 'The Director of the Berlin Museum, Mr. Lichtenstein, has allowed me to lay here the foundation for a Species Crustaceorum, a task that can only be undertaken in Berlin, because of its extensive collections, that are especially important through the presence of the original specimens mentioned in Herbst's work. I hope by the study of these to make an important contribution towards an accurate knowledge of this group' (free translation from the Dutch). Thus in 1826 he already intended to monograph the Crustacea. Therefore, it is quite understandable that when De Haan was charged with the study of the rich invertebrate collections brought together between 1823 and 1834 in Japan by Ph.F. von Siebold and H. Bürger he started with the Crustacea. The work was published in fascicles that were named decades, because each decas contained 10 plates (and a variable amount of text). The first decas of the Crustacea volume of *Fauna Japonica* was published in 1833, and until 1841 a decas appeared every other year. After the publication of Decas 5 in 1841 the regular order of publication was discontinued because of De Haan's illness, which began in 1842 and which prevented him from coming to the Museum. His illness was described as 'spinal consumption' and resulted in the paralysis of the lower part of the body. From his bed he directed the arrangement of the collections and did all that he possibly could do for the Museum. Director Temminck kept him on the payroll as long as possible, but in 1846 De Haan had to be pensioned. Around 1849 his condition improved somewhat and he managed to finish his work on *Fauna Japonica*. In 1844, 16 of the plates of Decas 6 and 7 had been published without text; the text followed in 1849 as well as the rest of the plates. The Crustacea volume, with an introductory chapter by Von Siebold, was completed in 1850. In his last years, De Haan, on his sickbed, studied the venation of the wings in Lepidoptera, a study he never finished.

On 11 November 1841 De Haan married Sophia E. van Vollenhoven, whose family, like his, belonged to the Dutch upper class (her brother was burgomaster of Amsterdam); no children were born to them. In 1847, the De Haans left Leiden and moved to Haarlem where they lived in the house 'Buitenzorg' at De Baan, a wide avenue in the southern part of the town. De Haan died there on 15 April 1855 and was buried in the General Cemetery of Haarlem; his grave lies in the oldest and nicest part of the cemetery and is well kept, the inscriptions on the tombstone are still clearly readable. De Haan's monograph of the Japanese Crustacea, in which he also gave an important contribution to the general classification of Crustacea, placed him in the foremost ranks of European carcinologists and many were the Academies and Societies that honored him with an honorary membership.

Apart from the first 9 months in office, when he shared the responsibilities for the invertebrate collections with J. van der Hoeven, De Haan throughout his tenure was in charge of all the invertebrate collections of the Museum. From 1824 to 1838 he was assisted by one preparator, Cornelis Overdijk, who left for the East Indies in 1838 when offered a more lucrative position with the East Indian government. At the Museum, Overdijk's annual salary in 1824 wa· Hfl.104, which by 1838 had increased to Hfl. 150. De Haan himself between 1823 and 1828 had a yearly income of Hfl.700, which doubled in 1828, and until 1846 his annual salary was Hfl.1400. His pension amounted to Hfl.548 a year.

There are no indications that another technician was appointed to succeed Overdijk. It was not until the beginning of the present century that the invertebrates again would have their own technical staff.

The dry Crustacea were mounted when large; when small, they were fastened to pieces of cardboard. Several smaller specimens would be placed flat on a sheet of cardboard in which, at either side of the carapace, a hole was drilled. A string put through these holes and around the carapace of the specimen was tied at the back of the cardboard and so kept the specimens in place. The labels were glued to the cardboard below the specimens. The cardboard sheets were placed at an angle to better display the specimens to the public. When the strings deteriorated and broke, the specimens would slither down the slope away from their labels, and this several times caused a mix-up in the labelling. Later, in the first two decades of the present century, each lot would be placed separately with its label in an open cardboard box, and around 1950 these open boxes were replaced with glass covered cardboard boxes, which had the advantage of keeping the dust out and still showing the specimens.

Alcohol material that arrived from Japan and the East Indies was preserved in arak, a strong alcoholic liquid made from rice and the Areca palm (hence its name). It seems that small specimens were placed in cork-stoppered vials, the large in jars. The difficulty in sealing these vials and jars properly allowed many samples to dry out and get lost; dry specimens fared much better than those preserved in spirit.

During De Haan's curatorship much material was received. Important collections were made by several scientists sent out by the government to collect and study zoological specimens. From 1820 to 1850 a small group of usually two scientists accompanied by an artist and a technician was sent to the East Indies for this purpose. Nine of the scientists died there, sometimes after a very short time, and had to be replaced. Of all these scientists only two returned to the Netherlands to publish the results of their investigations. Still this so-called Natural History Commission collected much material and information, and the Museum received large series of East Indian crustaceans from them. Important Mediterranean material was obtained by F.J. Cantraine who between 1827 and 1833 travelled for that purpose on a government grant all over Italy and the Adriatic. Several government officials were instructed to collect material in addition to their regular duties (like Ph.F. von Siebold and H. Bürger in Japan, 1823-1834; H.S. Pel in West Africa, 1840-1855). Also private citizens, interested in zoology, provided the Museum with interesting collections either as a present or against payment of the costs. Often, exchange with other museums and purchase from and exchange with dealers and others also provided important material. During his curatorship De Haan built up the Crustacea collection from one of 19 miscellaneous specimens to one well organized and of great scientific importance.

De Haan evidently was not a field biologist. The only Dutch specimens that the Museum received from him were some seashells, that he bought in the coastal town of Scheveningen. His contributions to the knowledge of the Dutch fauna are practically non-existent.

We know very little of De Haan's personality. In his obituary his steadfastness, persistency, and patience were mentioned as well as the fact that he accepted his illness so calmly and with so much resignation. He seemed to have loved to travel, judging by his visits to many parts of Germany in 1825 and 1826, his journey to Paris and Normandy with Temminck in 1828, and his plans to go to Scandinavia, which did not materialize because of the difficult financial position of the Netherlands at that time. From informal letters sent by the collectors for the Museum to the administrator Mr. J.A. Susanna, one gets the impression that De Haan was not insensible to female beauty and might even have been a bit of a womanizer, judging by expressions like 'our cockerel, alias the chicken-chaser' (the Dutch word 'haan' means cock or rooster).

2.1.2   *1846-1854, 1860-1872*

Jan Adrianus Herklots (born in Middelburg, province of Zeeland, 17 August 1820; died in Leiden, 31 March 1872), who had studied medicine and biology in Leiden, succeeded De Haan in 1846 as curator of the invertebrate collections. During the first years of his term he stayed in close contact with De Haan. He was curator of invertebrates from 1 July 1846 to 1 July 1854. In 1854, the Department of Invertebrates was split up into one of Articulata and one of Non-Articulata. Herklots remained curator of the latter, while the entomologist C.S. Snellen van Vollenhoven became curator of the new Department of Articulata, which included the Crustacea. Since Herklots had worked intensively with Crustacea and published on this group, this division was rather impractical and in July 1860 the two departments were redefined: C.S. Snellen van Vollenhoven's department became that of Entomology, while the Arthropoda-non-Insecta went to Herklots, and his department was then named Invertebrata-non-Insecta. Herklots remained curator of this department until his death in 1872. Although Herklots worked with several groups other than Crustacea and published fundamental papers on those (e.g., on the Pennatularia, and recent and fossil Echinodermata), he worked most intensively on the Crustacea and published important papers, like his thesis based on the West African Crustacean collections gathered by H.S. Pel and his catalogue of the Crustacea of the Leiden Museum. Herklots was not of robust health, and seems to have often been ill. He probably suffered from tuberculosis. In 1866, he went to Switzerland and returned in 1867 with his health seemingly much improved. However, his absences from the Museum became more and more frequent, and he died in 1872 'after a prolonged lingering disease.'

Herklots put the collection of Crustacea in order, following strictly De Haan's system. The annual report of 1862 remarked that finally enough shelving space was available to properly arrange the Crustacea collection. The annual reports give little information on Herklots's curatorial activities, but that of July 1863 said the cirripede collection and that of the Portunidae had been arranged and identified.

Herklots's (1861) catalog of the Crustacea of the Museum shows that the collection had steadily increased in numbers and importance and that it then contained 445 species in 512 lots of the Decapoda alone.

The most important carcinological work by Herklots is his (1851) thesis on the Decapoda and Stomatopoda of West Africa. Besides that he published a few shorter articles concerning new species of Decapoda and Isopoda, crustacean abnormalities, and the above mentioned catalog. Although important, his work did not have the impact that De Haan's crustacean volume of *Fauna Japonica* had, and Herklots, although his work was of good quality, certainly did not come up to the standard of his predecessor.

Herklots was much more interested in the Dutch fauna than De Haan was. While the Museum does not possess any crustacean labelled as having been collected by De Haan, there are several collected by Herklots on the North Sea shore at Noordwijk near Leiden. Herklots also wrote a popular book on Dutch invertebrates other than Arthropoda and edited the series *Bouwstoffen voor eene Fauna van Nederland* (*Materials for a Fauna of the Netherlands*), of which three volumes appeared. He married Miss Antoinetta Johanna Agatha Susanna (born 1831), daughter of Mr. J.A. Susanna, the administrator of the Museum. Very little is known of his personality.

2.1.3   *1854-1860*

Samuel Constant Snellen van Vollenhoven (born in Rotterdam, 18 October 1816; died in The Hague, 22 March 1880) from 1854 to 1860 had the supervision of the arthropod collection of the Museum, including the Crustacea. Snellen van Vollenhoven was first and foremost an entomologist (and a very good one at that), and apart from including, in 1859, the Crustacea in a popular book that he wrote on the Arthropoda of the Netherlands, he never published on that group. Herklots named the well known West African freshwater shrimp *Macrobrachium vollenhovenii* for him. A biography of Snellen van Vollenhoven has been published by Krikken et al. (1981).

2.1.4   *1872-1874*

Christiaan Karel Hoffmann (born in Heemstede, province of N.Holland, 16 July 1844; died in Leiden, 27 July 1903) studied medicine and biology in Amsterdam, Utrecht, and Göttingen. He received his M.D. in Utrecht, his Ph.D. in Göttingen. After having practiced as a physician and after having been prosector in the medical anatomy laboratories of the universities in Amsterdam and Leiden, he was appointed on 1 August 1872 as curator of invertebrates at the Leiden

Museum to succeed Herklots. He occupied this place only slightly more that 2 years and on 3 December 1874 left the Museum to become professor of zoology at Leiden University. In carcinology he is best known for his 1874 paper on Madagascar Crustacea collected by F.P. Pollen and D.C. van Dam. Later he became well known as a comparative anatomist and the 2.5 years spent at the Museum formed only an insignificant part of his scientific career.

2.1.5   *1874-1883*

Johannes Govertus de Man (born in Middelburg, province of Zeeland, 2 May 1850; died in Middelburg, 19 January 1930) was the only son of Dr. J.C. de Man, a prominent physician in Middelburg, and Neeltje Elisabeth Kamerman, daughter of a clergyman. He had two unmarried sisters, who like the rest of the family had a wide interest in science and history. Both had positions as curators at the local Museum of Middelburg. The older sister (Maria Goverdina, called Marie) was curator of the numismatic collection, the younger (Antoinette) of the Costume Department. Johannes Govertus (named Jan by his family) began his study of natural history at Leiden University in September 1868. His main zoology professor was Emil Selenka, a German zoologist who later became professor of zoology in Erlangen. Selenka was very popular among the Dutch zoology students and made a deep impression on them. De Man was influenced by Selenka and later kept in contact with him. De Man obtained his Ph.D. 29 September 1872 with a thesis on myological and neurological studies in amphibians and birds. A few days later (3 October 1872) he was appointed first assistant curator of the Leiden Museum. Although his thesis dealt with vertebrates, and De Man was more or less supposed to work on vertebrates in the Museum, he was more interested in invertebrates and in his new job devoted himself entirely to their study. This choice was not regarded favorably by the director, Schlegel, who was mostly interested in vertebrates. On 1 June 1875 De Man was appointed curator of invertebrates to replace Hoffmann, who had just left the Museum. De Man kept this post until November 1883. At first De Man studied Turbellaria and Nematoda. The latter group (together with the Decapod Crustacea) would have his interest up to his last years. He also published (in 1877) a paper on molluscs of Madagascar. His first carcinological papers appeared in 1879 and a continuous flow of publications on Decapoda and Stomatopoda followed. Apart from a paper on Sipunculida, two small notes on the mole (*Talpa europaea* L.), and a popular account of a visit to the Channel Islands, all his subsequent papers deal almost exclusively with decapod Crustacea and free-living Nematoda.

In 1876, he visited the Naples Zoological Station where he studied Nematoda, but after 6 weeks he contracted typhoid fever. His condition became so serious that his parents were requested to come to Naples. Fortunately, by the time they finally arrived, after a long and tedious journey, De Man's condition had sufficiently improved that he could return home by himself. From October 1881 to June 1882 he visited Erlangen and worked there with his former Leiden professor Selenka on Sipunculida, and on which they, together with C. Bülow, published a monograph. Between 1864 and 1905 De Man also made several pleasure trips to England, Germany, Denmark, and Belgium. His delightful accounts of these trips have been published in the biography that Van Benthem Jutting (1951) wrote of him.

The end of De Man's stay in Leiden was not a very pleasant one. It seems that from the beginning he had difficulties with Director Schlegel, who had succeeded Temminck in 1858. Their personalities were so different that clashes were not surprising. Schlegel, even though over 70 when De Man was appointed, was of robust health until very shortly before his death. He was rather extroverted and decisive, with not much consideration for his subordinates. De Man, on the other hand, was described as a shy and rather nervous person, entirely centered on his work, and very reserved in his contact with others. In November 1882, De Man's physician decided that the young man was under too much mental stress, and should have leave of absence from his work for a considerable time. He advised De Man to go back to his parents in Middelburg to recuperate, which was the more sensible because De Man's father was a physician. Schlegel agreed to request this leave from the trustees of the Museum, but after it was given he later objected to its extension, even when this was considered necessary by the two physicians. Schlegel then asked the trustees to discharge De Man from his post at the Museum. However, the Minister of Internal Affairs was more considerate and allowed De Man a leave with pay until 1 September 1883, followed by leave without pay until 1 January 1884. Schlegel then became bitter and abusive. He wrote De Man a harsh and cruel letter in which he cast severe doubt on De Man's abilities as a scientist and as a museum curator. Only then did De Man's father, on behalf of his son, tender a resignation, which the trustees granted in November 1883. One is inclined to lay the blame for this episode mainly with Schlegel, and the last letter that he wrote to De Man was unforgivable. However, one has to consider that Schlegel then was 79 (he died a few months later, on 23 January 1884), and was by then of poor health: he had diabetes and was practically blind. He had always been a fighter and evidently could not understand De Man who clearly was not. Tact and diplomacy were characteristics that were lacking in Schlegel.

Anyhow it was a good thing that De Man's father put an end to this sad situation and kept his son home. The De Mans were well to do and it was not necessary for the son to work for a living. For 10 years De Man stayed in his parents' home in Middelburg devoting himself entirely to his studies on freeliving Nematoda and decapod Crustacea. In 1893, he went to live in Ierseke, a small fishing town in the province of Zeeland, where a simple house was built for him on the dike near the harbor. A caretaker and his wife occupied the groundfloor and De Man lived on the second floor, where from his study he had a nice view over the harbor and the mudflats beyond. Here he lived, unmarried, until his death in 1930, still working on his animals and becoming one of the foremost authorities on Nematoda, and decapod and stomatopod Crustacea. He would go out on the mudflats to collect Nematoda, and also received material from the fishermen (he was called by them 'the mudflat doctor'). The greater part of his crustacean collection, however, consisted of duplicate specimens of collections that he had studied for museums all over Europe. He had a close relationship with professor Max Weber, the director of the Zoological Museum of Amsterdam and leader of the Siboga Expedition, who had entrusted De Man with the study of the Siboga Decapoda Macrura. Weber had a personality quite different from that of Schlegel: he had a great amount of tact and knew exactly how to handle De Man. The friendship with Weber and Mrs. Weber-van Bosse, must have been very dear to De Man. When De Man had finished a manuscript for the Siboga monographs, he did not want to entrust this precious document, written in his clear handwriting, to the mail, and would take it by train to Dordrecht (about halfway between Amsterdam and Ierseke) where he would meet Weber, who would take the parcel on to Amsterdam.

De Man's relations with the Leiden Museum remained good. He had close contacts with his successor, R. Horst, and identified and examined much material in the collection. His *Carcinological Studies in the Leyden Museum* continued to appear until 1892. De Man was not resentful as shown by the fact that in 1899 he contributed financially to a monument that was to be erected in Germany to commemorate Schlegel and two German zoologists (father and son Brehm). At the end of 1929, De Man became ill and was hospitalized in Middelburg where he died on 9 January 1930. He left his library and the larger part of his collection to the Amsterdam Museum. And, probably to show that he had no hard feelings towards the Leiden Museum, he donated to it his collection of Crustacea from the province of Zeeland. His two sisters survived him.

De Man's meticulous work in decapod and stomatopod Crustacea is still of the highest value. His Siboga papers became revisions of the groups concerned and are still consulted. Also his voluminous reports on collections made by J. Anderson, J. Brock, W. Kükenthal, Capt. Storm, and others are full of precise observations, lengthy descriptions, revisions of small groups, and illustrated by his magnificent and accurate drawings. Also his shorter papers contain a wealth of information. He clearly proved that Schlegel's accusations were entirely groundless. About De Man's curatorial activities we know little. It is said that he was the first to place vials with alcohol specimens in large glass-topped jars, to prevent evaporation. In 1882, he started a catalogue of the Brachyura preserved in alcohol and divided them into two categories: (a) those in the 'Gallerij,' i.e., the public exhibit, and (b) those in the 'Magazijn,' i.e., in the study collection not open to the public. Of each species one to four samples were on exhibit; the other samples were placed in the study collection.

In this period, important collections were obtained from private citizens, sometimes at the request of De Man, e.g., from his cousin P. Kamerman (Angola) and from Mr. J.A. Kruyt, Netherlands consul in Jidda, but also from expeditions like that by J. Büttikofer and C.F. Sala to Liberia and one to central Sumatra. Many of these collections were immediately studied by De Man and the results published without delay.

As to this curatorial work, Schlegel made the following remark in his letter of 17 September 1883 to De Man, 'After your departure (1 November 1882) I have inspected and made others inspect your Department, and it showed that the collection is in such a neglected condition that it needs immediately a different supervisor. Therefore I turned the care of this Department over to Dr. R. Horst.' It is possible, in view of what Schlegel said about De Man's scientific abilities, that he grossly exaggerated. On the whole, one gets the impression that De Man worked very accurately and precisely with the material. Or did he neglect the other invertebrates in order to give more time to the Crustacea? It seems unlikely. Van Benthem Jutting (1951: 242) in her biography of De Man mentioned 'his almost superhuman sense of order' and that his collection and library were in perfect shape, which seems better to describe the actual situation than do Schlegel's remarks.

2.2   Second period. Stagnation and recession. The period of R. Horst, J.J. Tesch, A.L.J. Sunier, H.C. Blöte and F.P. Koumans (1883-1933)

This period covers the administrations of the Museum's third and fourth directors: F.A. Jentink (1884-1913) and E.D. van Oort (1913-1933). During this period the interest of the government in the Museum declined. In addition, the trustees of the Museum (who were at the same time trustees of Leiden University) were not very helpful, and several attempts were made to incorporate the Museum directly into the University. In this period (in 1931), the official name of the museum was changed from 's Rijks Museum van Natuurlijke Historie to Rijksmuseum van Natuurlijke Historie.

The most important event during this period was the construction of a new museum building in the Raamsteeg to replace the old building on Rapenburg. Although the government had approved this new building in the 1870's, it was not until 1900 that the actual construction started. That building finally began was due to Director Jentink who continuously bombarded the government with complaints about the old building, complete with photographs of the fungus *Merulius lacrymans*, which grew luxuriantly under the floor 'providing the beautiful white view of a snowy landscape, weakening the floor of the spirit building to such an extent that it is a miracle that the thousands of jars did not crash through it.' Jentink must have greatly irritated the trustees and government with his complaints. He even published, at his own cost, a booklet entitled *A Visit to 's Rijks Museum van Natuurlijke Historie in the Autumn 1892* in which he, in the form of a guided tour through the Museum, pointed out all the deficiencies of the old building. He sent this to all members of parliament, to the ministers, and many influential people. Although effective, these tactics made him very unpopular with the trustees. Part of the building was finished in 1902, the rest in 1912; only the planned exhibition building was never even started. Jentink was most satisfied that the scientific collections would be adequately housed in spacious storage rooms with all possible safeguards against fire, light, and dust. Also the large and well arranged offices for the scientific, technical, and administrative personnel proved that his fight had not been in vain.

One gets the impression that the trustees almost from the start had a negative attitude towards Jentink, and they obstructed many of his other requests. Increase in staff and salaries was never willingly given, perhaps of necessity, and this frugality caused good workers to leave the Museum for more lucrative jobs elsewhere. Only those remained who either were really devoted to their jobs and willing to work for a small salary, or had little initiative and not likely to get a position elsewhere. Jentink, although energetic and a fighter, was not an inspiring leader, and the Museum, notwithstanding its beautiful building, became a rather sleepy institution.

E.D. van Oort, who succeeded Jentink, was not able to change this situation. He was honest and devoted to the Museum, but lacked the tact and subtleness to stem the tide. The trustees still wanted to reduce the Museum to a part of the Zoology Department of Leiden University, and because of the economic depression, they intended to drastically cut the museum expenditure by reducing staff and other costs. Van Oort's counter arguments were rather blunt and not always tactful or diplomatic, and he did not exploit his small victories. As a result, at Van Oort's death, the Museum was grossly understaffed and the staff underpayed. The scientific, technical, and administrative personnel were exactly the same number as at the end of Temminck's directorate in 1858.

This period was also one of stagnation in the Crustacea collection: The two curators (Tesch and Sunier) who really devoted time to the collection left after a few years because of the poor financial prospects. Horst, who supervised the Crustacea through most of this period, did his best, but was too occupied with Vermes, of which he was a world specialist. Blöte, who was in charge of the Crustacea at the end of this period, was mainly an entomologist, and left the care of the Crustacea to others.

During this period the Crustacea had four curators, although the curator of fishes, F.P. Koumans, as a sideline, devoted some time to the study of isopods.

2.2.1 *1883-1915, 1918-1923*

Rutgerus Horst (born in Angerlo, province of Gelderland, 16 August 1849; died in Leiden, 18 October 1930) studied biology at Utrecht University and in 1876 obtained his Ph.D. with a thesis on the anatomy of *Lumbricus terrestris*. On 19 September 1882 he was appointed the curator of fishes, reptiles, and amphibians at the Leiden Museum, and 1 year later became De Man's successor as the curator of invertebrates, a post he occupied until his retirement on 1 October 1923. Horst was first and foremost a specialist in polychaete and oligochaete worms, and his main interest remained with these groups. As supervisor of the Crustacea collection, he published only a few very short notes on the biology of *Birgus latro* and *Thalassina anomala*, based on field observations sent in by collectors; some notes on Dutch decapods (*Rhithropanopeus* and *Astacus*); and the description of one new bopyrid isopod (*Palaegyge buitendijki*). See also 2.2.3.

2.2.2   *1907-1908, 1915-1918*

Johan Jacob Tesch (born in Amsterdam, 7 February 1877; died in The Hague, 7 August 1954) was the son of J.W. Tesch, the director of an exclusive boys school in The Hague, and his wife T. Stoffel. J.J. Tesch studied biology in Leiden and Utrecht, obtaining his Ph.D. in Utrecht in 1906 with a thesis dealing with heteropod Mollusca, a group on which he had worked in the Zoological Station at Naples in 1904. On 1 May 1906, Tesch was appointed biological assistant at the Rijksinstituut voor Onderzoek der Zee (National Institute for Marine Research; later named National Institute for Biological Fishery Research). Except for two interruptions he was employed by this institute from 1906 until his retirement in 1942, at which time he was the head of the sea fisheries division.

The two interruptions (1 August 1907 to 18 May 1908 and 16 May 1915 to 1 September 1918) occurred when he accepted posts at 's Rijks Museum van Natuurlijke Historie. In 1907, he became assistant curator of invertebrates under Horst, but resigned the next year mostly for financial reasons. In 1915, he was appointed curator of Crustacea of the Leiden Museum, the Crustacea being split off from the rest of the invertebrates, which remained under Horst. It is likely that Tesch during his short first stay at the Museum occupied himself only with Mollusca, and that his interest in Crustacea was awakened after he went back to the Institute for Marine Research in 1908. There, he was charged with the study of the material collected by the R.V. *Wodan* in the southern North Sea. These collections were part of an international project of the Conseil International pour l'Exploration de la Mer to investigate the plankton and bottom fauna of the seas of northwest Europe. Under this project, which lasted from 1902 to 1911, the Rijks Instituut voor Onderzoek der Zee was responsible for the exploration of a section of the southern North Sea.

Tesch, who was foremost a systematist, identified part of the material and published reports on the Decapoda (1909), Mysidacea (1911), Cumacea (1912), Amphipoda (1911, 1912) as well as on Pycnogonida (1910), Echinodermata (1907) and Cephalopoda (1909), thereby providing an important contribution to knowledge of the North Sea fauna. In 1915, when there was a vacancy as curator in the Leiden Museum, Tesch applied for the job and obtained the post starting 16 May 1915. This was the first time that Crustacea would be a separate Department with its own curator. With much enthusiasm Tesch started his task and revised the very difficult genera *Macrophthalmus*, *Sesarma*, and related forms. He wrote two large and important monographs on crabs in the series of the Siboga Expedition, as well as some smaller notes on these animals. Notwithstanding the fact that he clearly enjoyed this work, he resigned on 1 September 1918 and returned to the Fisheries Institute, evidently again for financial reasons.

After his retirement from the Fisheries Institute, Tesch returned to the Museum in 1942 and was made an honorary research associate. He worked again on his Pteropoda of which he published five large papers, including his three reports on the pteropods of the Dana Expeditions. He also started a study of the grapsoid crabs of the 1929-1930 Snellius Expedition, but did not finish it. He was a regular visitor at the Museum until about 1952. He died in 1954.

The tragedy of Tesch's life was that he was a first class taxonomist and would have been happy to spend his entire career in the Museum. It was his responsibility to his family that made him leave the Museum twice, the first time a few months before his marriage (on 21 July 1908) to Miss Johanna de Wilde. This marriage was not a happy one: He left his wife in October 1932 and then lived separately from his family. After his wife's death on 23 June 1941 he married on, 3 September of the same year, Miss Aaltje Maria van der Hee, whom I remember as a very bossy woman who did not make his life particularly easy. He had two children, both from his first marriage.

Tesch was a man of small stature. He was rather shy and withdrawn, friendly but not easy to get close to. The short periods in which he was allowed to do taxonomic research showed his great qualities in this field. In his job at the Fisheries Institute he accomplished relatively little, and there, like at home, he seems to have been under the influence of personalities stronger than his own. One must be thankful for what he did accomplish for crustacean taxonomy but regret that he could not have made greater use of his talents in this field. Apart from his purely scientific carcinological papers, Tesch published accounts of the Dutch Decapoda in a popular Dutch journal. He also wrote a popular handbook on oceanography which is well written and was greatly appreciated by a wide public. For many of the younger generation of marine biologists it was the first solid basis for their knowledge of oceanography. Tesch was a very erudite person and had many interests outside biology, mainly in literature (he himself was quite an accomplished poet, who got several of his poems published) and history, especially medieval history. He was a good draftsman as shown by the illustrations he made for his scientific articles.

2.2.3 *1918-1923*

When Tesch left the Museum in 1918, the Crustacea collections again came under the care of Horst in the Invertebrate Department (see Section 2.2.1).

From 1919 to his retirement in 1923 Horst shared the Invertebrate Department with G. Stiasny, who was appointed in 1919 and was mostly interested in Coelenterata and Brachiopoda. After his retirement Horst came regularly to the Museum and continued his research and curational activities on the worms. His successor in the Invertebrate Department was A.L.J. Sunier.

2.2.4 *1923-1927*

Armand Louis Jean Sunier (born in The Hague, 17 December 1886; died in Baarn, province of Utrecht, 13 May 1974) studied biology in Leiden and obtained his Ph.D. on 10 April 1911 with a thesis dealing with the anatomy of fishes and Acrania. He left for the Netherlands East Indies in 1911 and became zoologist at the Department of Agriculture. In 1921, he was appointed head of the Laboratory of Marine Investigations in Batavia. He left the East Indies to take up a position at the Leiden Museum, where he was placed in charge of the Mollusca and Crustacea. At first he gave his main attention to the Crustacea, and in 1925, he published a short but interesting note on Palaemonidae. However, he was not charmed by the Crustacea and called the collection a 'cemetery of legs,' as so many specimens had broken off one or more legs on preservation, and he soon turned his attention to the Mollusca. On 1 October 1927 he left the Museum for good to accept the position of Director of the Amsterdam Zoological Gardens (Artis) where his considerable talents for organization and public relations could be used to much better advantage. He became a very popular and highly respected director of a well run zoo, and kept this position until his retirement on 31 March 1953. His influence on carcinology and on the Crustacea collection of the Museum was slight and negligible compared to his later accomplishments. When still in the East Indies he published a few carcinological papers on Stomatopoda and the fauna of marine fish ponds.

2.2.5 *1927-1940*

Hendrik Coenraad Blöte (born in Leiden, 12 February 1900; died in Voorschoten, near Leiden, 20 January 1990) studied biology at Leiden University and obtained his Ph.D. in 1935. He succeeded Sunier as curator of Crustacea and Mollusca. In September 1929, during a reshuffle of the Invertebrate Division of the Museum, Blöte became the second Curator of Arthropoda, to whom the Crustacea were again assigned after a separation of 70 years. The molluscs now got their own curator. Blöte was not particularly interested in the Crustacea, and apart from his curatorial duties did not bother with them. Blöte remained Curator of Entomology until his retirement in 1965.

2.2.6   *1928-1947*

Frederik Petrus Koumans (born in The Hague, 21 January 1905; died in Ede, province of Gelderland, 27 April 1977) was an eminent ichthyologist and Curator of Fishes at the Museum from 1 March 1928 to 1 November 1947. In his youth, he became interested in the study of the Dutch Isopoda and in 1928 published a paper on those in the Museum's collection. After that he seems to have lost his interest in this group, the fishes and molluscs (he was one of the founders of the Netherlands Malacological Society) taking his full attention. In 1947, he left the Museum to accept the position of director of a large medical library in The Hague, which position he occupied until his retirement in 1970.

Figure 1. Upper left, W. de Haan (1801-1855). From a portrait painted on ivory in 1826 by P. Daviot. Upper right, J.A. Herklots (1820-1872). From a photograph in the archives of the Leiden Museum. Lower left, J.G. de Man (1850-1931). From a photograph in the archives of the Leiden Museum. Lower right, J.J. Tesch (1877-1954). Photograph taken June 1948 by L.B. Holthuis. Mrs. Tesch's first reaction when told that this picture was taken: 'Was your tie straight?' Of course it was not.

Figure 2. Left, A.M. Buitendijk (1903-1950). Photograph, June 1948, by L.B. Holthuis. Right, H. Boschma (1893-1976). Photograph, 1961, by H.F. Roman.

2.3 *Third period. Resurrection, renewed activity and growth. The period of H. Boschma and Alida M. Buitendijk (1933-1950)*

The sad situation of the Museum at the time of E.D. Van Oort's death prompted the trustees to appoint a temporary director who was instructed to submit a report of the Museum's condition, indicate ways for improvement, and if possible curtail expenditures. To this end they appointed Prof. Dr. H. Boschma, the Ordinarius Professor of General Zoology of Leiden University and Director of the Zoological Laboratory. At first this seemed to be a new attack by the University (and the Zoology Department) on the independence of the Museum, and the Museum personnel must have had severe misgivings. However, Boschma, perfectly honest and straightforward, produced an extremely clear and lucid report showing that increased personnel at the Museum were not only desirable but imperative, and that other aspects of the Museum had to be strongly supported in order that it not be completely ruined. Boschma gave his arguments so convincingly and impartially, also providing a solution to end the feud between university and museum, that the trustees accepted most of his proposals. Here, I believe, Boschma's personality was crucial. He did not antagonize the trustees, something that Jentink and Van Oort, I am sure, secretly enjoyed doing. Boschma treated the trustees as equals and respected their view points when he could not change them. Boschma's personal charms played an important role, as did Temminck in his time. Boschma also got the full support of the government through the trustees.

On 18 May 1934, Boschma received his permanent appointment as director. When he started his directorate there were 4 curators on the Museum staff, and when he retired this number had risen to 13. In the technical department the growth was also impressive. Boschma awakened the Museum from its lethargy. Not only did the personnel increase, but there were also many more students studying the Museum collections for their Ph.D. theses. A great number of expeditions (e.g., to New Guinea, the West Indies, Surinam, the Red Sea, West Africa) in which the Museum took part, or which it organized, brought in interesting collections, as did excursions in the Netherlands and western Europe. Boschma was made Ordinarius Professor of Systematic Zoology and taught courses in taxonomy and nomenclature. The Museum blossomed as never before, notwithstanding the very difficult beginning for Boschma's directorship – economic depression of the 1930's followed by 4 years of German occupation during World War II. This blossoming and the congenial atmosphere created in the Museum was in a large part due to Boschma's inimitable personality.

During the period under consideration (1933-1950) there were two curators dealing with the Crustacea, Blöte already mentioned in the previous chapter and Buitendijk. Although not a curator, the most prominent carcinologist working at the Museum was the director, Boschma.

2.3.1 *1936-1950*

Alida M. Buitendijk (born in Leiden, 1 April 1903; died in Leiden, 12 September 1950) was the only child of Pieter Buitendijk (1870-1932), a ship's surgeon, and Marrigje de Graaf. Alida's father was very interested in natural history, and especially in fishes. During his journeys, mostly with mail-steamers from Amsterdam to Java and back, he collected extensive material, including marine invertebrates, which he deposited in the Leiden Museum. His intention to do research on the fishes after his retirement came to naught because of his early death. His daughter had inherited his interest in biology and after finishing girl's high school, she studied natural sciences in Leiden and obtained a degree allowing her to teach biology in high schools. Although she would have been an excellent teacher, her interest was more in taxonomic research. Therefore, on 1 October 1930, she accepted a post at the Museum as a typist in the Entomology Department. Before that time, in spare moments, she had begun the study of Dutch Collembola on which she published several papers (in 1929, 1930, 1933, 1941). Her scientific qualities were noted by Blöte and Boschma, who in 1936 entrusted her with the care of the collection of Crustacea, Chelicerata, and Collembola. On 1 January 1938, she became assistant curator for these groups. Boschma urged her to start with the study of the taxonomy of Paguridae, because one of his students (I. van Baal) studied the peltogastrid Rhizocephala of the 1929-1930 Snellius Expedition and it would be a great help if she could identify the hosts of these parasites. This lead to Buitendijk's first decapod publication, in 1937, 'The Paguridea of the Snellius Expedition.' Since Boschma worked mostly on rhizocephalan parasites of crabs, Buitendijk then devoted most of her time to the Brachyura. On 1 January 1940, she was appointed full curator of the Department of Crustacea and Chelicerata, which was split off from the Entomology Department. She worked mostly on Indo-West Pacific Brachyura and Anomura, and a number of publications appeared by her; she also published on crabs from the Pacific coast of Mexico. World War II made contacts with other countries difficult or impossible, and it was then that she started a handbook of the Dutch Decapoda Brachyura. Although the manuscript was almost finished before her death, unavoidable delays put publication off. After a complete revision of the manuscript with numerous additions and changes the book finally was published by J.P.H.M. Adema on 22 February 1991.

Buitendijk took excellent care of the Crustacea collection, helped by the technical assistant J.A.G. Delfos. Delfos, after having first worked for more than one division at the same time, now became the first technician who could fully concentrate on the Crustacea. His handwriting became well known to carcinologists all over the world, from the beautifully written labels of the material that was borrowed and the addresses on the packages. Buitendijk and Delfos kept the collection in first rate condition.

The pedagogical talents of Buitendijk were soon recognized by Boschma, and he let her supervise students working in the Museum for their master's or Ph.D. degree in zoological taxonomy. He did this often by suggesting to students, to their great benefit, that they select a subject for their theses in the field of Crustacea or Chelicerata. Many of these students still thankfully remember the time that they worked in Buitendijk's department and profited from her common sense advice and from the freedom she gave in the way the work should be done.

She could have a sharp tongue in defense of her division, but she could also use tears for the same purpose (she knew that Boschma could not stand tears). Her main interest outside Crustacea was working with the girl scouts of Leiden. Buitendijk, who never married, was rather small but very lively, with a good sense of humor. She was not particularly handsome, but her personality fully made up for that. She suffered from cancer, which she tried to ignore, in vain. She was only 47 years old when she died. She left her money to the Museum for a fund to be used to enable staff members to travel outside the country for study. The hope that she once entertained to go to America herself for the study of Crustacea never materialized.

2.3.2 *1934-1958*

Hilbrand Boschma (born in IJsbrechtum, province of Friesland, 22 April 1893; died in Leiden, 22 July 1976) was born and raised on a farm in the lake district of Friesland. He studied biology in Amsterdam and received his Ph.D. in 1920 on an anatomical subject, *Das Halsskelet der Krokodile*. The same year he went to the East Indies and got a job at the Treub Laboratory in Bogor. He returned to Leiden in 1922 to become chief assistant at the Zoological Laboratory. In 1928, he became lecturer at the University and in 1931, Professor Ordinarius of Zoology and Director of the Zoological Laboratory. As pointed out above, at the death of Dr. van Oort, he became temporary director of the Museum, and in 1934 came his permanent appointment.

During his directorship (1934-1958) Boschma not only brought the Museum to unprecedented prosperity, he also found the time to do research. He did this by almost completely ignoring unnecessary red tape and by not wasting his time on bureaucratic niceties; thereby, things went smoothly and there was time for useful activities. His research centered on a few widely diverse animal groups: Cetacea, Rhizocephala, corals (both Madreporaria and Stylasterida), and ellobiopsid protozoans. His interest in the Rhizocephala started in a roundabout way. Prof. Dr. P.N. van Kampen, Boschma's predecessor as Director of the Zoological Laboratory and his erstwhile chief, had promised Prof. M. Weber, the leader of the Siboga expedition, to study the Rhizocephala of the expedition. To this end Van Kampen asked the help of Boschma. Van Kampen and Boschma finished the study of this collection and a publication on it appeared in 1925 in the Siboga series. After this Van Kampen never looked at Rhizocephala again, but Boschma was intrigued by these crustaceans and continued working on them, becoming the world's foremost authority (he published over 120 articles on them).

After his retirement from the directorship in 1958, Boschma came regularly to the Museum and continued working on Rhizocephala (and Stylasterida). His last publication (on West Indian Sacculinidae) appeared in 1974.

Boschma bequeathed his carefully arranged slide-collection of Rhizocephala, his notes, and his drawings to the Leiden Museum, where they are in perfect order and can be easily consulted.

A biography and complete bibliography of Boschma has been published by W. Vervoort (1977).

REFERENCES

Benthem Jutting, W.S.S. van 1951. Dr. Johannes Govertus de Man, Middelburg 2 Mei 1850 - Middelburg 19 Januari 1930. Een Zeeuwse zoöloog van internationale vermaardheid. *Biologisch Jaarboek uitgegeven door het Koninklijk natuurwetenschappelijk genootschap Dodonaea te Gent* 18: 130-259.

Boeseman, M. 1970. The vicissitudes and dispersal of Albertus Seba's zoological specimens. *Zool. Med.* 44: 177-206.

Engel, H. 1986. Hendrik Engel's Alphabetical List of Dutch zoological cabinets and menageries, ed. 2. *Nieuwe Nederlandse Bijdragen tot de Geschiedenis der Geneeskunde en der Wetenschappen* 19: 1-340.

Herklots, J.A. 1861. Symbolae carcinologicae. I. Catalogue des Crustacés qui ont servi de base au système carcinologique de M. W. de Haan, rédigé d'après la collection du Musée des Pays-Bas et les Crustacés de la faune du Japon. *Tijdschr. Ent.* 4: 116-156.

Hoffmann, C.K. 1874. Crustacés et Echinodermes de Madagascar et de l'île de la Réunion. In F.P.L. Pollen & D.C. van Dam, *Recherches sur la faune de Madagascar et de ses dépendances*, 5(2): 1-58.

Krikken, J., C. van Achterberg, P.H. van Doesburg, R. de Jong & K.W.R. Zwart 1981. Samuel Constant Snellen van Vollenhoven (1816-1880) and his entomological work. *Tijdschr. Ent.* 124: 235-268.

Vervoort, W. 1977. Prof. Dr. Hilbrand Boschma, 22 April 1893 - 22 July 1976. Obituary and bibliography. *Zool. Bijdr.* 22: 1-28.

Carcinology in classical Japanese works

Eiji Harada
Seto Marine Biological Laboratory, Kyoto University Shirahama, Wakayama, Japan

1 INTRODUCTION

Classical Japanese works of natural history have been extensively studied by Uno Masuzo (1900-1989). Uno was a distinguished hydrobiologist as well as a leading carcinologist. He began his academic career at the Otsu Hydrobiological Station, later was director there, and retired as professor of biology at the College of Liberal Arts, Kyoto University. As a carcinologist he contributed chiefly to the taxonomy and ecology of cladocerans, amphipods, and bathynellids. During his early years but particularly after his retirement, he also devoted himself to the historical study of natural history in Japan. He published more than 100 papers on classical Japanese writings and their authors, as well as several comprehensive books on the history of biology in Japan. These publications established Uno as the eminent scholar of the history of Japanese natural history. Although he covered most major aspects of biological knowledge in his history papers, he included surprisingly little on early Japanese thinking on the systematic relationships of plant and animal groups, and nothing specifically on carcinology.

For this paper, I selected representatives of the major classical works and briefly reviewed their contents to demonstrate the position of crustaceans in the animal world of Japan before modern western biology was introduced.

The transliterations and translations to English of Chinese and Japanese titles and authors's names follow Needham (1986). The names of Japanese authors are written in the text in the order they are written in Japanese, family name first. When a work consists of two volumes, or when a volume or chapter is divided into two parts, these volumes or parts are traditionally called 'jo' (upper) and 'ge' (lower); and if in three parts, 'chu' (middle) is designated. I refer here to volumes or parts simply by 'U,' 'M,' or 'L' (see Table 1).

Many words in Japanese do not have corresponding words in English. The Japanese 'yebi' ('yibi', 'jebi', or now commonly 'ebi') denotes collectively those macrurous decapod crustaceans and allied forms that are referred to separately in English as either shrimp, prawn, or lobster. 'Mushi' is an interesting example that in various senses can mean almost any invertebrate and classically even snakes. Squids and cuttlefishes are collectively called 'ika.' In this paper I have arbitrarily adopted one English word for each of these kinds of Japanese terms, for instance, 'shrimp' for yebi, 'worm' for mushi and 'squid' for ika.

Table 1. Titles of volumes of *Honzo-wamyo* (897-929) with contents of volumes dealing with crustaceans. Letters in parentheses indicate multi-volume coverage of topic: (L) = Lower part; (M) = Middle part; (U) = upper part.

Volume	Title
1-2	No titles, general introduction
3	Gem & Stone (U)
4	Gem & Stone (M)
5	Gem & Stone (L)
6	Herb (U)
7	Herb (U)
8	Herb (M)
9	Herb (M)
10	Herb (L)
11	Herb (L)
12	Tree (U)
13	Tree (M)
14	Tree (L)
15	Beast & Bird
16	Worm & Fish
	'kani' (crab), 'kasame' (portunid crab), 'yibi' (shrimp), 'sei' (*Capitulum*), 'kamina' (hermit crab) [also includes] bee, oyster, abalone, sea turtle, eel, silkworm, slug, leech, squid, shark, toad, viper, spider mayfly, centipede, scorpion, firefly, earthworm, mussel, land snail, sea fish, whale, sea cucumber, jellyfish, octopus, sea urchin, and snail.
17	Cake
18	Vegetable
19	Rice & Cereal
20	Useless

2　CLASSICAL WORKS

Plants and animals were frequently mentioned by ancient writers. *Kojiki* and *Nihonshoki*, presumably written about 712 and 720, respectively, are the oldest records of the ancient history of Japan. According to Uno(1960), Chamberlain (1932) in his English translation of *Kojiki* enumerated 55 kinds of animals including two kinds of crustaceans, cirripedes and crabs. *Mannyoshu* is the oldest anthology of classical Japanese poetry and is believed to have been compiled about 900. It comprises more than 3500 long and short verses, composed mostly by educated nobles, officials, and priests, but also by commoners, in the years from about 645 to 759. Several of these verses were later included in the famous anthology *Hyakunin-isshu*, literally meaning 'one hundred poems by one hundred poets,' and which was translated into English by Porter (1909) as his *A Hundred Verses From Old Japan*. The verses of *Mannyoshu* give insight into daily life and culture of ancient Japanese society, as well as into Japanese scientific knowledge. Higashi (1935) identified and listed about 115 species or kinds of animals mentioned in about 1000 verses and accompanying notes of *Mannyoshu*; the majority of these were birds, mammals, and insects. Here again, the Crustacea is represented only by two kinds, crabs and possibly cirripedes. If the cirripedes are what are actually referred to, they are presumably *Capitulum* (= *Mitella*). The crabs are apparently *Eriocheir japonica*, and in one long verse, which is entitled, roughly, 'Ode to Crabs,' a little of their habits is presented, along with gratitude, and apology, to them for being food for man.

3 OLDEST ENCYCLOPEDIC WORKS

As has been stressed by Uno (1948), 'one of the characteristic features of Japanese culture is its nature of being imported,' and Chinese works had controlling influence on Japanese natural history until the 19th century. The schemes used in arranging and grouping animals in classical Japanese texts are indeed almost a complete replication of a non-hierarchical system found in the ancient Chinese works commonly called, in Chinese, the *Pên Tshao* books. These books were mostly encyclopedic compilations of existing knowledge of medicinal, beneficial, and edible natural things. Needham (1986) viewed the order mineral, plants, animals in the *Shen Nung Pên Tshao Ching* as 'the most primitively obvious scala naturae,' but also noted that Thao Hung-Ching, the author of one of the other *Pên Tshao* books, in about 500, was the first to depart from this primitive system of classifying natural objects, adopting instead a division into minerals, herbs and trees, fruits and vegetables, cereal grains, insects, and animals.

Honzo-wamyo (*Synonymic Materia Medica with Japanese Equivalents*), written around 923, is essentially a translation of the Chinese work *Hsin Hsin Pên Tshao* (*The New Pharmacopoeia*), better known as *Thang Pên Tshao* (*Pharmacopoeia of the Thang Dynasty*), completed about 660. In the *Honzo-wamyo*, the natural things tabulated in the *Thang Pên Tshao* are transcribed and explained briefly in Chinese, not in Japanese, but each item is given a phonetic Japanese name in Chinese characters (Fig. 1). The Japanese author made little attempt to add any contemporary information to the Chinese material.

The actual arrangement or classification of the entries in the entire *Thang Pên Tshao* is not

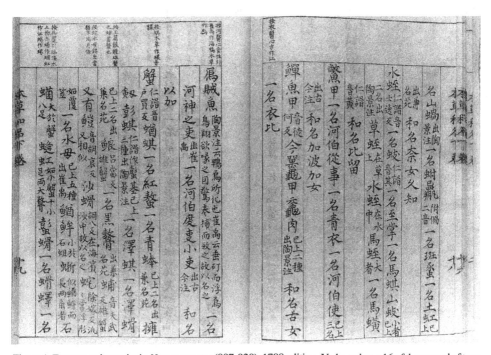

Figure 1. From woodcut print in *Honzo-wamyo* (897-929), 1798 edition, Yedo, volume 16 of the second of two books, showing the account of 'kani' (crab) on the sixth row from the left and continuing leftward, following 'ika' (squid).

exactly known, because the original texts and copies of the work no longer exist in China, and only 5 of the 20 volumes, copied from imported Chinese texts before 731, survive in Japan. However, the titles of the five extant volumes are identical to those of the corresponding volumes of *Honzo-wamyo*; therefore, the arrangement of materials in the 20 volumes of this Japanese work is presumed to be identical with that in *Thang Pên Tshao*.

The entries in *Honzo-wamyo* include inorganic minerals and living things, as well as natural phenomena, utensils, and imaginary creatures (Table 1). These are grouped into volumes, each entitled with a collective name. However, the entries in each volume are not divided into subgroups. Only some general terms, for example 'kani' meaning 'crab,' have examples of others of their kind listed under them. The volumes are arranged in a linear (non-hierarchical) system, with the volume titles not clearly distinguishing plants from animals, and living beings from non-living things. The words 'shokubutsu' for plant and 'dobutsu' for animal, used earlier (815) in the religious work *Henzyo-hokki-shourei-shu* by a great Buddhist priest, Kukai (See Iino et al. 1985, p. 918) are not found. In fact, these distinctions of living beings were not adopted in any of the classical works, except two in the 18th century: *Butsurui-shoko* (*Names of Things*, 1775) by Koshigaya Gozan and *Yamato-honzo* (1709) by Kaibara Ekken (see below).

The crustaceans of *Honzo-wamyo* are included in volume 16, entitled 'Worm and Fish' (Table 1). Five kinds of crustaceans are mentioned, scattered among other kinds of organisms. It is obvious that crustaceans are not recognized as a natural group, nor are, similarly, the fishes, the insects, the molluscs, the echinoderms, etc. This pattern of placing and treating natural

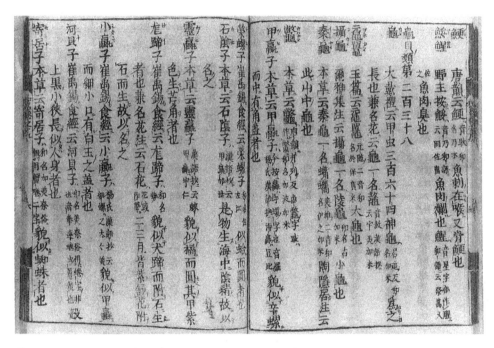

Figure 2. From woodcut print in *Wamyo-ruishusho* (ca. 930's), 1650 edition, Osaka, volume 19 of book 5, showing the account of 'sei' (*Capitulum*) on the eighth row from the left and continuing leftward, following 'uni' (sea urchin), and also showing the account of 'kamina' (hermit crab) on the left-most row, following 'nina' (freshwater snail).

Table 2. Titles of volumes of *Wamyo-ruishusho* (ca. 930) dealing with animals, and contents of those volumes that include crustaceans. Number in parentheses is chapter number.

Volume	Titles	
18	(28)	Feathered (bird)
	(29)	Haired (beast)
19	(30)	Scaled & Shelled
		Dragon & Fish: 'yebi' (shrimp), included with dragon, whale, fish, shark.
		Turtle & Shell: 'sei' (*Capitulum*), 'kamina' (hermit crab), 'kasame' (portunid crab), 'kani' (crab), 'inatsuki-kani' (fiddler crab), 'ashihara-kani' (seashore crab), 'ishikani' (seashore crab underneath stones), included with turtle, snail sea urchin, mussel, clam, abalone, oyster, squid, octopus, sea cucumber, sea squirt, jellyfish.
	(31)	Worm & Crawler (e.g. snake, earthworm, insect, toad)

history materials persisted in Japanese works until the 19th century, and indicates how profound was the Chinese influence on Japanese natural history.

Wamyo-ruishusho, compiled about 930, is a collection of Japanese names of things with explanations, based chiefly on *Thang Pên Tshao*, and is written in Chinese (Fig 2). The things entered and their arrangement are basically similar to those of the *Honzo-wamyo*, although the allocation of things to volumes is slightly different. The animals are together in volumes 18 and 19, which altogether comprise 4 chapters from the 28th to the 31st (Table 2). In each chapter, kinds are described and characteristic features of the group are explained separately. The shrimp is included in the section 'Dragon and Fish,' and *Capitulum*, hermit crab, and crab are mentioned in the section 'Turtle and Shell.' One can see some evidence of the concept of hierarchical arrangement for lower categories.

4 LATER NATURAL HISTORY WORKS

After the ancient works such as *Wamyo-ruishusho*, a dormant period in Japanese natural history began that lasted for centuries. Finally in 1590, a voluminous encyclopedic work was written in China, called *Pên Tshao Kang Mu* (*The Great Pharmacopoeia*, or preferably, *The Pandects of Natural History*) and was soon imported into Japan. This encyclopedia was more comprehensive and detailed than previous works, with much tighter groupings of things, and with detailed descriptions. However, the arrangement was fundamentally the same as in previous works. The crustaceans were described in three different volumes. In Japan, *Pên Tshao Kang Mu* became the long-term irrefutable standard. Almost without exception all native things of Japan were identified with things described in this Chinese work.

Kinmou-zuwi, published in 1666, is primarily an illustrated encyclopedia. The descriptions are rather brief, but each is accompanied by a figure (Figs. 3-6). These illustrations are in traditional Japanese style, showing animals as in life, rather than in European style, illustrating morphological characters. Crustaceans are included in two different volumes shrimps are placed with fish, and crabs with shell (Table 3). Descriptive information with the crustacean entries includes other names, uses as food, and where they are found.

Yamato-honzo (*The Medicinal Natural History of Japan*), written in 1709 (Fig. 7), is of particular significance to Japanese natural history because of its original, rather than imported, observations and descriptions. Uno(1960) affords this work an unequalled position because of its scientific approach, and Kimura (1974) regarded it as an unusually excellent variety of

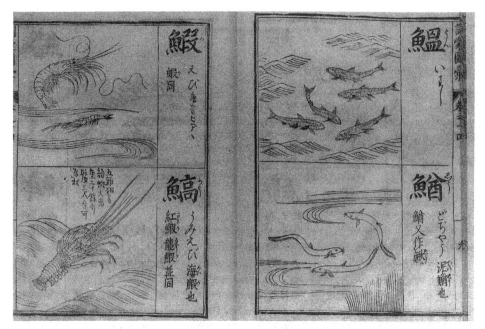

Figure 3. From woodcut print in *Kinmou-zuwi* (1666), first edition, Yedo, volume 14, showing the account of 'yebi' (shrimp) on the top, and that of 'umi-yebi' (spiny lobster) on the bottom.

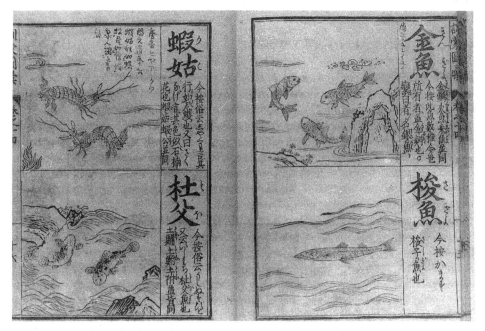

Figure 4. Also woodcut print from *Kinmou-zuwi* (1666) but showing the account (left page, top) of 'shaku' (mantis shrimp).

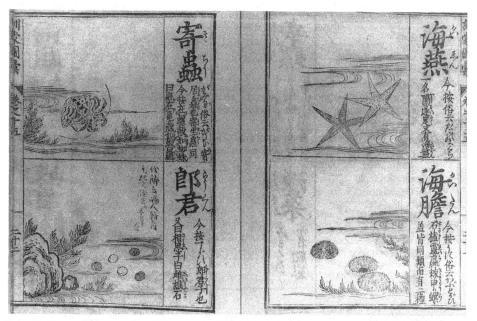

Figure 5. Also woodcut print from *Kinmou-zuwi* (1666) but showing the account (left page, top) of 'kamina' (hermit crab).

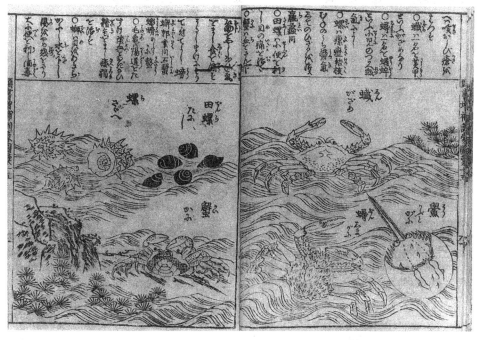

Figure 6. Woodcut print from *Kashiragaki-zouho-Kinmou-zuwi* (1695), an enlarged and revised edition of *Kinmou-zuwi*, showing the account, from volume 15, book 7, of 'kani' (crab), 'gasame' (portunid crab), 'shimagani' (crab), and 'kabutogani' (horseshoe crab).

Table 3. Titles of volumes from *Kinmou-zuwi* (1661) dealing with animals and plants, and the contents of those volumes that include crustaceans.

Volume	Title
12	Livestock & Beast
13	Bird
14	Dragon & Fish
	'awa-yebi' (river shrimp), 'yebi (shrimp), 'umi-yebi' (spiny lobster), 'ami' (opossum shrimp), included with dragon, whale, fish, jellyfish, octopus, squid.
15	Worm & Shell
	'kani' (crab, ?*Eriocheir*), 'shima-gani' (crab, ?*Erimacrus*), included with turtle, horseshoe crab, snail, mussel, clam, abalone, oyster, starfish, sea urchin, insect, frog, leech, centipede, toad, earthworm, spider, slug, snake, viper.
16	Rice & Cereal
17	Vegetable
18	Fruit
19	Tree & Bamboo
20	Flower & Grass

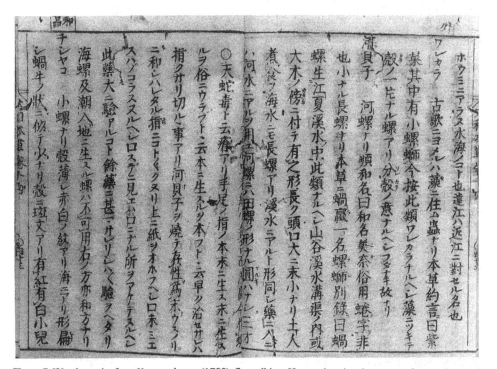

Figure 7. Woodcut print from *Yamato-honzo* (1709), first edition, Kyoto, showing the account from volume 14 of 'warekara' (skeleton shrimp or small sea snail) on the second row from the right and continuing leftward, between the snails *Charonia* and *Semisulcospira*.

genuine natural history. In the 'Discussion on Natural History Works' in volume 1 of *Yamato-honzo*, the author, Kaibara Ekiken (1630-1714), stresses the importance of using reason to arrive at truth and that simply learning from previous works is not natural history study. Thus, Kaibara points out questionable classifications of natural history things in *Pên Tshao Kang Mu* and proposes alternatives. For example, he questions why fishes in the Chinese work are not distinguished into freshwater and marine, and proposes that sea snakes, shrimps, and seahorses are not fishes and should be placed in the aquatic-worm grouping, which he has established distinct from his terrestrial worm and shell groupings (Table 4). These notions are an inevitable outcome of Kaibara viewing plants and animals ecologically. Also, unlike the author of *Pên Tshao Kang Mu*, Kaibara does not divide the worm groupings by types of reproduction, presumably because he thinks such types not exclusive to worm groupings but also applicable to fishes and even to plants. His descriptions of living organisms are detailed and original, and

Table 4. Titles of volumes of *Yamato-honzo* (1709), with contents of those volumes including crustaceans.Letters in parentheses indicate multi-volume coverage of topic: (L) = Lower part; (M) = Middle part; (U) = Upper part.

Volume	Tittle	
1	Introduction, Explanation, Discussion on NaturalHistory Works, Discussion on Natural Law	
2	Use of Medicine	
3	Water, Fire, Metal, Gem, Soil and Stone	
4	Cereal, Brewing and Processing	
5	Herb 1	Vegetable
6	Herb 2	Medicinal
7	Herb 3	Flowering, Garden leaf
8	Herb 4	Fruiting, Climbing, Fragrant, Freshwater, Marine
9	Herb 5	Weed, Fungi, Bamboo
10	Tree (U)	Four useful kinds, Fruiting
11	Tree (M)	Medicinal, Garden
12	Tree (L)	Flowering, Miscellaneous
13	Fish Freshwater, Marine (including whale, octopus,jellyfish, porpoise, mermaid, squid)	
14	Worm	
	Aquatic	
	'yebi' (freshwater shrimp *Macrobrachium*),'umi-yebi' (spiny lobster), 'shako' (mantis shrimp), 'ami' (opossum shrimp), includedwith sea cucumber, sea hare, sea squirt, seahorse, newt, insect v	
	Terrestrial	
	'funa-mushi' (*Ligia*), 'toko-mushi'(sowbug), included with insect, gecko, mite, silkworm, land snail, wire worm, toad, earthworm,spider, centipede.	
	Shell	
	'kamenote' (*Capitulum*), 'warekara'(skeleton shrimp or snail), 'kani' (crab), 'kasame' (portunid crab), 'shima-gani' (crab,?*Erimacrus*), 'gauna' (hermit crab), 'oni-gani' (*Nobilum*), 'tsu-gani' (*Eriocheir*), included with clam, mussel, pectin, argonaunt, snail, abalone, lamp shell, sea urchin,limpet, oyster, top shell, turtle, horseshoe crab.	
15	[no title] Water bird, Mountain bird, Small bird,Domestic bird, Miscellaneous bird, Foreign bird	
16	[no title] Beast, Human	
Appendix volume 1	Herb, Vegetable, Cereal, Fungi,Tree, Bamboo	
Appendix volume 2	Bird, Beast, Fish, Worm ['ami'(opossum shrimp) included with insect, seahorse], Shell ['ashi-kani' (seashore crab), 'shimamura-gani' (*Nobilum*), included with limpet, oyster, mussel, abalone], Water, Fire,Soil and Stone, Medicine	
Figures	Herb, Tree, Bird, Fish, Shell ['kobushi-gani' (*Leucosia*), 'kamenote' or 'shii' (*Capitulum*), 'kuzuma' (*Tetraclita*),vincluded with top shell, clam, mussel, limpet, horseshoe crab, pecten, snail, 'warekara' (smallsnail), chiton], Worm ['shakuge' or 'shako' (mantis shrimp) included with insect].	

include notes on external form and structure, habits, and uses in medicine.

Kaibara lists many different kinds of crustaceans, which he assigns to three groups under worm: Aquatic, terrestrial, and shell (Table 4). Such divisions seem illogical, and judging from the inclusion of crabs in the worm-shell, a traditional view, Kaibara's classification of things appears far from freed of the pervasive Chinese influence.

Wakan-sansai-zuye (*The Chinese and Japanese Universal Encyclopedia* based on the *San Tshai Thu Hui*), published in 1713, is also an illustrated encyclopedia. The presentation is similar to that of *Kinmou-zuwi*, and the descriptions of most entries have figures. Although this work appears much more comprehensive and detailed than *Kinmou-zuwi*, there is principally nothing new in the conception of animal classification, with only some differences in the order of presentation (Table 5). The crustaceans are placed with both shells and fishes, continuing the traditional view that crustaceans are not a separate group. Decapods are divided into three groups: Crabs with turtles into crust-shell; hermit crabs with cirripedes and various other invertebrates into a group with shell; and shrimps and spiny lobster into a grouping of marine, non-scaled fishes (Table 5).

For comparison, it is interesting to cite the descriptions given in Kaempfer's *The History of Japan*. This book was first published in 1727-1728, in London, as an English translation of the original German manuscript. Engelbert Kaempfer visited Japan in 1690, staying mostly in the

Table 5. Titles of volumes of *Wakan-sansai-zuye* (1713) dealing with animals, with contents of those volumes including crustaceans.

Volume	Title
37	Livestock
38	Beast
39	Mouse
40	Monster
41	Water bird
42	Field bird
43	Forest bird
44	Mountain bird
45	Dragon & Snake
46	Crust-shell
	'kani' (*Eriocheir*), 'ashihara-gani' (grapsid crab), 'houki' (grapsid crab), 'tumajiro' (fiddler crab, ?Uca), 'ishi-gani' (*Geothelphusa*), 'gasame' (portunid crab), 'shima-gani' (?*Erimacrus*), 'tebou-gani' (?ocypodid crab), 'takebun-gani' or 'shimamura-gani' (*Nobilum*), included with turtle, sea turtle, sea baldhead (giant)
47	Shell
	'kamenote' (*Capitulum*), 'kamina' or 'gouna' (hermit crab), included with abalone, pearl oyster, oyster, clam, mussel, pectin, snail, top shell, sea squirt, starfish, sea urchin, argonaut
48	Fish, Freshwater scaled
49	Fish, Marine scaled
50	Fish, Freshwater non-scaled
51	Fish, Marine non-scaled
	'yebi' (shrimp), 'kamakura-yebi' or 'ise-yebi' (spiny lobster), 'ami' (opossum shrimp), 'shako' or 'shakunage' (mantis shrimp) included with whale, fish, shark, porpoise, crocodile, ray, octopus, squid, sea cucumber, jellyfish
52	Worm, Oviparous birth
53	Worm, Spontaneous birth
54	Worm, Water birth

vicinity of Nagasaki for 2 years. During his stay, he travelled twice to Yedo (Tokyo) and had a chance to observe and collect plants and animals along the way, but seems not to have had the chance to meet Japanese naturalists. However, he was fairly knowledgeable of Japanese works of natural history, and in *The History of Japan* he includes many figures of animals in Japanese style seemingly copied from *Kinmou-zuwi*. In chapter 11, on aquatic animals, Kaempfer states, 'Both fish, crabs, and shells are known under one collective name 'kiokai' or 'iwokai' (p. 132) and he lists whales, fishes, squids, octopuses, jellyfishes, sea cucumbers, newts, cake urchins, turtles, crabs, shrimps, oysters and other shelled molluscs. All the crustaceans he mentions are in a paragraph under crabs and shrimps, and he includes Jebisako (*Crangon* or small shrimps), Si Jebi (a shrimp), Dakma Jebi (a shrimp), Kuruma Jebi (*Penaeus*), Umi Jebi (spiny lobster), Siakwa (mantis shrimp), Gamina or Koona (hermit crab), Kani (a crab, (?) *Eriocheir*), Kabuto-gani or Unkiu (horseshoe crab), Gadsame (portunid crab), and Simagani (a crab, (?) *Eri-macrus*). Obviously Kaempfer , unlike contemporary Japanese writers, regarded decapod crustaceans as a distinct group.

Although Kaempfer and his book had little influence on later Japanese works of natural history, this was not the case of Carl Peter Thunberg who visited Japan in 1775-1776. Many medical doctors and students of natural history, who knew his fame, came to offer assistance for his study, and to listen to his lectures on western biology. Some of them became convinced of and advocated its merits. Nevertheless, the western and Linnean system of classification of living beings did not immediately challenge the authoritative Chinese system in the leading Japanese writings and had to wait for another century to become established as the unchallenged principle.

Honzo-kouoku-keimou , written about 1803 by Ono Ranzan (1729-1810), is the detailed

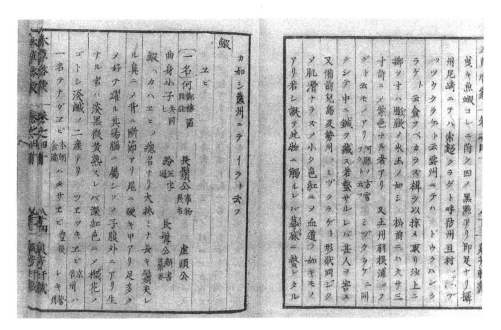

Figure 8. Woodcut print from *Honzo-Koumoku-keimou* (1811), Kyoto, reprint of first edition (1803), showing the account of 'yebi' (shrimp) on the 10th row from the left and continuing leftward, following 'kurage' (jellyfish).

exposition of *Pên Tshao Kang Mu*, containing also the descriptions of many Japanese things (Fig. 8). This is the last great work of traditional Japanese natural history. It was revised and enlarged (1847) with notes and comments from observations by pupils of Ono. Thus, the work actually became a large compilation of the knowledge of the time. For many plants and animals, various local names and names of varieties are given. However, as the titles of the volumes show (Table 6), the system of classifying things is principally a copy of that in *Pên Tshao Kang Mu* and descriptions follow the traditional Japanese way of identifying and equating things with those in Chinese works.

Traditional systems of classification of living things had been criticized and challenged in Japan before modernization in the second half of the 19th century. For example, there was Miura Baiyen, an imposing figure in Japanese philosophy and poetics, who lived on Kyushu in the late 18th century. His views on the living world were a corollary of his philosophical understanding of the universe, and he enunciated a system of classifying living beings into animal or plant, and then into terrestrial or aquatic, and so on, evidently in a dichotomous hierarchy. However, contemporary Japanese writers of natural history paid little attention to his work, probably because he was remote from the main cultural camps, and because he did not apply his system to actual animal and plant species.

Table 6. Titles of volumes of *Honzo-koumoku-keimou* (1803, Revised edition 1847) and contents of volumes dealing with crustaceans. Letters in parentheses indicate multi-volume coverage of topic: (L) = Lower (U) = Upper.

Volume	Title
1	Water
2	Fire
3	Soil
4	Metal & Stone
5-7	Stone
8-17	Herb
18-21	Cereal
22-24	Vegetable
25-29	Fruit
30-33	Tree
34	Cloth & Wear
35	Worm 1 – Oviparous birth (U)
36	Worm 2 – Oviparous birth (L)
37	Worm 3 – Spontaneous birth
38	Worm 4 – Water birth
39	Scaled 1 – Dragon Scaled 2 - Snake
40	Scaled 3 – Fish
	Scaled 4 – Non-scaled fish
	'yebi' (shrimp) and 'umi yebi' (spiny lobster), included with fish, mermaid, salamander, shark, squid, octopus, ray, jellyfish.
41	Shell 1 – Turtle
	'kani' (crab), included with sea turtles, turtles, horseshoe crab
42	Shell 2 – Mussel & Clam
	'sei' (*Capitulum*) and 'kamina' (hermit crab), included with oyster, mussel, pearl oyster, clam, cowrie, snail, cake urchin, starfish.
43-45	Bird
46-47	Beast
49	Human

By the mid-19th century, many Japanese writings and drawings dealing with specific groups of plants or animals were published. These included *Kaika rui sja sin*, a collection of illustrations of crabs and shrimps by Kurimoto Zuiken (1756-1834), published in 1820. Some of the drawings from this collection were later included in de Haan's volume on the Crustacea for von Siebold's *Fauna Japonica* (see Holthuis & Sakai 1970, Holthuis this volume). Kurimoto's work and similar publications show that some natural history work similar to that of the West occurred prior to 1868.

In an artistic picture entitled 'Kani-zukushi' (Crabs of All Kinds) by Katsushiks Hokusai (1760-1849), one of the most famous artists of the late-Yedo era, 101 crabs of different sizes and shapes are vividly drawn. (The original drawing is in the Freer Gallery of Art of the Smithsonian Institution, Washington, D.C.) Among the crabs the artist has also drawn a large shovel-nose lobster (*Scyllarides squamosus*) and a large horseshoe crab. One may assume the artist recognized these as crabs, too. Placing the horseshoe crab with true crabs occurs in all classical works of Japanese natural history.

5 CARCINOLOGY IN JAPAN

It seems doubtful whether the structure of crustaceans and most, if not all, other animal groups was systematically examined during the classical natural history period of Japan. Although

Table 7. Chronology of publications and events relevant to classical natural history in Japan and to the development of Japanese carcinology.

500-200	B.C.	*Erh Ya (Literary Expositor)* by Anonymous in China
500-200	B.C.	*Shan Hai Ching (Classic of the Mountains and Rivers)* by Anonymous in China.
ca. 100	A.D.	*Shen Nung Pên Tshao Ching (Pharmacopoeia of the Heavenly Husbandman)* by Anonymous in China.
ca. 290		*Po Wu Chin (Record of the Investigation of Things)* by Chang Hua in China.
ca. 660		*Thang Pên Tshao* by Su Kung in China
712		*Kojiki (Records of Ancient Things)* by Oh-no-Yasumaro in Japan.
720		*Nihon-shoki (Ancient History of Japan)* by Oh-no-Yasumaro and Toneri-Shinno in Japan
794		*Mannyo-shu* compiled by Ohtomo-no-Yakamochi in Japan
929		*Honzo-Wamyo* by Fukane-no Sukehito in Japan.
ca. 930		*Wamyo-ruishusho* by Minamoto-no-Shitagau in Japan.
1590		*Pên Tshao Kang Mu* by Li Shih-Chen in China.
1607		*San Tshai Thu Hui* by Wang Yin in China.
1666		*Kinmou-zuwi* by Nakamura Tekisai in Japan.
1690		Visit of E. Kaempfer to Japan.
1709		*Yamato-honzo* by Kaibara Ekken in Japan.
1713		*Wakan-sansai-zuye* by Terajima Ryoan in Japan.
1728		*The History of Japan* by E. Kaempfer in England.
1775		Visit of C.P. Thunberg to Japan
1803		*Honzo-koumoku-keimou* by Ono Ranzan in Japan.
1823		*Fauna Japonica* by C.P. Thunberg in Sweden.
1823		First visit of P.F. von Siebold to Japan.
1833		*Fauna Japonica*, the first part, by P.F. von Siebold in Holland.
1850		*Fauna Japonica*, the last part, by P.F. von Siebold in Holland.
1859		Second visit of P.F. von Siebold to Japan.
1868		Meiji Revolution.
1875		Visit of H.M.S. Challenger to Japan.

Table 8. The crustacean species presumed mentioned or referred to in the major classical works of natural history of Japan.

Anostraca
  Branchinella sp.
Conchostraca
  Caenestheriella sp.
Cirripedia
  Capitulum mitella   *Lepas anatifera*
  Lepas anserifera   *Coronula* sp.
  Balanus roseus   *Chthamalus challengeri*
  Tetraclita squamosa
Stomatopoda
  Oratosquilla oratoria
Mysidacea
  Neomysis sp.
Isopoda
  Ligia exotica   *Armadillidium vulgare*
Amphipoda
  Caprella sp.
Decapoda
  Dendrobranchiata
  Penaeus japonicus   *Penaeus semisulcatus*
  Metapenaeus ensis   *Metapenaeus joyneri*
  Metapenaeopsis palmensis
  Caridea
  Paratya compressa   *Neocaridina denticulata*
  Caridina leucostica   *Palaemon paucidens*
  Macrobrachium sp.   *Crangon* sp.
  Alpheus sp.
  Palinura
  Linuparus trigonus   *Panulirus japonicus*
  Panulirus versicolor   *Scyllarides squamosus*
  Parribacus japonicus
  Anomura
  Hippa pacifica   *Dardanus* sp.
  Pagurus sp.
  Brachyura
  Ranina ranina   *Nobilum japonicum*
  Calappa japonica   *Calappa hepatica*
  Matuta lunaris   *Macrocheira kaempferi*
  Parthenope sp.   *Leucosia anatum*
  Erimacrus isenbeckii   *Portunis* sp.
  Atergatis sp.   *Hypothalassia armata*
  Geothelphusa dehaani   *Pinnotheres* sp.
  Scopimera globossa   *Ocypode stimpsoni*
  Uca sp.   *Macrophthalmus* sp.
  Ilyoplax pusilla   *Goetice depressus*
  Sesarmops intermedium   *Chiromantes haematoceir*
  Helice tridens

profuse notions of animals are presented in the classical encyclopedic works, these descriptions are not based on any comparative analysis of morphological features. For example, with the collective names denoting a certain group of animals, for example, 'yebi' (shrimp) or 'kani' (crab), there is no statement giving the morphological characters defining the group and affording the basis for differentiating it from other groups. Instead descriptive information accompanying 'kani' includes: 'Our waters are too saline, so the taste is inferior. In summer it is less saline, and the taste is better.' There are writings on crabs. There are quite a few kinds of crabs, namely 'gasame,' 'shima-gami,' 'tsumashiro.' There is 'tsu-gani'...(from *Yamato-honzo*). For 'yebi' the following example is given: 'It is explained in *Pên Tshao Kang Mu* as follows. Those living in rivers and lakes are large and white, and those in streams and ponds small and green. They have whips and broad-axe nose, segmentation on their back, and hard scales on their tail. They have many legs and like dancing.... There are several kinds...' (from *Wakan-sansai-zuye*). For 'umi-yebi' (sea shrimp or spiny lobster) the following is reported: 'A collective name of shrimps living in the sea. There are many kinds. 'Ise-yebi' (=*Palinurus japonicus*), the same as 'xiakui' in a Chinese work.... One growing large is excellent. Its eyes are purplish black with yellow part in front and these are extending high up like tubercles. Its mouth bears four whips...' (from *Honzo-koumoku-keimou*).

So, some may argue that carcinology, at least in the sense of a biology of the Crustacea, did not exist in Japan before the Linnean system of classification was introduced in the mid-19th century. Or, that when there was no conception of Crustacea, carcinology could not exist. This is a terminology convention of purists and is a serious misunderstanding of language (Hull 1988). We can definitely see an accumulation of knowledge on crustaceans in the classical years in Japan. Nevertheless, after the Meiji Revolution, the classical systems of classification of living beings were undermined and simply replaced by the modern scientific system without much dispute, and unfortunately the native natural history of early years in Japan has become utterly disconnected from modern biology. A chronology of publications and events relevant to the development of carcinology in Japan is given in Table 7.

There is much difficulty in identifying the kinds of living beings mentioned in the classical works. For tentative reference the scientific names of crustaceans identified from the major classical works of Japan are listed in Table 8.

ACKNOWLEDGEMENTS

I directed my talk at the symposium and also this summary paper to those unfamiliar with the Japanese language. Those Japanese who read this paper probably are aware of the works I have introduced here, and I advise them to refer directly to them and to the relevant papers of those Japanese authors (listed below) who have studied the classical literature. All photographs are of pages from classical manuscripts collected by Uno Masuzo and placed in the Abe Memorial Library of Konan Women's University, whose permission to make photographs and publish them in this paper is gratefully acknowledged. I am especially grateful to F.M. Truesdale for making comments on the manuscript.

REFERENCES

Chamberlain, B.H. 1883. A translation of the *Ko-ji-ki*, or *Records of Ancient Matters*. Royal Asiatic Society 10: Appendix. (revised 2nd ed. 1932, Kobe).

Higashi, M. 1935. *The Animals of Mannyo-shu*. Kyoto: Jinmon-shoin (In Japanese).

Holthuis, L. B. & T. Sakai 1970. *PH. F. Von Siebold and Fauna Japonica*. Tokyo: Academic Press of Japan.

Hull, D.L. 1988. *Science as a Process. An Evolutionary Account of the Social and Conceptual Development of Science*. Chicago: University of Chicago.

Iino, T. (ed.) 1985. *Kagaku no Ziten* (*Encyclopedia of Science*), 3rd edition. Tokyo: Iwanami-shoten (In Japanese).

Kaempfer, E. 1727-1728. *The History of Japan*. London. Reprinted 1929, Kyoto: Kosei-kaku.

Kimura, Y. 1974. *Foundation of Japanese Natural History. Ran-gaku and Honzo-gaku*. Tokyo: Chuo-koron-sha (In Japanese).

Needham, J. 1986. *Science and Civilisation in China. Volume 6. Biology and Biological Technology*. Part I *Botany*. Cambridge: Cambridge University.

Uno, M. 1948. *A History of Japanese Natural History*. Kyoto: Hoshino-shoten (In Japanese).

Uno, M. 1960. *A Pre-Meiji History of the Japanese Biology. Volume 1*. Tokyo: Nihon-gakujutsu-shinkoukai (JSPS) (In Japanese).

Reflections on crab research in North America since 1758

Austin B. Williams
National Marine Fisheries Service Systematics Laboratory, National Museum of Natural History, Washington, D.C., USA

1 INTRODUCTION

History of crab research in North America is so broad a subject for coverage in the few pages allotted to it here that limits of some kind are necessary to give the treatment meaning. That record has already been published on in exacting detail in scattered accounts in a spectrum of publications. In a sense, these were summarized in Waterman's two-volume treatise, *Physiology of Crustacea* (1960-1961), containing contributions by 32 biologists. This was the first general review of crustacean research in English since the classic works on Crustacea of Calman (1909) in Lankester's multivolume *Treatise on Zoology*; Smith & Weldon (1909, reprinted 1923) in Harmer & Shipley's *The Cambridge Natural History*; and the review by Schmitt (1931, reprinted 1965). The massive summaries of crustacean morphology, ontogeny, physiology, and systematics in German by Balss (1940-1945, 1957) served the same purpose. Just 20 years after the Waterman volumes appeared, the field of crustacean biology had so advanced and broadened that need was perceived by Bliss for incorporating major breakthroughs of the science into a 10- volume treatise, and this was brought about through her initiative (1982-1985) and the cooperation of 8 editors and 75 authors. The busy period also saw the appearance of Glaessner's 'Decapoda' (1969) in the *Treatise on Invertebrate Paleontology*; the founding of the Crustacean Society in 1979 and its *Journal of Crustacean Biology*, which has flourished since its inauguration in 1981 (Manning 1990); the continuing presence of the well-established journal *Crustaceana*; the appearance of the several volumes of *Crustacean Issues* published by Balkema, Schram's (1986) *Crustacea*, other specialty volumes such as that on invertebrate fisheries edited by J.F. Caddy (1989), crustacean conferences and workshops too numerous to list in detail; and the continuing rain of publications on crustaceans scattered across the whole field of life sciences. In all this, brachyurans occupy a prominent place.

For practical reasons this review is limited to reflections on crab research in North America in the thought that this restriction may contain some representation of the whole. I did not believe it appropriate or practical to review extensively the research on North American crabs of commerce, for important as that activity has been the literature is almost unlimited in scope. In addition, most work accomplished in the 1980's seems too recent to be regarded as history.

A retrospective on crab research in North America during the years since its settlement by Europeans began, as of any research activity, is not simply condensation of results obtained, but a consideration of paths explored by persons and institutions, of conceptual strands woven into understanding, whose fabrication continues. After all, the activity began with the discovery of new territory, and subsequent developments paralleled the rise of modern science.

The central points considered here are exploration and discovery, some ramifications of selected investigations and economic development, research giants, and landmark papers. An outline that might apply to the general research history of many of the species can be conveyed by reading the subject headings in the table of contents of any FAO Fishery Synopsis volume on crustaceans (e.g., Milliken & Williams 1984).

2 DISCOVERY

From the earliest explorations of voyagers to North America, adventurers recorded observations on the aquatic fauna and flora. Among the easily observed biota were conspicuous crabs of interest either for fishing and food, or simply because they were natural curiosities that differed from familiar counterparts known to the explorers at home. From these quaint early accounts grew the recorded study on North American crabs. The results are remarkably similar for many species, illustrating perhaps what is an inherently natural pathway toward understanding.

3 EXPLORATION AND DESCRIPTION – NATURAL HISTORY

Early exploration of North America, including interactions with the American Indians, resulted almost immediately in a body of natural history information (e.g. Hariot 1588; Lawson 1714) including observations of crabs. Among conspicuous estuarine brachyurans, blue crabs were abundant, catchable, and highly palatable, and other crabs less abundant in nearshore waters were of interest as well.

It is hardly surprising that during the late 1700's following the appearance of *Systema Naturae* (Linnaeus 1758), only six naturalists published descriptions, still considered valid, of new species of North American crabs. All of these workers were Europeans. Of 27 brachyuran species (Table 1), Herbst described 10; Linnaeus, 7; O. Fabricius, 6; and J. C. Fabricius, G. Edwards, and Olivier, 1 each (see Williams et al. 1989). All these species were placed originally in the genus *Cancer*, and for well into the 1800's both European and American workers tended to use European names for North American species that appeared similar to those in Europe.

In the first half of the 1800's, a period during which pioneer American naturalists began making their contributions, work quickened to result in 56 descriptions of newly recognized crabs. Nearly half of that output was by French workers (H. Milne Edwards, including a coauthor, 16; Latreille, 7; Lamarck, 4), while Bosc, Brandt, Freminville, Herbst, Leach, Tilesius, and White collectively contributed 12 species (see Williams et al. 1989). In contrast, American authors described 17 species (Say, 10; Randall, 6; and Gould, 1).

Table 1. Number of valid brachyuran species occurring in North America that were described per half-century interval from 1758 to 1989.

1758-99	1800-49	1850-99	1900-49	1950-89
27	56	273	72	30

The stage was set for the great exploratory thrust achieved during the last half of the century when the noted French carcinologist A. Milne Edwards described 54 species of North American crabs and Saussure 6 species, far surpassing the contribution of other Europeans. However, the tide of this effort was turning toward the West. William Stimpson described 92 species of North American crabs during his brief, brilliant career, followed in productivity by other descriptive naturalists: M.J. Rathbun, 41; Dana, 16; S.I. Smith, 13; Lockington, 9; Gibbes, 6; and collectively 26 species by authors who described 3 or fewer (see Williams et al. 1989). Rathbun's effort (Schmitt 1973; Chace 1990) carried strongly into the first half of the 1900's, with her contributing an additional 58 new species, along with comprehensive biogeographic information in her great brachyuran monographs for the Western Hemisphere. Holmes named 3 species, and 11 others were described by various naturalists from the United States, Mexico, Brazil, and Europe.

The field of brachyuran studies then changed. As the main outlines of the North American brachyuran fauna became well established, descriptive efforts, always a mixture of several disciplines, became more analytical. More data were available, and biogeographic summaries along with ecological and experimental studies changed the emphasis that had long been placed on purely descriptive taxonomy. The result was that the more inclusive discipline of biosystematics offered reappraisals of brachyuran classification. During the late 1900's, Guinot (1978) has been a leader in this regard in brachyuran systematics, especially in treatment of categories above species level, while describing 6 new species from North America. Other contemporary workers extended that number (Williams, 5; Holthuis and Wass, 3 each; plus 13 collectively by still others; see Williams et al. 1989).

The fruits of exploration that are stored in museum research collections are grist for the descriptive and biogeographic mills and sources of information upon which all of the descriptive natural historians depend. Collecting and research collections form an inseparable linkage to which systematic effort is tied. Although individual collecting and collections traditionally have contributed information on brachyurans, biosystematic research on them has relied heavily on group effort for the gathering of data. The pioneer American naturalist Thomas Say (describer of *Menippe mercenaria*, for example) associated mainly with colleagues from the Academy of Natural Sciences of Philadelphia (Weiss & Ziegler 1931; Spamer & Bogan, this volume). He was followed by James Dana (1852) (Fig. 1) who was involved in the United States Exploring Expedition of 1838-42, which went around the world (Bartlett 1940, Appleman 1985). The American Philosophical Society and the Academy of Natural Sciences of Philadelphia were supporters of that expedition (Conklin 1940), and the embryonic National Museum of Natural History (NMNH), then being incorporated into the Smithsonian Institution, became the depository for those expedition specimens that survived the vicissitudes of sea transport, inadequate storage, and inexpert sorting in Washington, D.C. Still, this material was an important source of data for Dana's self-illustrated monographic volumes on invertebrates collected during the expedition, including decapods from the Oregon Territory and California. His (1851) designations for areolations on the crab carapace are followed to this day.

Many subsequent expeditions, public and private, amassed brachyuran research collections from North America. Leaders in research expeditions were the United States Commission of Fish and Fisheries and its successor agencies, the US Bureau of Fisheries, the Bureau of Commercial Fisheries, and the National Marine Fisheries Service that promoted exploration of the marine resources of contiguous waters of the United States and other regions for over 100 years. Congress in 1879 authorized appropriation of $45000 for construction of the Fish Hawk (Fig. 2), a 157-foot steam-powered vessel, designed primarily to aid in research on the hatching

Figure 1. James Dana by Daniel Huntington. Courtesy Yale University Art Gallery, bequest of Edward Salisbury Dana, B.A. 1870.

of eggs for propagation of fishes. Launched in 1880, Fish Hawk was utilized almost immediately in dredging on the continental shelf and slope off the eastern United States (Tanner 1884). Appropriations totalling $148000 followed in 1881 for construction of a larger vessel, the 234-foot Albatross (Fig. 2), launched in 1882 (Hedgpeth & Schmitt 1945). Data and specimens from thousands of stations sampled by workers on both of these vessels in the western North Atlantic, Gulf of Mexico, Caribbean, and on the Albatross in the South Atlantic, Strait of Magellan, eastern Pacific from Cape Horn to the Bering Sea, Hawaiian Islands, Micronesia, and the Philippines brought much material into the research collections of various museums, especially those of the NMNH, Washington, D.C.; the Yale University Peabody Museum, New Haven, Connecticut; and the Museum of Comparative Zoology, Harvard University, Cambridge, Massachusetts. It was largely upon data and collections in these institutions that Mary J. Rathbun based her voluminous output of research (Schmitt 1973).

In the 1930's and early 40's, a period of diminished activity occurred, but crab accessions into North American museum collections regained momentum following World War II as oceanographic research expanded. Brachyuran samples came from many vessels, but foremost among them again were exploratory fishing research vessels of the Bureau of Commercial Fisheries, now the National Marine Fisheries Service (Oregon, Oregon II, Silver Bay, Combat, and Pelican; see Springer & Bullis 1956, Bullis & Thompson 1965, Oregon II station lists, 1967-present). This resulted in generous deposits being made in the crustacean collection of the NMNH. To this must be added the considerable collections made by vessels in Canadian waters from the early 1800's onward, but especially from the Calanus expeditions of McGill University (Fisheries Research Board of Canada Eastern Arctic Investigations 1947-1961), reported by Squires (1965, 1990).

Figure 2. Top, United States Fish Commission Steamer Fish-Hawk (Tanner 1884). Bottom, United States Fish Commission Steamer Albatross. Reproduced from a photograph in the records of the Fish and Wildlife Service, Bureau of Fisheries, in the National Archives.

4 FISHERIES

Fisheries for crabs of the families Cancridae, Majidae, Portunidae, and Xanthidae, specifically the Dungeness (*Cancer magister*), snow (*Chionoecetes opilio*) and Tanner (*C. bairdi*), blue (*Callinectes sapidus*), and stone (*Menippe mercenaria*) crabs make up the bulk of brachyuran landings in North America. For example, United States landings for these species in 1988 were 197.3 million kg valued at $299.4 million. Comparative amounts for the American lobster were 22.1 million kg and $145.2 million, and for shrimps, 150.2 million kg and $506 million.

Fisheries for the blue and Dungeness crabs became organized activities in the mid-19th century (R. Rathbun 1884, 1887; Van Engel 1962; Methot 1989), and that for the stone crab was poorly organized until a century later (Erhardt & Restrepo 1989). The snow-Tanner crab fisheries, in deeper water than the others, developed quite differently in only the past 25 years (Brown 1971; Bailey & Elner 1989). All depend upon trapping as the primary method of capture.

The fishery for the Dungeness crab ranges from San Francisco, California, to Kodiak Island, Alaska, in depths usually less than 100 m and often less than 30 m (Methot 1989). A fishery at greater depths in the northwestern Atlantic, northern Pacific, and Sea of Japan for snow-Tanner crabs and relatives is the largest for brachyurans in the world (Brown 1971, Bailey & Elner 1989). Blue crabs are fished mainly from the middle Atlantic to southeastern and Gulf states of the United States, but also in other parts of the species-range from Massachusetts to Argentina in estuarine waters (Williams 1984). Stone crabs are fished mainly in the eastern Gulf of Mexico and Cuba in depths up to 50 m, but usually less (Williams & Felder 1986). Although blue crabs were caught by Native Americans, and after them by the early European settlers, an organized fishery did not develop in the Chesapeake Bay, for example, until the 1870's (Van Engel 1962); it was carried on with a variety of devices, but only pots, trotlines, and dredges proved suitable for catching hard crabs. Wire mesh pots dominate the fishery today. The Dungeness crab fishery in a sense had a parallel development; Indians hand speared and caught the crabs in baited circular frame and web traps, similar to ring nets still in use by sportsmen today, before a California fishery incidental to finfishing began in 1848 (Dahlstrom & Wild 1983). By 1880 the regular southern market was almost confined to San Francisco, but with the introduction of steam-powered vessels in 1885 the fishery expanded. Cylindrical, mesh-covered, baited traps typically 1 m in diameter dominate the fishery today. The stone crab fishery also began in the mid-1800's, mainly in south Florida (Ehrhardt & Restrepo 1989), but poorly developed marketing limited growth until around 1960 when a rapid expansion began that culminated in 1980, after which landings declined. Snow and Tanner crabbing in the northwest Atlantic and northern Pacific, respectively, began about 20 years ago as supplemental fisheries. The fishery for Tanner crabs began in Alaska during closed king crab seasons, and has continued (Brown 1971). Snow crabs were a nuisance to Canadian groundfishermen until they tired of cleaning them off their nets and realized the commercial potential as bycatch, and they initiated a trap fishery (Bailey & Elner 1989).

Fluctuations in catch characterized the fisheries for blue, Dungeness, and stone crabs once these resources were subjected to fully organized harvest (Pearson 1948; Methot 1989; Ehrhardt & Restrepo 1989); moreover, landings records for Dungeness crabs indicate more or less cyclic periods of abundance and scarcity. As these effects became apparent, questions concurrently arose concerning possible biotic and abiotic factors that might influence catch, and calls came for keeping records of catch and sales, for imposition of regulations, and for programs of research that might result in knowledge that could aid in management of the fisheries. An example of the latter is a regional group-study of the Dungeness crab, published under the editorship of Wild & Tasto (1983), that dealt with history of the fishery, stock identification, ocean and estuarine conditions during larval existence, larval development and staging, associated zooplankton, movement and growth of larva-adult, predation on crabs, effects of commercial trawling on crab survival, ocean climate and the fishery, ovarian development, influence of temperature on spawning, egg development and hatching success, effects of pollutants, and laboratory cultivation, including effects of substrate type on survival and growth. Similar efforts on Dungeness crab in other regions (Butler 1960, 1961; Cleaver 1949;

Snow & Nielson 1966), emphasize the recentness of much work of this kind. The larval studies especially had broad implications.

5 LARVAL DEVELOPMENT

The complex larval development of decapod Crustacea has long been studied (Gurney 1942; Epifanio 1979). Larval development of the Brachyura is of particular interest because of what it adds to our understanding of the distribution of a group of organisms that are largely benthic as adults. It also preserves morphological clues to the evolution of the group that bear on classification (Rice 1980, 1981). Earlier workers, intrigued as they were by the radical morphological changes from zoea to megalopa to juvenile crab, did not completely understand the number of stages in this transformation or their duration because the techniques and apparatus that are taken for granted in larval rearing studies today had not been developed. Historically, eggs of a species might have been hatched and reared through an incomplete series by makeshift means on inadequate diet, or development doubtfully pieced together from larval stages taken in plankton samples. Hyman (1924), for example, was able to rear the oyster pea crab, *Pinnotheres ostreum*, through only the first two zoeal stages, and over 20 years later Sandoz & Hopkins (1947) still only managed with difficulty to rear the species through four zoeal stages (one fragmentary), the megalopa, and first crab stage.

Soon afterward, however, Broad (1957a, b) worked out the development of a caridean shrimp, *Palaemonetes pugio*, by rearing the larvae in small containers that facilitated tracking the transformation of individuals fed experimental diets that included freshly hatched brine shrimp (*Artemia*) nauplii (generally called *A. salina* in some developmental studies to follow, but really a set of sibling species, see Bowen et al. 1980). Broad's technique provided for individual accounting of developmental stages by using a diet of small sized, nutritious prey that could be held indefinitely as dried resistant cysts ready to be hatched quickly when needed. That technique coupled with a method for developing brachyuran eggs in vitro (Costlow & Bookhout 1960), and use of other experimental diets (Sulkin & Epifanio 1975; Sulkin & Norman 1976), helped to foster a flood of decapod larval development studies. Descriptions of the larvae of *Callinectes sapidus*, *Pinnotheres maculatus*, *Menippe mercenaria*, *Cancer magister*, *C. productus*, *Bathynectes superba* (= *B. longispina*), *Ocypode quadrata*, *Hyas araneus*, *Chionoecetes bairdi* and *C. opilio*, and *Libinia emarginata*, respectively by Costlow & Bookhout (1959, 1966), Porter (1960), Poole (1966), Trask (1970), Roberts (1969), Christiansen (1971), Diaz & Costlow (1972), Haynes (1973), and Johns & Lang (1977), to cite only a few North American studies, followed along with many others in roughly 15 years, although some of these studies did not involve laboratory rearing.

Descriptions of larval stages reared from eggs of identified female crabs facilitated identification of planktonic zoeae and megalopae and yielded information on sequencing and duration of larval life. Most of the species reared were those easily available in estuarine and nearshore marine waters close to laboratories where the work was done, notably Duke University Marine Laboratory which was a leader in the rearing of crab larvae. Often the species reared were of primary economic or ecologic interest, but chance finds of ovigerous females from which rearings could stem were taken advantage of as well (*Homola*, Rice 1964; *Dromidia antillensis*, Rice & Provenzano 1966, for example). Developmental comparisons among species, genera, families, and other higher categories were initiated as soon as comparative data became available.

Relationships between environmental variables and larval development were of paramount interest, and the culturing methods allowed questions concerning these relationships to be rigorously tested with experiments. Effects of such factors as temperature, salinity, and photoperiod on larval developmental staging, duration, and rate of growth were monitored singly and in combination (*Callinectes sapidus*, Costlow 1965, 1967; *Sesarma cinereum, Panopeus herbstii, Rhithropanopeus harrisii*, Costlow et al. 1960, 1962, & 1966, respectively, for example).

Larvae in the real world, however, live and develop in dynamic systems influenced by climate, season, weather, currents, and tides, that offer circumstances far different from those in controlled laboratory compartments. Enlightening as were results from initial laboratory experiments, attempts were made to assess the more complex interrelationships of nature in the larval culturing systems. Much of this work was impelled by the attempt to explain linkages between dispersal, distribution, and survival of species and the dynamics of estuarine and coastal systems in which they live. Breeding, incubation, and hatching are linked to season; larvae released in systems with net seaward transport seemingly are able to capitalize on the dispersal abilities of those systems, either to remain within them, or return to the natal grounds after displacement to complete the cycle. Investigations were designed to study factors such as the time of appearance of phototaxis (Forward & Costlow 1974), effect of salinity and cyclic temperature on development (Christiansen & Costlow 1975), larval shadow response, depth regulation, and behavioral responses to rates of salinity change (Forward 1977). Aside from natural variables, none of the coastal aquatic waters where these studies were conducted are free from human influences, especially near urban centers. Possible effects of pollutants and contaminants on larvae could be assayed by manipulations analogous to those employed for testing their reactions to natural environmental processes (reviewed by Epifanio 1979; Williams & Duke 1979).

After identification of selected brachyuran larvae in cultures was accomplished, it was possible to plot their occurrence in nature with the aid of field sampling studies. The considerable evidence that larvae of estuarine invertebrates are found in coastal marine waters was reviewed by Sulkin & Van Heukelem (1982), and a question posed by them (p.472) and others was 'whether such larvae are lost from the adult habitat or whether they can be recruited back to the estuarine environment and if so, how significant is this source of recruitment?'

One of the seminal papers that led to this query was Christy's (1978) hypothesis that mating in *Uca* species is so synchronized with lunar tides that hatching of eggs and release of zoeae are timed to minimize downstream transport and loss from estuaries during development but maximize up-estuary return transport during the megalopal stage when this phase is ready to settle. The hypothesis involved vertical migration of larvae to mediate favorable transport. Papers ensued addressing the idea of estuarine retention of larvae (Cronin 1982), and tidally timed behavior and larval production (Cronin & Forward 1982, Christy & Stancyk 1982). It is evident that ability to assess larval adaptations to life in coastal waters has matured in recent decades.

Positive identification of larval crabs from offshore plankton samples soon revealed that species such as the commercially valuable blue crab, *Callinectes sapidus*, and Dungeness crab, *Cancer magister* are found there (Nichols & Keney 1963; Sulkin & Van Heukelem 1982; Reilly 1983; Hatfield 1983. Both of these species spawn in nearshore or estuarine waters. The hatching eggs of these species release zoeae that are swept to sea by oceanic currents in early stages of development. These undergo transformation through successive zoeal stages in offshore waters, sometimes near the continental slope. In late zoeal and megalopal stages, they apparently return to nearshore water by means of accommodation through vertical migration. They seem to concentrate in zones where conditions in the water column offer the best chance of shoreward transport by upwelling, intruding salt wedges, or longshore currents abetted by tidal sweeps, and thereby return to their natal waters. Spawning and larval development of penaeid shrimps in coastal waters, and subsequent inshore larval migration to estuaries where maturation continues before return as subadults to the sea had long been known of course (see for example Lindner & Anderson 1956), but the complexities of these cycles for crabs were neither dreamed of nor speculated upon in other than the sketchiest of ways prior to laboratory research on the larvae that allowed an understanding of the situation in nature.

6 GUIDEPOST PAPERS AND SOME OF THEIR SUCCESSORS

Certain publications on crabs established new lines of research, provided comprehensive comparative summaries, or gave significant support to the ongoing stream of studies. It seems appropriate to mention some of these as examples.

Any listing of guidepost papers on brachyurans of North America has to include the monographs of Rathbun (1918, 1925, 1930, 1937), still standards of reference. Aligned with these would surely be Garth's (1958) monograph on spider crabs of the Pacific coast of America. A series of no less important works that round out the working tools of anyone now attempting serious systematic study or routine identifications of brachyurans in North America are those of Guinot, among which are her preliminary revisions of the difficult Xanthidae-Goneplacidae (1967, 1968, 1969a, b, c, d, 1971), and her (1978) views of higher brachyuran categories already mentioned above that were opened to possible reinterpretation by de Saint Laurent (1980a, b). Some faunistic handbooks round out this list: California (Schmitt 1921), eastern Canada (Squires 1965, 1990), the Carolinian region (Williams 1965), and the eastern United States (Williams 1984), to say nothing of the West Indies (Chace & Hobbs 1969), subregional lists and keys that could be mentioned, and distributional analyses for decapods of the U.S. east coast (Wenner & Boesch 1979; Wenner & Read 1982). Manning & Holthuis' (1984) description of a new *Geryon* from Florida is an example of more finely drawn biosystematic reanalysis made possible by recent collecting activity.

Joseph Pearson's (1908) monograph on *Cancer pagurus*, the edible crab of Europe, is an early, thorough, artistically illustrated account of gross morphology, histology, larval development, economics, and bionomics, a model for similar studies that followed. For example, Churchill's (1919) treatment of the blue crab, *Callinectes sapidus*, provided similar, though less handsomely illustrated, coverage for that prominent species, and was followed by the work of Cochran (1935) on skeletal musculature and Cronin (1947) on anatomy and histology of the male reproductive system of the species. Van Engel's (1958) review of reproduction, early development, growth, and migration of the blue crab returned to Pearson's comprehensive concept while directed more toward fishery application. Johnson's (1980) histology of the blue crab gave added dimension to such analytical summaries. In this vein, a set of studies on reproduction, molting, and growth of *Cancer borealis*, *C. irroratus*, *Geryon quinquedens*, and *Acanthocarpus alexandri* combined experiment with observation to gain better understanding of the processes studied (Haefner 1976, 1977a, b, 1981; Haefner & Van Engel 1975). Finally, study of visual and accoustical signals in fiddler crabs produced morphological, behavioral, and ecological evidence for reproductive isolation among species of the genus *Uca* (Salmon 1965, 1971; Salmon & Atsaides 1968a, b, 1969; Salmon et al. 1978).

7 SUMMARY

Research on North American crabs began with the exploration and descriptive work following the publication of Linnaeus's *Systema Naturae* in 1758. However, accounts of Native American knowledge and usage of crabs preceded this along with early comparisons of the new fauna to that known in homelands of the explorers. European domination of systematic research in the 18th and early 19th centuries gave way to increasing efforts by Americans as the age of great exploration passed. That effort, along with a search for fisheries resources and the related growth of systematic collections, provided the basis for the research that continues today.

Purely descriptive effort has given way to more analytical biosystematic study as the resource base has grown. Fisheries developed for economically valuable species underwent innovations in exploitation followed by introduction of managed limits as pristine conditions changed. Those changes produced a need for greater understanding and study of the fished species and related organisms. The full array of contemporary biological research on crabs was not necessarily initiated by fisheries needs, but a close connection is evident.

Crustacean research as a whole flowered in the last 50 years, and the part concerned with crabs of North America is an important part of the movement. Growth in numbers of publications, conferences, and workshops expanded exponentially with the information explosion that accompanied our increased ability to analyze data. While we can afford some perspective on the history of earlier periods, analysis of the present is better left to topical comment.

ACKNOWLEDGMENTS

The manuscript benefitted from critical reading by F.A. Chace, Jr., B.B. Collette, R.B. Manning, and anonymous reviewers.

REFERENCES

Appleman, D.E. 1985. James Dwight Dana and Pacific geology. In H.J. Viola & C. Margolis (eds), *Magnificent Voyagers. The US Exploring Expedition, 1838-1842*: pp. 88-117. Washington: Smithsonian Inst. Press.

Bailey, R.F.J. & R.W. Elner. 1989. Northwest Atlantic snow crab fisheries: Lessons in research and management. In J.F. Caddy,(ed.), *Marine Invertebrate Fisheries: Their Assessment and Management*: pp. 261-280. Wiley: New York.

Balss, H. 1940-1945. Decapoda. In *H.G. Bronn's, Klassen und Ordnungen des Tierreichs* (Bd. 5, Abt. 1, B. 7, Lf. 1-7,): pp. 1-1006. Leipzig: Akad. Verlag.

Balss, H. 1957. Decapoda. VIII. Systematik. In *H.G. Bronn's, Klassen und Ordnungen des Tierreichs* (Bd. 5, Abt. 1, B. 7, Lf. 12):pp. 1505-1672. Leipzig: Akad. Verlag.

Bartlett, H.H. 1940. The reports of the Wilkes Expedition, and the work of the specialists in science. *Proc. Amer. Phil. Soc.* 82: 601-719.

Bliss, D.E. (ed. in chief). 1981-1985. *The Biology of Crustacea*, vols. 1-10. New York: Academic Press.

Bowen, S.T., M.L. Davis, S.R. Fenster & G.A. Lindwall. 1980. Sibling species of *Artemia*. In G. Persoone, P. Sorgeloos, O. Roels & E. Jaspers (eds), *The Brine Shrimp Artemia*, Vol. 1: pp. 155-167. Belgium: Universa.

Broad, A.C. 1957a. Larval development of *Palaemonetes pugio. Biol. Bull.* 112: 144-161.

Broad, A.C. 1957b. The relationship between diet and larval development of *Palaemonetes. Biol. Bull.* 112: 162-170.

Brown, R.B. 1971. The development of the Alaskan fishery for Tanner crab, *Chionoecetes* species, with particular reference to the Kodiak area, 1967-1970. *Alaska Dep. Fish Game, Inform. Leafl.* 153: 1-26.

Bullis, H.R., Jr. & J.R. Thompson. 1965. Collections by the exploratory fishing vessels Oregon, Silver Bay, Combat, and Pelican made during 1956-1960 in the southwestern North Atlantic. *US Fish Wildl. Serv. Spec. Sci. Rep. Fish.* No. 510: 1-130.

Butler, T.H. 1960. Maturity and breeding of the Pacific edible crab, *Cancer magister* Dana. *J. Fish. Res. Bd. Can.* 17: 641-646.

Butler, T.H. 1961. Growth and age determination of the Pacific edible crab *Cancer magister* Dana. *J. Fish. Res. Bd. Can.* 18: 873-891.

Caddy, J.F. (ed.). 1989. *Marine Invertebrate Fisheries: Their Assessment and Management.* New York: Wiley.

Calman, W.T. 1909. Crustacea. In R. Lankester (ed.) *A Treatise on Zoology, Part 7, Appendiculata. Third fascicle*: 1-346. London: Adam & Charles Black.

Chace, F.A., Jr. 1990. Mary J. Rathbun (1860-1943). *J. Crust. Biol.* 10: 165-167.

Chace, F.A., Jr. & H.H. Hobbs Jr. 1969. The freshwater and terrestrial decapod crustaceans of the West Indies with special reference to Dominica. Bredin-Archbold-Smithsonian Biological Survey of Dominica. *US Nat. Mus. Bull.* 292: 1-258.

Christiansen, M.E. 1971. Larval development of *Hyas araneus* (Linnaeus) with and without antibiotics (Decapoda, Brachyura, Majidae). *Crustaceana* 21: 307-315.

Christiansen, M.E. & J.D. Costlow Jr. 1975. The effect of salinity and cyclic temperature on larval development of the mud-crab *Rhithropanopeus harrisii* (Brachyura: Xanthidae) reared in the laboratory. *Mar. Biol.* 32: 215-221.

Christy, J.H. 1978. Adaptive significance of reproductive cycles in the fiddler crab *Uca pugilator*: A hypothesis. *Science* 199: 453-455.

Christy, J.H. & S.E. Stancyk. 1982. Timing of larval production and flux of invertebrate larvae in a well-mixed estuary. In V.S. Kennedy (ed.), *Estuarine Comparisons*: pp. 489-503. New York: Academic Press.

Churchill, E.P.J. 1919. Life history of the blue crab. *Bull. Bur. Fish.*, 36: 95-128, pls. 47-55.

Cleaver, F.C. 1949. Preliminary results of the coastal crab (*Cancer magister*) investigation. *Wash. Dep. Fish. Biol. Rep.* 49A: 47-82.

Cochran, D.M. 1935. The skeletal musculature of the blue crab, *Callinectes sapidus* Rathbun. *Smithson. Misc. Collect.* 92(9):1-76.

Conklin, E.G. 1940. Connection of the American Philosophical Society with our first national exploring expedition. *Proc. Amer. Phil. Soc.* 82: 519-549.

Costlow Jr., J.D. 1965. Variability in larval stages of the blue crab, *Callinectes sapidus*. *Biol. Bull.* 128: 58-66.

Costlow Jr., J.D. 1967. The effect of salinity and temperature on survival and metamorphosis of megalops of the blue crab *Callinectes sapidus*. *Helgolaender Wiss. Meeresunters.* 15: 84-97.

Costlow Jr., J.D. & C.G. Bookhout. 1959. The larval development of *Callinectes sapidus* Rathbun reared in the laboratory. *Biol. Bull.* 116: 373-396.

Costlow Jr., J.D. & C.G. Bookhout 1960. A method for developing brachyuran eggs in vitro. *Limnol. Oceanogr.* 5: 212-215.

Costlow Jr., J.D. & C.G. Bookhout 1966. Larval stages of the crab, *Pinnotheres maculatus*, under laboratory conditions. *Chesapeake Sci.* 7: 157-163.

Costlow Jr., J.D. C.G. Bookhout & R. Monroe 1960. The effect of salinity and temperature on larval development of *Sesarma cinereum* (Bosc) reared in the laboratory. *Biol. Bull.* 118: 183-202.

Costlow Jr., J.D. C.G. Bookhout & R. Monroe 1962. Salinity-temperature effects on the larval development of the crab, *Panopeus herbstii* Milne Edwards, reared in the laboratory. *Physiol. Zool.* 35:79-93.

Costlow Jr., J.D. C.G. Bookhout & R.J. Monroe 1966. Studies on the larval development of the crab, *Rhithropanopeus harrisii* (Gould). 1. The effect of salinity and temperature on larval development. *Physiol. Zool.* 39: 81-100.

Cronin, L.E. 1947. Anatomy and histology of the male reproductive system of *Callinectes sapidus* Rathbun. *J. Morph.* 81: 209-239.

Cronin, T.W. 1982. Estuarine retention of larvae of the crab *Rhithropanopeus harrisii*. *Estuarine Coastal Shelf Sci.* 15: 207-220.

Cronin, T.W. & R.B. Forward Jr. 1982. Tidally timed behavior: effects on larval distributions in estuaries. In V.S. Kennedy (ed.), *Estuarine Comparisons*: pp. 505-520. New York: Academic Press.

Dahlstrom, W.A. & P.W. Wild 1983. A history of Dungeness crab fisheries in California. In P.W. Wild & R.N.

Tasto, (eds),MI Life History, Environment, and Mariculture Studies of the Dungeness Crab, *Cancer magister*, with Emphasis on the Central California Fishery Resource: pp. 7-23, *Calif. Dep. Fish Game Fish Bull.* 172.

Dana, J.D. 1851. On the markings of the carapax of crabs. *Amer. J. Sci. Arts* (2)11: 95-99.

Dana, J.D. 1852. Crustacea. In *United States Exploring Expedition During the Years 1838, 1839, 1840, 1841, 1842, Under the Command of Charles Wilkes, U.S.N. Vol. 13. With a folio atlas (1855) of 96 plates.* Philadelphia: C. Sherman.

De Saint-Laurent, M. 1980a. Sur la classification et la phylogénie des Crustacés Décapodes Brachyoures. I. Podotremata Guinot, 1977, et Eubrachyura sect. nov. *C.R. Hebd. Seances Acad. Sci. (Paris)* ser. D, 290: 1265-1268.

De Saint-Laurent, M. 1980b. Sur la classification et la phylogénie des Crustacés Décapodes Brachyoures. Heterotremata et Thoracotremata Guinot, 1977. *C.R. Hebd. Seances Acad. Sci. (Paris)* ser. D, 290: 1317-1320.

Diaz, H. & J.D. Costlow 1972. Larval development of *Ocypode quadrata* (Brachyura: Crustacea) under laboratory conditions. *Mar. Biol.* 15: 120-131.

Epifanio, C.E. 1979. Larval decapods (Arthropoda: Crustacea: Decapoda). In C.W. Hart & S.L.H. Fuller (eds), *Pollution ecology of estuarine invertebrates*: pp. 259-292. New York: Academic Press.

Ehrhardt, N.M. & V.R. Restrepo 1989. The Florida stone crab fishery: a reusable resource? In J.F. Caddy (ed.), *Marine Invertebrate Fisheries: Their Assessment and Management*: pp. 225-240. New York: Wiley.

Forward Jr., R.B. 1977. Occurrence of a shadow response among brachyuran larvae. *Mar. Biol.* 39: 331-341.

Forward Jr., R.B. & J.D. Costlow 1974. The ontogeny of phototaxis by larvae of the crab *Rhithropanopeus harrisii. Mar. Biol.* 26: 27-33.

Garth, J.S. 1958. Brachyura of the Pacific coast of America, Oxyrhyncha. *Allan Hancock Pac. Exped.* 21(1):1-499, (2):501-854, pls. A-Z, 1-55.

Glaessner, M.F. 1969. Decapoda. In R.C. Moore (ed.), Treatise on Invertebrate Paleontology, Pt. R, Arthropoda 4, Vol. 2: pp. R399-R533, R626-R628. Lawrence: Univ. of Kansas & Geol. Soc. Amer.

Guinot, D. 1967. Recherches préliminaires sur les groupements natureles chez les Crustacés Décapodes Brachyoures. II. Les anciens genres *Micropanope* Stimpson et *Medaeus* Dana. *Bull. Mus. Nat. Hist. Natur. Paris* (2)39:345-374.

Guinot, D. 1968. Recherches préliminaires sur les groupements natureles chez les Crustacés Décapodes Brachyoures. IV. Observations sur quelques genres de Xanthidae. *Bull. Mus. Nat. Hist Natur. Paris* (2)39(4, for 1967): 695-727.

Guinot, D. 1969a. Sur divers Xanthidae notamment sur *Actaea* de Haan et *Paractaea* gen. nov. (Crustacea Decapoda Brachyura). *Cah. Pac.* 3: 223-267.

Guinot, D. 1969b. Recherches préliminaires sur les groupements natureles chez les Crustacés Décapodes Brachyoures. VII. Les Goneplacidae. *Bull. Mus. Nat. Hist. Natur. Paris* (2)41: 241-265.

Guinot, D. 1969c. Recherches préliminaires sur les groupements natureles chez les Crustacés Décapodes Brachyoures. VII. Les Goneplacidae (suite). *Bull. Mus. Nat. Hist. Natur. Paris* (2)41: 507-528.

Guinot, D. 1969d. Recherches préliminaires sur les groupements natureles chez les Crustacés Décapodes Brachyoures. VII. Les Goneplacidae (suite et fin). *Bull. Mus. Nat. Hist. Natur. Paris* (2)41: 688-724.

Guinot, D. 1971. Recherches préliminaires sur les groupements natureles chez les Crustacés Décapodes des Brachyoures. VIII. Synthèse et Bibliographie. *Bull. Mus. Nat. Hist. Natur. Paris* (2)42(5, for 1970): 1063-1090.

Guinot, D. 1978. Principes d'une classification évolutive des Crustacés Décapodes Brachyoures. *Bull. Biol. Fr. Belg.* n.s. 112: 211-292.

Gurney, R. 1942. *Larvae of Decapod Crustacea.* London: Ray Society.

Haefner, P.A., Jr. 1976. Distribution, reproduction and moulting of the rock crab, *Cancer irroratus* Say, 1817, in the mid-Atlantic bight. *J. Nat. Hist.* 10:377-397.

Haefner Jr., P.A. 1977a. Reproductive biology of the female deep-sea red crab, *Geryon quinquedens*, from the Chesapeake bight. *Fish. Bull. US* 75:91-102.

Haefner Jr., P.A. 1977b. Aspects of the biology of the Jonah crab, *Cancer borealis* Stimpson, 1859 in the mid-Atlantic bight. *J. Nat. Hist.* 11: 303-320.

Haefner Jr., P.A. 1981. Morphometry, reproductive biology, and diet of *Acanthocarpus alexandri* Stimpson, 1871 (Decapoda, Brachyura) in the Middle Atlantic bight. *J. Crust. Biol.* 1: 348-357.

Haefner Jr., P.A. & W.A. Van Engel 1975. Aspects of molting, growth and survival of male rock crabs, *Cancer irroratus*, in Chesapeake Bay. *Chesapeake Sci.* 16: 253-265.

Hariot, T. 1588. *A Briefe and True Report of the New Found Land of Virginia: ... etc.* London. (Pages unnumbered, but 46 by actual count in facsimilie No. 278 of 315 in the Clements copy printed April 1931, Ann Arbor, Michigan.)

Hatfield, S.E. 1983. Intermolt staging and distribution of Dungeness crab, *Cancer magister*, megalopae. In P.W. Wild & R.N. Tasto (eds), *Life History, Environment, and Mariculture Studies of the Dungeness Crab*, Cancer magister, *With Emphasis on the Central California Fishery Resource*: pp. 85-96. *Calif. Dep. Fish & Game, Fish Bull.* 172.

Haynes, E. 1973. Descriptions of prezoeae and Stage I zoeae of *Chionoecetes bairdi* and *C. opilio* (Oxyrhyncha, Oregoniinae). *Fish. Bull. US* 71:769-775.

Hedgpeth, J.W., with appendix by W.L. Schmitt 1945. The United States Fish Commission steamer Albatross. *Amer. Neptune* 5: 5-26.

Hyman, O.W. 1924. Studies on larvae of crabs of the family Pinnotheridae. *Proc. US Nat. Mus.* 64(7)(for 1925): 1-7, 6 pls.

Johns, D.M. & W.H. Lang 1977. Larval development of the spider crab *Libinia emarginata* (Majidae). *Fish. Bull. U.S.* 75: 831-841.

Johnson, P.T. 1980. *Histology of the Blue Crab,* Callinectes sapidus, *a Model for the Decapoda.* New York: Praeger Special Studies.

Lawson, J. 1714. *History of North Carolina ... etc.* London: Taylor and Baker, 3rd ed., 1960, F.L. Harriss, ed., Richmond: Garrett & Massie.

Lindner, M.J. & W.W. Anderson 1956. Growth, migrations, spawning and size distribution of shrimp *Penaeus setiferus*. *Fish. Bull. US* 56: 553-645.

Linnaeus, C. 1758. *Systema Naturae*, ed. 10, vol. 1, Stockholm.

Manning, R.B. 1990. History of the Crustacean Society 1979-1989. *J. Crust. Biol.* 10: 735-750.

Manning, R.B. & L.B. Holthuis 1984. *Geryon fenneri*, a new deep-water crab from Florida (Crustacea: Decapoda: Geryonidae). *Proc. Biol. Soc. Wash.* 97: 666-673.

Methot, R.D. 1989. Management of a cyclic resource: the Dungeness crab fisheries of the Pacific coast of North America. In J.F. Caddy (ed.), *Marine Invertebrate Fisheries: Their Assessment and Management*: pp. 205-223. New York:Wiley.

Milliken, M.R. & A.B. Williams 1984. Synopsis of biological data on the blue crab, *Callinectes sapidus* Rathbun. *NOAA Tech. Rep. NMFS 1, FAO Fish. Synop.* 138: 1-39.

Nichols, P.R. & P.M. Keney 1963. Crab larvae (*Callinectes*), in plankton collections from cruises of M/V *Theodore N. Gill*, south Atlantic coast of the United States, 1953-54. *US Fish Wildl. Serv. Spec. Sci. Rep. Fish.* 448: 1-14.

Oregon II. 1967. Station lists. U.S. Fish Wildl. Serv., Bur. Comm. Fish, et seq., Pascagoula, MS.

Pearson, J.C. 1948. Fluctuations in the abundance of the blue crab in Chesapeake Bay. *US Fish Wildl. Serv. Res. Rep.* 14: 1-26.

Pearson, J. 1908. *Cancer* (The edible crab). (Liverpool Marine Biological Committee Memoir No. XVI). *Proc. Trans. Liverpool Biol. Soc.* 22: 291-499, pls. I-XIII.

Poole, R.L. 1966. A description of laboratory-reared zoeae of *Cancer magister* Dana, and megalopae taken under natural conditions (Decapoda, Brachyura). *Crustaceana* 11: 83-97.

Porter, H.J. 1960. Zoeal stages of the stone crab, *Menippe mercenaria* Say. *Chesapeake Sci.* 1:168-177.

Rathbun, M.J. 1918. The grapsoid crabs of America. *US Nat. Mus. Bull.* 97: 1-461, 161 pls.

Rathbun, M.J. 1925. The spider crabs of America. *US Nat. Mus. Bull.* 129: 1-613, 283 pls.

Rathbun, M.J. 1930. The cancroid crabs of America of the families Euryalidae, Portunidae, Atelecyclidae, Cancridae and Xanthidae. *US Nat. Mus. Bull.* 152: 1-609, 230 pls.

Rathbun, M.J. 1937. The oxystomatous and allied crabs of America. *US Nat. Mus. Bull.* 166: 1-278, 86 pls.

Rathbun, R. 1884. Crustaceans. In G. B. Goode, *The Fisheries and Fishery Industries of the United States. Sec. I, Pt. 5, Natural History of Useful Aquatic Animals*: pp. 759-830, pls. 260-275. Washington, D.C.: US Comm. Fish and Fisheries.

Rathbun, R. 1887. Part XXI. The crab, lobster, crayfish, rock lobster, shrimp, and prawn fisheries. 1. The crab fisheries. In G.B. Goode et al., *Section V. History and Methods of the Fisheries, Vol. II. The Fisheries and*

Fishery Industries of the United States: pp. 629-658, atlas of plates. Washington, D.C.: US Comm. Fish and Fisheries.

Reilly, P.N. 1983. Dynamics of Dungeness crab, *Cancer magister*, larvae off central and northern California. In P.W. Wild & R.N. Tasto (eds), *Life History, Environment, and Mariculture Studies of the Dungeness Crab, Cancer magister, With Emphasis on the Central California Fishery Resource*: pp. 57-84. *Calif. Dep. Fish Game, Fish Bull.* 172.

Rice, A.L. 1964. The metamorphosis of a species of *Homola* (Crustacea, Decapoda: Dromiacea). *Bull. Mar. Sci. Gulf Carib.* 14: 221-238.

Rice, A.L. 1980. Crab zoeal morphology and its bearing on the classification of the Brachyura. *Trans. Zool. Soc. Lond.* 35: 271-424.

Rice, A.L. 1981. Crab zoeae and brachyuran classification: A re-appraisal. *Bull. Brit. Mus. (Nat. Hist.) Zool.* 40:287-296.

Rice, A.L. & A. J. Provenzano Jr. 1966. The larval development of the West Indian sponge crab *Dromidia antillensis* (Decapoda: Dromiidae). *J. Zool. Soc. Lond.* 149:297-319.

Roberts Jr., M.H. 1969. Larval development of *Bathynectes superba* (Costa) reared in the laboratory. *Biol. Bull.* 137:338-351.

Salmon, M. 1965. Waving display and sound production in the courtship behavior of *Uca pugilator*, with comparisons to *U. minax* and *U. pugnax*. *Zoologica* 50: 123-150, pls. 1-5.

Salmon, M. 1971. Signal characteristics and acoustic detection by the fiddler crabs, *Uca rapax* and *Uca pugilator*. *Physiol. Zool.* 44: 210-224.

Salmon, M. & S.P. Atsaides 1968a. Visual and acoustical signalling during courtship by fiddler crabs (genus *Uca*). *Amer. Zool.* 8: 623-639.

Salmon, M. & S.P. Atsaides 1968b. Behavioral, morphological and ecological evidence for two new species of fiddler crabs (genus *Uca*) from the Gulf coast of the United States. *Proc. Biol. Soc. Wash.* 81: 275-290.

Salmon, M. & S.P. Atsaides 1969. Sensitivity to substrate vibration in the fiddler crab, *Uca pugilator*. *Anim. Behav.* 17: 68-76, pl. 7.

Salmon, M., G. Hyatt, K. McCarthy & J.D. Costlow Jr. 1978. Display specificity and reproductive isolation in the fiddler crabs, *Uca panacea* and *U. pugilator*. *Z. Tierpsychol.* 48:251-276.

Sandoz, M. & S.H. Hopkins 1947. Early life history of the oyster crab *Pinnotheres ostreum* (Say). *Biol. Bull.* 93:250-258.

Schmitt, W.L. 1921. The marine decapod Crustacea of California. *Univ. Calif. Pub. Zool.* 23:470 pp., 50 pls.

Schmitt, W.L. 1931. Crustaceans. *Smithsonian Scientific Series* 10:87-248, 40 pls.

Schmitt, W.L. 1965. *Crustaceans*. Ann Arbor: University of Michigan Press.

Schmitt, W.L. 1973. Mary J. Rathbun, 1860-1943. *Crustaceana* 24:283-297.

Schram, F.R. 1986. *Crustacea*. New York: Oxford Univ. Press.

Smith, G. & W.F.R. Weldon 1909, reprinted 1923. Crustacea. In S.F. Harmer & A.E. Shipley (eds.), *The Cambridge Natural History* 4: 1-217. London: McMillan.

Snow, C.D. & J.R. Nielsen 1966. Premating and mating behavior of the Dungeness crab (*Cancer magister* Dana). *J. Fish. Res. Bd. Can.* 23: 1319-1323.

Springer, S. & H.R. Bullis Jr. 1956. Collections by the Oregon in the Gulf of Mexico. List of crustaceans, mollusks, and fishes identified from collections made by the exploratory fishing vessel *Oregon* in the Gulf of Mexico and adjacent seas 1950 through 1955. *US Fish Wildl. Serv. Spec. Sci. Rep. Fish.* 196: 1-134.

Squires, H.J. 1965. Decapod crustaceans of Newfoundland, Labrador and the Canadian eastern Arctic. *Fish. Res. Bd. Can., Manuscr. Rep. Ser. (Biol.)* 810: 1-212.

Squires, H.J. 1990. Decapod Crustacea of the Atlantic coast of Canada. *Can. Bull. Fish. Aquat. Sci.* 221: 1-532.

Sulkin, S.D. & C.E. Epifanio 1975. Comparison of rotifers and other diets for rearing early larvae of the blue crab, *Callinectes sapidus* Rathbun. *Estuarine Coastal Mar. Sci.* 3: 109-113.

Sulkin, S.D. & K. Norman 1976. A comparison of two diets in the laboratory culture of the zoeal stages of the brachyuran crabs *Rhithropanopeus harrisii* and *Neopanope* sp. *Helgoländer Wiss. Meeresunters.* 28: 183-190.

Sulkin, S.D. & W. Van Heukelem 1982. Larval recruitment in the crab *Callinectes sapidus* Rathbun: An amendment to the concept of larval retention in estuaries. In V.S. Kennedy, (ed.), *Estuarine Comparisons*: pp. 459-475. New York: Academic Press.

Tanner, Z.L. 1884. Report on the construction, and work in 1880, of the Fish Commission steamer Fish Hawk. *US Comm. of Fish and Fisheries. Part IX . Rept. of the Comm.* 1881: 3-53, pls. 1-18.

Trask, T. 1970. A description of laboratory-reared larvae of *Cancer productus* Randall (Decapoda, Brachyura) and a comparison to larvae of *Cancer magister* Dana. *Crustaceana* 18: 133-146.

Van Engel, W.A. 1958. The blue crab and its fishery in Chesapeake bay. Part I. Reproduction, early development, growth and migration. *Comm. Fish. Rev.* 20: 6-17.

Van Engel, W.A. 1962. The blue crab and its fishery in Chesapeake bay. Part 2. Types of gear for hard crab fishing. *Comm. Fish. Rev.* 24: 1-10.

Waterman, T.H. (ed.). 1960. The Physiology of Crustacea. Vol. I. Metabolism and Growth. New York: Academic Press.

Waterman, T.H. (ed.). 1961. The Physiology of Crustacea. Vol. II. Sense Organs, Integration, and Behavior. New York: Academic Press.

Weiss, H.B. & G. M. Ziegler 1931. Thomas Say, Early American Naturalist. Springfield: Charles C. Thomas.

Wenner, E.L. & D.F. Boesch 1979. Distribution patterns of epibenthic decapod Crustacea along the shelf-slope coenocline, Middle Atlantic Bight, USA. *Bull. Biol. Soc. Wash.* 3:106-133.

Wenner, E.L. & T. Read 1982. Seasonal composition and abundance of decapod crustacean assemblages from the South Atlantic Bight, USA. *Bull. Mar. Sci.* 32: 181-206.

Wild, P.W. & R.N. Tasto (eds) 1983. Life history, environment, and mariculture studies of the Dungeness crab, *Cancer magister*, with emphasis on the central California fishery resource. *Calif. Dep. Fish Game, Fish Bull.* 172: 1-352.

Williams, A.B. 1965. Marine decapod crustaceans of the Carolinas. *Fish. Bull. US* 65: 1-298.

Williams, A.B. 1984. Shrimps, Lobsters, and Crabs of the Atlantic Coast of the Eastern United States, Maine to Florida. Washington, D.C.: Smithsonian Inst. Press.

Williams, A.B. & T.W. Duke. 1979. Crabs (Arthropoda: Crustacea: Decapoda: Brachyura). In C.W. Hart & S.L.H. Fuller (eds), *Pollution Ecology of Estuarine Invertebrates*: pp. 171-233. New York: Academic Press.

Williams, A. B. & D.L. Felder. 1986. Analysis of stone crabs: *Menippe mercenaria* (Say), restricted, and a previously unrecognized species described (Decapoda: Xanthidae). *Proc. Biol. Soc. Wash.* 99: 517-543.

Williams, A.B., L.G. Abele, D.L. Felder, H.H. Hobbs Jr., R.B. Manning, P.A. McLaughlin, & I. Pérez Farfante. 1989. Common and scientific names of aquatic invertebrates from the United States and Canada: Decapod crustaceans. *Amer. Fish. Soc. Spec. Pub.* 17: 1-77.

Highlights in the history of research on freshwater Anostraca Crustacea in North America

Ralph W. Dexter (deceased)
Department of Biological Sciences, Kent State University, Kent, Ohio, USA

1 INTRODUCTION

From the great bulk of published research on the Anostraca Crustacea of North America, I have selected references that to me indicate significant additions to our knowledge, realizing that other authors would likely make a somewhat different selection. Time and space do not permit complete coverage, and giving details is virtually impossible. This review, then, is merely an outline sketch and hopefully a guide to future research and the literature. References in the works quoted here lead to the greater part of publications on North American Anostraca.

2 GENERAL STUDIES, MANUALS, AND KEYS

A.S. Packard Jr. (1883) published a monograph on the phyllopod Crustacea of North America, giving an outline of their general biology and reviewing all North American species to date. Daday (1910) produced the first comprehensive, systematic monograph for the Anostraca, including all North American species known up to that time. Pearse (1918) published a synopsis of all North American phyllopods as chapter 21 in the volume on *Freshwater Biology* edited by Ward & Whipple (Dexter (1959) revised the Anostraca in the second edition edited by Edmondson). Creaser (1935) likewise prepared a chapter on the phyllopods for *A Manual of the Common Invertebrate Animals* by Pratt (revised in 1948).

F. Linder (1941) included North American species in his often cited general review of the morphology and taxonomy of the Anostraca. Pennak (1953) included a section on anostracans in his volume on the *Freshwater Invertebrates of the United States*. Dexter (1961) prepared a section on the Branchiopoda for the *Encyclopedia of the Biological Sciences*. Belk (1975) published an updated key to the Anostraca of North America, and a review of the Branchiopoda (Belk 1982) in Parker's *Synopsis and Classification of Living organisms*.

3 TAXONOMY AND DESCRIPTION OF NEW SPECIES

A total of 24 authors have described species of Anostraca from the US and Canada. The following ten have described more than one species: A.S. Packard (4); S.A. Forbes (2); J.A. Ryder (2); E.P. Creaser (2); J.E. Lynch (4); R.W. Dexter (3); W.G. Moore (2); L.L. Eng, D. Belk, and C.H. Erikson (4, plus another 1 for Belk).

Some of the major taxonomic works have been published by Verrill (1869, 1870); Packard (1874a, b, 1877); Forbes (1876); Ryder (1879a, b), Holmes (1910); Creaser (1930a, b); Dexter (1953, 1956); Lynch (1937, 1958, 1960, 1964, 1972); Moore (1966); Sissom (1976); Belk (1979); and Eng et al. (1990) who recently revised the classification and distribution of records for California. Brtek (1964, 1965, 1966a, b) revised the classification of several species from North America and introduced nine new taxa. Mackin (1952) unfortunately introduced some taxonomic problems that were cleared up by Lynch (1964).

4   GEOGRAPHICAL DISTRIBUTION AND LOCALITY RECORDS

In addition to the general treatises already mentioned, many writers have added new localities and extensions of range for a wide variety of species throughout North America. Following are 25 selected examples: Underwood (1886); Hay (1892); Pearse (1912); Dodds (1915); Johansen (1921, 1922); Van Cleave (1928); Creaser (1929); Ferguson (1935); Mackin (1939); Coopey (1946); Leonard & Ponder (1949); Daggy (1952); Dexter (1953); Wiese (1957); Cole (1959); McCarraher (1959); Reed (1963); Prophet (1963a); Moore & Young (1964); Hartland-Rowe (1965); Sissom (1976); Belk (1977a); Belk & Milne (1984); and Eng et al. (1990). The latter is the most comprehensive zoogeographic study of a regional anostracan fauna detailing records for the state of California wherein occur 17 species, (including 6 endemic), equaling nearly 40% of all species known from North America.

5   LIFE HISTORY AND ECOLOGY

Many authors have described in more or less detail the life history and general ecology of freshwater Anostraca. Some of the more notable ones are as follows. Packard (1878) gave an early account of *Eubranchipus vernalis* in ponds of eastern Massachusetts in the wintertime, Shantz (1905) recorded habits of North American species of *Branchinecta* in Colorado, Weaver (1943) described the life cycle of *E. vernalis* in Ohio, while Dexter & Ferguson (1943) did the same for *E. serratus* in Illinois. Dexter & Sheary (1943), Dexter (1946), and Dexter & Kuehnle (1948, 1951) published a series of papers on *E. vernalis* in Ohio.

Coopey (1950) described the life history of *E. oregonus* while Moore (1951, 1955a) did the same for *Streptocephalus seali* in Louisiana. *B. coloradensis* was included in Needer & Pennak's (1955) study of an alpine pond in Colorado, while Moore (1959) gave field observations of *E. moorei* Brtek, 1966, in Louisiana. Prophet (1959) studied a winter population of *S. seali* in Kansas. Later, Prophet (1963b) presented information on physical-chemical characteristics of habitats and seasonal occurrence for *Branchinecta lindahli, Eubranchipus serratus, Streptocephalus seali, S. texanus,* and *Thamnocephalus platyurus* in Oklahoma and Kansas. Moore (1963) discussed interspecific relationships between *E. moorei* and *S. seali* in Louisiana. Proctor & Malone (1965) gave evidence of dispersal of fairy shrimps, along with other aquatic organisms, through the intestinal tract of birds. Hartland-Rowe (1966, 1967) studied fairy shrimps and their ecology in Canada. Dexter (1967) reviewed his longterm study of seasonal and annual population changes of the Anostraca over many years in temporary ponds of Illinois and Ohio. Horne (1967) in Wyoming and Sublette & Sublette (1967) in New Mexico and Texas studied physio-chemical factors related to distribution and occurrence of phyllopods, and McCarraher (1970) showed ecological relationships of fairy shrimps in alkaline ponds in

Nebraska. Moore (1970) made limnological studies of anostracan ponds in Louisiana, and Hartland-Rowe (1972) did the same for Canada. Daborn (1975, 1976a, b, 1977a, b) studied the life history of five species in Alberta, while Retallack & Clifford (1980) did the same for three species, also in Alberta. Modlin (1982) compared *E. vernalis* with *E. holmani* in Alabama, while Donald (1983) studied five species inhabiting a single pond over 14 years in Alberta.

6 MORPHOLOGY, PHYSIOLOGY, AND REPRODUCTIVE BIOLOGY

Morphological and physiological studies began with early work by Herrick (1885) who worked out metamorphosis of a species that Moore (personal communication) believes was *E. bundyi* rather than *E. holmani*, as reported by Herrick. McGinnis (1911) studied the reactions of *E. serratus* to light, heat, and gravity. Pearse (1913) studied the behavior of the same species (which was reported as *E. dadayi*). The most thorough morphological study, which was on external development of *Linderiella occidentalis*, was made by Heath (1924). Baker & Rosof (1927) described in detail spermatogenesis in *E. vernalis*. Heat death for *S. seali* was investigated by Moore (1955b). H.J. Linder (1959, 1960) reported on the structure and histochemistry of the reproductive system of *E. bundyi*. Gaudin (1960) studied egg production of streptocephalids, Moore and Ogren (1962) studied the breeding behavior of *E. moorei*, and Baqai (1963) studied the postembryonic development of *S. seali*. Broch (1965) described the mechanism of hatching for *E. bundyi* as this relates to environmental parameters. He also investigated the osmotic adaptation of three species to saline water (Broch 1969, 1988). Moore & Burn (1968) worked on lethal oxygen thresholds, and Knight et al. (1975) studied the effect of temperature on oxygen consumption. Wiman (1979a, b, 1981) reported on hybridization and mating behavior in streptocephalids. McLaughlin (1980) included the Anostraca in her study of comparative morphology of the Crustacea, as did Eriksen & Brown (1980) in their study of comparative respiratory physiology and Schram (1986) in his survey of Crustacea. Modlin (1983) and Hazelwood & Hazelwood (1985) studied the effect of temperature on oxygen consumption of several species. Daborn (1979) reported on limb structure and sexual dimorphism, and Belk (1984) showed the relation of antennal appendages to reproductive success and reviewed (Belk 1991) reproductive behavior in the Anostraca.

7 HATCHING STUDIES

Related to life history and ecological studies of the Anostraca is the intriguing problem of hatching of the eggs (actually the cysts). Hay & Hay (1889) showed at an early date the difficulty of hatching cysts of *E. vernalis* and the importance of periodic drying as a stimulant for a successful hatch, but contrary to common belief, showed that freezing was not necessary, although helpful. Norton (1909) also succeeded in getting a hatch, although many attempts by others before and after that time were not very successful. Creaser (1931) pointed out that a resting period of unknown length is required before hatching in phyllopods, although in some species drying was not necessary. Kelly (1956) made a comparative study of natality in eight species of phyllopods, including four Anostraca that with one exception were all native to Missouri. He reviewed the literature on this matter thoroughly and from his own studies concluded there were many differences within and between species in regard to hatching results. Two species produced two types of eggs, one hatching immediately and the other after drying. Freezing or low oxygen tension were also effective parameters. While some did not require either drying or freezing, others gave a better hatch with either, or both drying and freezing. Most investigators have found that drying following a long period of rest is beneficial, if not necessary for hatching, and that low oxygen pressure resulted in a better hatch. In general, drying was thought to be the most effective agent, but cysts of *E. serratus* and of *E. vernalis* seldom hatched, regardless of treatment in the laboratory. Moore (1957) found cysts of *S. seali* to hatch in laboratory cultures without drying. Gaudin's (1960) report on the large number of cysts produced by that species may explain its wide geographic distribution. Prophet (1963c, d) studied egg production and hatching in laboratory cultures of five species with reference to temperature, drying, dilution of culture medium, and aging. Broch (1965) also studied the hatching process in *E. bundyi* with reference to the ephemeral habitat and concluded that prehatching of the egg occurred, and that through physiological mechanisms, phases of the life cycle have become synchronized with seasonal changes in the temporary pond habitat. His approach was evolutionary adaptation to the temporary pond, and his study is one of the most thorough involving embryological development and the relation of life cycle of a single species to its ephemeral environment. Moore (1967) studied hatching of *S. seali* in reference to various environmental factors and related them to seasonal differences in the natural life cycle. Hutchinson's (1967) review of the literature lead him to conclude that changes in osmotic pressure are responsible for the hatch of phyllopod eggs. Horne (1971) studied the effects of temperature and oxygen concentration on *Branchinecta packardi* and *S. texanus*. Brown & Carpelan (1971), in studying *B. mackini*, concluded that salinity and oxygen concentration control hatching in arid regions, whereas temperature and oxygen tension control hatching in humid regions. Dexter (1973), working with mud collected in New Mexico and California, found that repeated soaking and drying and re-soaking of the same sample of mud over a long period of time would bring out a succession of hatchings immediately following almost every period of drying (4 spp. Anostraca, 4 spp. Conchostraca, 1 sp. Notostraca). While aging and drying were stimuli, they did not result in a complete hatch at any one time of all of the cysts present in each sample of mud. It is obvious that we do not yet have a clear and thorough understanding of the mechanism of hatching for all species of Anostraca. Variation among the species is certain, and the matter is probably a complex of environmental and physiological parameters and their interactions.

8 EVOLUTION

Relatively little has been published on the evolution of the Anostraca. Gissler (1881) was a pioneer in the field, but Daday (1910) did the first detailed study involving the evolution of the group, followed by the work of F. Linder (1941) and Tasch (1963), who both presented detailed interpretations of branchiopod evolution. Belk (1977b), Belk et al. (1990) traced the evolution of egg size strategies for the Anostraca, while Wiman (1979b) was concerned with speciation in the genus *Streptocephalus*. The latest general contribution is from Spicer (1985), who presents a phylogenetic analysis of streptocephalids from North America. Much remains to be done.

REFERENCES

Baker, R.C. & J.A. Rosof 1927. Spermatogenesis in *Branchipus vernalis*. *Ohio J. Sci.* 27: 175-186; 28: 50-68; 28: 315-328.

Baqai, I.U. 1963. Studies on the postembryonic development of the fairy shrimp *Streptocephalus seali* Ryder. *Tulane Stud. in Zool.* 10: 91-120.

Belk, D. 1975. Key to the Anostraca (fairy shrimps) of North America. *Southwest Nat.* 20: 91-104.

Belk, D. 1977a. Zoogeography of the Arizona fairy shrimps (Crustacea: Anostraca). *J. Ariz. Acad. Sci.* 12:70-78.

Belk, D. 1977b. Evolution of egg size strategies in fairy shrimps. *Southwest Nat.* 22: 99-105.

Belk, D. 1979. *Branchinecta potassa* new species (Crustacea: Anostraca) a new fairy shrimp from Nebraska, U.S.A. *Southwest Nat.* 24: 93-96.

Belk, D. 1982. Branchiopoda. In: S.P. Parker (ed.), *Synopsis and Classification of Living Organisms. Vol. 2*: pp. 174-180. New York: McGraw Hill.

Belk, D. 1984. Antennal appendages and reproductive success in the Anostraca. *J. Crust. Biol.* 4: 66-71.

Belk, D. 1991. Anostracan mating behavior: a case of scramble- competition polygyny. In R.T. Bauer & J.W. Martin (eds), *Crustacean Sexual Biology*: pp 111-125. New York: Columbia Univ. Press.

Belk, D., G. Anderson, & S. Hsu 1990. Additonal observations on variations in egg size among populations of *Streptocephalus seali* (Anostraca). *J. Crust. Biol.* 10: 128-133.

Belk, D. & W.D. Milne Jr. 1984. Anostraca in Alabama. *J. Alabama Acad. Science* 55: 245-247

Broch, E.S. 1965. Mechanism of adaptation of the fairy shrimp *Chirocephalus bundyi* Forbes to the temporary pond. *Cornell Univ. Agr. Exper. Sta. Memoir* 392: 1-48.

Broch, E.S. 1969. The osmotic adaptation of the fairy shrimp *Branchinecta campestris* Lynch to saline astatic waters. *Limnol. Oceanogr.* 14: 485-492.

Broch, E.S. 1988. Osmoregulatory patterns of adaptation to inland astatic waters by two species of fairy shrimp, *Branchinecta gigas* Lynch and *Branchinecta mackini* Dexter. *J. Crust. Biol.* 8: 383-391.

Brown, L.R. & L.H. Carpelan 1971. Egg hatching and life history of a fairy shrimp *Branchinecta mackini* Dexter (Crustacea: Anostraca) in a Mohave Desert playa (Rabbit Dry Lake). *Ecology* 52: 41-54.

Brtek, J. 1964. A new genus and family of Order Anostraca. *Annotationes Zoologicae et Botanicae, Slovenske Narodne Museum islave Brat* 7: 1-7.

Brtek, J. 1965. Eine neue Art und eine neue Gattung der Ordnung Anostraca. *Annotationes Zoologicae et Botanicae, Slovenske Narodne Museum islave Brat* 21: 1-7.

Brtek, J. 1966a. Einiga Notizen zur Taxonomie der Familie Chirocephalidae Daday 1910. *Annotationes Zoologicae et Botanicae*, Slovenske Narodne Museum islave Brat No. 33: pp. 1-65.

Brtek, J. 1966b. *Eubranchipus (Creaseria) moorei. Annotationes Zoologicae et Botanicae, Slovenske Narodne Museum islave Brat* 36: 1-8.

Cole, G.A. 1959. A summary of our knowledge of Kentucky crustaceans. *Trans. Kentucky Acad. Sci.* 20: 66-79.

Coopey, R.W. 1946. Phyllopods of southeastern Oregon. *Trans. Amer. Micros. Soc.* 65: 338-345.

Coopey, R.W. 1950. The life history of the fairy shrimp *Eubranchipus oregonus. Trans Amer. Micros. Soc.* 69: 125-132.

Creaser, E.P. 1929. The phyllopoda of Michigan. *Papers Michigan Acad. Sci. Arts, and Letters* 11: 381-388.

Creaser, E.P. 1930a. Revision of the phyllopod genus *Eubranchipus*, with the description of new species. *Occ. Pap. Mus. Zool., Univ. Mich.* 208: 1-13.

Creaser, E.P. 1930b. The North American phyllopods of the genus *Streptocephalus. Occ. Pap. Mus. Zool., Univ. Mich.* 217: 1-10.

Creaser, E.P. 1931. North American phyllopods. *Science* 74: 267-268.

Creaser, E.P. 1935. Phyllopoda. In H.S. Pratt (ed.), *A Manual of the Common Invertebrate Animals*: pp. 373-377. Philadelphia: Blakiston.

Daborn, G.R. 1975. Life history and energy relations of the giant fairy shrimp *Branchinecta gigas* Lynch 1937 (Crustacea: Anostraca). *Ecology* 56: 1025-1039.

Daborn, G.R. 1976a. The life cycle of *Eubranchipus bundyi* (Forbes) (Crustacea: Anostraca) in a temporary vernal pond of Alberta. *Can. J. Zool.* 54: 193-201.

Daborn, G.R. 1976b. Occurrence of an arctic fairy shrimp *Polyartemialla hazeni* (Murdock) 1884 (Crustacea: Anostraca) in Alberta and Yukon Territory. *Can. J. Zool.* 54: 2026-2028.

Daborn, G.R. 1977a. The life history of *Branchinecta mackini* Dexter (Crustacea: Anostraca) in an argillotrophic lake of Alberta. *Can J. Zool.* 55: 161-168.

Daborn, G.R. 1977b. On the distribution and biology of an arctic fairy shrimp *Artemiopsis stefanssoni* Johansen, 1921 (Crustacea: Anostraca). *Can J. Zool.* 55: 280-287.

Daborn, G.R. 1979. Limb structure and sexual dimorphism in the Anostraca (Crustacea). *Can. J. Zool.* 57: 894-900.

Daday, E. 1910. Monographie systematique des Phyllopodes Anostraces. *Annales des sciences naturelles, Zoologie* (9)11: 91-489.

Daggy, T. 1952. Records of fairy shrimps in North Carolina. *J. Elisha Mitchell Sci. Soc.* 68: 143.

Dexter, R.W. 1946. Further studies on the life history and distribution of *Eubranchipus vernalis* (Verrill). *Ohio J. Sci.* 46: 31-44.

Dexter, R.W. 1953. Studies on North American fairy shrimps with the description of two new species. *Amer. Midl. Nat.* 49: 751-771.

Dexter, R.W. 1956. A new fairy shrimp from western United States with notes on other North American species. *J. Wash. Acad. Sci.* 46: 159-165.

Dexter, R.W. 1959. Anostraca. In W.T. Edmondson (ed.), *Ward and Whipple's Fresh-Water Biology,* 2nd ed.: 558-571. New York: Wiley.

Dexter, R.W. 1961. Branchiopoda. In P. Gray (ed.), *The Encyclopedia of the Biological Sciences*: pp. 175-176. New York: Reinhold.

Dexter, R.W. 1967. Annual Changes in Populations of Anostraca Crustacea. *Proc. Symposium on Crustacea Mar. Biol. Ass. India.* Part 2: 568-576.

Dexter, R.W. 1973. Persistence of viability in the eggs of certain phyllopod Crustacea and its ecological significance. *Amer. Zool.* 13: 1341-1342.

Dexter, R.W. & M.S. Ferguson 1943. Life history and distributional studies on *Eubranchipus serratus* Forbes (1876). *Amer. Midl. Nat.* 29: 210-222.

Dexter, R.W. & C.H. Kuehnle 1948. Fairy Shrimp populations of northeastern Ohio in the seasons of 1945 and 1946. *Ohio J. Sci.* 48: 15-26.

Dexter, R.W. & C.H. Kuehnle 1951. Further studies on the fairy shrimp populations of northeastern Ohio. *Ohio J. Sci.* 51: 73-86.

Dexter, R.W. & L.E. Sheary 1943. Records of anostracan phyllopods in northeastern Ohio. *Ohio J. Sci.* 43: 176-179.

Dodds, G.S. 1915. Descriptions of two new species of Entomostraca from Colorado with notes on other species. *Proc. US Nat. Mus.* 49: 97-102.

Donald, D.B. 1983. Erratic occurrence of anostracans in a temporary pond: Colonization and extinction or adaptation to variations in annual weather? *Can. J. Zool.* 61: 1492-1498.

Eng. L.L., D. Belk, & C.H. Eriksen 1990. Californian Anostraca: Distribution, habitat, and status. *J. Crust. Biol.* 10: 247-277.

Eriksen, C.H. & R.J. Brown 1980. Comparative respiratory physiology and ecology of phyllopod Crustacea. II. Anostraca. *Crustaceana* 39: 11-21.

Ferguson, M.S. 1935. Three species of *Eubranchipus* new to Canada. *Can. F. Nat.* 49: 47-49.

Forbes, S.A. 1876. List of Illinois Crustacea with descriptions of new species. *Bull. Illinois State Lab. Nat. Hist.* 1: 3-25.

Gaudin, F.A. 1960. Egg production of *Streptocephalus seali* Ryder, with notes on the distinctions between certain North American Streptocephalidae. *Southwest Nat.* 5: 61-65.

Gissler, C.F. 1881. Evidences of the effect of chemiophysical influences in the evolution of branchiopod crustaceans. *Amer. Nat.* 15: 280-282.

Hartland-Rowe, R. 1965. The Anostraca and Notostraca of Canada with some new distribution records. *Can. Field. Nat.* 79: 185-189.

Hartland-Rowe, R. 1966. The fauna and ecology of temporary pools in western Canada. *Verh. Int. fur Limnol.* 16: 577-584.

Hartland-Rowe, R. 1967. *Eubranchipus intricatus* n. sp., a widely distributed North American fairy shrimp, with a note on its ecology. *Can. J. Zool.* 45: 663-666.

Hartland-Rowe, R. 1972. The limnology of temporary waters and the ecology of Euphyllopoda. In R.B. Clark & R.J. Wooton (eds), *Essays in Hydrobiology*: pp. 15-31. Exeter: University of Exeter Press.

Hay, O.P. & W.P. Hay 1889. A contribution to the knowledge of the genus *Branchipus. Amer. Nat.* 23: 91-95.

Hay, W.P. 1892. The Crustacea of Indiana. *Proc. Indiana Acad. Sci.* 1891: 147-151.

Hazelwood, D.H. & S.E. Hazelwood 1985. The effect of temperature in four species of freshwater fairy shrimp (Crustacea: Anostraca). *Freshwater Invert. Biol.* 4: 133-137.

Heath, H. 1924. The external development of certain phyllopods. *J. Morph.* 38: 453-483.

Herrick, C.L. 1885. Metamorphosis and morphology of certain phyllopod Crustacea. *Bull. Sci. Lab., Denison Univ.* 1: 16-24.

Holmes, S.J. 1910. Description of a new species of *Eubranchipus* from Wisconsin with observations on its reactions to light. *Trans. Wisconsin Acad. Sci.* 16: 1252-1254.

Horne, F.R. 1967. Effects of physical-chemical factors on the distribution and occurrence of some southeastern Wyoming phyllopods. *Ecology* 48: 472-477.

Horne, F.R. 1971. Some effects of temperature and oxygen concentration on phyllopod ecology. *Ecology* 52: 343-347.

Hutchinson, G.E. 1967. *A Treatise on Limnology*, Vol. 2: pp. 552-567. New York: Wiley.

Johansen, F. 1921. The larger freshwater Crustacea from Canada and Alaska. III. Euphyllopoda. *Can. Field. Nat.* 35: 21-30; 45-47; 88-94.

Johansen, F. 1922. Euphyllopoda Crustacea of the American arctic. *Report Can. Arctic Exped. 1913-l8.* 7(G): G10-G34.

Kelly, R.W. 1956. Natality factors of selected species of phyllopods. Ph.D. Dissertation, Univ. Missouri. Ann Arbor, Michigan: Univ. Microfilms.

Knight, A.W., R.L. Lippson, & M.A. Simmons 1975. The effect of temperature on the oxygen consumption of two species of fairy shrimps. *Amer. Midl. Nat.* 94: 236-240.

Leonard, A.B. & L.H. Ponder 1949. Crustacea in Eastern Kansas. *Trans. Kan. Acad. Sci.* 52: 168-204.

Linder, F. 1941. Contributions to the morphology and the taxonomy of the Branchiopoda Anostraca. *Zoologiska Bidrag Fran Uppsala* 20: 101-302.

Linder, H.J. 1959. Studies on the freshwater fairy shrimp *Chirocephalopsis bundyi* (Forbes). I. Structure and 19 histochemistry of the ovary and accessory reproductive tissues. *J. Morph.* 104: 1-60.

Linder, H.J. 1960. Studies on the freshwater fairy shrimp *Chirocephalopsis bundyi* (Forbes). II. Histochemistry of egg-shell formation. *J. Morph.* 107: 259-280.

Lynch, J.E. 1937. A giant new species of fairy shrimp of the   genus *Branchinecta* from the state of Washington. *Proc. US Nat. Mus.* 84: 555-562.

Lynch, J.E. 1958. *Branchinecta cornigera*, a new species of anostracan phyllopod from the state of Washington. *Proc. US Nat. Mus.* 108: 25-37.

Lynch, J.E. 1960. The fairy shrimp *Branchinecta campestris* from northwestern United States (Crustacea: Phyllopoda). *Proc. US Nat. Mus.* 112: 549-561.

Lynch, J.E. 1964. Packard's and Pearse's species of *Branchinecta*: An analysis of a nomenclatural involvement. *Amer. Midl. Nat.* 71: 466-488.

Lynch, J.E. 1972. *Branchinecta dissimilis* n. sp., a new species of fairy shrimp with a discussion of specific characters in the genus. *Trans. Amer. Micros. Soc.* 91: 240-243.

Mackin, J.G. 1939. Key to the species of Phyllopoda of Oklahoma and neighboring states. *Proc. Oklahoma Acad. Sci.* 19: 45-47.

Mackin, J.G. 1952. On the correct specific names of several North American species of the phyllopod genus *Brachinecta* Verrill. *Amer. Midl. Nat.* 47: 61-65.

McCarraher, D.B. 1970. Some ecological relationships of fairy shrimps in alkaline habitats of Nebraska. *Amer. Midl. Nat.* 84: 59-68.

McGinnis, M.O. 1911. Reactions of *Branchipus serratus* to light, heat, and gravity. *J. Exper Zool.* 10: 227-240.

McLaughlin, P.A. 1980. *Comparative morphology of recent Crustacea*. San Francisco: W.H. Freeman.

Modlin, R.F. 1982. A comparison of two *Eubranchipus* species (Crustacea: Anostraca). *Amer. Midl. Nat.* 107: 107-113.

Modlin, R.F. 1983. Effect of temperature and body size on oxygen comsumption of two species of *Eubranchipus* (Crustacea: Anostraca). *Amer. Midl. Nat.* 109: 55-62.

Moore, W.G. 1951. Observations on the biology of *Streptocephalus seali*. *Proc. Louisiana. Acad. Sci.* 14: 57-65.

Moore, W.G. 1955a. The life history of the spiny-tailed fairy shrimp in Louisiana. *Ecology* 36: 176-184.

Moore, W.G. 1955b. Observations on heat death in the fairy shrimp, *Streptocephalus seali*. *Proc. Louisiana Acad. Sci.* 18: 5-12.

Moore, W.G. 1957. Studies on the laboratory culture of Anostraca. *Trans. Amer. Micros. Soc.* 76: 159-173.

Moore, W.G. 1959. Observations on the biology of the fairy shrimp, *Eubranchipus dolmani*. *Ecology* 40: 398-403.

Moore, W.G. 1963. Some interspecies relationships in Anostraca populations of certain Louisiana ponds. *Ecology* 44: 131-139.

Moore, W.G. 1966. New world fairy shrimps of the genus *Streptocephalus* (Branchiopoda, Anostraca). *Southwest. Nat.* 1: 24-48.

Moore, W.G. 1967. Factors affecting egg-hatching in *Streptocephalus seali* (Branchiopoda, Anostraca), p. 724-735 In *Proc. Symp. Crustacea Pt. II*. Mar. Biol. Assoc. India

Moore, W.G. 1970. Limnological studies of temporary ponds in southeastern Louisiana. *Southwest Nat.* 15: 83-110.

Moore, W.G. & A. Burn 1968. Lethal oxygen threshold for certain temporary pond invertebrates and their application to field situations. *Ecology* 49: 349-351.

Moore, W.G. & L.H. Ogren 1962. Notes on the breeding behavior of *Eubranchis holmani* (Ryder). *Tulane Stud. in Zool.* 9: 315-318.

Moore, W.G. & J.B. Young 1964. Fairy shrimps of the genus *Thamnocephalus* (Branchiopoda, Anostraca) in the United States and Mexico. *Southwest Nat.* 9: 68-77.

Needer, K.H. & R.W. Pennak 1955. Seasonal faunal variations in a Colorado alpine pond. *Amer. Midl. Nat.* 53: 419-430.

Norton, A.H. 1909. Some aquatic and terrestrial crustaceans of the state of Maine. *Proc. Portland Soc. Nat. Hist.* 2: 245-255.

Packard Jr., A.S. 1874a. Descriptions of new North American Phyllopoda. *6th Annual Rept. Peabody Acad. Sci.* 1873: 54-57.

Packard Jr., A.S. 1874b. Synopsis of the freshwater Phyllopoda of North America. *US Geol. & Geograph. Survey Rept.* 1873: 613-622.

Packard Jr., A.S. 1877. Descriptions of new phyllopod Crustacea from the west. *Bull. US Geol. & Geograph. Survey* 3: 171-179.

Packard Jr., A.S. 1878. Occurrence of the phyllopod *Eubranchipus* in winter. *Amer. Nat.* 12: 186.

Packard Jr., A.S. 1883. A monograph of the phyllopod Crustacea of North America. *Annual Rept. US Geol. & Geograph. Survey of the Territories for 1878*. 12 (1): 295-592.

Pearse, A.S. 1912. Notes on phyllopod Crustacea. *Rept. Michigan Acad. Sci.* 14: 191-197.

Pearse, A.S. 1913. Observations of the behavior of *Eubranchipus dadayi*. *Bull. Wisconsin Nat. Hist. Soc.* 10: 109-117.

Pearse, A.S. 1918. The fairy shrimps (Phyllopoda). In H.B. Ward & G.C. Whipple (eds), *Fresh-water Biology*: pp. 661-670. New York: Wiley.

Pennak, R.W. 1953. *Fresh-water Invertebrates of the United States*: pp. 326-349. New York: Ronald Press.

Proctor, V.W. & C.R. Malone 1965. Further evidence of the passive dispersal of small aquatic organisms via the intestinal tract of birds. *Ecology* 46: 728-729.

Prophet, C.W. 1959. A winter population of *Streptocephalus seali* Ryder inhabiting a roadside ditch in Lyon County, Kansas. *Trans. Kansas Acad. Sci.* 62: 153-161.

Prophet, C.W. 1963a. Distribution of Anostraca in Oklahoma. *Proc. Oklahoma Acad. Sci.* 43: 144.

Prophet, C.W. 1963b. Physical-chemical characteristics of habitats and seasonal occurrence of some Anostraca in Oklahoma and Kansas. *Ecology* 44: 798-801.

Prophet, C.W. 1963c. Egg production by laboratory cultured Anostraca. *Southwest. Nat.* 8: 32-37.

Prophet, C.W. 1963d. Some factors influencing the hatching of anostracan eggs. *Trans. Kansas Acad. Sci.* 66: 150-159.

Reed, E.B. 1963. Records of freshwater Crustacea from arctic and subarctic Canada. *Bull. Nat. Mus. Can.* 199: 29-62.

Retallack, J.T. & H.F. Clifford 1980. Periodicity of crustaceans in a saline prairie stream of Alberta, Canada. *Amer. Midl. Nat.* 103: 123-132.

Ryder, J.A. 1879a. Description of a new species of *Chirocephalopsus*. *Proc. Acad. Nat. Sci. Philadelphia* 1879: 148-149.

Ryder, J.A. 1879b. Description of a new branchiopod. *Proc. Acad. Nat. Sci. Philadelphia* 1879: 200-202.

Schram, F.R. 1986. *Crustacea*. New York: Oxford Univ. Press.

Shantz, H.L. 1905. Notes on the North American species of *Branchinecta* and their relatives. *Biol. Bull.* 9: 249-264.

Sissom, S.L. 1976. Studies on a new fairy shrimp from the playa lakes of west Texas (Branchiopoda, Anostraca, Thamnocephalidae). *Crustaceana* 30: 39-42.

Spicer, G.S. 1985. A new fairy shrimp of the genus *Streptocephalus* from Mexico with a phylogenetic analysis of the North American species (Anostraca). *J. Crust. Biol.* 5: 168-174.

Sublette, J.E. & M.S. Sublette 1967. The limnology of the playa lakes on the Llano Estacado, New Mexico and Texas. *Southwest. Nat.* 12: 369-406.

Tasch, P. 1963. Evolution of the Branchiopoda. In H.B. Whittington & W.D.I. Rolfe (eds), *Phylogeny and Evolution of Crustacea*: pp. 145-157. Cambridge: Mus. Comp. Zool.

Underwood, L.M. 1886. List of the described species of freshwater crustacea from America, north of Mexico. *Bull. Illinois State Lab. Nat. Hist.* 2: 323-384.

Van Cleave, H.J. 1928. The fairy shrimps of Illinois. *Trans. Illinois State Acad. Sci.* 20: 130-132.

Verrill, A.E. 1869. Description of some new American phyllopod Crustacea. *Amer. J. Sci. & Arts* (2) 48: 244-254.

Verrill, A.E. 1870. Observations on phyllopod Crustacea of the family Branchipidae, with description of some new genera and species from America. *Proc. Amer. Ass. Adv. Sci.* 1869: 230-247.

Weaver, C.R. 1943. Observations of the life cycle of the fairy shrimp *Eubranchipus vernalis*. *Ecology* 24: 500-502.

Weise, J.G. 1957. A new state and size record for the spinytailed fairy shrimp, *Streptocephalus seali* Ryder. *Trans. Kentucky Acad. Sci.* 18: 75-77.

Wiman, F.H. 1979a. Hybridization and detection of hybrids in the fairy shrimp genus *Streptocephalus*. *Amer. Midl. Nat.* 102: 149-156.

Wiman, F.H. 1979b. Mating patterns and speciation in the fairy shrimp genus *Streptocephalus*. *Evolution* 33: 172-181.

Wiman, F.H. 1981. Mating behavior in the *Streptocephalus* fairy shrimps (Crustacea: Anostraca). *Southwest Nat.* 25: 541-546.

Two centuries of larval crab papers: A preliminary analysis

A.L. Rice
Institute of Oceanographic Sciences Deacon Laboratory, Wormley, Surrey, UK

1 INTRODUCTION

Bibliometric analysis of the scientific literature has become very fashionable in recent years; whole teams of bibliometricians make a living from it and the results of their analyses, particularly citation indices, are used to assess the success or otherwise of individual scientists, research institutions, or major programs. It has even been suggested recently (de Bruin et al. 1991) that a bibliometric analysis of the scientific output of Iraq and its neighbors during the 1980's might have reinforced other indications of the tensions leading to the Gulf War. This paper has no such pretensions, but simply attempts to identify some of the general trends in the published literature on the larval stages of brachyuran crabs over the past 200 years.

In a recent paper, Young (1990) undertook a rather broader task, reviewing developments in the study of larval ecology of marine invertebrates. Young's study included a detailed analysis of all larval papers published in six major marine biological journals, two older ones established around the turn of the century, and four newer journals begun within the last 25 years. The data revealed an exponential increase in larval studies over the last century, with a particularly dramatic increase since the early 1960's. Some of the general trends were thought to reflect changes in the productivity patterns in most fields of science. Young, however, was able also to relate specific changes to particular historical events. For instance, the fluctuating interest in larval biology in the *Biological Bulletin* seemed to be correlated with major historical or economic events affecting the United States, such as the two World Wars, the depression, and the Vietnam War. Similarly, many of the laboratory studies on larvae published in the *Bulletin* in the late 1960's and early 1970's dealt with responses to pollutants, reflecting increasing public concern for the environment. Nevertheless, by focussing on a small number of journals Young acknowledged that certain biases were inevitably introduced with the newer journals being dominated by laboratory studies. Even within this broad category, however, the analysis of different journals suggest different trends. The journal *Ophelia*, for instance, had a major emphasis on morphological papers until very recently, whereas *Marine Ecology Progress Series* accepts no papers on morphology but has generally concentrated on papers dealing with laboratory experiments on larvae.

In this paper, I attempt a broader chronological coverage, but a much simpler analysis of all (or almost all) of the larval literature on one relatively small group, the brachyuran crabs. As far as I am aware, such a review has never been carried out before.

2   METHODS

The analysis is based on the bibliography of crab larvae published recently by Soltanpour-Gargari et al. (1989), which attempts to be exhaustive for the period between Linnaeus's (1767) description of *Cancer germanicus* and the mid-1980's. Although this list is not complete, it misses relatively few papers and probably has no particular bias that would affect my analysis significantly. Moreover, it has the advantage of being readily accessible to anyone with the interest (or temerity) to wish to check my findings. To bring the analysis as up-to-date as possible, I have also extracted data from the *Zoological Record* up to and including volume 126 (1989/90).

The resulting data set contains 1194 entries of which all but a handful were included in this analysis. Each publication was assigned a classification based on its geographical and subject area coverage. In the geographical classification, each paper received a single entry based on a simple hierarchy: first, the ocean from which the studied material came (Atlantic, Pacific, Indian (including the Red Sea) or Mediterranean); second, in the case of the Atlantic and Pacific, eastern, western or central areas. Coarse though this classification is, a small number of publications, such as some expedition reports, could not be readily accommodated within it and were ignored.

The subject classification was a little more complicated. First, each paper was assigned to one of eleven broad subject areas that I judged to represent the main purpose of the study. Second, I assigned each paper up to two further levels of subject classification, using the same categories. Thus, a paper describing the morphology of the larval stages of a species that was reared in the laboratory and also reporting ancillary data on the larval survival in various temperature and salinity conditions would be classified primarily under 'morphology,' with a secondary classification under 'ecology,' and finally under 'rearing.' The subject areas ranged from very broad ones like morphology and ecology, to those of more narrow focus such as physiology, biochemistry, and responses to toxic substances. In practice, however, only four of the categories – morphology, ecology, rearing, and aquaculture – showed interesting distributions, and the remaining ones were therefore grouped together as 'others.' Although the main subject categories may seem to be self-explanatory, my use of them was somewhat idiosyncratic so that some clarification is needed.

Morphology: Used for any paper concerned primarily with the description of zoeas or megalopas, whether the material was obtained from the plankton or from laboratory culture.

Ecology: Used for all papers concerned with the distribution of larvae or the study of the effects of variations of environmental variables on their survival. As with 'morphology,' this category was used whether the study was based on field or laboratory observations.

Rearing: In the primary level of classification, used only for those papers concerned with aspects of laboratory culture technique, but not specifically directed towards commercial cultivation. At the secondary and tertiary levels this category was used for all papers dealing with reared larvae whatever the primary objective of the work.

Aquaculture: Used for papers concerned specifically with the commercial cultivation of exploitable species.

3 RESULTS

3.1 Total publications

Figure 1 shows the total number of larval crab papers appearing in each five year period up to 1989. Until the 1940's there is little of interest in these data. Over the previous 100 years or so the publication rate had increased gradually to an average of three or four papers each year during the 1920's and 1930's, but with considerable variations even between the 5-year totals. A significant fall in the late 1940's presumably reflects the effects of World War II, though no similarly convincing fall is apparent following World War I.

From the mid-1950's, however, there has been a dramatic increase in the mean yearly publication rate up to almost 50 per year in the early 1980's, but with an apparent decrease during the last 5 years. When the period since 1950 is analyzed by the year (Fig. 2), considerable inter-annual variations are apparent, but the general increase is clear, as is the decreased publication rate since about 1985.

3.2 Geographical distribution

Figure 3 shows the distribution of larval crab papers according to the major ocean regions with which they were concerned. Prior to 1950 the literature dealt predominantly with the eastern Atlantic, presumably reflecting the major influence of European workers. The exceptions include the sporadic appearance of western Atlantic papers, particularly in the 1880's due to the efforts of authors such as Faxon at the Harvard Museum of Comparative Zoology, Birge associated with Brooks at the Chesapeake Zoological Laboratory of Johns Hopkins University,

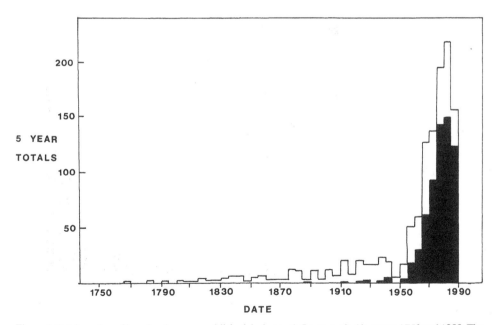

Figure 1. Total number of larval crab papers published during each 5-year period between 1750 and 1989. The blacked-in areas represent papers based on laboratory rearing.

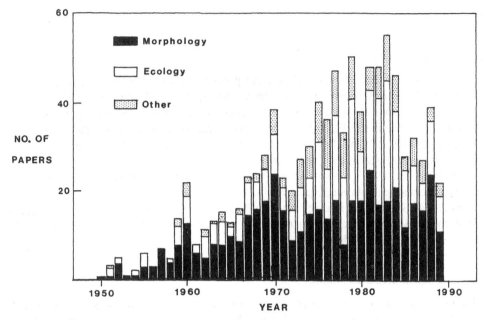

Figure 2. Yearly production of larval crab papers between 1950 and 1989 according to the broad area of their main subject.

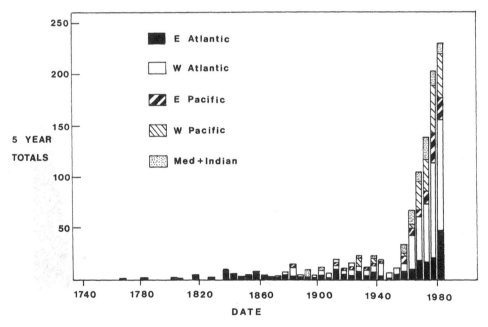

Figure 3. Five-year totals for larval crab papers from 1750 to 1985 according to the main oceanic areas with which they dealt. Note that a small number of the papers included in Figure 1 are not represented in this figure.

and in the 1940's Sandoz and her co-workers at the Virginia Fisheries Laboratory. Similarly, the brief flurry of Mediterranean papers in the late 1880's and early 1890's resulted from the activities of Cano in Italy, while Aikawa produced a small peak in western Pacific (Japanese) papers in the 1920's and 1930's. Because of the more even spread of eastern Atlantic papers, the efforts of individual European workers such as H.C. Williamson, M.V. Lebour, and R. Gurney are not so obvious, though their contributions to the brachyuran larval literature were considerably greater.

With the exception of the Mediterranean, all of the ocean areas experienced the general increase in attention since the 1950's, but this increase is far more marked for the western Atlantic (Fig. 3) than for the other areas.

3.3 *Subject analysis*

Over the whole period considered, the majority of larval crab papers were concerned primarily with the morphological description of the larval stages (some 650 out of a total of almost 1200 papers). Indeed, prior to the 1950's, papers not dealing with morphology were rather rare. Morphology continued to dominate the larval crab literature numerically even into the early 1960's. It continues to be an important category today, being the primary subject of about 20 papers each year (Fig. 2). Since the early 1970's, however, morphological papers are frequently outnumbered by publications concentrating on other topics, and particularly those falling in my 'ecology' category. In fact, the growth in these ecological papers is largely responsible for the dramatic increase in the total publication rate over the last 30 years.

All of the other primary subject categories (rearing and aquaculture, physiology, toxicity, etc.) waxed and waned rather irregularly over the past 40 years, but rarely formed a significant proportion of the total larval publications. Under the secondary and tertiary level classifications, however, the most dramatic change is noted in the number of papers involving laboratory rearing as a technique irrespective of the primary objective. Figure 1 shows that prior to the 1950's such papers were rare. From the mid-1950's onwards the number of papers based on rearing grew rapidly; but since the early 1970's almost half of all larval crab publications are based on laboratory cultivation.

4 DISCUSSION

The above simple analysis probably reveals nothing that was not already apparent in general terms to anyone familiar with the larval crab literature. Nevertheless, at least for me it has helped to crystallize these general impressions into the following specific points: 1) The rapid increase in the number of papers published each year since the mid-1950's; 2) the increasing predominance of papers dealing with the western Atlantic and eastern Pacific, mainly originating from the USA; 3) the predominance of morphological papers prior to this period and the increasing importance of other studies subsequently; and 4) the predominance of papers based on laboratory rearing during the last 3 decades.

The dramatic increase in the number of larval crab papers published since World War II undoubtedly reflects the general increase in scientific activity during this period. Similarly, the increasing proportion of papers originating from North America reflects the increasing post-war dominance of the USA on the world stage, in science as in other areas, resulting, at least partly, from its relatively unscathed economy. These factors, however, provide by no means the whole explanation; they would not explain, for instance, the change in emphasis from 'morphology' towards 'ecology' and the other disciplines. In turn, these changes were certainly influenced partly by fashion in the sense that the last four decades have seen an increasing emphasis on the study of species in an environmental context rather than in isolation. However, in the study of decapod crustacean larvae in general, and of brachyuran crab larvae in particular, I believe that many of the changes owe their origins to the relatively simple, but unprecedented, improvements in the techniques for the laboratory culture of the larval stages in the late 1950's and early 1960's. Prior to this time there were numerous attempts to rear decapod larvae, some (e.g. Lebour 1927,1928; Hart 1935) met with considerable success; but the results were inconsistent. With the pioneering work of Knudsen in California (Knudsen 1958), Chamberlain in Chesapeake Bay (Chamberlain 1957), and particularly of Bookhout, Costlow, and their co-workers at the Duke University laboratory at Beaufort, North Carolina (Costlow & Bookhout 1959, 1964), this situation changed dramatically. By maintaining the larvae in small batches or even individually; by feeding them on fresh food, particularly *Artemia* nauplii; and by applying simple but careful hygiene measures, success was more or less assured. By the early 1960's some 30 local decapod species had been reared at Beaufort (Costlow & Bookhout 1964). Within a few years the same general techniques, modified to suit local conditions or requirements, were being used successfully in a variety of other laboratories in the US and soon spread to other countries. In 1962, for instance, I was introduced to the technique myself at the University of Miami marine laboratory where not only brachyurans but also macrurans and anomurans were reared (e.g. Provenzano & Rice 1964; Dobkin 1965; Yang 1967). In turn, I introduced the techniques to the British Museum (Natural History) in London when I joined the staff in 1966, so beginning a rearing program that continued with considerable success until the mid-1980's. Similar programs were begun over the next few years in a number of centers around the world.

For many workers, the rearing techniques simply provided them with abundant well-preserved material of known parentage on which to base their morphological descriptions. The result was a spate of papers describing the reared zoeas and megalopas of crabs from a variety of environments and largely accounting for the increase in morphological papers. In turn, this led to the increased use of larval characters in attempts to unravel the complexities of brachyuran relationships, but a realization at the same time that many of the earlier accounts, including those of the subject's 'gurus' such as Lebour and Gurney, were woefully inadequate. A direct consequence of the use of laboratory culturing techniques therefore was a significant improvement in the quality of larval descriptions. Further improvements are still required, but these will depend upon the availability of appropriate microscope technology and both the will and expertise to use it. By using the well-established rearing techniques, anyone can now obtain excellent material with which to work.

The advent of reliable rearing techniques also opened up possibilities for experimental work, attracting the attention of scientists who would never otherwise have worked on crab larvae. Costlow, Bookhout, and their co-workers, for example, exploited the technique extensively to investigate the effects of salinity and temperature on larval growth and survival, the control of larval molting, and the tolerance of larval stages to toxic substances. These studies, and similar ones taken up by other groups (see Epifanio 1979), were largely responsible for the growth of the non-morphological categories in my classification. Thus, in my opinion the impact of the development of successful laboratory rearing techniques on the study of crab larvae was enormous, contributing significantly not only to its overall growth, reflected in the total output in terms of publications, but also encouraging its diversification.

Why then is the period of rapid growth apparently over, with indications of a decrease in the publication rate during the last few years? Several factors probably are involved. First, the general reduction in science funding, experienced in many western countries during the late 1970's and 1980's, has had an effect. Moreover, even where the total available funding has not fallen significantly, there has been an increasing emphasis on accountability, often expressed as increased support for 'applied' at the expense of 'pure' research. Students of brachyuran crabs are not well-placed to survive such a change. Relatively few crab species are exploited commercially, and even fewer support important fisheries compared with fish or even other crustaceans. Furthermore, the economic value of crabs is in general too low to make them attractive subjects for commercial cultivation. Consequently, under the aquaculture category there has been no significant growth in brachyuran larval studies.

Second, larval crab studies suffered from the decreased support for, and interest in, taxonomy over recent decades. Studies of larval morphology, including those dealing with phylogeny, fall within this general category and therefore have become more difficult both to fund and to publish. Moreover, the improved standards expected of such papers also contributed to their reduced numbers; for example, as a referee I have recommended the rejection of morphological papers on these grounds more frequently in recent years than previously. On the other hand, Figure 2 suggests that the number of morphological papers published each year has not fallen dramatically, but that the decrease in the overall publication rate resulted from a fall in the other categories, including ecology.

Finally, there may be an even simpler explanation for the decreased publication rate. When laboratory rearing techniques are first introduced to a new geographical region, almost any local crab is worth culturing since it was almost certainly not reared before. After a few years of active rearing work, all the most common and easily obtained crabs have been cultured, so that acquiring ovigerous females of unreared species becomes more and more difficult. Coupled with the other reasons for reduced support for larval work, these logistic difficulties may be the last straw leading to the final demise of a rearing program; such a scenario certainly contributed to the cessation of the program at the British Museum (Natural History).

What does this analysis forecast for the future of larval crab work? First, there is clearly still much to be done. Of the almost 5000 known crab species, larvae of less than 10% have been described and less than half of these have been reared through all of the larval stages. For some years past, the larvae of between about 10 and 20 'new' species have been described each year. In the present climate, this rate of publication of morphological papers is likely to be reduced in the future. However, it will probably continue at a reduced level for many years to come, mainly as a result of the efforts of students working in geographical regions in which the larval stages of the common crabs are still undescribed. Such students are likely to work only temporarily on crab larvae, perhaps to complete an M.Sc. or Ph.D., before moving onto more fashionable or applied areas of research. In developed countries, larval crab work is likely to concentrate on ecological, behavioral, physiological, and biochemical studies, with morphological papers produced only as a byproduct when attention turns to species of which the larvae are still undescribed. Hopefully, nevertheless, even in these areas the potential value of larval studies in unravelling crab interrelationships will encourage specialists to study material of particularly interesting or difficult groups as material becomes available. In all cases, however, it seems certain that laboratory rearing will continue to dominate the subject for the foreseeable future.

REFERENCES

Chamberlain, N.A. 1957. Larval development of the mud crab *Neopanope texana sayi* (Smith). *Biol. Bull.* 113: 338.

Costlow, J.D. & C.G. Bookhout 1959. The larval development of *Callinectes sapidus* Rathbun reared in the laboratory. *Biol. Bull.* 116: 373-396.

Costlow, J.D. & C.G. Bookhout 1964. An approach to the ecology of marine invertebrate larvae. In *Symp. Exp. Mar. Ecol. Grad. Sch. Oceanogr., Univ. Rhode Island, Occas. Publ.* 2: 69-75.

de Bruin, R.E., R.R. Braam & H.F. Moed 1991. Bibliometric lines in the sand. *Nature* 349: 559-562.

Dobkin, S. 1965. The early larval stages of *Glyphocrangon spinicauda* A. Milne Edwards. *Bull. Mar. Sci.* IS: 872-884.

Epifanio, C.E. 1979. Larval decapods (Arthropoda: Crustacea: Decapoda). In C.W. Hart Jr., & S.L.H. Fuller (eds), *Pollution Ecology of Estuarine Invertebrates*: 259-292. New York: Academic Press.

Hart, J.F.L. 1935. The larval development of British Columbia Brachyura. I. Xanthidae, Pinnotheridae (in part) and Grapsidae. *Can. J. Res.* 12: 411-432.

Knudsen, J.W. 1958. Life cycle studies of the Brachyura of western North America, I. General culture methods and the life cycle of *Lophopanopeus leucomanus leucomanus* (Lockington). *Bull. So. Calif. Acad. Sci.* 57: 51-59.

Lebour, M.V. 1927. Studies on the Plymouth Brachyura. I. The rearing of crabs in captivity, with a description of the larval stages of *Inachus dorsettensis, Macropodia longirostris* and *Maia squinado. J. Mar. Biol. Ass. U.K.* 14: 795-813.

Lebour, M.V. 1928. The larval stages of the Plymouth Brachyura. *Proc. Zool. Soc. Lond.* 1928: 473-560.

Linnaeus, C. 1767. *Systema Naturae*, Ed. 13, Bd. 1, Pars 2. Vindobonne.

Provenzano, A.J. & A.L. Rice 1964. The larval stages of *Paguristes marshi* Benedict (Decapoda, Anomura). *Crustaceana* 7: 217-235.

Soltanpour-Gargari, A., R. Engelmann & S. Wellershaus 1989. Development and rearing of zoea larvae in Brachyura (Crustacea Decapoda). A bibliography. *Crustaceana* Suppl. 14: 1-173.

Yang, W.T. 1967. A study of zoeal, megalopal, and early crab stages of some oxyrhynchous crabs (Crustacea: Decapoda). Ph.D. Diss., Univ. Miami. Florida.

Young, C.M. 1990. Larval ecology of marine invertebrates: A sesquicentennial history. *Ophelia* 32: 1-48.

History of North American decapod paleocarcinology

Gale A. Bishop
Department of Geology & Geography and Institute of Parasitology and Arthropodology, Georgia Southern University, Statesboro, Georgia, USA

1 INTRODUCTION

The crabs have a significant recent (Holocene) record of over 434 genera and 4500 species (Moore & McCormick 1969: p. R79) occupying most marine and many non-marine habitats (Schmitt 1965), forming significant parts of communities, and occupying important positions in food webs. Many species of crabs support important commercial fisheries. The diversity and abundance of crabs in easily accessible waters along the coasts of continents has made them one of the most collected and best understood groups of invertebrates; accessibility makes it easy to procure material for comparison and dissection. Crab abundance, diversity, and ecologic and commercial importance has led many biologists to spend their lives studying these interesting organisms, producing a large literature on their systematics, ecology, physiology, embryology, and other aspects of their biology.

From a paleontologist's point of view, crabs are an exciting group because they are an important, rapidly evolving, extant group of complex organisms having great diversity and abundance and a moderate fossil record. The most recent compilation (Glaessner 1929) of our knowledge of fossil decapods includes about 700 literature citations including 230 genera and 300 species of crabs from rocks of Cretaceous and Early Tertiary Age (since publication these numbers are estimated to have increased about 50%). The moderate fossil record not only allows the paleontologist to maintain an overview of all decapods, but also, because the decapods are an under-studied group, provides great opportunities to further our knowledge of this important group. This historical note attempts to provide an overview of major themes in fossil decapod research, a chronology of this research in North America, a regional analysis of its development, and a brief stratigraphic review. The reader can enter the paper by any of these thematic tracks. I was constrained in this small history by several factors, particularly, limited time and limited library materials. Also, I was unable to examine primary archival documents such as many of the described specimens and museum labels, field notes, correspondence, and manuscript notes of individual researchers. This has led me to write only a general overview based on my perceptions of the history of North American decapod paleocarcinology.

2   MAJOR THEMES OF DECAPOD HISTORY

2.1   *Geologic setting*

The evolutionary history of any group of organisms must be set in the context of changing environments and the stage of development of contemporaries in the past. These historical constraints are in turn a result of natural selection and contingency in a changing physical geography of the past (see Gould 1989: 284). The major portion of the history of crabs was developed after the latest breakup of the continents. The crabs of the margins and epicontinental seas of North America must be studied in the context of the opening of the Atlantic, the destruction of the circumequatorial Tethys Sea, and the progradation of continental margins.

As the North American continent was carried away from Laurasia, the Atlantic Ocean basin began opening. This movement caused compressional features along the leading edge of the plate (the Rocky Mountains and metamorphism of the active Pacific Slope), a 'standing wave trough' (the Western Interior Seaway) in the continental interior and along the plate's trailing edge (Gulf and Atlantic coasts). The high standing Rockies began rapidly eroding and the derived sediment transported to the continental margins, eventually filling in the Western Interior Seaway and extending the Atlantic and Gulf coastal plains hundreds of kilometers beyond the old edge of the North American continent. The changing environmental conditions apparently snuffed out most reefs after the Early Cretaceous until reestablishment of this habitat later in the Paleocene or Eocene (Kauffman 1979: 30), and certainly provided a series of dynamic habitats, in the clastic and carbonate environments of the continental margins, in which the evolving decapods played a major (but largely unrecognized) role as prey, predators, and sediment movers. This period saw the crabs develop from primitive forms (e.g., prosoponids, dromiids, and homolids) into taxa more recognizable as modern types (xanthids, portunids, grapsids, etc.). This change from 'primitive' to 'modern' crabs was played out against the dynamic changes of the Late Cretaceous and Early Tertiary of the world's oceans.

2.2   *Systematics*

Martin F. Glaessner (Fig. 1A), an active student of the crabs for half a century, provided us with the *Fossilium Catalogus*, the *Treatise of Invertebrate Paleontology* volume on decapods, and many systematic and evolutionary syntheses (Glaessner 1929, 1960, 1969, 1980). Two of these syntheses center on the fossil evidence for the evolution of the crabs (Glaessner 1960, 1980), the most recent includes a phylogenetic chart summarizing the hypothesized brachyuran relationships and highlighting three times of adaptive radiation: The Late Jurassic, the 'mid-Cretaceous,' and the Late Cretaceous-Early Paleogene. The systematic arrangement of taxa closely follows the systematic revision of Guinot (1977, 1978), developed using primarily Recent forms but including paleontologic data. This acknowledgment of Guinot's systematics by the leading paleontologist of the crabs has initiated a dramatic revision of our picture of crab systematics and evolution. This last adaptive radiation is of particular significance because it is the source of many of our recent taxa including numerous commercial species (*Callinectes, Cancer*, etc.), predators, sediment movers, and prey species. Particularly rewarding have been the recent collaborations (Fig. 1E) between paleontologists and systematists such as those by myself and Austin B. Williams of the National Marine Fisheries Systematics Laboratory (Bishop & Williams 1986), mentioned here primarily as a model for future decapod paleontology, which should integrate paleontology and neontology wherever possible.

Figure 1. Photographs of paleocarcinologists important to the development of the study of fossils of North American decapod crustaceans. A. Martin F. Glaessner (1906-1989). Passport photograph (1976) sent to G. Bishop in 1981. B. Henryk B. Stenzel (1899-1980) at Little Campus, University of Texas at Austin on 19 January, 1971 (Photograph by G. Bishop). C. Henry B. Roberts (1910-1979) at his desk in the National Museum of Natural history 3 November 1977, examining a specimen of *Dakoticancer australis* Rathbun. D. Rodney M. Feldmann (1939-), collecting fossil decapods at the Heart Tail Ranch Locality (see Bishop 1985c) on 27 June 1982. E. National Geographic Expedition members Michael Klug, Gale A. Bishop (1942-), Reinhard Förster (1935-1987), and Austin B. Williams (1919-) on the summit of Harney Peak, Black Hills, South Dakota, taking a break from collecting crabs from South Dakota's Tepee Buttes, Cretaceous Pierre Shale, 4 July 1987.

2.3   *Ecology*

The complex bodies of crabs lend themselves to functional morphologic analyses (Schäfer 1954, 1972; Wharton 1942) that will eventually allow paleontologists to confidently (as the functional morphology hypotheses are tested in the Recent) assign fossil species to certain locomotory and feeding groups. Our knowledge of described decapod communities in modern environments is limited (Thorson 1957, Crane 1975), but the paleontologic data indicates that although many fossil crab faunas occur as parts of molluscan-dominated communities (Bishop 1973, 1976, 1983b), other faunas probably represent decapod-dominated communities (Feldmann et al. 1977; Bishop 1981a, 1983c). As more data become available it will be possible to study the evolution of community structure in these decapod communities (as Schram has done for the Carboniferous). The position of decapods in food webs of the past is not well understood but evidence does exist of crabs as prey (Bishop 1972a, 1975) and as molluscan predators (Vermeij 1977: 249; LaBarbera 1981: 522). The evolution of such predators is thought (Vermeij 1977: 251, LaBarbera 1981: 524) to have had a negative effect on reclining bivalves, a niche which became nearly empty at the end of the Cretaceous. The rapid evolution of the crabs, with adaptive bursts at the Jurassic-Cretaceous boundary, during the Mid-Cretaceous, and near the Cretaceous-Tertiary boundary (Glaessner 1969: fig. 22) could lead, as the record becomes more complete, to using fossil decapods as biostratigraphic markers.

2.4   *Decapod communities and paleocommunities*

Because decapods are mobile animals that usually feed as carnivores, omnivores, or scavengers, they generally constitute relatively minor parts of communities that are dominated by other organisms. In spite of this, some communities rich in decapods exist in the Recent. Thorson (1957) described three crab communities: the *Pinnixa rathbunae* community from Ise Wan and Mikawa Wan, Japan; The *Xenophthalamus pinnotheroides* community from the Persian Gulf; and the *Diogenes costatus* hermit crab community from from the Madras coast. Thorson went on (1957: 518) to suggest, 'Also the sand crab (*Emerita*) communities described from the California beaches by MacGinitie (1949) (1939) may probably be regarded isocommunities to the crab and hermit crab community just mentioned, but since the *Emerita* community is intertidal only, it is not discussed in detail here.' Camp et al. (1977: 48) described, but did not name, a polychaete-decapod community that occurs off the coast of Florida. Even though they may not have been formally described as 'communities,' such decapod-rich assemblages might be appropriately considered communities. Included should be the faunas of the northeastern United States and Canada dominated by the American Lobster, *Homarus americanus*, the marsh faunas of the southeastern United States dominated by the fiddler crabs (*Uca pugnax*, *U. minax*, and *U. pugilator*), and the sandy beach faunas of the southeastern United States dominated by the Carolinian ghost shrimp, *Callichirus major* (Say). In each of these cases, decapod crustaceans constitute a significant, if not dominant, proportion of the biomass and can be thought to be indicators of the other organisms that inhabit each environment, thus, in one sense, constituting a community.

The concept of decapod communities has not often been cited in the literature, probably because fossil decapods often are scarce and constitute only a minor element of the total fauna. Bishop (1986c: 349) has shown that about 42% of the described Cretaceous crabs from North America are known from single specimens. These taxa represent randomly preserved sparse elements in molluscan-dominated paleocommunities. The occurrence of decapod-dominated or decapod-rich, apatite-preserved faunas (Bishop 1981a, 1986b, 1987b; Bishop & Williams

1986) have been cited as preserved decapod community fractions (Fig. 2F-H). Their repeated occurrence in the North American Cretaceous has lead to the ability to decipher some of the complex relationships of the evolution of decapod communities in the Cretaceous.

2.5 *Decapod ichnology*

Trace fossils of decapod crustaceans are numerous and diverse, mostly consisting of burrows used for domiciles or for feeding, but including tracks and trails, and feeding traces. Frey et al. (1984) reviewed the tracemaking activities of crabs, finding that these were under represented in the fossil record. The bioturbation of sediment by decapods is significant (Frey et al. 1984: 333, Pryor 1975: 1244) in many environments. Two burrowing behaviors were discerned by Frey et al. (1984: 335): 1) Burrowing backwards to shelter in the substrate (as is done by the beach-dwelling mole crab, the blue crab, and the deeper-water frog crabs (raninids); and 2) burrowing sideways into the substrate to excavate a 'normal' domicile as is done by the beach-dwelling ghost crab (*Ocypode*), the marsh-dwelling fiddler crabs (*Uca*), the American lobster (*Homarus*), and numerous other decapods. These behaviors would lead to the production of resting traces and burrows preservable in the fossil record.

Decapod burrows are variable in morphology and are assigned to various ichnogenera such as *Thalassinoides* Ehrenberg, 1944; *Ophiomorpha* Lundgren, 1891, (see Fig. 2J); *Spongeliomorpha* Saporta, 1887; *Macanopsis* Macsotay, 1967; *Gyrolithes* Saporata, 1884; *Chagrinichnites* Feldmann et al., 1978; *Ardelia* Chamberlain & Baer, 1973; and *Psilonichnus* Ffrsich, 1981, (Frey et al. 1984: 333). A few body fossils of decapods have been reported *in situ* in burrows (Richards 1975, Jenkins 1975, and Bishop 1988c: Fig. 3D), although these traces have not been formally assigned to trace fossil genera (Fig. 2I). The complexities of decapod traces and trace makers are beautifully developed by Frey et al. (1978) in a review paper on *Ophiomorpha*. In it they document the production of similar burrow morphologies by diverse burrowers as well as the variation of burrows made by a single burrower in sediment with different physical characteristics.

Because burrowers are often tied to discrete habitats, their traces tend to be sediment dependent and in some cases useful for delineation of depositional environments. Frey & Mayou (1971) described the vertical zonation of decapod burrows across the strandline on Georgia barrier islands, with ghost shrimp burrows in the lower foreshore, ghost crab burrows on the backbeach, and fiddler crab burrows in the adjacent marshes. Thus a facies tract of ichnofossils can be useful in interpreting ancient burrowed sediment. Such an interpretation was made by Howard & Scott (1983) on an exposed sequence of an ancient barrier island along the St. Mary's River on the Georgia-Florida line. In addition to using trace fossils to decipher the facies tract, they also discussed the overprinting of deep-burrowing fabrics on superjacent and subjacent beds.

The description and utilization of decapod trace fossils is still in its infancy, but holds great promise as an alternative methodology for documenting the presence and behavior of decapods in lithosomes that commonly do not preserve body fossils.

2.6 *Biogeography and paleobiogeography*

A plethora of neontological biogeographic papers, monographs, and books document the modern distribution of decapod crustaceans. Many of these have been produced by authors describing faunas of discrete areas (e.g. Sakai (1965) *The Crabs of Sagami Bay* or Schmidtt

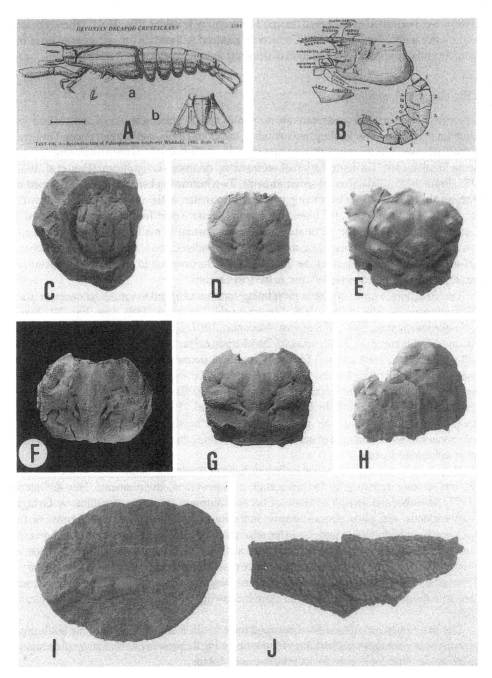

Figure 2. Historically significant North American decapod crustaceans. A. Reconstruction of *Palaeopalemon newberryi* Whitfield, 1880, from Schram et al. (1978) illustrating the oldest known decapod. B. Reconstruction of *Homarus brittonestris* Stenzel, 1945, illustrating the clarity and completeness with which Stenzel executed his work. C & D. *Tetracarcinus subquadratus* Weller, 1905, showing differences in quality between steinkerns similar to those described and figured by Weller (C. Specimen USNM 73716) and well-preserved, intact, specimens which result from continued, aggressive collecting (D. Specimen GAB 37-1113). E. *Homolopsis*

(1921) *Marine Decapod Crustacea of California*), discrete taxonomic bundles (e.g. Crane (1975) *Fiddler Crabs of the World*), or specimens collected by oceanographic expeditions.

Parallel studies of decapod paleobiogeography have been produced occasionally, but tend to emphasize the fossil decapods of an author's country or area of interest. Such monographic works include Via Boada's (1969) *Crustaceos Decapodos de Eoceno Español*, Bell's monograph (1857) of the *Malacostracous Decapods of Great Britain*, and Wright & Collin's work (1972) on *British Cretaceous Crabs*. These works, although not on North American faunas, directly impinged on North American studies by providing easy access to known faunas and stimulating further research in North America.

Regional monographs were produced for North America by Mary Rathbun: for the West Coast, *The Fossil Stalk-eyed Crustacea of the Pacific Slope of North America* (1926b), and for the East and Gulf coasts, *Fossil Crustacea of the Atlantic and Gulf Coastal Plain* (1935b). Both these monographs were important systematic inventories of known fossil crustaceans consistent with expectations of scientists at the time they were produced, but neither presented an adequate analysis of preservation, taphonomy, community structure, paleoecology, or paleobiogeography.

In 1975, Nations published a very detailed monograph, *The Genus Cancer (Crustacea:Brachyura): Systematics, Biogeography and Fossil Record*. In it, the biogeographic ranges of extant species of *Cancer* were detailed and paleobiogeographic ranges of fossil subgenera were summarized. Feldmann (1981) produced a summary, 'Paleobiogeography of North American lobsters and shrimps (Crustacea, Decapoda),' citing 98 species from North America, the greatest number being from the Cretaceous. Bishop (1986b) published a comprehensive summary, 'Occurrence, preservation, and biogeography of the Cretaceous crabs of North America,' which placed all known North American Cretaceous crabs into a paleobiogeographic context, drew attention to their patterns of endemism, and chronicled their occurrence by locality.

2.7 Taphonomy of the Decapoda

Recognition of the importance taphonomy plays in the preservation of decapod crustaceans was first recognized by the Europeans Hans Mertin (1941) and Wilhelm Schäfer (1951) and

---

atlantica Roberts, 1962, showing the anterior of a partial carapace to compare with Roberts (1962, Plate 89, Fig. 4). Notice the difference in quality of reproduction which has hindered the usefulness of Roberts's otherwise significant contribution of 1962. F-H. *Dakoticancer overanus* Rathbun, 1917, described in one of Rathbun's earliest paleontological papers and named in honor of W.H. Over who founded a natural history museum at the University of South Dakota. Over collected and submitted the type suite (including the holotype (USNM 32055) in Figure F. Subsequent collecting by Bishop has made this taxon (Figure G) the best known fossil decapod in the world. The significance of this decapod and its assemblage lies in its documentation as a fraction of a decapod paleocommunity, in which thousands of specimens were preserved allowing further definition of biological processes (such as molting, Fig. H). I. *Callianassa whiteavesi* Woodward, 1896, named in honor of J.F. Whiteaves. This specimen (RTMP 91.112.140), showing a mold of the outer face of the right major cheliped and the outer face of the left minor cheliped, was collected at Sounding Creek, Alberta, from a locality previously collected by J.B. Tyrrell in 1880, yielding the hypotypes (Feldmann & McPherson 1980: 18). J. Section of a pelleted burrow, (*Ophiomorpha nodosa*), from *Notopocorystes* Assemblage, Britton Formation, Eagle Ford Group, Cretaceous of north-central Texas. Although not mentioned by Stenzel (1945), it is probably the burrow of *Upogebia rhachiochir* Stenzel, 1945, with which it occurs. Figures F, H, I & J, x 1.0; E & G, x1.5.

	1850	1860	1870	1880	1890	1900	1910
Pacific Slope	Dana 1849		Gabb 1869 Gabb 1864		Whiteaves 1885	Woodward 1896 Whiteaves 1895	Rathbun 1908 Whiteaves 1903 Woodward 1900	Rathbun 1917 Rathbun 1916 Dickerson 191
Western Interior		Hall & Meek 1855		Cope 1871	Packard 1881b Packard 1881a Packard 1880 Whitfield 1880	Ortmann 1897 Whitfield 1891	Whitfield 1907	Rathbun 191
Gulf Coastal Plain					Roemer 1887			
Atlantic Coastal Plain	Lyell 1844 Hitchcock 1841?		Stimpson 1863				Weller 1907 Cushman 1905 Weller 1905a Weller 1905b Martin 1904 Pilsbry 1901	Pilsbry 1916

Figure 3. Publication time line for publications on fossil decapod crustaceans of North American. Monographic or seminal works are shown with CAPITAL LETTERS, those of particular importance are shown at right angles and span several regions.

refined by a series of papers such as Glaessner (1929), Bachmayer & Mundlos (1968), and Mundlos (1975). Mertin (1941) described the preservational phenomena that biased the fossil record of decapods, while Schäfer in his *Fossilizations-Bedingungen Brachyurer Krebse* (1951) established the foundation for future experimental work on decapod taphonomy.

Taphonomy of North American decapods has been directly addressed in only a few works such as those by Bishop (1981a, 1981b), Feldmann et al. (1977), and Plotnick (1986). Feldmann et al. (1977) described the taphonomy of the Ft. Peck lobster assemblage, tabulating their final burial position. Bishop (1981a) described the taphonomic history of the *Dakoticancer* Assemblage from the Late Cretaceous Pierre Shale of South Dakota as leading to apatite-preserved decapod community fractions. Important experimental work was initiated by Plotnick (1986) as he studied the decomposition of penaeid shrimp under laboratory conditions, finding that integrity of the remains virtually disappeared within two weeks. Bishop (1986c) gave a holistic review of the taphonomy of North American decapods in a paper presenting concepts generally applicable to decapod taphonomy. Bishop & Williams (1986) applied those concepts to a lobster assemblage from the Cretaceous Carlile Shale of the Black Hills of South Dakota and hypothesized a worm-decapod short cycle that leads to preservation of apatite-preserved decapod assemblages. This theme was documented in a subsequent paper (Bishop 1987a) that postulated a positive taphonomic feedback cycle was generally operable and periodically lead to preservation of decapod community fractions as apatite-preserved, decapod assemblages.

Walker (1988) studied the taphonomic imprints left by hermit crabs inhabiting gastropod shells. This line of study may allow a much better analysis of the presence and abundance of hermit crabs in ancient faunas as the gastropod shell are preferentially preserved over the more easily destroyed decapod cuticle.

	1920	1930	1940	1950	1960	1970	1980	1990	2000
Rathbun 1925b: Fossil Stalk-eyed Crustacea of the Pacific Slope	Withers 1924	Rathbun 1932c Rathbun 1932b Rathbun 1932a		Menzies 1951	Nations 1968	Richards 1975 NATIONS 1975 Jensen 1975 Feldmann 1974 Kooser & Orr 1973 Orr & Kooser 1971	Feldmann 1989 Berglund & Feldmann 89 Bishop 1988c Miller & Ash 1988 Squires 1980	Feldmann et al 1991 Tucker & Feldmann 1990	
Glaessner 1929: Fossilium Catalogue / *Glaessner 1969: Treatise on Invertebrate Paleontology*		Van Straelen 1936 Rathbun 1930a		Holland & Cvancara '58	Waage 1968	Feldmann 1979 Feldmann & West 1978 Herrick & Schram 1978 Schram et al 1978 Feldmann et al 1977 Feldmann et al 1976 Feldmann 1974, 1979 BISHOP 1975 Bishop 74, 76, 77, 78 Bishop 72a, 72b, 73 Feldmann & Holland 1971 Schram 1971	Feldmann & Copeland 88 Bishop 1988b Tucker et al. 1987 Bishop 1987a, 1987b BISHOP 84b, 86b, 86c Bishop & Williams 1986 Bishop 1985c, 1986a Bishop 83d, 84a, 85a Bishop 81a, 82, 83a, Feldmann 1981 Feldmann et al 1981 Kues 1980 FELDMANN & McPHERSON 1980	Bishop & Williams 1991 Bishop 1992b Bishop 1991	
Rathbun 1935b: Fossil Crustacea of the Atlantic and Gulf Coastal Plain	Rathbun 1928 Rathbun 1926a	Rathbun 1936 Stenzel 1935 Stenzel 1934 Rathbun 1930b	STENZEL 1945 Stenzel 1944b Stenzel 1944a Stenzel 1941	Kesling & Reimann '57a Richardson 1955 Stenzel 1951	Davidson 1966 Davidson 1963		Bishop 1988a Bishop 1988d Bishop 1985b Schram & Mapes 1984 Bishop 1983c Bishop 1983b Whetstone & Collins 1982 Bishop 1981b	Bishop 1992a Bishop & Brannen 1992 Vega & Feldmann 1991 Kensley & Williams 1990	
	Rathbun 1929 Davis & Long 1927 Rathbun 1923	Rathbun 1936 Rathbun 1935a Palmer 1935 Rathbun 1931	Kindle 1949	Roberts 1956 Roberts 1955 Roberts 1953	Kesling & Reimann 67 Lewis & Ross 1965 Ross & Scolaro 64 Ross et al. 1964 Roberts 1962		Bishop & Portell 1989 Bishop & Whitmore 1986		

3 CHRONOLOGY OF NORTH AMERICAN PALEOCARCINOLOGY

Although it is clearly impossible to completely document all the papers and authors that have contributed to our knowledge of the North American fossil decapods, I believe a chronological review of the history of the literature is a reasonable way to approximate that effort and begin to draw this information together.

3.1 *Early North American work*

'Dr. Edward Hitchcock was the first definitely to publish the occurrence of the crab remains at Gay Head,' according to Cushman (1905). Hitchcock reported the occurrence of fossil crabs in several earlier reports and, in 1841, published a description of the broken remains for which he '...thought that drawings would not be of use.' Charles Lyell (1844) published two papers on the Gay Head fossil decapods at Martha's Vineyard, Massachusetts. William Stimpson (1863) described *Archaeoplax signifera* from the Gay Head fauna and, subsequently, Alpheus Packard described (1900) *Cancer proavitus* and provided photographs of *A. signifera*. Gabb (1867) described *Callianassa stimpsoni* from the Cretaceous of California and in (1869) described *Cancer brewerii* and synonymized it with *C. productus*. E.D. Cope (1871) described three new species of *Astacus* from Idaho. Roemer (1887) described a crab, *Graptocarcinus texanus*, from the Upper Cretaceous of Texas. Packard (1880, a, 1881b) described a crayfish, *Cambarus primaevus*, from the Eocene Green River Formation of Wyoming, subsequently reassigned to *Procambarus* (*Austrocambarus*) *primaevus* by Feldmann et al. (1981). This body of literature remains obscure because it is difficult for most researchers to procure.

3.2   *The Americans circa 1900*

At the turn of the century there was a small spurt of activity describing decapods from the Western Interior and from the Northern Atlantic Coastal Plain. Ortmann (1897) described two specimens collected by H.F. Wells in the headwaters of Cotton-Wood Creek, Mead (sic) Co., South Dakota, which were rebuilt from fragments by Mr. Gidley at the Princeton Museum. This fauna was restudied and further documented by Bishop & Williams (1986). H.A. Pilsbry (1901) described a number of decapods from the Cretaceous of New Jersey, including *Hoploparia gabbi, H. gladiator, Callianassa mortoni,* and *C. conradi.* Stuart Weller (1905a, 1905b) published *The Fauna of the Cliffwood (N.J.) Clays* in which he described *Tetracarcinus subquadratus,* and said, 'One of the most notable features of the fauna from these Cliffwood nodules is the great number of crustacean remains. Nearly every one of the concretions, when broken, yields the remains, more or less fragmentary and crushed, of one of these creatures; indeed, a crab of some sort seems to have been the nucleus around which every one of these concretionary nodules in the clay has been formed.' Weller (1907) followed his initial decapod work with a review of all the then known Cretaceous decapods of New Jersey, including Pilsbry's taxa.

Cushman (1905) described the decapod fauna collected from the greensand at Gay Head (Martha's Vineyard, Massachusetts) in a paper that is remarkably modern in all aspects. H.A. Pilsbry (1916) described fossil decapods from the Cretaceous of Maryland.

3.3   *The Canadians circa 1900*

Early geological investigations of the Rocky Mountains allowed geologists and paleontologists the opportunity to amass large collections of fossils. One of these collectors, J.B. Tyrrell, collected on Sounding Creek in southern Alberta in 1880 and 1886. The Royal Tyrrell Museum of Paleontology is named after this early worker. The extensive callianassid fauna collected by Tyrrell is currently being reexamined by my colleagues Nancy Brannen and Paul Johnston, and me. In the late 19th century and early 20th century, three workers, Whiteaves, Woodward, and Whitfield, described 10 species of fossil decapods from all the decapod material collected from the Canadian Cretaceous until that time (Feldmann & McPherson 1980). J.F. Whiteaves (1885) described a lobster, *?Hoploparia canadensis,* from the Cretaceous of the Highwood River, and later (Whiteaves 1895) described *Podocrates vancouverensis* from the Nanaimo Group of Vancouver Island. Henry Woodward (1896) described *Callianassa whiteavsii, Homolopsis richardsoni, Palaeocorystes harveyi,* and *Plagiolophus vancouverensis* from the Cretaceous of British Columbia. Later, he (1900) reassigned *P. vancouverensis* to *Linuparus,* and described *Eryma dawsoni, Hoploparia westoni, H. bennetti, Enoploclytia minor,* and *?Meyeria harveyi.* Whiteaves (1903) described an unnamed species of *Glyphaea* and synonymized *Linuparus atavus* Ortmann with *L. canadensis.* R.P. Whitfield (1907) described a new lobster, *Hoploparia browni,* from the Cretaceous Bearpaw Shale of Montana, a taxon subsequently referred to *Palaeonephrops browni* (Whitfield) by Mertin (1941) and emended by Feldmann et al. (1977). No additional decapod taxa were described from Canada until 1960 (Feldmann & McPherson 1980: 1). This brief spurt of research typified so much of the early study of decapod paleocarcinology: A few workers turning on to opportunities afforded by collections, exposures, or interest, yielding a sporadic literature centered about a limited area, perhaps because of travel limitations of the times.

3.4   *The Rathbun years*

Mary Jane Rathbun began her publishing career in 1891 (Schmitt 1973), publishing a total of

166 papers, monographs, and books, 27 of these (16%) on fossil decapods of North American. Rathbun's involvement with fossil material began in 1908 with a publication describing a small Pliocene fauna, including *Loxorhynchus grandis*, *Cancer fissus* (new species), *Branchiolambrus altus*, and *Archaeopus antennatus* (new genus, new species), from the Northwest; followed in 1916 by a description of an Eocene crab, (new genus, new species) from Port Townsend, Washington; and in 1917 (Rathbun 1917a, b) by descriptions of a California Pliocene fauna and a South Dakota Cretaceous fauna. Sporadic papers on fossil material appeared in her bibliography until the years from about 1924 to 1932 when a burst of publication on fossil taxa occurred roughly balanced with publication on Recent taxa (see bibliography in Schmitt 1973). After this period, Rathbun largely focused again on Recent taxa.

During the latter part of her career, Mary Rathbun significantly impacted paleocarcinology with two monographs on fossil North American decapods, *The Fossil Stalk-Eyed Crustacea of the Pacific Slope of North America* (1926b) and *Fossil Crustacea of the Atlantic and Gulf Coastal Plain* (1935b). Modern critics often question the validity of so many taxa based on very fragmentary remains, particularly taxa in the monograph on the Atlantic and Gulf Coastal Plain. Although true, this criticism ought to be offered humbly, as the bulk of Rathbun's work with fossil taxa is basically sound. During her later years (while actively writing her Atlantic and Coastal Plain monograph), she apparently suffered from severe memory loss, and perhaps this combined with her naming such fragmentary remains led contemporary colleagues and subsequent students to doubt not only much of her later work, but by association, the validity of her paleontological work.

3.5 *Regional analysis*

As with so many aspects of science, the great depression and World War II seem to form a natural and distinct boundary in paleontology between classical science and new scientific methodologies and attitudes, aided by rapidly developing technologies. Not only was scientific technology rapidly advanced by the war, but the ability and opportunity to travel was enhanced by modern automobiles and airplanes. Coupled to these advances was a burgeoning geologic exploration and knowledge fueled by the search for petroleum. At this boundary I will also change my attack on history and approach our knowledge of North American decapod paleocarcinology from a regional perspective. This change from a chronological analysis to regional analysis is in keeping with the rapidly expanding technology and ability to travel within regions, but will necessitate some redundancy in each region for the sake of continuity.

3.5.1 *Pacific Slope*
The historical development of paleocarcinology on the Pacific Slope includes a spate of very early publications including Dana (1849), Gabb (1864, 1867, 1869) followed by publication of several papers at the turn of the century by the Canadians, Whiteaves (1895, 1900, 1903) and Woodward (1896, 1900). This was followed by Rathbun's initial paleocarcinological contribution in 1908, which culminated in her 1926 monograph on Pacific Slope decapods. That publication initiated an hiatus of almost half a century interrupted only by publication of a few papers. Since 1971 decapod research on the Pacific Slope has seen a resurgence with the appearance of papers by Nations (1971), Orr & Kooser (1971), Kooser & Orr (1973), Richards (1975), Berglund & Feldmann (1980), Bishop (1988c), Feldmann (1989), Tucker & Feldmann (1990), and Feldmann et al.(1991).

In 1849 Dana described a new fossil decapod, *Callianassa oregonensis* from the Miocene of Oregon. Gabb (1864) described *Callianassa stimpsoni* and subsequently cited it, along with other taxa, in 1867 and 1869. Professor Bruce L. Clark (University of California) sent collections made by Harold Hannibal in Washington and on Vancouver Island to Mary Rathbun for study. J.F. Whiteaves (1900) described *Zygastrocarcinus richardsoni* from the Queen Charlotte Islands of British Columbia and subsequently (1903) described a new glypheid and reassigned *Linuparus atavus* Ortmann, 1897, to *L. canadensis*. In (1908), Mary Jane Rathbun described two new species, *Loxorhynchus grandis* and *Cancer fissus*, and two new genera and species, *Branchiolambrus altus* and *Archaeopus antennatus*, from the Miocene of California, launching the paleontological aspect of her career. In 1916 she described *Branchioplax washingtoniana*, from Port Townsend, Washington, and later (1917) described another new species, *Cancer urbanus*, from the California Pliocene. Thomas H. Withers (1924) described *Callianassa clallamensis* and *Ranina americana* from the Oligocene of Washington. In 1926, Rathbun published her monograph *The Fossil Stalk-eyed Crustacea of the Pacific Slope of North America*, which reviewed all previously described taxa, including 54 new species.

Current research includes efforts by Feldmann, Tucker, Bergland, and Richards that should soon result in numerous papers describing the beautifully preserved faunas of the Olympic coastlines.

3.5.2   *Eastern Margin*

The history of paleocarcinology on the Atlantic and eastern Gulf coastal plains include the earliest notes of fossil decapods by Hitchcock (1841), Lyell (1844), and Stimpson (1863). These early notes were followed by several papers published at the turn of the century: Packard 1900; Pilsbry 1901, 1916; Martin 1904; Cushman 1905; and Weller 1905, 1907 (see Fig. 2C, D). Beginning with her 1923 paper, nearly all research on fossil decapods of the East Coast was done by Rathbun at the US National Museum (Rathbun 1923, 1926a, 1929, 1931, 1935a, 1935b, and 1936), all but culminating her career with her 1935 monograph (Rathbun 1935b). Subsequently, Henry Roberts (1953, 1955, 1956, 1962) published on fossil decapods of the East Coast (Figs 1C, 2E), and Ross & Scolaro (1964), Ross et al. (1964), and Lewis & Ross (1965) did some Florida work. A little additional work has since been done: Bishop & Whitmore (1986), Bishop & Portell (1989), and Portell (1991).

Rathbun (1923) described a Cretaceous crab, *Avitelmessus grapsoideus*, from the Cretaceous of North Carolina. She (Rathbun 1926a) described the decapods of the Ripley Formation of Coon Creek, Tennessee, listing ten species of decapods, including *Linuparus canadensis* (Whiteaves), *Protocallianassa mortoni* (Pilsbry), *Dakoticancer overana* Rathbun, and *Avitelmessus grapsoideus* Rathbun, and describing six new species, including *Peneus wenasogensis*, *Hoploparia mcnaieyensis*, *Hoploparia tennesseensis*, *Enoploclytia sculpta*, *Eryma flecta*, and *Raninella testacea*. In 1935 Rathbun published her monograph summarizing knowledge of fossil crustaceans of the Atlantic and Gulf coastal plains, which included descriptions of three new genera (*Prehepatus*, *Ophthalmoplax*, *Scyllarella*) and 88 new species.

3.5.3   *Gulf Coastal Plain*

The early paleocarcinological history of the Gulf Coastal Plain is surprisingly sparse. Roemer (1887) described *Graptocarcinus texanus* from the Upper Cretaceous of Texas, but the next published research was produced by Rathbun (1926a, 1928, 1935b). In the Gulf Coastal Plain, the pattern seen elsewhere of a decline of fossil decapod research after publication of Rathbun's monographs was thwarted by overlapping of research activity by Henryk Stenzel (Figs 1B, 2B). Stenzel began publishing (1934, 1935) before Rathbun's monograph was published and continued his activity into the war years (1941, 1944a, b, 1945) and beyond (1951). Except for a couple of papers during the 1950's and 1960's (Richardson 1955; Davidson 1963, 1966) very little work was done on the Gulf Coastal Plain until the 1980's (Bishop 1981b, 1983b, c, 1985b, 1986d, 1988a; Whetstone & Collins 1982).

Stenzel's publishing presents an interesting anomaly, because if it had not been for his work during the war years, the hiatus in fossil decapod work following publication of Rathbun's 1935 monograph would have been continent wide! The Gulf Coastal Plain remains an interesting and active research objective for fossil decapods as is evidenced by recent work by me and my colleagues (e.g. Bishop & Brannen 1992).

3.5.4 *Western Interior*

Early paleocarcinological work in the Western Interior consisted of description of freshwater decapods (crayfishes) by Hall & Meek (1855), Cope (1871), and Packard (1880, 1881a, b). Ortmann described a new lobster from South Dakota (1897) and Whitfield described a lobster from Montana (1907). Rathbun (1917b) described several new species of Cretaceous decapods from the Pierre Shale of South Dakota including *Dakoticancer overanus* (Figs 2F-H), *Homolopsis punctata*, and *Campylostoma pierrense*; later (1930a) she described *Callianassa cheyennense* from the same formation. Essentially a major hiatus in research on fossil decapods of the Western Interior occurred from 1917 to 1972; except for Rathbun (1930a) and a short paper by Holland & Cvancara (1958) describing an extremely interesting crab fauna from the Cannonball Formation of North Dakota, little work was published. In 1971, a burst of publication began by Feldmann and colleagues (1971, 1976, 1977, 1979, 1981, 1988); Bishop (1972a, b, 1973, 1974, 1975, 1976, 1977, 1978, 1981a, 1982, 1983a, d, 1984a, b, 1985a, c, 1986a, b, c, 1987a, b, 1988b, 1991, 1992b,); Herrick & Schram (1978); Kues (1980); and Bishop & Williams (1986, 1991). This body of publications may represent the most coherent research on fossil decapods of any region, ranging from the purely descriptive to almost the purely speculative and interpretive.

The richly fossiliferous beds of the Western Interior present virtually unlimited possibilities for discovering additional taxa and resolving significant questions of decapod paleontology such as rates of evolution, ancestral relationships of modern taxa, and paleocommunity structure.

3.6 *The torch is passed*

Henry B. Roberts (Fig. 1C) joined the staff of the National Museum of Natural History and in effect should have carried on the traditions of an active program in the study of fossil decapods forged by Mary Rathbun. He published several small papers and a small monograph (1962) on the decapods of the Northern Atlantic Coastal Plain. Unfortunately this definitive work was published with poor quality plates, virtually erasing its usefulness to subsequent workers. In 1971, I visited Roberts at his office in the National Museum and found him to be very scholarly, in the words of one of my reviewers, 'He was an absolute encyclopedia of knowledge!' Roberts impressed me as being very conservative, in person, in earlier correspondences, and in reviews of my early papers. It would be interesting to speculate on the reasons for Roberts's inability to capitalize on his eminent potential to become a leader in paleocarcinology. Perhaps his conservatism and perfectionism had a deleterious effect on his efforts to publish.

Henryk Bronislaw Stenzel (1899-1980) entered the paleocarcinological stage (1934) with a redescription of *Harpactocarcinus americanus* Rathbun and descriptions of *H. rathbunae,*

Zanthopsis peytoni, *Calappilia diglypta*, *Callianassa brazoensis*, and *C. wechesensis* from the Middle Eocene of Texas. He later (1935) described *Lobonotus natchitochensis* and *L. bra-zosensis* from the Middle Eocene of Louisiana and Texas, emended the description of *H. americanus*, described *?Portunus vicksburgensis* from the Oligocene of Mississippi, and named and provided a key for several species of *Callianassa*. In 1941 Stenzel named and illustrated, and, in 1945, described most of the known Cretaceous decapods from the Eagle Ford including *Linuparus grimmeri*, *L. watkinsi*, *Homarus brittonestris*, *Notopocorystes dichrous*, *Upogebia rhacheochir*, *Necrocarcinus ovalis*, and *Cenomanocarcinus vanstraeleni*. In 1944 he published papers describing *Graptocarcinus muiri* from Mexico (1944a) and *Tehuacana tehuacana* from the Paleocene of Texas (1944b). In 1945, Stenzel's monographic work, *Decapod Crustaceans from the Cretaceous of Texas*, appeared. This work includes sections on comparative morphology (Fig. 2B) and stratigraphy, which are clearly presented, as well as individual descriptions, which are definitive and complete with data on locality, preservation, and associated fauna. Stenzel's final work on decapods was a description of a fauna from the Woodbine Formation of Texas published in 1951, in which he described fragments questionably assigned to a palinurid and callianassid, emended his description of *Cenomanocarcinus vanstraeleni*, and described *Woodinax texanus* (new genus, new species). Although Stenzel's publications on decapods only consist of six papers (Young 1982) and a monograph, the work is clearly of superior quality and is significantly focused. It has provided a model to be emulated. At a meeting in Keith Young's lab at the Department of Geosciences, University of Texas in Austin in 1971, Stenzel stopped by to examine collections I had made from the South Dakota *Dakoticancer* Assemblages; he carefully examined the specimens in my cabinet, silently walked to the door, opened it, turned and said, 'You have my blessing,' and left.

3.7　*The big monographs*

The history of the study of fossil decapods in North America has been strongly affected by the appearance of several large monographic works, not all dealing directly with North American faunas. In 1929, Martin F. Glaessner published the *Decapoda* volume of the *Fossilium Catalogus* that taxonomically summarized all work to that time, making the work of subsequent paleontologists much easier. This comprehensive monograph not only listed all described species with the related literature, but also summarized the stratigraphic and geographic distribution of decapod crustaceans. This data base is still being used today as it has not been superseded by a revision (which was being compiled by Reinhard Förster prior to his untimely death in 1987).

In 1962, the second volume (Richards 1958) of the *Cretaceous Fossils of New Jersey* appeared with a section on *Crustacea* by Henry B. Roberts. In his section, Roberts reviewed the known Cretaceous decapod crustaceans from the Northern Atlantic Coastal Plain and added many new species. Sylvie Secretan (1964) produced a dissertation publication on the Upper Jurassic and Cretaceous decapods of Madagascar that remains little known. However, the descriptions, illustrations, morphology, and systematics are presented very clearly. In 1969, Luis Via Boada produced a most impressive and comprehensive 479 page monograph, *Crustáceos Decápodos del Eoceno Español*, that described and reviewed the morphology and systematics of the decapods from the Spanish Eocene. The year 1969 also saw the publication of the *Treatise on Invertebrate Paleontology, Part R* , the *Decapoda*, by Martin F. Glaessner. In this 252 page monograph Glaessner presented the general morphology, ecology, and systematics of the decapod crustaceans as well as a systematic diagnosis of all valid genera illustrated by species-level taxa from each genus. This treatment differed from that of his *Decapoda* in the *Fossilium Catalogus* by illustrating taxa and addressing aspects of preservation, morphology, ecology, paleoecology, as well as systematics. However, synonymies of species, presented so nicely in the *Fossilium Catalogus*, are missing from the *Treatise on Invertebrate Paleontology*. Glaessner's *Decapoda*, above all other monographs, stabilized paleocarcinology, allowing paleontologists to rapidly assess the significance of the decapod crustaceans and setting an example for the synthesis of subsequent research.

C.W. Wright & J.S.H. Collins (1972) monographed British Cretaceous crabs in a handsome volume published by The Palaeontographical Society of London. In this important work Wright & Collins reviewed 26 genera (4 new) and 62 species or subspecies (27 new), providing locality data, stratigraphic ranges, concise diagnoses and comparisons, clearly drawn illustrations, and high quality halftone plates. This volume which updated previous generations of monographs on British fossil decapods (e.g. Bell 1857) will remain a standard far into the future. Often cited, this monograph has stimulated research on Cretaceous decapods and served as a scholarly data base for those working on North American Cretaceous decapods.

3.8 *The modern chronicles*

Much of the history of research on fossil decapods depended on the slow accumulation of knowledge and experience by individuals who happened across fossil decapods, found them intriguing, and published small, descriptive papers on their discoveries. This process often resulted in the researchers developing a deeper interest in decapod fossils and subsequently searching for additional remains, and often publishing a string of additional descriptive papers. In a few instances, interest built to a level of activity that eventually resulted in the publication of a major work on fossil decapods, perhaps a monograph or book.

The accumulation of inventory-level papers seemed to be consistently pursued through the last 150 years at a pace that did not significantly change until the 1970's, except in cases where individuals acquired experience and expertise that resulted in periods of increased activity within their careers (see Section 3.4). A relatively new phenomenon in the study of fossil decapods is the dedication of nearly entire careers to the study of paleocarcinology by members of the current generation of paleontologists including the late Reinhard Förster of Germany, Luis Via Boada of Spain, Rodney Feldmann and myself of the United States. These paleotologists have not only significantly increased the inventory of known fossil decapods, but have acted as foci of activity for colleagues, stimulating and assisting the further study of decapod crustaceans in collaborative papers and indirect support. It now appears that a second generation of such dedicated workers is evolving in teacher-student lineages (e.g. Rod Feldmann (Fig. 1D) and his student Ann Tucker (see Tucker & Feldmann 1990, Feldmann et al. 1991). The discussion presented in section 4 will try to define the contribution of these scientists who have documented the fossil decapods of North America.

A group deserving special recognition for their service to decapod paleocarcinology is the paraprofessionals or serious amateur collectors. This group of interesting and interested persons accumulates decapod remains and often unselfishly share their specimens and localities with decapod paleontologists. The contribution of these persons to our understanding of the geologic history of the decapods is extremely significant and pervasive. Some have been acknowledged in publications resulting from their labors or even given coauthorship of specific papers (e.g. Feldmann et al. 1991); others have been honored by the assignment of patronyms (e.g. *Zygastrocarcinus mendryki* (Bishop 1982), after Harry Mendryk; *Notopocorystes (Euco-*

rystes) eichhorni (Bishop 1983d) after Larry Eichhorn, and *Dromiopsis kimberlyae* (Bishop 1987a) after Kimberly Bishop).

4 STRATIGRAPHY OF THE NORTH AMERICAN DECAPODS

4.1   *The North American Paleozoic record*

The earliest recognized decapod crustacean is *Palaeopalaemon newberryi* Whitfield, 1880, from the Devonian of North America (Schram et al. 1978; see Fig. 2A). This ancestral decapod is known from the stable interior in Iowa, Ohio, New York, and Kentucky. Its assignment to the decapods, which pushes the range of decapods back in time almost 100 million years, is questionable, as *P. newberryi* shows an unusual array of characters. The lack of conformance with more modern decapod characters is understandable because of the extreme age of *Palaeopalaemon*; until its recognition a taxonomic scheme for decapods was based largely on a Mesozoic-Cenozoic record. The extreme age of *Palaeopalaemon* also indicates a significant gap in our knowledge of the early record (Late Paleozoic) of decapods (Schram et al.1978: 1386).

Schram & Mapes (1984) described *Imocaris tuberculata*, from the Upper Mississippian of Arkansas. This remarkable taxon is extremely crab-like and was assigned to the brachyuran Subsection Dromiacea (an extant group), extending its range back from the Early Jurassic to the middle of the Carboniferous (Fig. 4).

4.2   *The North American Mesozoic decapod record*

The chronicling the North American Mesozoic record of the decapod crustaceans illustrates the general development of decapod paleocarcinology. The Triassic and Jurassic have poorly known decapod faunas and the Cretaceous has well documented faunas (Fig. 4). The future will reveal if the apparent lack of Triassic and Jurassic fossil decapods is real or an artifact of collecting, preservation, or habitat. The documentation of the European record, which is almost the same as saying the North American record because the North Atlantic Ocean was only beginning to open at that time, would seem to suggest a diverse and substantial record is present. I believe that the sparse North American record, at least in the Jurassic, which is well represented by marine sedimentary rocks, is due mostly to lack of decapod collecting.

4.2.1   *Triassic*
The Triassic rocks of North America are restricted in the east to limited fillings of fault block basins and are often buried beneath younger coastal plain sediments, providing little opportunity and usually the wrong habitat for the predominantly marine decapods. Triassic rocks of the West consist largely of terrestrial rocks with limited marine sedimentary rocks. Only a few taxa have been documented from the North American Triassic. These include *Litogaster turnbullensis* Schram, 1971, from the Thaynes Formation near Hot Springs, Idaho; *Pseudoglyphaea* Mulleri (Van Straelen 1936) from the upper Triassic of Nevada; *Erymastacus* from the Lower Jurassic of Canada; and *Enoploclytia porteri* Miller & Ash, 1988, from the Late Triassic Chinle Formation of the Petrified Forest National Park, Arizona. *Enoploclytia porteri* is especially interesting because it is the oldest known freshwater crayfish.

Litogaster turnbullensis exemplifies a common decapod scenario, the fortuitous discovery of an isolated specimen, its recognition as a rarity, and subsequent description. When discovered by vertebrate paleontologists, this unique specimen was lying exposed on a slab of rock, in talus, on a ridge east of Hot Springs, Idaho. The specimen went undescribed from 1950 until Schram's paper in 1971.

4.2.2 *Jurassic*

The Jurassic rocks of North America are virtually restricted to the western half of the continent or are deeply buried beneath younger sediments on the Gulf Coastal Plain. During the Jurassic the North Atlantic Ocean was just beginning to open while marine deposition was controlled by the distribution of the circumequatorial Tethys Seaway. The Rocky Mountain region was beginning to undergo sporadic orogeny resulting in emplacement of large batholiths and rapid tectonic fluctuations of sea level. As sea level fluctuated, marine lithosomes were deposited in the western areas of the continent. These rocks, some of them marine, have not yet been seriously searched for included decapod faunas. In spite of this, a few decapods have already been described as they have turned up in mapping or paleontological exploration.

Herrick & Schram (1978) described a marine decapod fauna from the Stockade Beaver member of the Sundance Formation of Wyoming, which included *Antrimpos* sp., *Bombur* sp., *Mecochirus* sp., a glypheid, and unidentifiable anomuran remains. Feldmann (1979) described another taxon, *Eryma foersteri*, from the same fauna.

Feldmann & McPherson (1980) reviewed most of the known Canadian decapods including two Jurassic taxa, *Eryma bordensis* (Copeland 1960) and *Glyphaea robusta*, new species. Feldmann & Copeland (1988) described a new erymid lobster, ?*Eryma ollerenshawi*, from the Lower Jurassic of Southwestern Alberta, a specimen preserved as a compression film.

This sparse Jurassic fauna, when compared to the extensive European record, is quite obviously severely limited, probably by lack of collecting. As North American decapod workers begin to search for Jurassic decapods in earnest, this record will almost surely become as robust as the European.

4.2.3 *Cretaceous*

The Cretaceous rocks of North America were deposited largely under marine conditions and carry robust marine faunas. Important sequences of little-deformed marine sedimentary rocks are known from the Atlantic and Gulf Coastal plains, including the Mississippi Embayment, outcropping as a narrow, continuous band from New England into old Mexico. Cretaceous marine and non-marine sedimentary rocks of the Western Interior Seaway are sporadically exposed from northern Texas into the Canadian Arctic, often tectonically modified and occasionally covered by younger sediments. Cretaceous marine sedimentary rocks are present on the West Coast as scattered outcrops. These widely distributed marine rocks carry abundant decapod faunas that have received significant and long-standing attention from numerous workers. The decapod record of the Lower Cretaceous is sparse and restricted to a few specimens from the Neocomian of Oregon (Feldmann 1974), a lobster from the Middle Albian Kiowa Shale of Kansas (Feldmann & West 1978), and a diverse decapod fauna consisting of 11 species from the Lower Albian Glen Rose Limestone of Central Texas (Bishop 1983b). The decapod record of the Late Cretaceous is robust and constantly increasing in quantity and quality, represented by numerous papers by several authors. It is these Late Cretaceous faunas that launched decapod paleontology in North America beyond the inventory level and set the stage for analyses of evolution, biostratigraphy, paleoecology, functional morphology, and taphonomy.

Age (Ma)	Period	Epoch	Pacific Slope	Western Interior	Gulf Coastal Plain	Atlantic Coastal Plain
0–10	Tertiary	Pleistocene / Pliocene	Nations 1968 Gabb 1869 Rathbun 1917a, 1932a			
10–20	Tertiary	Miocene	Dana 1849 Rathbun 1908, 1932b,	Cope 1871		Bishop & Portell 1989 Cushman 1905 Martin 1904 Packard 1900 Lyall 1844 Hitchcock 1841 Stimpson 1863
20–30	Tertiary	Oligocene	Withers 1924			
30–50	Tertiary	Eocene	Feldmann et al. 1991 Tucker & Feldmann 1990 Benglund & Feldmann 89 Jensen 1975 Kooser & Orr 1973 Kooser & Orr 1973 Orr & Kooser 1971 Dickerson 1916 Rathbun 16, 17a, 30b, 32c Rathbun 1908	Feldmann et al. 1981 Packard 80,81a, 81b	Rathbun 1928 Stenzel 34, 35 Rathbun 1930b	Kensley & Williams 1990 Lewis & Ross 1965 Ross & Scolaro 1964 Ross et al. 1964 Roberts 1953, 1955 Palmer 1935 Rathbun 1929
50–65	Tertiary	Paleocene	Squires 1980 Feldmann 1989	Feldmann & Holland 1971 Feldmann et al. 1976 Holland & Cvancara 1958	Stenzel, 1944b Davidson 1966	Roberts 1956 Davidson 1966 Bishop & Whitmore 1986
65–72	Cretaceous	Maastrichtian	Gabb 1864 Richards 1975	Feldmann et al. 1976 Tucker et al. 1987 Bishop 72a, 72b, 73, 74, 76, 77, 78, 81a, 82, 83a, 84a, 91	Bishop 1986d Bishop 1985b Bishop 1986a	Kindle 1949 Rathbun 31, 35a
72–82	Cretaceous	Upper Campanian / Lower Campanian	Bishop 1986c Woodward 96, 1900 Rathbun 1908 Whiteaves 1895, 1900, 1903	Rathbun 1917b, 1930a BISHOP 75, 84b, 96b, 86c, 87b Feldmann et al. 1977 Bishop 76, 83, 85a,85c, 84d, 87a, 89b Feldmann 1991, Whitfield 1907	Bishop 81b, 83c Rathbun 1926a Whetstone & Collins 1982 Kesling & Reimann 1957	Rathbun 1926 Rathbun 1923 Pilsbry 1901, 1916 Davis & Lang 1927 Weller 1905a, 1905b, 1907
82–87	Cretaceous	Santonian		FELDMANN & McPHERSON 1980	Roemer 1887	
87–90	Cretaceous	Coniacian / Turonian		Bishop & Williams 1996 Ortmann 1897 Kues 1980 Hall 1865	Richardson 55 Davidson 63 STENZEL 1941, 1944a, 1945, 1961	
90–100	Cretaceous	Cenomanian				
100+	Cretaceous	Upper Albian	Feldmann 1974			

Rotated/marginal annotations: Nations 1975; Rathbun 1926b; McPherson 1980; Feldmann & McPherson 1980; Rathbun 1935; Rathbun 1835; Roberts 1962

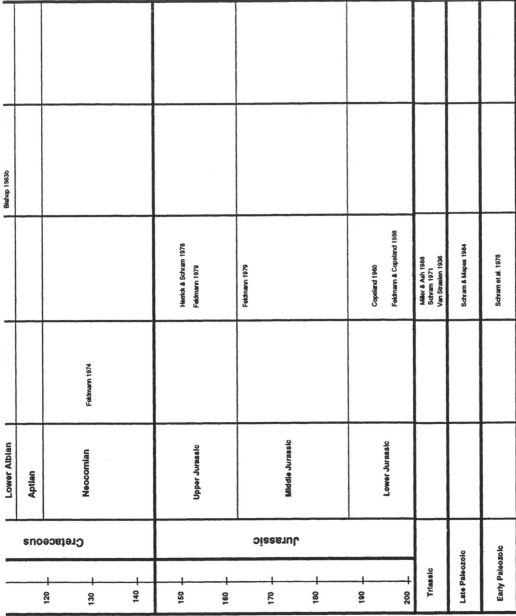

Figure 4. Stratigraphic summary of literature on fossil decapod crustacea of North American. Monographic or seminal works are shown with CAPITAL LETTERS, those of particular importance are shown at right angles and span several stratigraphic units. Approximate stratigraphic position is depicted by position on the diagram.

Numerous papers provide an excellent inventory of Cretaceous lobsters and crabs. The shrimp, as is typical, remain virtually non-represented in the North American fossil record. The earlier literature on the Cretaceous decapods is quite regional. At a time when transportation was slow and cumbersome most workers were relatively restricted in their travel opportunities. Local or regional faunas were studied by most workers, including Pilsbry and Weller in the Northern Atlantic Coastal Plain and Stenzel on the Gulf Coastal Plain, or local faunas were collected by local geologists and passed along to decapod paleontologists for description, such as the case of Mary Rathbun at the National Museum of Natural History. Rathbun seldom, if ever, saw the decapod fossils in place or in their geological context. This in turn led to a plethora of descriptive papers with scant stratigraphic documentation and often incorrect stratigraphic information. Data on the collected decapods were set in the time of the description, and must always be adjusted for new data and concepts.

Special notice needs to be given to the work of Stenzel, which, although relatively limited in quantity, remains truly superior in quality and significance. Even taken in the context of today, his papers and monograph, *Decapod crustaceans from the Cretaceous of Texas*, represent classical models for the presentation of paleontological data in scientific publication. His small monograph addressed functional morphology, systematics, and morphology of an entire fauna, placed the fauna in geological context, and provided subsequent workers with a firm foundation upon which to build.

4.3   *The North American Cenozoic record*

The Cenozoic rocks of North America were deposited largely on prograding continental margins, except for a remnant of the Western Interior Seaway preserved as the Cannonball Formation of south-central North Dakota. This Paleocene marine remnant has yielded both lobsters (Feldmann & Holland 1971) and crabs (Holland & Cvancara 1958). Terrestrial lacustrine sediments of the Green River Basin have yielded freshwater decapods (Feldmann et al. 1981).

The decapod faunas of the continental margins have been described in numerous descriptive papers, often of a regional flavor. Rathbun (1928, 1930a) and Stenzel (1934, 1935, 1944b) described the Paleocene and Eocene taxa of the western Gulf Coastal Plain. The decapod faunas of Florida were described by Rathbun (1929), Roberts (1953, 1955, 1956), Ross & Scolaro (1964), Ross et al.(1964), and Lewis & Ross (1965). Kensley & Williams (1990) described a unique silicified thalassinoid from the middle Eocene of South Carolina.

The northern Atlantic Coastal Plain has yielded a decapod fauna that must take a special place in North American decapod paleontology. The decapods from Gay Head at Martha's Vineyard, Massachusetts, were the first decapod fauna from our continent described and published on. The fauna was reported as early as 1841 by Hitchcock, studied by Charles Lyell (1844), and published on by J. A. Cushman (1905) who went on to help establish the study of Foraminifera.

The Eocene, Oligocene, Miocene, and Pliocene rocks of the Pacific Slope have yielded a well described and beautifully preserved series of decapod faunas presented in papers by Dana (1849), Dickerson (1916), Withers (1924), Rathbun (1908, 1916, 1917a, 1930b, 1932a, b, c), Nations (1968, 1975), Orr & Kooser (1971), Kooser & Orr (1973), and Tucker & Feldmann 1990). From among these small papers have come two monographic works: Rathbun's (1926b) monograph, *The Fossil Stalk-eyed Crustacea of the Pacific Slope of North America*, and a very nice little monograph by Dale Nations (1975), *The Genus Cancer (Crustacea:Brachyura): Systematics, Biogeography, and Fossil RecordD*.

The North American record of Eocene decapods will eventually become one of the world's significant faunas. Robust faunas exist in carbonate lithosomes of the Florida Platform and the southern Atlantic Coastal Plain. These faunas have been hinted at by Bishop & Whitmore (1986) and, because of their strategic position within the last major adaptive radiation of decapod crustaceans, will allow us to decipher much of the early history of extant taxa.

5 CONCLUSIONS

This review of the history of decapod paleocarcinology of North America illustrates that the efforts of many persons have built a significant body of knowledge (Fig. 3). The literature indicates there was first a slow accumulation of papers on local faunas from near where early paleontologists lived. This was followed by monographic efforts of Rathbun, Stenzel, and Roberts which produced summaries of decapod faunas for various regions of North America. Glaessner's monograph (1969) in *Treatise on Invertebrate Paleontology*, as well as a systematic monograph he published in 1929 in the *Fossilium Catalogus*, provided a firm conceptual foundation in decapod paleocarcinology for subsequent workers. Since 1970, there has been an explosion of publication as workers were able to easily travel from area to area documenting widely separated faunas.

This later literature also indicates another major change in decapod paleontology in North America. The work of earlier scientists who were only occasionally involved with fossil decapods, although in some cases (i.e. Rathbun and Stenzel) responsible, none the less, for major pulses of research, has been partly replaced by that of those dedicating most of their careers to decapod paleontology, a tradition that has probably existed for a long time in Europe (e.g. the careers of Van Straelen, Woods, Bell, and Glaessner). The recognition of the abundance of fossil decapods in North America has led Rodney Feldmann and me to dedicate most of our professional careers to studying them.

Decapod paleocarcinology has progressed from the production of inventory-type papers describing new species, through a more holistic approach focusing on broader areas and time intervals, to today's eclectic integration of paleontology, ecology, taphonomy, biogeography, and phylogeny. I anticipate that many present and future students of the fossil decapod crustaceans will build on this foundation, gradually increasing our knowledge until we have a nearly complete record of the geological history of the decapods in North America.

If any lessons are to be gained from a historical perspective such as this, I guess an important one should be that we encourage and nurture each and every student interested in decapod paleontology, for she, or he, may be the next scientist to become enraptured by the mystery and intrigue of solving the myriad problems associated with the preservation of fossil decapods and develop into the next Mary Jane Rathbun or Martin F. Glaessner.

REFERENCES

Bachmayer, F. & R. Mundlos 1968. Die tertiären Krebse von Helmstedt bei Braunschweig Deutchland. *Ann. Naturhistor. Mus. Wien* 72:649-692.

Bell, T. 1857. A monograph of the fossil malacostracous Crustacea of Great Britain. Pt I. Crustacea of the London Clay. *Palaeontograph. Soc. Lond.* 10(40):10-40.

Berglund, R.E. & R.M. Feldmann 1989. A new crab, *Rogueus orri*, n. gen., n. sp. (Decapoda: Brachyura), from the Lookingglass Formation (Ulatisian Stage: Lower Middle Eocene) of southwestern Oregon. *J. Paleont.* 63: 69-73.

Bishop, G.A. 1972a. Crab bitten by a fish from the Upper Cretaceous Pierre Shale of South Dakota. *Geol. Soc. Amer. Bull.* 83:3823-3826.

Bishop, G.A. 1972b. Moults of *Dakoticancer overanus*: An Upper Cretaceous crab from the Pierre Shale of South Dakota. *Paleontology.* 15:631-636 and 122-123.

Bishop, G.A. 1973. *Homolopsis dawsonensis*: A New Crab (Crustacea, Decapoda) from the Pierre Shale (Upper Cretaceous, Maastrichtian) of Cedar Creek Anticline, Eastern Montana. *J. Paleont.* 47(1):19-20.

Bishop, G.A. 1974. A sexually aberrant crab (*Dakoticancer overanus* Rathbun, 1917) from the Upper Cretaceous Pierre Shale of South Dakota. *Crustaceana* 26:212-219.

Bishop, G.A. 1975. Traces of Predation. In R.W. Frey (ed.), *The Study of Trace Fossils*: pp. 261-281. New York: Springer Verlag.

Bishop, G.A. 1976. *Ekalakia lamberti* n. gen., n. sp. (Crustacea, Decapoda) from the Upper Cretaceous Pierre Shale of Eastern Montana. *J. Paleont.* 50:398-401.

Bishop, G.A. 1977. Pierre Feces: A scatological study of the *Dakoticancer* Assemblage, Pierre Shale (Upper Cretaceous) of South Dakota. *J. Sediment. Petrol.* 47:129-136.

Bishop, G.A. 1978. Two new crabs: *Sodakus tatankayotankaensis* n. gen., n. sp. and *Raninella oaheensis* n. sp., from the Pierre Shale (Maastrichtian) of South Dakota. *J. Paleont.* 52:609-617.

Bishop, G.A. 1981a. Occurrence and fossilization of the *Dakoticancer* Assemblage, Upper Cretaceous Pierre Shale, South Dakota. In J. Gray, et al. (eds), *Communities of the Past*: pp. 383-413. Stroudsburg, PA.: Hutchinson Ross.

Bishop, G.A. 1981b. The lobster *Linuparus* preserved as an attachment scar on the oyster *Exogyra costata*, Ripley Formation (Late Cretaceous), Union County, Mississippi. *Miss. Geol.* 2(1):2-5.

Bishop, G.A. 1982. *Homolopsis mendryki*: A new fossil crab (Crustacea, Decapoda) from the Late Cretaceous *Dakoticancer* Assemblage, Pierre Shale (Maastrichtian) of South Dakota. *J. Paleont.* 56:221-225.

Bishop, G.A. 1983a. A second sexualy aberrant specimen of *Dakoticancer overanus* Rathbun, 1917, from the Upper Cretaceous *Dakoticancer* Assemblage, Pierre Shale, South Dakota (Decapoda, Brachyura). *Crustaceana* 44:23-26.

Bishop, G.A. 1983b. Fossil decapod crustaceans from the Lower Cretaceous Glen Rose Limestone of Central Texas. *Trans. San Diego Nat. Hist. Soc.* 20:27-55.

Bishop, G.A. 1983c. Fossil decapod crustaceans from the Late Cretaceous Coon Creek Formation, Union County, Mississippi. *J. Crust. Biol.* 3:417-430.

Bishop, G.A. 1983d. Two new species of crab, *Notopocorystes* (*Eucorystes*) *eichhorni* and *Zygastrocarcinus griesi* (Decapoda, Brachyura) from the Bearpaw Shale (Campanian) of north-central Montana. *J. Paleont.* 57:900-910.

Bishop, G.A. 1984a. Orbital bulla reticularis : A new orbital structure of a Cretaceous crab. *J. Crust. Biol.* 4:514-517.

Bishop, G.A. 1984b. Paleobiogeography and evolution of the Late Cretaceous crabs of North America, 1976-78. *Nat. Geog. Res..* 17:189-201.

Bishop, G.A. 1985a. A new crab, *Eomunidopsis cobbani* n. sp. (Crustacea, Decapoda), from the Pierre Shale (Early Maastrichtian) of Colorado. *J. Paleont.* 59:601-604.

Bishop, G.A. 1985b. A new crab, *Prehepatus harrisi* (Crustacea, Decapoda) from the Coon Creek and Prairie Bluff formation, Union County, Mississippi. *J. Paleont.* 59:1028-1032.

Bishop, G.A. 1985c. Fossil decapod crustaceans from the Gammon Ferruginous Member, Pierre Shale (Early Campanian), Black Hills, South Dakota. *J. Paleont.* 59:605-624.

Bishop, G.A. 1986a. A new crab, *Zygastrocarcinus cardsmithi* (Crustacea, Decapoda) from the lower Pierre Shale, southeastern Montana. *J. Paleont.* 60:1097-1102.

Bishop, G.A. 1986b. Occurrence, preservation and biogeography of the Cretaceous crabs of North America. *Crustacean Issues* 4:111-142.

Bishop, G.A. 1986c. Taphonomy of the North American decapods. *J. Crust. Biol.* 6:326-355.

Bishop, G.A. 1986d. Two new Crabs, *Parapaguristes tuberculatus* and *Palaeoxantho libertiensis*, from the Prairie Bluff Formation (Middle Maastrichtian) Union County, Mississippi. *Proc. Biol. Soc. of Wash.* 99:602-609.

Bishop, G.A. 1987a. *Dromiopsis kimberlyae*: A new Late Cretaceous crab from the Pierre Shale of South Dakota *Proc. Biol. Soc. Wash.* 100:35-39.

Bishop, G.A. 1987b. Positive taphonomic feedback in North American Tethyan Cretaceous decapod-worm associations In K. McKenzie (ed.), *Shallow Tethys 2*: pp. 319-329. Rotterdam: Balkema.

Bishop, G.A. 1988a. A new crab, *Seorsus wadei*, from the Late Cretaceous Coon Creek Formation, Union County, Mississippi. *Proc. Biol. Soc. Wash.* 101:72-78.

Bishop, G.A. 1988b. New fossil crabs, *Plagiophthalmus izetti*, *Latheticocarcinus shapirous*, and *Sagitta carabus* (Crustacea, Decapoda), from the Western Interior Cretaceous, USA *Proc. Biol. Soc. Wash.* 101:375-381.

Bishop, G.A. 1988c. Two crabs, *Xandaros sternbergi* (Rathbun, 1926) n. gen., and *Icriocarcinus xestos* n. gen., n. sp., from the late Cretaceous of San Diego County, California, USA, and Baja California Norte, Mexico. *Trans. San Diego Soc. Nat. Hist.* 21:245-257.

Bishop, G.A. 1991. *Xanthosia occidentalis* Bishop, 1985, and *Xanthosia spinosa*, new species, two Late Cretaceous crabs from the Pierre Shale of the Western Interior. *J. Crust. Biol.* 11:305-314.

Bishop, G.A. 1992a. The Coon Creek decapod assemblages of northern Mississippi. *Miss. Geol.* 12(1): 8-17.

Bishop, G.A. 1992b. Two new crabs, *Homolopsis williamsi* and *Homolopsis centurialis* (Crustacea: Decapoda), from the Western Interior Cretaceous of the United States. *Proc. Biol. Soc. Wash.* 105: 55-56.

Bishop, G.A. & N.A. Brannen 1992. A new crab, *Homolopsis pikeae* (Crustacea: Decapoda), from the Cretaceous of Texas. *J. Crust. Biol.* 12:317-323.

Bishop, G.A. & R.W. Portell 1989. Pliocene crab-sea star association from southwest Florida. *J. Crust. Biol.* 9:453-458.

Bishop, G.A. & J.L. Whitmore 1986. The Paleogene crabs of North America: Occurrence, preservation and distribution. In W.B. Harris et al.(eds), *Eocene Carbonate Facies of the North Carolina Coastal Plain*: pp. 297-306, 325-332. Field Trip Guidebook 8, Society of Economic Paleontologists and Mineralogists Third Midyear Meeting.

Bishop, G.A. & A.B. Williams 1986. The fossil lobster *Linuparus canadensis*, Carlile Shale (Cretaceous, Turonian), Black Hills. *Nat. Geogr. Res.* 2:372-387.

Bishop, G.A. & A.B. Williams 1991. *Necrocarcinus olsonorum*, new species, a crab (Decapoda: Calappidae) from the Cretaceous Carlile Shale (Turonian), Western Interior United States. *J. Crust. Biol.* 11:451-459.

Camp. D.K., N.H. Whiting, & R.E. Martin 1977. Nearshore marine ecology at Hutchinson Island, Florida: 1971-1974, V. Arthropods. *Fla Mar. Res.* 25:1-63.

Cope, E.D. 1871. On three extinct Astaci from the Fresh-Water Territory Idaho. *Proc. Amer. Philos.* Soc. 11(1870):605-608.

Copeland, M.H. 1960. *Erymastacus bordenensis*, a new Mesozoic decapod from the Canadian Arctic. *Geol. Surv. Can. Bull.* 60: 55-57.

Crane, J. 1975. *Fiddler Crabs of the World (Ocypodidae: Genus Uca)*. Princeton: Princeton Univ. Press.

Cushman, J.A. 1905. Fossil crabs of the Gay Head Miocene. *Amer. Natur.* 39(462): 381-387.

Dana, J.D. 1849. *United States Exploring Expedition. During the Years, 1838, 1839, 1840, 1841, 1842. Under the Command of Charles Wilkes, USN. Geology*. Philadelphia: C. Sherman.

Davidson, E. 1963. New linuparid crustaceans from the Upper Cretaceous of Texas. *Bull. Amer. Paleont.* 206:69-76.

Davidson, E. 1966. A new Paleocene crab from Texas. *J Paleont.* 40: 211-213.

Davis, W.T., & C.W. Leng. 1927. Cretaceous fossils from Staten Island. *Proc. Staten Is. Instit. Arts and Sci.* 4:47-50.

Dickerson, R.E. 1916. Stratigraphy and fauna of the Tejon Eocene of California. *Univ. Calif. Geol. Pub.* 9(17):434.

Feldmann, R.M. 1974. Hoploparia riddlensis, a new species of lobster (Decapoda: Nephropidae) from the Days Creek Formation (Hautervian, Lower Cretaceous) of Oregon. *J. Paleont.* 48:586-593.

Feldmann, R.M. 1979. *Eryma foersteri*, a new species of lobster (Decapoda) from the Jurassic (Callovian) of North America. *Amer. Mus. Nat. Hist. Novitates* 2668:1-5.

Feldmann, R.M. 1981. Paleobiogeography of North American lobsters and shrimps (Crustacea, Decapoda). *Geobios* 14:449-468.

Feldmann, R.M. 1989. *Lyreidus alseanus* Rathbun from the Paleogene of Washington and Oregon, USA *Ann.Carnegie Mus.* 58:61-70.

Feldmann, R.M. & M.J. Copeland 1988. A new species of erymid lobster from Lower Jurassic strata (Sinemurian/Pliensbachian), Fernie Formation, Southern Alberta. In Contributions to Canadian Paleontology, *Geol. Surv. Can. Bull.* 379:93-101.

Feldmann, R.M. & F.D. Holland Jr. 1971. A new species of lobster from the Cannonball Formation (Paleocene) of North Dakota. *J. Paleont.* 45:838-843.

Feldmann, R.M. & C.B. McPherson 1980. Fossil decapod crustaceans of Canada. *Geol. Surv. Can. Pap.* 79-16:1-20.

Feldmann, R.M. & R.R. West 1978. *Huhatanka*, a new genus of lobster (Decapoda: Mecochiridae) from the Kiowa Formation (Cretaceous: Albian) of Kansas. *J. Paleont.* 52:1219-1226.

Feldmann, R.M., E.E. Awotua, & J. Welshenbaugh 1976. *Necrocarcinus siouxensis* A new species of calappid crab (Crustacea: Decapoda) from the Fox Hills Formation (Cretaceous: Maastrichtian) of North Dakota. *J. Paleont.* 50:985-990.

Feldmann, R.M., G.A. Bishop & T.W. Kammer 1977. Macrurous decapods from the Bearpaw Shale (Cretaceous: Campanian) of Northeastern Montana. *J. Paleont.* 51:1161-1180.

Feldmann, R.M., A.B. Tucker, & R.E. Berglund 1991. Fossil crustaceans. *Nat. Geog. Res.* 7:352-363.

Feldmann, R.M., L. Grande, C.P. Birkhimer, J.T. Hannibal, & D.L. McCoy 1981. Decapod fauna of the Green River Formation (Eocene) of Wyoming. *J. Paleont.* 55:788-799.

Frey, R.W. & T.V. Mayou 1971. Decapod burrows in Holocene barrier island beaches and washover fans, Georgia. *Senckenbergiana maritima* 3:53-77.

Frey, R.W., H.A. Curran, & S.G. Pemberton 1984. Tracemaking activities of crabs and their environmental significance: The Ichnogenus *Psilonichnus. J. Paleont.* 58(2):333-350.

Frey, R.W., J.D. Howard, & W.A. Pryor 1978. *Ophiomorpha*: Its morphologic, taxonomic, and environmental significance. *Palaeo.* 23:199-229.

Gabb, W.M. 1864. Description of the Cretaceous Fossils. *Geol. Surv. Calif. Palaeont.* 1:55-236.

Gabb, W.M. 1867. On the subdivisions of the Cretaceous Formation in California. *Proc. Calif. Acad. Nat. Sci.* 3:301-306.

Gabb, W.M. 1869. Cretaceous and Tetiary Fossils. *Geol. Surv. Calif. Palaeont.* 2.

Glaessner, M.F. 1929. Crustacea Decapoda. In J.F. Pompeckj (ed.), *Fossilium Catalogus* Pt. 41. Berlin: W.Junk.

Glaessner, M.F. 1960. The fossil decapod Crustacea of New Zealand and the Evolution of the Order Decapoda. *NZ Geol. Surv. Paleont. Bull.* 310:1-63.

Glaessner, M.F. 1969. Decapoda. In R.C. Moore (ed.), *Treatise on Invertebrate Paleontology*, Part R, Arthropoda 4, volume 2. Lawrence: University of Kansas.

Glaessner, M.F. 1980. New Cretaceous and Tertiary Crabs from Australia and New Zealand. *Trans. R. Soc. S. Aust.* 104:171-192.

Gould, S.J. 1989. *Wonderful Life.* New York: Norton.

Guinot, D. 1977. Propositions pour une nouvelle classification des Crustacés Décapodes. *C. R. Acad. Sc. Paris* 285:1049-1052.

Guinot, D. 1978. Principles of a new classification of the Crustacea Decapoda Brachyura. *Bull. Biol. Fr. Belg.* 3:211-292.

Hall, J. & F.B. Meek. 1855. Descriptions of new species of fossils from the Cretaceous formations of Nebraska. *Mem. Amer. Acad. Arts and Sci.* N. S. 17:356-360.

Herrick, E.M. & F.R. Schram 1978. Malacostracan crustacean fauna from the Sundance Formation (Jurassic) of Wyoming. *Amer. Mus. Nat. Hist. Novitates* 2652:1-12.

Hitchcock, E. 1841. *Final Report on the Geology of Massachusetts.* 2 volumes. Northhampton.

Holland Jr. F.D. & A.M. Cvancara 1958. Crabs from the Cannonball Formation (Paleocene) of North Dakota. *J. Paleont.* 32:495-505.

Howard, J.D. & R.M. Scott 1983. Comparison of Pleistocene and Holocene barrier island beach-to-offshore sequences, Georgia and Northeast Florida coasts, USA *Sed. Geol.* 34:167-183.

Jenkins, R.J.F. 1975. The fossil crab *Ommatocarcinus corioensis* (Cresswell) and a review of related Australasian species. *Mem. Nat'l Mus. Vict.* 36:33-62.

Jensen, D.E. 1975. Fossil crabs of the Northwest: A summary of the life and geologic events of the Oligocene in northwest Oregon and southwest Washington. *Ward's Geol. Newletter* 46:6-7.

Kaufman, E.G. 1979. The ecology and biogeography of the Cretaceous-Tertiary extinction event. In W.K. Christensen & T. Birkelund (eds), *Cretaceous-Tertiary Boundary Events Symposium*: pp. 29-37. Copenhagen: University of Copenhagen.

Kensley, B. & A.B. Williams 1990. *Axiopsis eximia*, a new thalassinidean shrimp (Crustacea, Decapoda, Axiidae) from the middle Eocene of South Carolina. *J. Paleont.* 64:798-802.

Kesling, R.V. & I.G. Reimann 1957. An Upper Cretaceous crab, *Avitelmessus grapsoideus* Rathbun. *Contrib. Mich. Mus. Paleont.* 14: 1-15.

Kindle, C.H. 1949. The Cretaceous crab *Raninella testacea* in New Jersey. *Trans. N. Y. Acad. Sci.* Ser. II, 12(1): 17.

Kooser, M.A. & W.N. Orr. 1973. Two new decapod species from Oregon. *J. Paleont.* 47:1044-1046.

Kues, B.S. 1980. A fossil crab from the Mancos Shale (Upper Cretaceous) of New Mexico. *J. Paleont.* 54:862-864.

LaBarbera 1981. The ecology of Mesozoic *Grypaea*, *Exogyra*, and *Ilymatogyra* (Bivalvia: Mollusca) in a modern ocean. *Paleobiology*. 7:510-526.

Lewis, J.E. & A. Ross 1965. Notes on the Eocene Brachyura of Florida. *Quart. J. Fla. Acad. Sci.*28: 233-244.

Lyell, C. 1844. On the tertiary strata of the island of Martha's Vineyard in Massachusetts. *Amer. J. Sci.* 46:318-320.

MacGinitie, G.E. 1939. Littoral marine communities. *Amer. Mid. Nat.* 21: 28-55.

Martin, G.C. 1904. Systematic Paleontology of the Miocene deposits of Maryland. Malacostraca. *Md. Geol. Surv., Miocene*:1-94.

Menzies, R.J. 1951. Pleistocene Brachyura from the Los Angeles area: Cancridae. *J. Paleont.* 25:165-170.

Mertin, H. 1941. Decapode Krebse aus dem subhercynen und Braunschweiger Emscher und Untersenon, sowie Bemerkungen uber winige verwandte Formen in der Oberkreide. *Abhandl. Kaiserl. Leopold.- Carolin. Deutsch. Akad. Naturforsch.* 10:152-257.

Miller, G.L. & S.R. Ash 1988. The oldest freshwater decapod crustacean, from the Triassic of Arizona. *Palaeontology* 31:273-279.

Moore, R.C. & L. McCormick 1969. General features of Crustacea. In R.C. Moore (ed.), *Treatise of Invertebrate Paleontology*, Part R, Arthropoda 4, volume 2: pp. 57-120. Lawrence: Univ. of Kansas Press.

Mundlos, R. 1975. Okologie, Biostratinomie, und Diagenese brachyurer Krebse aus dem Alt-Tertiär von Helmstedt (Neidersachsen, BRD). *N. Jb. Geol. Palaont. Abh.* 148:252-271.

Nations, J.D. 1968. A new species of cancroid crab from the Pliocene of California. *J. Paleont.* 42:33-36.

Nations, J.D. 1975. The Genus *Cancer* (Crustacea: Brachyura): Systematics, biogeography, and fossil record. *Bull. Los Ang. Cy. Mus. Nat. Hist.: Sci.* 23:1-104.

Orr, W.N. & M.A. Kooser 1971. Oregon Eocene Decapod Crustacea. *Ore Bin* 33:119-129.

Ortmann, A.E. 1897. On a new species of the palinurid genus *Linuparus* found in the Upper Cretaceous of South Dakota. *Amer. Jour. Sci.* ser. 4, 4:290-296.

Packard Jr., A.S. 1880. Fossil crawfish from the tertiaries of Wyoming. *Amer. Nat.* 14:222-223.

Packard Jr., A.S. 1881a. A fossil Tertiary crayfish. *Amer. Nat.* 15:832-834.

Packard Jr., A.S. 1881b. On a crayfish from the lower Tertiary beds of western Wyoming. *US Geol. Geog. Surv. Bull.* 6:391-397.

Packard Jr., A.S. 1900. A new fossil crab from the Miocene greensand bed of Gay Head, Martha's Vineyard, with remarks on the phylogeny of the genus *Cancer*. *Proc. Amer. Acad. Arts. & Sci.* 36:1-9.

Palmer, E.M. 1935. Preliminary report on a possible new species of fossil crab from the Miocene of Maryland. *Md. Nat. (Nat. Hist. Soc. Md.)* 6(2): 7-8.

Pilsbry, H.A. 1901. Crustacea of the Cretaceous formation of New Jersey. *Proc. Acad. Soc. Phila.* 53:111-118.

Pilsbry, H.A. 1916. Systematic paleontology of the Upper Cretaceous deposits of Maryland (Arthropoda). *Md. Geol. Survey, Upper Cretaceous*:361-370.

Plotnick, R.E. 1986. Taphonomy of a modern shrimp: Implications for the arthropod fossil record. *Palaios* 1:286-293.

Portell, R.W. & K.S. Schindler 1991. *Menippe mercenaria* (Decapoda: Xanthidae) from the Pleistocene of Florida. *Pap. Fla. Paleont.* 3:1-8.

Pryor, W.A. 1975. Biogenic sedimentation and alteration of argillaceous sediments in shallow marine environments. *Geol. Soc. Amer. Bull.* 86:1244-1254.

Rathbun, M.J. 1908. Descriptions of fossil crabs from California. *Proc. US Nat. Mus.* 35:341-349.

Rathbun, M.J. 1916. Description of a new genus and species of fossil crab from Port Townsend, Washington. *Amer. J. Sci.* 41:344-346.

Rathbun, M.J. 1917a. Description of a new species of crab from the California Pliocene. *Proc. US Nat. Mus.* 53:451-452.

Rathbun, M.J. 1917b. New species of South Dakota Cretaceous crabs. *Proc. US Nat. Mus.* 52:385-391.

Rathbun, M.J. 1923. Decapod crustaceans from the Upper Cretaceous of North Carolina. *N. C. Geol. Survey* 5:1-407.

Rathbun, M.J. 1926a. Decapoda. In B. Wade (ed.), *The fauna of the Ripley Formation On Coon Creek, Tennessee. Prof. Pap. US Geol. Surv.* 137:184-191.

Rathbun, M.J. 1926b. The fossil stalk-eyed Crustacea of the Pacific Slope of North America. *Bull. US Nat. Mus.* 138:1-155.

Rathbun, M.J. 1928. Two new crabs from the Eocene of Texas. *Proc. US Nat. Mus.* 73(6):1-6.

Rathbun, M.J. 1929. A new crab from the Eocene of Florida. *Proc. US Nat. Mus.* 75(15): 1-4.

Rathbun, M.J. 1930a. A new *Callianassa* from the Cretaceous of South Dakota. *J. Wash. Acad. Sci.* 20: 1-3.

Rathbun, M.J. 1930b. Fossil decapod crustaceans from Mexico. *Proc. US Nat. Mus.* 78(8): 1-10.

Rathbun, M.J. 1931. A new fossil palinurid from Staten Island. *Proc. Staten Is. Inst. Arts & Sci.* 5:161-162.

Rathbun, M.J. 1932a. A new species of *Cancer* from the Pliocene of the Los Angeles Basin. *J. Wash. Acad. Sci.* 22(1): 19.

Rathbun, M.J. 1932b. Fossil pinnotherids from the California Miocene. *J. Wash. Acad. Sci.* 22:411-413.

Rathbun, M.J. 1932c. New species of fossil Ranninidae from Oregon. *J. Wash. Acad. Sci.* 22:239-242.

Rathbun, M.J. 1935a. A new xanthid crab from the Cretaceous of New Jersey. *Proc. Acad. Nat. Sci. Phila.* 87:165-166.

Rathbun, M.J. 1935b. Fossil Crustacea of the Atlantic and Gulf Coastal Plain. *Geol. Soc. Amer. Spec. Pap.* 2:1-160.

Rathbun, M.J. 1936. Corrections of names of fossil decapod crustaceans. *Proc. Biol. Soc. Wash.* 49:37.

Richards, B.C. 1975. *Longusorbis cuniculosus*: A new genus and species of Upper Cretaceous crab; with comments on Spray Formation at Shelter Point, Vancouver Island, British Columbia. *Can. J. Earth Sci.* 12: 1850-1863.

Richards, H.G. 1958. *The Cretaceous Fossils of New Jersey* Pt. 2. Trenton: N. Jers. Geol. Sur.

Richardson, E.S. 1955. A new variety of Cretaceous decapod from Texas. *Fieldiana: Zoology* 37:445-448.

Roberts, H.B. 1953. A new species of decapod crustacean from the Inglis member. *Bull. Fla. Geol. Surv.* 35: 64-67.

Roberts, H.B. 1955. New xanthid crab from the Claiborne Eocene of New Jersey. *Bull. Wagner Free Instit. Sci.* 30(1): 9-12.

Roberts, H.B. 1956. Early Tertiary decapod crustaceans from the Vincentown Formation in New Jersey. *Bull. Wagner Free Instit. Sci.* 31(2): 5-12.

Roberts, H.B. 1962. The Upper Cretaceous decapod crustaceans of New Jersey and Delaware. In H.G. Richards (ed.), *The Cretaceous fossils of New Jersey*: pp. 163-191. Bull. N. Jers. Bur. Geol. & Topo. 61.

Roemer, F.A. 1887. *Graptocarcinus texanus*, ein Brachyure aus der oberen Kreide von Texas. *Neues Jb. Miner. Geol. Palaont.* 3-176.

Ross, A. & R.J. Scolaro 1964. A new crab from the Eocene of Florida. *Quart. J. Fla. Acad. Sci.* 27:98-106.

Ross, A., J.E. Lewis, & R.J. Scolaro 1964. New Eocene decapods from Florida. *Quart. J. Fla. Acad. Sci.* 27: 187-196.

Sakai, T. 1965. *The Crabs of Sagami Bay*. Honolulu: East-West Center Press.

Schäfer, W. 1951. Fossilisations-Bedingungen brachyurer Krebse. *Abh. senckenb. naturforsch. Ges.* 485:221-238.

Schäfer, W. 1954. Form und Funktion der Brachyuren-Schere. *Abh. senckenb. naturforsch. Ges.* 489:1-65.

Schäfer, W. 1972. *Ecology and Palaeoecology of Marine Environments*. Chicago: Univ. Chicago Press.

Schmitt, W.L. 1921. *The Marine Decapod Crustacea of California*. Berkeley: Univ. Calif. Press.

Schmitt, W.L. 1965. *Crustaceans*. Ann Arbor: Univ. of Michigan Press.

Schmitt, W.L. 1973. Mary J. Rathbun. *Crustaceana* 24:283-297.

Schram, F.R. 1971. *Litogaster turnbullensis* (sp. nov.): A Lower Triassic glypheid decapod crustacean from Idaho. *J. Paleont.* 45: 534-537.

Schram, F.R. & R.H. Mapes 1984. *Imocaris tuberculata* n. gen., n. sp. (Crustacea: Decapoda) from the upper Mississippian Imo Formation, Arkansas. *Trans. San Diego Nat. Hist. Soc.* 20:165-168.

Schram, F.R., R.M. Feldmann, & M.J. Copeland 1978. The Late Devonian Palaeopalaemonidae and the earliest decapod crustaceans. *J. Paleont.* 52:1375-1387.

Secretan, S. 1964. Les Crustacés Décapodes du Jurassique Supérieur et du Crétacé de Madagascar. *Mém. Mus. Natn. Hist. Natn., Paris* 14C:1-144.

Squires, R.L. 1980. A new species of brachyuran from the Paleocene of California. *J. Paleont.* 54:472-476.

Stenzel, H.B. 1934. Decapod Crustaceans from the Middle Eocene of Texas. *J. Paleont.* 8:38-56.

Stenzel, H.B. 1935. Middle Eocene and Oligocene Decapod Crustaceans from Texas, Louisiana, and Mississippi. *Amer. Mid. Nat.* 16:379-400.

Stenzel, H.B. 1941. Crustaceans. In *Geology of Dallas County, Texas*: pp. 35-39. Dallas Petroleum Geologists.

Stenzel, H.B. 1944a. A new Cretaceous crab *Graptocarcinus muiri*, from Mexico. *J. Paleont.* 18: 550-552.

Stenzel, H.B. 1944b. A new Paleocene catometope crab from Texas, *Tehuacana tehuacana*. *J. Paleont.* 18:546-552.

Stenzel, H.B. 1945. Decapod crustaceans from the Cretaceous of Texas. *Univ. of Tex. Bur. of Econ. Geol.* 4401:401-476.

Stenzel, H.B. 1951. Decapod crustaceans from the Woodbine formation of Texas In L.W. Stephenson (ed.), *Larger Invertebrate fossils of the Woodbine formation (Cenomanian) of Texas*: pp. 212-217. 7: 583-589, *US Geol. Surv. Prof. Pap.* 242.

Stimpson, W. 1863. On the fossil crab of Gay Head (Martha's Vineyard, Mass.). *Boston J. Nat. Hist.* 7:583-589.

Thorson, G. 1957. Bottom Communities (Sublittoral or Shallow Shelf), In J.W. Hedgpeth (ed.), *Treatise on Marine Ecology and Paleoecology*, Vol. 1: pp.461-534. *Mem. Geol. Soc. Amer.* 67.

Tucker, A.B. & R.M. Feldmann 1990. Fossil decapod crustaceans from the lower Tertiary of the Prince William Sound region, Gulf of Alaska. *J. Paleont.* 64:73-80.

Tucker, A.B., R.M. Feldmann, F.D. Holland, & K.F. Brinster 1987. Fossil crab (Decapoda: Brachyura) fauna from the Late Cretaceous (Campanian-Maastrichtian) Pierre Shale in Bowman County, North Dakota. *Ann. Carnegie Mus.* 56:275-288.

Van Straelen, V. 1936 Sur les crustacés Décapoda Triassiques du Nevada. *Bull Mus. r. Hist. nat. Belg.* 12(29): 1-7.

Vega, F.J. & R.M. Feldmann 1991. Fossil crabs (Crustacea: Decapoda) from the Maastrichtian Difunta Group, northeastern Mexico. *Ann. Carnegie Mus.* 60: 163-177.

Vermeij, G.J. 1977. Patterns in crab claw size: The geography of crushing. *Sys. Zool.* 16:138-151.

Via Boada, L. 1969. Crustaceos decapodes del Eocene Español. *Institute de Estudies Pirenaicos* 91-94:1-479.

Waage, K.M. 1968. The type Fox Hills Formation, Cretaceous (Maastrichtian), South Dakota, Part 1. Stratigraphy and paleoenvironments. *Peabody Mus. Nat. Hist. Bull.* 27:1-175.

Walker, S.E. 1988. Taphonomic significance of hermit crabs (Anomura: Paguridea): Epifaunal hermit crab-Infaunal gastropod example. *Palaeo., Palaeo., Palaeo.* 63:45-71.

Weller, Stuart. 1905a. The fauna of the Cliffwood Clays. *A. Rep. N. Jers. State Geol.* 1904:133-144.

Weller, Stuart. 1905b. The fauna of the Cliffwood (N.J.) Clays. *J. of Geol.* 13: 324-337.

Weller, Stuart. 1907. A report on the Cretaceous paleontology of New Jersey. *N. Jers. Geol. Surv. Paleont.* Ser.4: 846-853.

Wharton, G.W. 1942. A typical sand beach animal, the mole crab, *Emerita talpoida* (Say). In A.S. Pearse et al. Ecology of Sand Beaches of Beaufort, N. C. *Ecol. Monogr.* 12:35-190.

Whetshone, K.N. & J. S.H. Collins 1982. Fossil crabs (Crustacea: Decapoda) from the Upper Cretaceous Eutaw Formation of Alabama. *J. Paleont.* 56:1218-1222.

Whiteaves, J.F. 1885. Note on a decapod crustacean from the Upper Cretaceous of Highwood River, Alberta, Northwest Territory. *Trans. Roy. Soc. Can.* 2:237-238.

Whiteaves, J.F. 1895. On some fossils from the Nanaino group of the Vancouver Cretaceous. *Trans. R. Soc. Can.* Ser 2. 1: 119-133.

Whiteaves, J.F. 1900. Mesozoic fossils. On some additional or imperfectly understood fossils from the Cretaceous rocks of the Queen Charlotte Islands with a revised list of species from these rocks. *Geol. Surv. Can.* 1:263-307.

Whiteaves, J.F. 1903. On some additional fossils from the Vancouver Cretaceous with a revised list of the species therefrom. Mesozoic fossils. *Geol. Surv. Can.* 1(5):309-409.

Whitfield, R.P. 1880. Notice of new forms of fossil crustaceans from the Upper Devonian rocks of Ohio, with a description of new genera and species. *Amer. J. Sci.* (3)19:33-42.

Whitfield, R.P. 1891. Contributions to invertebrate paleontology. *Ann. N.Y. Acad. Sci.* 5: 505-620.

Whitfield, R.P. 1907. Notice of an American species of the genus *Hoploparia* McCoy from the Cretaceous of Montana. *Bull. Amer. Mus. Nat. Hist.* 23:459-462.

Withers, T.H. 1924. Some Decapod Crustaceans (*Callianassa* and *Ranina*) from the Oligocene of Washington, USA *Ann. Mag. Nat. Hist.* 14: 121-127.

Woodward, H. 1896. On some Podophthalmatous Crustacea from the Cretaceous Formation of Vancouver and Queen Charlotte Islands. *Quart. J. Geol. Soc.* 52:221-228.

Woodward, H. 1900. Further note on podophthalmous crustaceans from the Upper Cretaceous formation of British Columbia. *Geol. Mag.* 7: 392-401.

Wright, C.W. and J.S.H. Collins 1972. British Cretaceous Crabs. *Paleontograph. Soc. Monogr.* 533: 1-114.

Young, K.P. 1982. Memorials: Henryk Bronislaw Stenzel (1899-1980). *Newslet. Dept. Geol. Sci., Univ. Texas at Austin.*

The British School: Calman, Cannon, and Manton and their effect on carcinology in the English speaking world

Frederick R. Schram

Instituut voor Systematiek en Populatiebiologie, Amsterdam, Netherlands

1 INTRODUCTION

Three people in 20th century Britain dominate crustacean studies: William Thomas Calman (1871-1952), Herbert Graham Cannon (1897-1963), and Sidnie Milana Manton (1902-1979). Obituaries of all three (Gordon 1952; Cannon 1953; Peacock 1953; Smith 1963; Yonge 1963; Fryer 1980) provide a basis for the biographic information that follows. A study of archived personal papers and letters could form the nucleus of a thorough historical analysis of their work. This paper presents only a preliminary examination of these people and their achievements. Calman, Cannon, and Manton had profound impacts on carcinology. They completely determined the course of the 'British School,' indeed they and their associates were the British School.

2 HISTORICAL ROOTS

The publication of Darwin's *On the Origin of Species*, while controversial, had a muted impact on biology in the late 1800's. By the time Darwin published his book considerable evolutionary speculation had appeared in the literature of the early half of the 19th century, especially among English authors. Much of this literature focused on the writings of Lamarck. For example, Robert Grant, one of the most prominent zoologists in the first half of the century and Professor of Zoology at University College, London, had studied in Paris with Cuvier, Goeffroy St. Hilaire, and Lamarck and came under the influence of the latter (Desmond 1982). Grant spread Lamarckian ideas into Britain well before Sir Charles Lyell found it necessary to come out against them in his 1832 edition of *Principles of Geology*.

In addition to the writings of professional scientists, a considerable popular literature on the subject of evolution appeared in the early decades of the 19th century. The contentious *Vestiges of the Natural History of Creation* appeared in 1844. This originally anonymous work of Robert Chambers, while founded on the idea of ultimate divine creation of the universe, viewed the world as the result of the workings of universal, natural law and did not advocate a necessary, direct, divine intervention in the evolutionary process. Chambers often made strange comparisons between animals, such as stating that turtles formed the sole source for birds [a statement Darwin found especially amusing (di Gregorio 1984)]. Although at times naive and at other times simply incorrect in his assertions, Chamber's *Vestiges* had considerable public impact, going through several editions.

Chamber's approach, while subject to great criticism, possessed more scientific merit than the more deistic works of this same period, such as Paley's *Natural Theology* and the *Bridgewater Treatises*. The approach in these volumes grew out of the 'clockwork' concept of the universe so common in the 18th century, which held that the universe contained many examples of purpose and design that clearly manifested proof for the existence of 'beneficent Providence' (Livingston 1986). This compelling view grew out of the religious revivalism of the age, which focused on 'self evident argument' that rational men could examine 'if they cared.' As Buckland stated in his volumes of the *Bridgewater Treatises*, 'To attribute all this harmony and order to any fortuitous causes that would exclude Design, would be to reject conclusions founded on a kind of evidence on which the human mind reposes with undoubting confidence in all the ordinary business of life, as well as physical and metaphysical investigations.' (Buckland 1869:483).

At mid-century, Darwin effectively argued that anatomical features arose as the result of strict cause and effect without necessary recourse to explanations of design. Nevertheless, this materialistic or realistic approach to science had many difficulties. The problem hinged on the fact (e.g. Macbeth 1971) that the concepts of selection and fitness carry considerable teleologic connotations. While Darwin attracted a visible and vocal array of supporters largely centered about Thomas Huxley (Desmond 1982) other equally prominent men of science took exception, men such as the anatomist Richard Owen, the zoologist St. George Mivart, and the paleontologist Harry Seeley. In response to Darwin, these latter figures effectively reformulated the Paleyian argument of design into a more idealist approach; 'God's creation' became the 'workings of natural law.' This viewpoint gained wide acceptance, evidenced by the fact that demand persisted for further editions of Buckland's *Treatise* even after his death, the fourth edition appearing in 1869 well after the first edition of Darwin's *Origin*.

The main opposition to the idea of evolution by natural selection came from Darwin's contemporary, Richard Owen, founder of the British Museum of Natural History. Owen, a prominent anatomist, became the principal advocate of an idealist view of life, the intellectual origins for which came from the continent. German 'Naturphilosophie' interpreted the variety evident in living things as mere manifestations of basic plans or archetypes. These archetypes in turn could be linked to a 'plan' whereby all the possible patterns evident in life, whether of orders, families, species, or whatever, appeared.

Although not all idealists advocated Lamarckianism (nor did all Lamarckians subscribe to idealism), men like Owen and Grant had much in common. The Lamarckians believed in a positive relationship of organic form with environmental constraints and this nicely dovetailed with the idealist overview based on archetypes. Both approaches modulated functional adaptations through the workings of natural law, i.e., the 'design' arose from some 'need.'

While we in the late 20th century have come to view the triumph of evolutionary theory in terms of Thomas Huxley and his associates defending Darwin in Britain against the likes of Bishop Wilberforce and Professor Owen (e.g., Desmond 1982; di Gregorio 1984), most rank and file scientists in the later half of the 19th century had serious problems with trying to explain the origins of specific adaptations. The best example of this occurred in America where the principal scientific figure of the age, Louis Agassiz, an idealist (but personally a non-evolutionist), founded an American school that came to effectively advocate and defend essentially Lamarckian thought (Livingston 1986). Figures such as Nathaniel Southgate Shaler, Alpheus Hyatt, and Alpheus Packard, all students of Agassiz, and Edward Drinker Cope came to not only teach a generation of American scientists, but by their writings exerted influence in other parts of the English speaking world, especially Britain.

Idealists and romantics flourished in the late 19th century. The natural theologians and religious transcendentalists focused on the importance of subjective experience over rational empiricism. To the transcendentalist and romantic people, contemplation of Nature lead to spiritual insight and thus to ultimate truth. However, the scientific counterparts of these, the idealists, believed in objective observations as not only possible but as necessary and empirical. To the idealist, study of nature led to an understanding of adaptations and a realization of the significance of design.

A central figure in our story matured in this climate, D'Arcy Wentworth Thompson. Thompson, born in Edinburgh in 1860, started his studies as a medical student in the university there but came under the influence of Sir Wyville Thompson (no relation) of Challenger Expedition fame. Publishing some papers on hydroids as well as fossil seals, with a scholarship in classics, young D'Arcy went on to take his degree at Trinity College, Cambridge. There he worked under Francis Balfour and Michael Foster, prominent figures in the then growing field of experimental zoology.

At the age of 24, Thompson became professor of biology at University College, Dundee. He matured into one of the leading figures in experimental biology, especially with the publication in 1917 of his benchmark treatise, *On Growth and Form* – a remarkable volume. Although Thompson designed the book to inject the rigor of mathematics and physics into biology, he did not write it in a materialist or realist vein. In fact, Thompson labeled others who produced 'physicomathematical' research as too materialistic. To Thompson matter could not function as an operant element but only as the location of energy and the focus of forces. As he said in the epilogue of the second edition of *On Growth and Form*, '...the harmony of the world is made manifest in Form and Number...(for)...the perfection of mathematical beauty is such... that whatsoever is most beautiful and regular is also found to be most useful and excellent... This is the teaching of Plato and Pythagoras...the living and the dead...are bound alike by physical and mathematical law (Thompson 1942:1096-1097).' I can conceive of no clearer pronunciation of scientific idealism.

Thompson personally preferred to play in his mind with mathematical descriptions so as to tie them to physical analogies. Experiments did not figure as a prime factor in this. In fact, in the first edition of *On Growth and Form*, Thompson refused to cite experiments that had direct bearing on his work. His conviction that the interplay of physical forces could explain biologic structure resulted in a causal morphology, something quite distinct from the historical morphology of Darwin (Chambers 1949).

With the 20th century and the rediscovery of Mendel's work, the science of genetics began. At first, however, its applicability to Darwinian selection theory seemed rather vague. Many scientists viewed De Vries's theory to explain the origin of variation by mutation as something quite separate from what Mendel had documented, and strong idealist and Lamarckian currents continued to exist in evolutionary theory. The Lamarckian's had a coherent world view and covertly came to have profound influence on experimentalists seeking to understand the nature of functional adaptations. The realists and Darwinians found themselves in a weak position, hindered at that time without an effective mechanism to explain the origin of change until well into the 20th century, and even then, the efficacy of totally relying on genes to produce both micro- and macroevolutionary events found no universal acceptance.

We must remember that our three principals, Calman, Cannon, and Manton, all matured as scientists before formulation of the Synthetic Theory of Evolution. Any understanding of their work must keep this fact in view.

3 WILLIAM THOMAS CALMAN

Calman was an only son born on December 29, 1871 in Dundee, Scotland. His father, Thomas Calman, a music teacher blind since birth, came from a long line associated with shipping, either as builders, owners, or masters of ships. His grandfather, John Calman, had some fame not only as a shipbuilder, but also as a prominent member of the local government in Dundee.

Calman's mother, Agnes Beatts MacLean, came from a line of cabinet makers. A few years after William Thomas' birth she bore a daughter, but her husband died soon afterward. She never remarried and supported her children as a seamstress. Thomas Calman had actively participated in the Free Church of Scotland, a pious and puritanical sect, but Agnes Calman managed to raise her son and daughter in a manner that instilled in them a strong element of humor and common sense. William Thomas, by all later accounts, gained a reputation for his wry sense of humor with a strong streak of Scottish pride.

Agnes Calman also conveyed to her son a strong belief in the efficacy of education for education's sake, a hallmark of people of the Scottish middle class. Calman himself confessed to having read in his boyhood most of the books on science in the Dundee Free Library. A quiet and shy boy, he enjoyed reading and scholarly activities and expressed little interest in the traditional schoolboy athletics. His retiring behavior probably grew from a bad stammer. In later years, Calman blamed this on efforts to force him as a child to write with his right hand; he claimed he cured the stammer after shifting in middle age to writing with his left.

A small scholarship allowed him to attend the Dundee High School. There he fell under the influence of a teacher of science who placed a strong emphasis on classroom experiments. Calman learned a great deal of his physics and chemistry in this way, but a microscope given to him as a teenager by one of his relatives introduced him to the wonders of biology. He began a study of pond rotifers that eventually lead to his first published scientific papers a few years later.

His graduation at age 16 and his family's financial condition forced him to take an apprenticeship as an insurance clerk. Nevertheless, he indulged his scientific interests by joining the Dundee Working Men's Field Club. In the era before television, radio, and other forms of mass entertainment, people sought educational as well as entertaining diversions. Organizations such as the Field Club, common in Victorian Britain, had memberships characterized by the seriousness with which they pursued their interests. Often times, as with the Dundee Club, these societies published their own magazines or journals. During the four years Calman worked as an insurance clerk, the comraderie and activities of the Dundee Working Men's Field Club provided him with the intellectual stimulation he needed.

During this time, Calman also joined the Dundee Naturalist's Society. This crucial event in Calman's life allowed him to meet D'Arcy Thompson, the Professor of Zoology at University College in Dundee. At the time, Calman's superiors, because of his speech impediment, discouraged him from pursuing a career in business. Yet, ironically, a lecture Calman gave before the Naturalist's Society on his rotifer studies attracted Thompson's interest and prompted Thompson to offer Calman a post as junior laboratory assistant in the College. Thompson actually paid him out of his own pocket, and this began a mentor/student relationship that persisted for the rest of both their lives.

Calman's assistantship entailed preparing specimens for the student laboratories and helping Thompson to sort material referred for study from the Antarctic and Arctic. By working at the college, Calman could take classes for free. This enabled him to earn a B.S. in 1895 with a first class honors tripos in zoology, botany, and physiology. By assisting with Thompson's work, Calman gained some time to continue his own research projects, and his first two papers in 1892 and 1893 dealt with the rotifers he had studied since his high school days. In addition, Calman worked on some of the polar material that came into the laboratory, and this resulted in a paper in 1894 describing a new genus of tunicates.

Thompson had a crucial effect in forming Calman's perspective and approach to science. As indicated above, Thompson did not want to advocate materialistic experimentalism when he wrote *On Growth and Form* (Bonner 1966) for he had essentially an idealist outlook. Thompson believed one could use mathematics and physical analogy to focus on the details of form and interpret its meaning. This required a tremendous capacity to target observations and develop the relevant comparisons and analogies.

In Calman, the effect of Thompson's instruction manifested itself in a tremendous ability to deal with details. Calman made meticulous observations. For example, early in his career (1908) Calman published a short note on stridulating organs in river crabs. This paper comprised four and one-half pages of description and comparative analysis of a tiny bit of anatomy encountered while '...examining a collection of river-crabs (Potamonidae).' While sorting this material Calman discovered, with his characteristic acuity, that these crabs stridulated not by means of the ridges like those found on other crabs but rather by means of special setae akin to those found on arachnids. This little paper forms a bit of frosting effectively skimmed from the vast confection of crustaceans from all over the world that passed through Calman's laboratory. It is the sort of thing D'Arcy Thompson thoroughly approved of.

Perhaps Calman's early instruction in the uses of analogy may account for another hallmark of his work, his free use of the fossil record, best illustrated in Calman's research on syncarids. He read his first paper in this regard (1896) before the Royal Society of Edinburgh in 1895 while still a student. The first species described and eventually attributed to the syncarids was a little Permian fossil, *Gampsonyx fimbriatus* (Jordan) in 1847. For the next 45 years, the relationships of this species and similar fossils continued in a constant state of flux (e.g., see Schram 1984). Only after G.M. Thomson (1893, 1894) described the first living species, *Anaspides tasmaniae*, could Calman place the various fossils forms into perspective uniting them with *Anaspides* into the taxon Syncarida.

This fusion of information from fossil and living forms to generate new insights into animal relationships formed a remarkable achievement for the time, but it became only the first in a series of papers on this group that continued in this vein. In 1899, Calman properly discerned the relationships of *Bathynella natans* Vejdovsky, 1882 to the syncarids. Vejdovsky had claimed superficial similarities between this species and gammarid amphipods. Calman went through the features of the anatomy of *Bathynella* and systematically analyzed them, noting the form of the legs, correcting the number and identity of the trunk segments, and clarifying the location of the gonopores. By checking each character, Calman established a clear connection of *Bathynella* to *Anaspides*, and from there to the Paleozoic 'Gampsonychidae.' Indeed, he observed that in some respects *Bathynella* seemed more like the fossils than *Anaspides*, given the free first thoracomere and the single series of epipodites on the thoracic limbs. Calman did not make any distinction at that time between primitive and advanced characters as do present day workers; however, this *Bathynella* paper clearly reveals that in his own mind there existed qualitative differences between anatomical features, differences tempered by whether features appeared in fossil or living forms.

He displayed this same clarity of thought and attention to analyzing detail in his 1902 paper attacking Anton Fritsch's contention that the fossil syncarid *Uronectes* had no relationship at all to *Anaspides*. Calman acted like a Scottish terrier when he caught someone in error, and he went after Fritsch with a vengeance. Fritsch had completely misinterpreted the anatomy of his fossils by reconstructing seven abdominal segments and two sets of maxillipedes. Calman called attention to these matters as being '...so exceedingly peculiar as to preclude direct comparison with any known crustacean (p.66).' Fritsch had reconstructed a uniramous uropod, ignoring the work of Jordan and von Meyer's (1854) observations of distinctly biramous uropods in *Gampsonyx/Uronectes*, which in Calman's mind had strong priority since it accorded with Packard's (1885, 1886) conclusions about another fossil crustacean, *Palaeocaris*. Calman rejustified his affiliation of the syncarid fossils with *Anaspides*, pointing out that the possible absence of exopods on the thoracic limbs, which Fritsch claimed for his *Uronectes* fossils, would not necessarily preclude their relationship to the living syncarids, just as their possession of exopods would not sufficiently place them among the 'schizopods,' i.e., in close relation to mysids and krill. Calman effectively argued in this paper that one cannot focus just on one or two features when assessing animal relationships, but one must examine and weigh all the evidence and perceive animals as a whole.

Calman continued his interests in syncarids for the rest of his career, publishing another nine papers on the subject up until 1934. In all of these, he treated fossils as integral elements in the analysis of living forms and vice-versa, continued his careful attention to detail, and insisted that animals (whether they be fossil or living) be looked at as integrated wholes.

Upon receiving his B.S. degree, Thompson appointed Calman as Assistant Lecturer and Demonstrator in zoology at Dundee. Although some papers appeared on non-arthropods over the next few years, Calman by this point had definitively settled on crustaceans as his prime field of interest. His expertise, and undoubtedly the patronage of Thompson as well, insured a steady rise among zoologists of the time. By 1901 he had attracted the attention of Sir E. Ray Lankester, then at the British Museum, who invited him to collaborate on the *Treatise on Zoology*. Lankester came from a distinguished background. His Father, Edwin Lankester, became a well-known microscopist in the circle of Darwin, Hooker, and Huxley. Lankester himself developed a strictly realist and staunchly anti-teleological outlook (Desmond 1982). Lankester planned his *Treatise* as a massive, multivolume compendium in the English language to match Bronn's *Das Tierreich* in German. He never completed the project, but the finished volumes become classics in their fields. Lankester invited Calman to prepare the volume on Crustacea.

Lankester enlisted the help of R.I. Pocock in recruiting Calman, and in the course of the correspondence between Calman and Pocock the latter mentioned the possibility of a position for Calman at the British Museum. Indeed, Lankester invited Calman to become a 'temporary assistant' on the museum staff in 1903. The following year Calman accepted a permanent position in charge of crustaceans. The transfer to London with its access to the greatest collection of crustaceans at the time provided Calman a secure position in British science for the rest of his career. It certainly insured that his crustacean volume for Lankester's *Treatise* would have as thorough a coverage as possible.

Calman's 1904 paper in the *Annals and Magazine of Natural History*, 'On the classification of the Crustacea Malacostraca,' presaged the appearance of the Treatise. In this paper, Calman advanced the now well-known subdivision of the Malacostraca into superorders. Up until then, personal whim pretty much determined the classification of malacostracans. Debates in the literature of that time commonly exhibited vagueness and subjectivity, such as whether or not particular malacostracan groups were 'schizopods.' In this 1904 paper, Calman set out a carefully defined set of characters that delineated major groups of malacostracans. The taxonomic synthesis of Calman brought much needed stability to crustacean systematics. Again, his attention to detail enabled Calman to achieve this breakthrough.

The 1909 volume on Crustacea in Lankester's *Treatise* turned out as a tour d'force. It summarized knowledge about crustacean biology to that date, clearly organized it, and stabilized the higher taxonomy. So well constructed, *Crustacea* served as the standard general reference in the English language on crustaceans for almost 80 years, though remarkably a relatively slim volume. Gordon (1952: 780) felt that the book's usefulness arose directly from Calman's ability at synthesis, 'Calman possessed, to a degree that has rarely been equaled, a gift for winnowing the significant from masses of detail.' An ability to analyze details could be a curse if not coupled with an ability to make judgements about which details are, and which are not, important.

The Life of Crustacea soon followed the crustacean treatise in 1911. This book surveyed crustacean natural history, with chapters covering the major habitat types that Crustacea occupy. Much of this book reorganized material from the crustacean treatise, but it does contain some additional material on ecology and functional morphology. Calman demonstrated the importance of the fossils again by devoting a whole chapter at the end of the book to the fossil record, much of its material based on Calman's own research. Together, the two books form a state of the art summary of what people knew about crustacean biology at the opening of the 20th century.

Calman's ability as a synthesizer rested upon another of his talents; he had a photographic memory. For years, he compiled the arachnid and crustacean sections of the *Zoological Record*, and did this without benefit of any index cards. Not only that, he remembered references 20 or 30 years after the fact.

Despite his imposing intellect, Calman struck everyone as the most genial of people. Gordon recalls that he made it a point to meet every visitor to the department, whether a student or a professional. His mild character probably acted to his disadvantage on at least two occasions. In 1908, a few years after Calman arrived at the museum, Ray Lankester, both Director and Keeper of Zoology, departed. This necessitated filling the keepership. However, since Calman had first entered museum service on a temporary appointment before promotion into a permanent position without benefit of a competitive procedure usually required by the Civil Service, some people objected to Calman's succession to the position. Instead, Calman had to defer in favor of Tate Regan, who had less time in museum service than Claman. Calman eventually succeeded to the keepership, but only after Regan himself moved up to become Director in 1927.

In another incident in 1919, Calman tried to succeed his mentor D'Arcy Thompson in the professorship at Dundee. Thompson had taken the chair of zoology at the University of St. Andrews. Cannon records that Calman did all the proper things one did in those days to qualify one's candidacy. He had his carefully drawn application printed and accompanied with letters of support from 21 Fellows of the Royal Society. However, Calman neglected to travel to Dundee and personally lobby the electors, or so called Curators of Patronage. He lost the election.

A less genial, more aggressively ambitious personality might have overcome either of these situations and advanced their careers. Calman could not do so. Of course, in considering the British system, one can never completely disregard the role that social origin and position might play in matters of promotion and advancement.

The zoology department under Regan and then Calman, grew into a stimulating place indeed. Its proximity to Imperial College insured a steady stream of students out of the laboratory of Prof. E.W. MacBride, students like H. Graham Cannon and Sidnie Manton, who freely interacted with museum staff, which included in the crustacean section Isabella Gordon. Calman, although aware of his responsibilities as museum taxonomist, had concerns beyond classification of organisms. His books certainly demonstrate his wide interests extending into the natural history and functional morphology of crustaceans. So he exerted considerable influence on younger workers who preferred more experimental research, students like Cannon who reflected the vision of D'Arcy Thompson in *On Growth and Form*.

Calman himself, however, effectively parted company with his idealist mentor by this time. Calman nicely illustrated the materialist nature of his outlook on science with two papers published near the end of his career (Calman 1931, 1935), actually derived from public addresses. The first paper dealt with the role of taxonomy, and reveals Calman as a realist concerning systematic work. To him, the 'complete description' of a species did not exist. As an example, he pointed out species of the mantis shrimp, *Squilla*, where he had found from experience that about six characters could serve to clearly delineate species within the genus. He found, when comparing his experience with the published results of an unnamed German colleague who had 'completely' described the very same species, that not one of the features Calman found useful appeared in the latter work. To Calman, this difference did not cast aspersions on anyone's ability as an observer. Rather, it demonstrated the limits of observational science. Calman believed it impractical to demand too much of qualified observers; no one ideal or archetype exists to which nature conforms.

The above issue touches upon issues still current today, viz., the characterization and polarization of anatomical features. Calman displayed concern with the role characters play in establishing an evolutionary view of zoology. To Calman, the idealistic demands placed on taxonomists results in a denigration of their work and an unwarranted emphasis on the 'validity of experimentalism.' Calman believed this resulted in a loss of an evolutionary perspective. These demands by nontaxonomists lead to an obsession with the differences among animal species. Calman chided people such as F.A. Bather who claimed that all phyla were 'riddled through and through with polyphyly and convergence,' or the contention of Przibram's that every metazoan species developed independently from different species of protozoans. To Calman, these contentions grew from the consequences of a reductionist approach to science, i.e., the experimentalist approach of breaking the whole into its parts for easier study. Calman felt that good taxonomists synthesized, that, although they did not have all the answers, they could frame the great evolutionary questions properly in order to gain a truer perspective on the history of life.

His 1935 paper tried to demonstrate the meaning of taxonomy for all biologists. He felt compelled to point out that taxonomy functioned in a way different from stamp collecting, rather it acted as a means to an end. In a display of his affinities with Thompson, Calman believed that experimentalists replace what he called 'historical morphology' with 'causal morphology' in their efforts to understand function. He observed that experimentalists often seek to link function with 'need.' This leads, in Calman's mind, to a penchant for linking similar forms to similar needs, the needs then become the causes, and to completely loosing site of the historical nature of species lineages.

Calman reacted to a tendency among his colleagues towards a Lamarckian view of evolution; indeed, Cannon forcefully advocated Lamarckism a few years later. Calman approached his science as a realist and a Darwinian, although he had little sympathy with the experimental genetics of his day. What Calman dealt with in these 'taxonomy' papers arose from his continuing problem, as we have already seen, with character assessment. As an example, he cited the work of his colleague Tate Regan (1930) on primitive primates. Regan believed that lemurs and lorises had converged, yet Calman, while he recognized differences between these groups, demanded to know the significance of apparently important characters like procumbent and slender lower incisors and canines, important features that would seem to unite these groups. Regan wanted to separate these two groups of primates because of their purported differences. However, Calman believed the similarities of animals more significant, for as he wrote, 'Where do we stop? Is structure any indication of phylogenetic affinity? Is community of descent ever a cause of organic similarity? Is blood ever, in fact, thicker than water?' (1935, p. 153).

While Calman instinctively believed that characters could show qualitative differences, he never could clearly conceptualize how issues of primitive and derived could affect the analysis of lineages. He did recognize, however, that the obsession with differences and implied convergence carried to its logical conclusion effectively obliterated the evidence of its own existence. Quoting Shakespeare, Calman left the issue open, 'Treason doth never prosper. What's the reason?/Why if it prosper, none dare call it treason.'

Calman's questioning of the use of 'need' to assess function had grown from his early association with Lankester. Calman's old mentor, Thompson, did not believe that selection could function as a creative force in evolution, that selection only served to eliminate the unfit. Classic views of natural selection of course arose from the philosopher Spencer's dictum of 'survival of the fittest.' If true, then to survive a species 'needed' to possess 'fitness.' Yet Calman, while agreeing up to a point, eventually parted company with Thompson. Although both he and Thompson rejected untoward emphasis on experiments, their views of evolution differed. Thompson, with his concept of physics as an analogy, saw structures arising as the result of direct forces, molecular in small cases, mechanical in the large instances (a prescient view of the dichotomy between micro- and macroevolution). Thompson's idealism, thus ignored phylogeny in favor of a concept of form as a consequence of function, not of blood. Calman on the other hand never gave up his Darwinian and Lankesterian reliance on the importance of lineage in studying animal characters; he never wavered in his belief that blood mattered.

I could speculate what might have happened if Calman had crystalized his views on character assessment and use of the fossil record into a coherent system? Might cladistics have appeared 30 years earlier, with a British rather than a Germanic cast? Instead, Calman and colleagues like Gordon continued to grapple as taxonomists with the problems of character assessment while not seriously constrained by Thompson's penchant for physical analogy. Functional morphologists like Cannon and Manton, despite Calman's admonitions, became strict experimentalists. While they accepted Calman's systematic synthesis, they ignored his admonition on the importance of similarities and instead focused exclusively on the differences between species.

Calman retired from the British Museum in 1936. He had garnered a long list of honors during his career, including the presidency of the Linnean Society and fellowship in the Royal Society. When he left museum service, he returned with his wife to his beloved Scotland, specifically to Tayport in Fife, Alice Calman's childhood home. He rejoined the Dundee Naturalist's Society, and during World War II took up teaching again at University College as well as St. Andrews. In 1948, he and Alice returned south for health reasons and settled in Surrey, where Calman died suddenly, September 29, 1952.

4 HERBERT GRAHAM CANNON

Cannon born in Whimbledon April 14, 1897, came of working class parents. His father, David William Cannon, worked as a compositor in a printing firm. His mother, the daughter of Charles Graham, the owner and operator of one of the first horse-drawn buses in south London, bore

Herbert Graham as the third of four children. With only a printer's wages to support the family, the Cannons eventually had to relocate to Brixton in south London with some decent council schools in the area. Young Cannon did well enough to win a scholarship to Wilson's Grammar School, Camberwell. His education there allowed him to specialize in science in the higher grades or 'forms.'

Like Calman, Cannon had little interest in sports, except for rifle marksmanship. He enjoyed chess, did well at arts, and excelled at music. A most impressionable youth, he had a vivid imagination. His memorializer, J.E. Smith, recalled that at age 16, Cannon had seen a particularly exciting stage play wherein one of the characters jumped from an airplane as it descended across the stage. That night he had a vivid dream wherein he took the part of the actor, walked in his sleep, and jumped from the window of his bedroom. He broke both wrists and severely injured his pelvis and lumbar vertebrae. This resulted in a permanent disability, plaguing him with acute and often crippling pain for the rest of his life.

His injuries exempted him from service in World War I, and he entered Cambridge in 1916 as a Choral Scholar at Christ's College. He earned a first class honor tripos in zoology, chemistry, and physiology, and then applied for a term leave in 1918 to take up a temporary post as naturalist with the Board of Fisheries. He served for awhile at Conway on the coast and then in London.

At war's end, he returned to Cambridge and fell under the tutelage of Leonard Doncaster. Most of the rest of the Cambridge faculty were too busy getting their research restarted after war service to bother with students, so Cannon came to work with Doncaster, a Quaker and conscientious objector, almost by default. Doncaster had established expertise in the then budding field of cytogenetics, and he put Cannon to work on a problem studying spermatogenesis in the louse, *Pediculus*. This resulted in Cannon's first paper in 1920 (and Doncaster's last – he died in 1919).

As already noted, the role of experimentation in biology had received a great boost with the publication of D'Arcy Thompson's *On Growth and Form*, and Cannon decided to continue working in the field of cytology. Upon receiving his B.S., he took a position as a demonstrator in zoology at Imperial College in London, working under E.W. MacBride a leading figure in experimental zoology of the time. MacBride ran his laboratory in a stimulating way , and he consistently attracted animal researchers from all over London to Imperial College for discussions on form, function, and phylogeny and evolution. MacBride, besides his devotion to experimental science, ardently advocated a Lamarckian outlook and greatly influenced his students in this regard, especially Cannon.

At this time Cannon established an association with Calman. Calman's office and lab in back of the British Museum lay across the road from MacBride's laboratory. Cannon took to frequenting Calman's lab, and the two often stepped outside away from the alcohol collections to enjoy a smoke. After an abortive attempt at studying nudibranch embryology undertaken as the result of a stay at the Marine Biological Laboratory at Plymouth, Cannon let Calman convince him to direct his attentions toward crustaceans, and after 1922 Cannon shifted his research interests entirely to that group. Calman provided Cannon with some mud sent to him from Mesopotamia that contained dormant eggs of the choncostracan *Estheria*. This prompted Cannon to begin a series of studies on organ development in the branchiopods that definitively launched his career.

Cannon soon discovered that some muscles in the anostracan *Chirocephalus* actually differentiated from ectoderm – those associated with the presumptive proctodeum fold inward with proctodeal invagination and function as hindgut dilators. In addition, Cannon noted that the ventral muscles attached to the endoskeleton also arose from involuted material derived from ectoderm. Cannon's subsequent studies on the development of segmental structures, especially those associated with transient coelomic pouches, earned him a solid reputation as an experimental zoologist, and garnered for him in quick succession a D.Sc. from the University of London and election to the chair of zoology at the University of Sheffield in 1926, and fellowship in the Royal Society and the awarding of the Linnean Society Crisp Medal in 1927 – all by the age of 30.

Cannon had met, while in MacBride's laboratory, Sidnie Manton. The two of them established a collaboration and carried on research on the feeding mechanisms of *Hemimysis lamornae*, as well as studies of segmental excretory organs in crustaceans (1927) and feeding in syncarids (1929). These articles illustrate an ironic twist to Cannon's reputation as an experimental zoologist; he often engaged in study of what some have called 'the functional morphology of the dead.' The syncarid work offers an excellent example of this. Without any available living material, Cannon and Manton relied on the examination of preserved specimens. D'Arcy Thompson would not have objected, since he preferred analogy as an analytical technique over actual experimentation. Cannon and Manton inferred conclusions about syncarid functional morphology from the position and form of fixed material.

However, Cannon's experimental techniques displayed another weakness of his work. When he employed living specimens, as in the Cannon/Manton *Hemimysis* study, he often observed animals under very abnormal conditions. In that particular case, he confined the mysids between two glass slides in a condition just short of being squashed, and drew conclusions about feeding currents around their bodies by observing the motion of carmine particles moved by frantically vibrating appendages. Observations by other workers (e.g., Attramadal 1981) made under more natural conditions came to decidedly different conclusions about how mysids feed.

Although these kinds of studies refected 'state of the art' techniques for those times, they did lead to some misconceptions about crustacean functional morphology. Certain assumptions concerning feeding were taken as givens, such as the presumption of the universal existence of filter feeding among crustaceans not obviously carnivores. For example, Cannon carried on his work (1928) on feeding in the copepods *Calanus* and *Diaptomus* with the same confined setup as in the mysid studies. Cannon's interpretation of the observations lead him to confirm the assumption of filter feeding. This 'discovery' gained acceptance and wide quotation in books and papers for decades until the work of Koehl & Stickler (1981) who, using different techniques, revealed that study of feeding in copepods involved complex issues and did not necessarily encompass true filter feeding at all.

Study of the biology of crustaceans up until the work of the German Carl Claus in the late 1800's focused largely on recording animal distributions and describing new species. Although Calman made some quantum leaps in crustacean studies by using information gained from the study of fossils and carefully observing the details of crustacean structure, Cannon tried to use observations and experiments to move the study of crustaceans outside the confines of museum collections and descriptive anatomy. That he didn't always succeed because of the shortcomings of technique and material does not negate the value of his work. On the other hand, we should recognize the limits of his studies.

In 1931, Cannon became Beyer Professor of Zoology and head of the department at the University of Manchester. The larger department and better facilities enabled him to deal with more students and a more stimulating staff than at Sheffield. Although he did not establish a long line of crustacean workers from his students there, he did have a few successes (Blower 1989).

Chiefly among these was Ralph Dennell who published studies on feeding in the cumacean *Diastylis* and the tanaid *Apseudes* (Dennell 1934, 1937). Dennell's research closely paralleled the work of Cannon (1933) on feeding in the branchiopods, a study which earned Cannon his fellowship in the Royal Society.

Subsequently, Cannon turned his attentions to more purely anatomical studies focusing on barnacles, ostracodes, and *Nebalia*. The latter culminated in his preparation of a volume on leptostracans in *Bronn's Tierreich*. However, in his later years, like Calman, Cannon turned his attentions to more theoretical matters, in his case a defense of Lamarckism. Indeed, he became a vocal critic of what he considered a lamentable emphasis on genetics in modern evolutionary theory for, as he wrote, 'Mendelism is a panacea for evolutionary troubles' (Cannon 1956:1).

Cannon did not totally reject out of hand the science of genetics, he just questioned whether genes controlled all characters. Species characters, to him, had undoubted Mendelian origins, but characters useful in defining higher taxonomic categories seemed '...due to some entirely different mechanism... some type of inheritance not particulate but organismal' (Cannon 1956:3-4).

Cannon devised an ingenious metaphor to make his point. In the automobile industry, many permits and machines went into making a particular car. If some crucial bit of machinery or a particular government permit disappeared, the car production stops. This analogy mimics the genetic control of development. However, the 'evolution' or development of the automobile happened quite separate from the permits and machines necessary to make any particular car. Cannon then asked whether the evolution of the eye occurred quite independently of any genes necessary for the presence or absence of eyes.

Cannon believed the question of gene reality irrelevant. He believed it more important to ask whether the changes necessary for evolution resulted from chance or some directed design. As an idealist and a Lamarckian, he believed that changes arose from functional need, and being functional they had to be directed toward 'a' design. He failed to see that the element of chance could play a critical role in evolution with or without design.

Cannon believed that need leads to appearance of form or structure, and use leads to degrees of development. Like Calman, he ascribed to a holistic view of nature, that organisms lived in total equilibrium with their environment. However, Cannon saw evolution as a kind of biological Le Chatelier's Principle, i.e., biological systems adjusted to external changes in the environment so as to remain in equilibrium with them.

At the time he composed the above ideas, Cannon delivered a lecture to the Linnean Society in London (Nov. 5, 1955), 'What Lamarck really said.' He advanced at that time the idea that a misunderstanding of Lamarck's real meaning arose in translating his ideas. Lamarck used the word *besoin*, meaning need, as opposed to *desir*, meaning desire. *Besoin* carried the connotation of unconscious need, and to Cannon this conveyed a more acceptable idea as opposed to the distorted translation of Cuvier and other Lamarckian critics who claimed that Lamarck placed in animals a willful desire.

Of course, both words suffer from the problems associated with teleology. This interpretation of Lamarck in turn led Cannon to ask the wrong question of evolution, i.e., he sought to know the 'why' of form and structure rather than being satisfied with trying to understand the 'how.' Cannon blamed Darwin for Lamarck's reputation in English-speaking biology. However, Cannon revels in pointing out that by the sixth edition of *The Origin of Species* Darwin clearly took a Lamarckian position when he concluded that the peculiar form of the giraffe occurred 'no doubt in an important manner with the inherited effects of the increased use of parts' (Darwin 1878:178).

Cannon found a receptive audience for his ideas. The above lecture actually triggered a conference with the Systematics Association on the inheritance of acquired characters that occurred on Nov. 1, 1956 at the Linnean Society (published in the Proceedings of the Linnean Society, 1958) with the participation of several notable figures including among others Sidnie Manton, Richard Melville, and C.H. Waddington. In 1959, Cannon published a small book on the subject, *Lamarck and Modern Genetics*. In his spirited defense of Lamarck, Cannon tried to deal with a problem that modern evolutionary theoreticians still debate. In effect, while he believed that genes controlled little things, Cannon could not understand how genes controlled major adaptive features. Essentially, this mirrors modern discussions of the problem of micro- versus macroevolution. Cannon's view in this regard stemmed from what Calman had called the problem of causal versus historical morphology. As a functional morphologist, Cannon believed in experimentation as a way of discerning form and function. All form had to have function in his mind, and so the function became the determinant of form. Although Cannon preached holism, his inclination to experimentally manipulate systems in order to fanthom form predisposed him to reductionism. This in turn caused him to focus on the differences between animal types rather than the similarities.

How might Cannon's ideas figured in late 20th century debates on the subject if he had lived into his 70's, or even 80's? As it was, he died rather suddenly and at his peak. Early in 1962 he suffered an attack of acute appendicitis. His recovery from surgery was very slow, and he took a leave that summer with his wife to visit one of his children in West Africa. He seemed well, but on his way home he collapsed in Spain and rushing back to London doctors diagnosed liver disease. He died soon thereafter early in 1963. His effect on arthropod studies lived on, though, in the person of his most famous student, Sidnie Manton.

5   SIDNIE MILANA MANTON

Manton, born in London on May 4, 1902, came from a professional family; her father, George Sidnie Frederick Manton, practised as a dental surgeon, and her mother, Milana d'Humy, of Scottish-French ancestry, had artistic inclinations. Sidnie had a younger sister, and both girls developed a strong interest in the natural sciences from their holiday stays at the family country cottage. (Her sister Irene also entered the sciences and became an accomplished student of flagellates, and both sisters eventually became Fellows of the Royal Society.)

At age 4, Sidnie began attending the Froebel Educational Institute, Kensington, where Rosalie Lulham, a teacher and author of an introductory zoology text, encouraged Sidnie's interests in science with the study of live animals and excursions into the field for fungi. Her later education at St. Paul's Girls School, Hammersmith, and repeated visits to the Natural History Museum reinforced her interests in biology, and she did well enough in her studies to earn a scholarship to Girton College, Cambridge. Unlike Calman and Cannon, Manton liked athletics, excelling at swimming and hockey, both before and during her Cambridge years.

A brilliant student, Manton held scholarships all the way through Cambridge, attaining her first class honors tripos in zoology, botany, and physiology while winning the Montefiore Prize. With the completion of her zoology degree she used a Alfred Yarrow Research Studentship in 1924 to work in the laboratory of E.W. MacBride at Imperial College, London. There she met W.T. Calman and H. Graham Cannon, beginning a collaboration with the latter that went on for several years. In 1926, she returned to Cambridge to continue her studies and became in 1927 the first woman demonstrator in comparative anatomy in the university. She joined the staff at Girton College when she received her Ph.D. in 1928.

Her year in London determined the direction of her career. She published her first two papers in 1927 with Cannon as co-author (their studies of mysid feeding and crustacean excretory organs noted above). Parallel to these projects, she carried out her own work on what became her classic study of the embryology of *Hemimysis lamornae* (Manton 1928a), as well as publishing some observations relevant to the feeding and natural history of lophogastrids (Manton 1928b) and collaborating with Cannon on a study of feeding in syncarids (Cannon & Manton 1929). All these papers, aside from their voluminous and careful observations of form and inferred function, saw publication in highly conspicuous places either by the Royal Society of London, the Royal Society of Edinburgh, or the Linnean Society of London. Combining this publication record with her Cambridge academic achievements and Manton could not help but progress as a rising star in British zoology in the 1920's.

In 1928, with the completion of her degree, she sailed for Australia to participate in the Great Barrier Reef Expedition. On the way, influenced by Calman's studies on fossil and recent syncarids and, wishing to augment her feeding studies with Cannon on *Anaspides* and relatives, she detoured to Tasmania to collect and observe living 'mountain shrimp.' Two short papers resulted (Manton 1929, 1930), based largely on field observations, that corrected some misconceptions arrived at during her functional studies with Cannon. These discoveries demonstrated to her the importance of making observations on living material while studying preserved specimens. When she began her studies of arthropod locomotion some years later, she definitely preferred to include observations on living material.

Returning to Cambridge, Manton resumed her duties as anatomy demonstrator, publishing with J.T. Saunders a very successful textbook, *A Manual of Practical Vertebrate Morphology* (1931), which went through several editions. [Interestingly, the two great English speaking women invertebrate zoologists of this period, Sidnie Manton and Libbie Hyman, both published successful vertebrate comparative anatomy texts.] However, Manton's research interests began to shift. She did complete an important study on the embryology on *Nebalia* in 1934, but this effectively completed her purely crustacean research. From the mid-1930's onward, she directed her research within the larger framework of arthropod biology. At that time she began her series 'Studies on the Onychophora,' which extended (see appendix) from her contribution on feeding in 1937, to the last and eighth paper in the series on placental development in 1972 (with D.T. Anderson).

A strong phylogenetic slant characterises Mantonian papers, evident in the very first paper she published with Cannon on mysid feeding. Greatly influenced by Calman, a substantial portion of the text of that paper presented a detailed analysis of the comparative anatomy of fossil and recent arthropod limbs as they related to inferred feeding patterns. Thus a fine paper on functional morphology also became a seminal contribution on the theory of arthropod evolution. Indeed, I would date the beginning of serious modern discussion of crustacean evolution to this paper. Certainly its impact exceeded the related and contemporaneous contributions of Borradaile (1926), which did not deal with fossils, and Hansen (1925, 1930), which made only passing reference to past life.

The phylogenetic focus of her early work carried over into her onychophoran studies and grew into the motive force to Manton's greatest research effort, on 'The evolution of arthropod locomotory mechanisms,' a series of papers begun in 1950 and completed with number 11 in 1973 (see appendix). This research effort culminated in her magnum opus, *The Arthropoda. Habits, Functional Morphology, and Evolution* (1977). This series resulted as a consequence of her onychophoran studies when her colleague at Cambridge, Sir James Gray, provided her with some movie film of a walking *Peripatus*. The film so intrigued her that she used Gray's footage in the first of a series of studies that employed film to record live action for later analysis in detail.

Manton's work on locomotory mechanisms may never be surpassed. The shear volume of detail overwhelms, almost 1000 pages in the main series of 11 papers. Her relocation from Cambridge to London facilitated progress on the project. In 1937, Manton married John Harding, a noted copepod authority who eventually succeeded Calman as Keeper of Zoology at the British Museum. Manton took up a succession of positions at several colleges of the University of London, the most enduring being a readership at King's beginning in 1949 (where she gained her most productive student, D.T. Anderson). Her academic positions culminated in 1960 with an honorary fellowship at Queen Mary College. However, she spent most of her time in residence at the British Museum carrying on her locomotion experimental work and related dissections. She systematically proceeded through study of the arthropods using movie film of animals walking under various experimental conditions, focusing her attentions mostly on the myriapods and insects. This extensive work, augmented by the influence of the ideas of O.W. Tiegs, caused Manton to recognize these arthropods as a distinct lineage quite separate from the crustaceans, for which she coined the term Uniramia. This stood as an unusual conclusion at the time, since the majority of arthropod researchers subscribed to the conclusions of the great insect researcher, Robert Snodgrass, who created the taxon Mandibulata to unite the myriapods and hexapods with the crustaceans. Manton believed that the difference in locomotory anatomy formed one of two fundamental reasons why uniramians and crustaceans must remain separate.

Her other reason for separating these groups grew from her research on mandibular mechanisms (1964), which elucidated for the first time in great detail the variety expressed in arthropod mandibular anatomy. Both Manton's locomotory and mandibular work, however, linked considerations of form to function, and her analyses fixed on the differences between various arthropod groups at all levels. This achieved an emphasis on causal rather than historical factors in arthropod evolution. As a result, her phylogenetic conclusions rejected all the traditional arthropod family trees, disassociated lines of evolution from each other, and produced a series of phylogenetic grasses.

The basically idealist and Lamarckian underpinnings to her research emerged into print only once to my knowledge. Following the published remarks of the scheduled participants in Cannon's conference at the Linnean Society on 'The Inheritance of Acquired Characters,' Manton made some comments (Manton 1958). She offered that anatomical characters are 'intimately correlated' with each other to form a complex harmonious whole, no single feature of anatomy seemed sufficient in and of itself to totally achieve some functional end. She cited the tergite heteronomy seen in fast moving centipedes. This prominent and readily observed character would have minimal impact on the locomotory abilities of these animals without simultaneous changes in tergite size, the development and orientation of the 300 muscles used by these animals in locomotion, and the structure of the leg joints. Manton concluded that to rely on genes alone to produce such complex adaptations both by chance and simultaneously 'may need reconsideration.' Her position in this matter clearly approaches that of Cannon already discussed above, and indeed echoes an overview first voiced by Paley.

The philosophic foundations of Manton's work should interest us because her immense effect on arthropod studies. Not only the shear volume of her work impressed her colleagues, but she also spread her ideas on arthropod functional morphology and evolution with missionary zeal. Beginning with the well known and often cited review with O.W. Tiegs (1958) on 'The evolution of the Arthropoda,' she took every available opportunity to spread her point of view. These efforts included contributed sections in the *Treatise on Invertebrate Paleontology*

(1969), review chapters in Ramsay & Wigglesworth's *The Cell & the Organism* (1961), Whittington and Rolfe's *Phylogeny and Evolution of Crustacea* (1963), Gray's *Arthropod Locomotion* (1968), Florkin's *Chemical Zoology* (vol. 5, 1970), Gupta's *Arthropod Phylogeny* (1979), and, with D.T. Anderson, House's *The Origin of Major Invertebrate Groups* (1979). She wrote papers in the *Proceedings of the 15th International Congress of Zoology* (1959), *Biological Review* (1960), *Journal of Natural History* (1967), and the *Journal of Zoology* (1973). All these articles essentially repeat the same theme, that the major groups of arthropods differed so much from each other that they evolved separately – that arthropods constituted a polyphyletic taxon.

Indeed, by the end of her career she had promulgated the message of arthropod polyphyly so thoroughly that it became a virtual mantra among many arthropod workers. With so commanding a status few workers published anything dealing with arthropod locomotion without consulting with her first. For example, the Burgess Shale paleontologists who worked on arthropods tended to preface their remarks at scientific meetings or in private conversations with the statement, 'Of course, we've passed this by Manton and ...'

Manton herself became too certain of her own opinions and once, towards the end, fell victim to her own methods. Her 1978 paper on pycnogonids utilized a dissection of a single preserved specimen, and her conclusion that sea spiders evolved from true spiders stood in stark opposition to the observations of Schram & Hedgpeth (1978) who based their analysis on movie films of living material as well as dissections of several species.

However, she did not always take herself too seriously. She described her book *The Arthropoda* to me in a letter saying, 'It is not intended to be a high-powered work but something that students can read, and there will be an enormous number of figures in it' (letter of early Dec. 1974). Her temper could be short, nevertheless, when she believed she perceived muddled reasoning on someone's part. The conclusions of a young trilobite paleontologist in a series of papers he wrote so annoyed her that through several letters to me in 1974 she constantly dwelt on his conclusions. 'How he invents such rubbish I don't know' (Dec. 20, 1974), and 'You can come down on (him) like a ton of bricks' (Dec. 23, 1974).

In the 1970's, her health declined. Letters to me complained of cataract and arthritic problems. Indeed, she underwent extensive surgery to replace joints and remodel her wrists. Although she went about her work while in great pain, she continued speaking and writing. Yet her performance at lectures could mesmerize. She would hobble to the podium, barely able to move, and within minutes into her talk she would seemingly shed her ailments, blossom, and become an animated and dynamic speaker. Only after she sat down did the pain of her afflictions return. She worked almost to the end, suffering a physical collapse in late 1978 from which she died in the hospital January 2, 1979.

6   SUMMARY

From the above discussion, it would appear that the field of British systematic carcinology effectively divided between two groups. One party derived from W.T. Calman and took root in a materialist tradition skeptical of experimentalism. It developed from an essentially selectionist belief, but did not hold necessarily to strict Darwinism in the modern sense. While the role of genes in the process of developing 'new' traits seemed often vague and allowed some cynicism, the line of thought expressed by this group had a strong commitment to lineage and history as essential elements in understanding the record of life.

The other approach, that of H.G. Cannon and S.M. Manton, took root in an idealist tradition strongly committed to the efficacy of experiments. It developed from an essentially, though not necessarily overtly, Lamarckian approach. Environmental perturbation loomed as the source for 'new' traits, and this outlook grew from the notion that structures have functions, functions serve purposes, purposes arise from environmental constraints, and therefore environment determines structure. Historical lineage seemed not as important (indeed Manton viewed it as something impossible to arrive at) as functional causes in trying to understand the past events of life.

These divergent approaches towards understanding crustacean biology did not exclude each other. Indeed, the mutual friendship and respect that existed between these three people and their colleagues obscured the existence of these two very distinct methods. However, the fact remains that the materialists, Calman and his associates, retreated into essentially a strict taxonomic realm of activity, while the idealists, Cannon and Manton and their students, excelled at experimental work. I use these verbs deliberately because even today the connotation persists in the minds of many that taxonomy lacks in productivity or significance for the advancement of biological research as opposed to experimental work. Calman and his colleagues did not possess an effective notion as to how major adaptive features could arise, nor could they make definitive qualitative decisions about differences between characters.

To non-taxonomists, the conclusions of taxonomy (i.e., classifications) often appear as somehow eternal, and fluxes in taxonomy (i.e., chaging classifications) stand somehow as a sign of bad science. Experimental science on the other hand appears as dynamic and productive when new experiments overturn old ideas. Of course, taxonomy can be dynamic, and experiments once well done can stabilize our view of life. We need to meld somehow the two approaches, i.e., enliven our materialism with a sense of creativity and dampen our idealism with due regard for reality. Only when we recognize the limits of these viewpoints can we build on the potential of each towards advancing our understanding of nature.

APPENDIX

Certain errors and inconsistencies occur in the published bibliographies of Calman, Cannon, and Manton. To correct these and for ease of reference, I present bibliographies for these people here.

William Thomas Calman
1892   On certain new or rare rotifers from Forfarshire. *Ann. Scot. Nat. Hist.* 4: 240-245, pl. 8.
1893   A new *Pedalion. Ann. Mag. Nat. Hist.* (6)11: 332-333.
1894   On *Julinia*, a new species of compound ascidians from the Antarctic Ocean. *Quart. J. Micro. Sci.* 37: 1-17, pls. 1-3.
1895   Note on rotifers from County Galway, pp. 697-698. In J. Hood, On the Rotifera of the Country Mayo. *Proc. Roy. Irish Acad.* (3)3: 664-706, pls. 21, 22.
1896   On species of *Phoxicephalus* and *ApherusaD. Trans. Roy. Irish Acad.* 30: 743-754, pls. 31, 32. On deep-sea Crustacea from the south-west of Ireland. *Trans. Roy. Irish Acad.* 31: 1-20, pls. 1,2.
1897   On the genus *Anaspides* and its affinities with certain fossil Crustacea. *Trans. Roy. Soc. Edinb.* 38: 787-802, 2 pls.
1898   The progress of research on the reproduction of Rotifera. *Nat. Sci.* 8: 43-51.
        Rotifera in Lake Bassenthwaite. *Nature* 58:271.
        On a collection of Crustacea from Puget Sound. *Ann. N. Y. Acad. Sci.* 9: 259
1899   On the characters of the crustacean genus *Bathynella* Vejdovsky. *J. Linn. Soc. (Zool.)* 27: 338-344, pl.20.

On the British Pandalidae. *Ann. Mag. Nat. Hist.* (7)3:2 7-39, pls. 1-4.

On two species of macrurous crustaceans from Lake Tanganyka. *Proc. Zool. Soc.* 1899: 704-712, pls. 39,40.

1900   On a collection of Brachyura from the Torres Strait. *Trans. Linn. Soc. Lond. (Zool.)* 8: 1-50, pls. 1-3.

1902   *Uronectes* and *Anaspides.* A reply to Prof. Anton Fritsch. *Zool. Anz.* 25: 65-66.

On the occurrence of terrestrial planarians of Scotland. *Ann. Soc. Nat. Hist.* 1902: 231-233.

1903   On macrurous Crustacea obtained by Mr. George Murray during the cruise of the 'Oceanea' in 1898. *Ann. Mag. Nat. Hist.* (7)11: 416-420.

Note, p. 463. In E.R. Lankester, On the modification of the eye peduncles in the crabs of the genus *Cymonomus. Quart. J. Micro. Sci.* 47: 439-463.

1904   On the classification of the Crustacea Malacostraca. *Ann. Mag. Nat. Hist.* (7)13: 144-158.

On *Munidopsis polymorpha* Koelbel, a cave-dwelling marine crustacean from the Canary Islands. *Ann. Mag. Nat. Hist.* (7)14: 213-218.

Report on the Cumacea collected by Prof. Herdman at Ceylon in 1902. *Rept. Pearl Oyster Fish. Gov. of Manaar, Part 2, Suppl. Rept.* 12: 159-180, 5 pls.

1905   Note on a genus of euphausid crustacean. *Rept. Fish. Ireland*, 1902-1903, Part 2, App. iv(2): 153-155, pl. 26.

The marine fauna of the west coast of Ireland. Part. 4. Cumacea. *Sci. Invest. Fish. Brit. Ireland*, 1904. 52, pp., 5 pls.

On a new species of river-crab from Yunnan. *Ann. Mag. Nat. Hist.* (7)16: 154-158.

The Cumacea of the 'Siboga' Expedition. *Siboga Exped.* 36: 1-23. 2 pls.

1906   Notes on some genera of the crustacean family Hippolytidae. *Ann. Mag. Nat. Hist.* (7)17: 29-34.

The Cumacea of the 'Puritan' Expedition. *Mitt. Zool. Stat. Neapel.* 17: 411-432, pls. 27, 28.

Zoological results of the Third Tanganyka Expedition, conducted by Dr. W.A. Cunnington, 1904-1905.

Report of the macrurous Crustacea. *Proc. Zool. Soc.* 1906: 187-206.

On a lobster with symmetrical claws. *Proc. Zool. Soc.* 1906: 633.

1907   On a freshwater decapod crustacean collected by W.J. Burchell at Parà in 1829. *Ann. Mag. Nat. Hist.* (7)19: 295-299.

Sur quelques Cumacés des côte de France. *Bull. Mus. Paris.* 1907: 116-124.

On new or rare Crustacea of the Order Cumacea from the collection of the Copenhagen Museum. Part. 1. The families Bodotriidae, Vauntompsoniidae, and Leuconidae. *Trans. Zool. Soc.* 18: 1-58, pls. 1-9.

1908   Cumacea. *Natl. Antarctic Exped. Nat. Hist.* 2: 1-6.

Decapoda. *Natl. Antarctic Exped. Nat. Hist.* 2:1-7.

Notes on a small collection of plankton from New Zealand. *Ann. Mag. Nat. Hist.* (8)1:232-240.

On a stridulating organ in certain African river-crabs. *Ann. Mag. Nat. Hist.* (8)1:469-473.

On a parasitic copepod from *Cephalodiscus. Trans. S. Afr. Phil. Soc.* 17:177-184, pls. 28, 29.

Notes on some characters of *Koonunga* and *Anaspides*, p.15. In A. O. Sayce, On *Koonunga cursor*, a remarkable new type of malacostracous crustacean *Trans. Linn. Soc. Lond. (Zool.)* 11:1-16, pls. 1, 2.

An early figure on the king-crab (*Limulus polyphemus*). *Science* 27:669.

1909   A new river-crab of the genus *Gecarcinus* from New Guinea. *Proc. Zool. Soc.* 1908:960-963.

The fauna of the Cocos-Keeling Atoll, collected by F. Wood-Jones. Crustacea. *Proc. Zool. Soc.* 1909:159-160.

On decapod Crustacea from Christmas Island, collected by Dr. C. W. Andrews, F.R.S., F.Z.S. *Proc. Zool. Soc.* 1909:703-713, pl. 72.

Ruwenzori Expedition Reports. 5. Crustacea. *Trans. Zool. Soc.* 19:51-56.

On a blind prawn from the Sea of Galilee (*Typhlocaris galilea* g. et sp. n.) *Trans. Linn. Soc. Lond. (Zool.)* (2)11:93-97, pl. 19.

On a new crab from a deep-sea telegraph cable in the Indian Ocean. *Ann. Mag. Nat. Hist.* (8)3:30-33.

The genus *Puerulus* Ortmann, and the postlarval development of the spiney lobsters (Palinuridae). *Ann. Mag. Nat. Hist.* (8)3:441-446.

The problem of the Pycnogonida. *Sci. Progr.* 3:687.

Gigantocypris and the 'Challenger.' *Nature* 80:248.

A Treatise on Zoology, Part 7: Appendiculata. Third Fascicle. Crustacea. London: Adam & Charles Black.

1910   *Guide to the Crustacea, Arachnida, Onychophora, and Myriapods.* London: British Museum (Natural History).

On two new species of wood-boring Crustacea from Christmas Island. *Ann. Mag. Nat. Hist.* (8) 5:181-186, pl. 5.

Note on two species of *Pandalus. Ann. Mag. Nat. Hist.* (8)5:524-527.

On *Heterocuma sarsi* Miers. *Ann. Mag. Nat. Hist.* (8)5:612-616.

The researches of Bouvier and Bordage on mutations in Crustacea of the family Atyidae. *Quart. J. Micr. Sci.* 55:785-797.

Les Cumacés des expéditions du 'Travailleur' et du 'Talisman.' *Bull. Mus. Paris* 1910:180-182.

1911   *The Life of Crustacea.* London: Methuen & Co.

Crustacea. *Encyclopaedia Britannica,* 11th edit. 7:552-561.

On new or rare Crustacea of the order Cumacea from the collection of the Copenhagen Museum. Part 2. The families Nannastacidae and Diastylidae. *Trans. Zool. Soc.* 18:341-398, pls. 32-37.

On the transference of names in zoology. *Science* 33:219; *Nature* 85:406.

Note on a crayfish from New Guinea. *Ann. Mag. Nat. Hist.* (8)8:366-368.

An epizoic hydroid on a crab from Christmas Island. *Ann. Mag. Nat. Hist.* (8)546-550.

On *Pleurocaris,* a genus from the English Coal Measures. *Geol. Mag.* 8:156-160.

Transference of the term 'genotype.' *Science* 34:685 (with F.A. Bather)

On some Crustacea of the Division Syncarida from the English Coal Measures. *Geol. Mag.* 8:488-495.

1912   On *Dipteropeltis,* a new genus of the crustacean order Branchiura. *Proc. Zool. Soc.* 1912:763-766. pl. 84.

On a terrestrial amphipod from Kew Gardens. *Ann. Mag. Nat. Hist.* (8)10:132-137.

The Crustacea of the order Cumacea in the collection of the United States National Museum. *Proc. US Nat. Mus.* 41:603-676.

Patrick Matthew of Gourdiehill. *J. Bot.* 50:193.

1913   Eucrustacea, pp. 730-769. In C.R. Eastman (ed.), K.A. von Zittel's *Textbook of Paleontology.* London: MacMillan.

Note on the brachyuran genera *Micippioides* and *Hyastenus. Ann. Mag. Nat. Hist.* (8)11:312-314.

A new species of the crustacean genus *Thaumastocheles. Ann. Mag. Nat. Hist.* (8)12:229-233.

Two cases of abnormal appendages in crabs. *Ann. Mag. Nat. Hist.* (8)11:399-404.

On freshwater decapod Crustacea (families Potamonidae and Palaemonidae) collected in Madagascar by the Hon. Paul A. Methuin. *Proc. Zool. Soc.* 1913:914-932, pls. 91, 92.

On *Aphareocaris* nom. nov. (*Aphareus*), a genus of the crustacean family Sergestidae. *J. Linn. Soc. (Zool.)* 32:219-223.

1914   On the crustacean genus *Sicyonella* Borradaile. *Ann. Mag. Nat. Hist.* (8)13:358-260.

A new crab of the genus *Calappa* from West Africa. *Ann. Mag. Nat. Hist.* (8)14:493-494.

Report on the river-crabs (Potamonidae) collected by the British Ornithologists' Union Expedition and the Wollaston Expedition in Dutch New Guinea. *Trans. Zool. Soc. Lond.* 20:307-313.

Arthropleura moyseyi n. sp. from the Coal Measures of Derbyshire. *Geol. Mag.* 1:541-544.

1915   The holotype of *Ammothea carolinensis* Leach (Pycnogonida). *Ann. Mag. Nat. Hist.* (8)15:310-314.

The holotype of *Nymphon gracilipes* Miers (Pycnogonida). *Ann. Mag. Nat. Hist.* (8)15:584-588.

Pycnogonida. *Brit. Antarct. ('Terra Nova') Exped., 1910. Nat. Hist. Rept. Zool.* 3(1):1-74.

A parasite of the flying fish. *West Ind. Bull.* 15:120.

1916   A new species of the crustacean genus *Squilla* from West Africa. *Ann. Mag. Nat. Hist.* (8)18:373-376.

Lobster fisheries and their conservation. (3 articles in *Country Life,* January-February).

1917   Notes on the morphology of *Bathynella* and some allied Crustacea. *Quart. J. Micro. Sci.* 62:489-514.

Crustacea. Part 4. Stomatopoda, Cumacea, Phyllocarida, and Cladocera. *Brit. Antarct. ('Terra Nova') Exped., 1910. Nat. Hist. Rept. Zool.* 3(5):137-162.

Cumacés. *Deuxiéme Exped. Antarct. Française (1908-1910). Documents Scientifiques.* Arthropoda 1917:1p.

1918   Cumacea and Phyllocarida. *A. A. Expedit. Sci. Rept.* Ser. C. V, Part 6:1-11, pls. 19, 20.
          On barnacles of the genus *Scalpellum* from deep-sea telegraph cables. *Ann. Mag. Nat. Hist.* (9)1:
          96-124.
          A new crab from the Transvaal. *Ann. Mag. Nat. Hist.* (9)1:234-236.
          The type specimen of *Poecilasma carinatum* Hoek (Cirripedia). *Ann. Mag. Nat. Hist.* (9)1:401-408.
          Review: A Triassic isopod crustacean from Australia. *Geol. Mag.* 5:227-280.
1919   Dr. Walcott's researches on the appendages of the trilobites. *Geol. Mag.* 6:359-363.
          On barnacles of the genus *Megalasma* from deep-sea telegraph cables. *Ann. Mag. Nat. Hist.* (9)
          4:361-374.
          Marine Boring Animals Injurious to Submerged Structures. *BM(NH) Econ. Series* 10:1-35.
1920   A new crab of the genus *Sesarma* from Basra. *Ann. Mag. Nat. Hist.* (9)5:62-65.
          A whale-barnacle of the genus *Xenobalanus* from Antarctic seas. *Ann. Mag. Nat. Hist.* (9)6:165-166.
          On a collection of Pycnogonida from the South Orkney Islands. *Ann. Mag. Nat. Hist.* (9)6:244-247.
          A new isopod of the genus *Serolis. Ann. Mag. Nat. Hist.* (9)6:299-304.
          On marine boring animals. In P.M. Crosthwaite & G.R. Redgrave (eds), *Deterioration of Structures
          in Sea-water*, pp. 63-78. London: Comm. Inst. Civil Engineers.
          Notes on marine wood-boring animals. 1. The shipworms (Teredinidae). *Proc. Zool. Soc.* 1920:391-
          403.
          Cumacea of the Canadian Arctic Expedition, 1913-1918. *Rept. Can. Arct. Exped.* 7:1-4.
1921   Notes on marine wood-boring animals. 2. Crustacea. *Proc. Zool. Soc.* 1921:215-220.
1922   The holotype of *Parazetes auchenicus* Slater (Pycnogonida). *Ann. Mag. Nat. Hist.* (9)9:199-203.
1923   Pycnogonida of the Indian Museum. *Rec. Ind. Mus.* 25:265-299.
1924   An abnormal specimen of the edible crab (*Cancer pagurus*). *Ann. Mag. Nat. Hist.* (9)14:326-328.
1925   A new crab of the genus *Sesarma* from New Guinea. *Ann. Mag. Nat. Hist.* (9)15:454-456.
          Crustacea Decapoda collected in Korinchi and on the west Sumatran coast by H.C. Robinson and C.
          Baden Kloss. *J. Fed. Malay States Mus.* 8:166-167.
          Growth stages of a crustacean. *Nature* 115:783-784.
1926   The Rhynie crustacean. *Nature* 118:89-90.
          On freshwater prawns of the family Atyidae from Queensland. *Ann. Mag. Nat. Hist.* (9)17:241-246.
          On macrurous Decapoda collected in South Africa waters by S.S. 'Pickle' with a note on specimens
          on the genus *Sergestes* by H.J. Hansen. *Fish. Mar. Biol. Survey, Cape Town* 4:1-22, 4 pls.
1927   Zoological results of the Cambridge Expedition, Suez, 1924: Crustacea Decapoda (Brachyura).
          Trans. Zool. Soc. Lond. 22:211-217.
          Zoological results of the Cambridge Expedition, Suez, 1924: Phyllocarida, Cumacea, and Stoma-
          topoda. *Trans. Zool. Soc. Lond.* 22:399-401.
          Zoological results of the Cambridge Expedition, Suez, 1924: Pycnogonida. *Trans. Zool. Soc. Lond.*
          22:403-410.
          A blind prawn from the river of Lethe. *Nat. Hist. Mag.* 1:53-55.
          T.R.R. Stebbing 1835-1926. *Proc. Roy. Soc. Lond.* 101B:xxx-xxxii, portr.
1928   On prawns of the family Atyidae from Tanganyka. *Proc. Zool. Soc.* 1928:737-741.
          Subterranean Crustacea. *J. Quekett Micro. Club* 16:1-8.
1929   The Pycnoginida. *J. Quekett Micro. Club* 16:95-106.
1931   The taxonomic outlook in Zoology. *Brit. Assc. Adv. Sci., Rept.* 1930:83-91; *Nature* 126:440-444;
          Science 72:279-282.
1932   A cave dwelling crustacean of the family Mysidae from the Island of Lanzarote. *Ann. Mag. Nat. Hist.*
          (10)10:127-131.
          Notes on *Palaeocaris praecursor* (H. Woodward), a fossil crustacean of the Division Syncarida. *Ann.
          Mag. Nat. Hist.* (10)10:537-541.
          Occurrence of *Bathynella* in England. *Nature* 130:62.
          Survivors of a dead world. *The Times*, 18 July 1932.
          Remarks. *Proc. Linn. Soc.* 145:37.
          Taxonomy in the Museum. *Nat. Hist. Mag.* 31:431.
          William Carmichael M'Intosh 1838-1931. *Proc. Roy. Soc. Lond.* 110B:xxiv-xxviii, portr.
1933   On *Anthracocaris scotica* (Peach), a fossil crustacean from the Lower Carboniferous. *Ann. Mag. Nat.
          Hist.* (10)11:562-565.

1934   Notes on *Uronectes fimbriatus* (Jordon), a fossil crustacean of the Division Syncarida. *Ann. Mag. Nat. Hist.* (10)13:321-330, pls. 12 & 13.

The mitten crab (*Eriocheir*), a Chinese immigrant into Europe. *Proc. Linn. Soc.* 146:69-70.

1935   The meaning of biological classification. *Proc. Linn. Soc.* 147:145-158; *Nature* 136:9-10.

1936   The origin of insects. *Proc. Linn. Soc.* 148:193-204.

Marine boring animals injurious to submerged structures, 2nd ed. (with G.I. Crawford). *BM(NH) Econ. Series* 10:1-38.

1937   James Eights, a pioneer Antarctic naturalist. *Proc. Linn. Soc.* 149:171-184, pl. 4.

1938   Pycnogonida. *Sci. Rept. 'John Murray' Expedit.* 5:147-166.

1939   Crustacea: Caridea. *Sci. Rept. 'John Murray' Expedit.* 6:183-224.

1940   A museum zoologist's view of taxonomy. In J.H. Huxley (ed), *The New Systematics*: pp. 455-459. Oxford: Clarendon Press; *Nature* 146:42.

1949   *The Classification of Animals: An Introduction to Zoological Taxonomy.* London: Methuen's Monographs on Biol. Subjects.

Herbert Graham Cannon

1920   On the spermatogenesis of the louse (*Pediculus corporis* and *P. capitis*) with some observations on the maturation of the egg. *Quart. J. Micro. Sci.* 64:303-328, pl. 15. (with L. Doncaster).

1921   The early development of the summer egg of the cladoceran (*Simocephalus vetelus*). *Quart. J. Micro. Sci.* 65:627-642, pl. 25.

1922   On the labral glands of a cladoceran (*Simocephalus vetelus*) with a description of its mode of feeding. *Quart. J. Micro. Sci.* 66:213-234, pls 9, 10.

A further account of the spermatogenesis of lice. *Quart. J. Micro. Sci.* 66:657-667.

1923   On the nature of the centrosomal force. *J. Genet.* 13:47-78.

Spermatogenesis of the Lepidoptera. *Nature* 111:670-671.

A note on the zoea of a land-crab *Cardisoma armatum*. *Proc. Zool. Soc.* 1923:11-14.

1924   On the development of an estherid crustacean. *Phil. Trans. Roy. Soc. Lond.* (B)212:395-430, pls. 18-24.

A new trematode from the grass-snake. *Ann. Mag. Nat. Hist.* (9)13:194-196, pl. 6. (with H.A. Baylis).

Further note on a new trematode from the grass-snake. *Ann. Mag. Nat. Hist.* (9)13:558-559. (with H.A. Baylis).

1925   Ectodermal muscles in a crustacean. *Nature* 115:458-459.

On the segmental excretory organs of certain fresh-water ostracods. *Phil. Trans. Roy. Soc. Lond.* (B)214:1-27, pls. 1, 2.

1926   On the feeding mechanisms of a freshwater ostracod, *Pinocypris vidua* (O.F. Müller). *J. Linn. Soc. (Zool.)* 36:325-335, pl. 11.

On the post-embryonic development of the fairy shrimp *Chirocephalus diaphanus*. *J. Linn. Soc. (Zool.)* 36:401-416, pls. 22, 23.

1927   On the feeding mechanism of a mysid crustacean *Hemimysis lamornae*. *Trans. Roy. Soc. Edinb.* 55(1):219-253, pl. 1-4. (with S.M. Manton).

On the feeding mechanism of *Nebalia bipes*. *Trans. Roy. Soc. Edinb.* 55(2):355-369.

Notes on the segmental excretory organs of Crustacea. Parts 1-4. *J. Linn. Soc. (Zool.)* 36:439-456. (with S.M. Manton).

An aerating and circulating apparatus for aquaria and general use. *J. Roy. Micro. Soc.* 47:319-322. (with A.J. Grove).

1928   On the feeding mechanism of the fairy shrimp *Chirocephalus diaphanus* Prevost. *Trans. Roy. Soc. Edinb.* 55:807-822.

On the feeding mechanisms of the copepods *Calanus finmarchicus* and *Diaptomus gracilis*. *Brit. J. Exp. Biol.* 6:131-144.

1929   On the feeding mechanism of the syncarid Crustacea. *Trans. Roy. Soc. Edinb.* 56(1):175-189. (with S.M. Manton).

1931   On the anatomy of a marine ostracod *Cypridina (Dolaris) levis* (Skogsberg). *Discovery Repts.* 2:435-482, pls. 6 & 7.

On the blood system of *Parabathynella malaya* G.O. Sars. *Ann. Mag. Nat. Hist.* (10)8:109-114.

Nebaliacea. *Discovery Repts.* 3:199-222, pl. 32.

1933   On the feeding mechanism of certain marine ostracods. *Trans. Roy. Soc. Edinb.* 57(3):739-764.
        On the feeding mechanism of the Branchiopoda (HGC), with an appendix on the mouth parts of the
        Branchiopoda (HGC & FL). *Phil. Trans. Roy. Soc. Lond.* (B)222:267-352. (with F.M.C. Leak).
1934   Feeding mechanism of the fairy shrimp. *Nature* 133:329.
1935   On the rock-boring barnacle *Lithotrya valentiana. Sci. Rept. Gr. Brit. Barrier Reef Exped. 1928-
        1929.* 5(1):1-17, 2 pls.
        Function of the labral glands in *Chirocephalus. Nature* 136:758.
        A further account of the feeding mechanism of *Chirocephalus diaphanus. Proc. Roy. Soc. Lond.*
        (B)117:455-470, pl. 24.
1936   *A Method of Illustration for Zoological Papers.* Norwich: Assc. Brit. Zool.
1937   A new biological stain for general purposes. *Nature* 139:549.
        On the term 'gnathobase' (Lankester), a correction. *Proc. Zool. Soc.* (B)107:539-541.
1940   Ostracoda. *Sci. Rept. 'John Murray' Exped.* 1933-1934. 6:319-325.
        On the anatomy of *Gigantocypris mülleri. Discovery Repts.* 19:185-244, pls.39-42.
1941   A note on fine needles for dissection. *J. Roy. Micro. Soc.* 61:58-59.
        On chlorazol black E and some other new stains. *J. Roy. Micro. Soc.* 61:88-94.
1946   *Nebaliopsis typica. Discovery Repts.* 23:213-222, pl. 15.
1947   On the anatomy of the pedunculate barnacle *Lithotrya. Phil. Trans. Roy. Soc.* (B)233:89-136.
        More graduates. *Manchester Guardian,* 16 May.
1948   Undergraduate zoology. *Advancement of Science* 5:194-204.
        The place of biology in a curriculum. *Nature* 162:1006.
        New certificate examination-university entrance requirements. *The Times,* 16 Oct.
        University entrance requirements. *The Times,* 11 Dec.
1949   The teaching of biology in schools. *Nature* 163:577.
        University entrance requirements and the new examinations. *J. Educ.,* April.
        The teaching of biology in schools. *Nature* 163:846.
        University awards. *J. Educ.,* May.
        University education in natural science. *Nature* 163:882.
1951   University standards. *The Times,* 20 Nov.
1953   Disadvantages of specialization. *The Times,* 7 Jan.
        William Thomas Calman. 1871-1952. *Obit. Not. Roy. Soc.* 8:355-372.
1954   The perfect athlete. *Manchester Guardian,* 26 May.
        Is the problem of evolution solved? *School Sci. Rev.* 35:232-236.
1955   The history of teaching. *The Times,* 7 Jan.
        Illiterates. *Manchester Guardian,* 10 Oct.
        Scientific language. *Manchester Guardian,* 21 Nov.
1956   The shortage of science masters. *Manchester Guardian,* 10 Feb.
        An essay on evolution and modern genetics. *J. Linn. Soc. (Zool.)* 43:1-17.
        Recovery from poliomyelitis – courage alone not the answer. *The Times,* 25 May.
        Cambridge science tripos. *Manchester Guardian,* 30 June.
        The science tripos. *Manchester Guardian,* 9 July.
1957   Specialization at school. *Manchester Guardian,* 8 Jan.
        Specialization in the sixth form. *The Times,* 25 Jan.
        Specialization at school. *Manchester Guardian,* 25, Jan.
        What Lamarck really said. *Proc. Linn. Soc. Lond.* 168:70-85.
        Britain's two cultures. *Sunday Times,* 31 March.
        The most specialized creature. *The Times,* 10 April.
1958   The misuse of penicillin. *Manchester Guardian,* 27 Jan.
        Use of penicillin. *Manchester Guardian,* 4 Feb.
        Opening address. Joint Discussion with the Systematics Association on 'The Inheritance of Acquired
        Characters.' *Proc. Linn. Soc. Lond.* 169:41-45.
        Origin of species. *The Times,* 21 May.
        The idea of evolution. *Manchester Guardian,* 16 June.
        The idea of evolution – Darwin and Lamarck. *Manchester Guardian,* 26 June.

How the giraffe got a good long neck. *Manchester Guardian*, 10 June.
The Evolution of Living Things. Manchester: Manchester Univ. Press.
1959   *Lamarck and Modern Genetics*. Manchester: Manchester Univ. Press (Amer. edit., Springfield: Charles C. Thomas Publ.)
Darwin and Lamarck, their roles in the history of science as seen in the centenary year 1959. *Main Currents in Modern Thought* 1959:106-110.
Language requirements. *The Times*, 8 June.
Oxford and Cambridge scholarships. *Manchester Guardian*, 23 June.
Quiz programmes and education. *Manchester Guardian*, 27 Oct.
Balance sheet. *The Times*, 21 Nov.
Basis of life. *Manchester Guardian*, 10 Dec.
1960   The future of man. *The Listener*, 21 Jan.
Leptostraca. In *Bronn's Klassen und Ornungen des Tierreichs* Bd. 5, Abt. 1, Buch 4, Teil 1:1-81. Leipzig: Akad. Verlag.
The myth of the inheritance of acquired characters. *The New Scientist* 7:798-800.
The professor's job. *The Guardian*, 4 Feb.
Specialization in the sixth. *The Guardian*, 22 Feb.
Wider view of learning. *The Times*, 27, May.
Mass graduation. *The Times*, 28 Dec.
1961   Genetics at the universities. *The Times*, 1 Aug.
Animals on Tristan da Cunha. *The Times*, 19 Oct.
Sir Bertram Jones – an adventurous life. *The Times*, 20 Oct.
The universities' function. *The Guardian*, 15, Dec.
1962   Education in depth. *The Times*, 7 Spet.

Sidnie Milana Manton
1927   On the feeding mechanism of a mysid crustacean *Hemimysis lamornae*. *Trans. Roy. Soc. Edinb.* 55(1):219-253, pl. 1-4. (with H.G. Cannon).
Notes on the segmental excretory organs of Crustacea. Parts 1-4. *J. Linn. Soc. (Zool.)* 36:439-456. (with H.G. Cannon).
1928   On the embryology of the mysid crustacean *Hemimysis lamornae*. *Phil. Trans. Roy. Soc. Lond.* (B)216:363-463.
On some points in the anatomy and habits of the lophogastrid Crustacea. *Trans. Roy. Soc. Edinb.* 56:103-119.
1929   On the feeding mechanisms of the syncarid Crustacea. *Trans. Roy. Soc. Edinb.* 56:175-189. (with H.G. Cannon).
Observations on the habits of some Tasmanian Crustacea. *Victorian Nat.* 45:298-300.
1930   Notes on the habits and feeding mechanisms of *Anaspides* and *Paranaspides* (Crustacea: Syncarida). *Proc. Zool. Soc.* 1930:791-800.
1931   *A Manual of Practical Vertebrate Morphology*. Oxford: Oxford Univ. Press. (with J.T. Saunders)
Notes on the segmental excretory organs of Crustacea. 5. On the maxillary glands of the Syncarida. *J. Linn. Soc. (Zool.)* 37:467-472.
Photograph of a living *Anaspides tasmaniae*. *Proc. Zool. Soc.* 1930:1079.
1932   On the growth of the adult colony of *Pocillopora bulbosa*. *Sci. Rept. Grt. Barrier Reef Exped. 1928-1929* 3:157-166.
1934   On the embryology of the crustacean *Nebalia bipes*. *Phil. Trans. Roy. Soc. Lond.* (B)223:163-238.
1935   Ecological surveys of coral reefs. *Sci. Rept. Grt. Barrier Reef Exped. 1928-1929* 3:273-312.
1937   Studies on the Onychophora. 2. The feeding, digestion, excretion, and food storage of *Peripatopsis*. *Phil. Trans. Roy. Soc. Lond.* (B)227:411-464. (with N.G. Heatley).
Studies on the Onycophora. 3. The control of water loss in *Peripatopsis*. *J. Exp. Biol.* 14:470-472. (with J.A. Ramsay).
1938   Studies on the Onychophora. 4. The passage of spermatozoa into the ovary in *Peripatopsis* and the early development of the ova. *Phil. Trans. Roy. Soc. Lond.* (B)228:421-441.
Studies on the Onychophora. 5. Onychophora found in Cape Colony. *Ann. Mag. Nat. Hist.* (11) 1:476-480.

        Studies on the Onychophora. 6. The life-history of *Peripatopsis. Ann. Mag. Nat. Hist.* (11)1:515-529.

1940   On two new species of the hydroid *Myriothela. Sci. Rept. Br. Graham Ld. Exped. 1934-1937*
        1:255-294.

1941   On the hydrorhiza and claspers of the hydroid *Myriothela cocksi* (Vigurs.) *J. Mar. Biol. Assc. U.K.*
        25:143-150.

1945   The larvae of the Ptinidae associated with stored products. *Bull. Ent. Res.* 35:341-365.

1949   *A Manual of Practical Vertebrate Morphology*, 2nd. edit. Oxford: Oxford Univ. Press. (with J.T.
        Saunders).
        Studies on the Onychophora. 7. The early embryonic stages of *Peripatopsis* and some general
        considerations concerning the morphology and phylogeny of the Arthropoda. *Phil. Trans. Roy. Soc.*
        (B)233:483-540.

1950   The evolution of arthropodan locomotory mechanisms. 1. The locomotion of *Peripatus. J. Linn. Soc.*
        (Zool.) 41:529-570.

1952   The evolution of arthropodan locomotory mechanisms. 2. General introduction to the locomotory
        mechanisms of the Arthropoda. *J. Linn. Soc. (Zool.)* 42:93-117.
        The evolution of arthropodan locomotory mechanisms. 3. The locomotion of the Chilopoda and the
        Pauropoda. *J. Linn. Soc. (Zool.)* 42:118-167.
        A biologist in Soviet Central Asia. *Geogr. Mag.* 25:255-264.
        How the USSR is developing Central Asia. *The Listener* 1952:118-167.
        Developments and science in the Soviet Union. *Nature* 169:729-731.
        The Soviet Union Today, a Scientist's Impression. London: Lawrence & Wishart.

1953   Locomotory habits and the evolution of the large arthropodan groups. *Symp. Soc. Exp. Biol.* 7:339-
        376.

1954   Biological research and the productive transformation of steppe and desert in the Soviet Union. In J.
        L. Cloudsley-Thompson, *Proceedings of a Symposium on the Biology of Hot and Cold Deserts*, pp.
        148-155. London: Inst. Biol.
        The evolution of arthropodan locomotory mechanisms. 4. The structure, habits, and evolution of the
        Diplopoda. *J. Linn. Soc. (Zool.)* 42:299-368.

1955   The changing geography of Soviet Asia. *Indian J. Pwr. Riv. Vall. Dev.* 4:1-10.

1956   The evolution of arthropodan locomotory mechanisms. 5. The structure, habits, and evolution of the
        Pselaphognatha. *J. Linn. Soc. (Zool.)* 43:153-187.

1958   The evolution of arthropodan locomotory mechanisms. 6. Habits and evolution of the Lysiope-
        taloidea (Diplopoda, some principles of leg design in Diplopoda and Chilopoda, and limb structure
        of Diplopoda. *J. Linn. Soc. (Zool.)* 43:487-556.
        Habits of life and evolution of body design in Arthropoda. *J. Linn. Soc. (Zool.)* 44:58-72.
        The evolution of the Arthropoda. *Biol. Rev.* 33:255-337. (with O.W. Tiegs).
        Hydrostatic pressure and leg extension in arthropods, with special reference to arachnids. *Ann. Mag.
        Nat. Hist.* (13)1:161-182.
        Embryology of the Pogonophora and the classifications of animals. *Nature* 181:748-751.

1959   Functional morphology and the evolution of diagnostic characters of arthropodan groups. *Proc. 15th.
        Int. Congr. Zool.* 1:390-393.
        Functional morphology and taxonomic problems of Arthropoda. *Publs. Syst. Assc.* 3:23-32.
        A manual of Practical Vertebrate Morphology, 3rd edit. Oxford: Oxford Univ. Press. (with J.T.
        Saunders).

1960   Concerning head development in the arthropods. *Biol. Rev.* 35:265-282.

1961   Experimental zoology and problems of arthropod evolution. In J.A. Ramsay & V.B. Wigglesworth
        (eds), *The Cell and the Organism*, pp. 234-255. Cambridge: Cambridge Univ. Press.
        The evolution of arthropodan locomotory mechanisms. 7. Functional requirements of body design
        in Colobognatha (Diplopoda), together with a comparative account of diplopod burrowing tech-
        niques, trunk musculature, and segmentation. *J. Linn. Soc. (Zool.)* 44:383-461.

1963   Obituary. H.G. Cannon. *Proc. Linn. Soc. Lond.* 175:83-86.
        Jaw mechanisms of Arthropoda with particular reference to the evolution of Crustacea. In H.B.
        Whittington & W.D.I. Rolfe (eds), *Phylogeny and Evolution of Crustacea*, pp. 111-140. Cambridge:
        Mus. Comp. Zool.

Arthropod segmental organs and malpighian tubules, with particular reference to their function in the Chilopoda. *Ann. Mag. Nat. Hist.* (13)5:545-556. (with D. S. Bennett).

1964   Mandibular mechanisms and the evolution of arthropods. *Phil. Trans. Roy. Soc. Lond.* (B)247:1-183.

1965   The evolution of arthropodan locomotory mechanisms. 8. Functional requirements and body design in Chilopoda, together with a comparative account of their skelatomusculature systems and an appendix on a comparison between burrowing forces of annelids and chilopods and its bearing upon the evolution of the arthropodan haemocoel. *J. Linn. Soc. (Zool.)* 46:251-483.
Locomozione. *Galileo* 92:140-144.

1966   The evolution of arthropodan locomotory mechanisms. 9. Functional requirements of body design in Symphyla and Pauropoda and the relationships between Myriapoda and pterygote insects. *J. Linn. Soc. (Zool.)* 46:103-141.

1967   The polychaete *Spinther* and the origin of the Arthropoda. *J. Nat. Hist.* 1:1-22.

1968   Terrestrial Arthropoda (II). In J. Gray (ed.), *Animal Locomotion*, pp. 333-376. London: Weidenfeld & Nicholson.

1969   Introduction to classification of Arthropoda. In R.C. Moore (ed.), *Treatise on Invertebrate Paleontology*, Part. R, Arthropoda 4, pp. 3-15. Lawrence: Univ. of Kansas Press & Geol. Soc. Amer.
Evolution and affinities of Onychophora, Myriapoda, Hexapoda, and Crustacea. In R.C. Moore (ed.), *Treatise on Invertebrate Paleontology*, Part R, Arthropda 4, pp. 15-56. Lawrence: Univ. of Kansas Press & Geol. Soc. Amer.
A Manual of Practical Vertebrate Morphology, 4th edit. Oxford: Oxford Univ. Press. (with M.E. Brown).

1970   Introduction. In M. Florkin & B.T. Scheer (ed.), *Chemical Zoology*, vol. 5, Arthropoda, part A, pp. 1-34. London: Academic Press.

1971   *Colourpoint, Longhair, and Himalayan Cats*. London: Allen & Unwin.

1972   The evolution of arthropodan locomotory mechanisms. 10. Locomotory habits, morphology, and evolution of hexapod classes. *Zool. J. Linn. Soc.* 51:203-400.
Studies on the Onychophora. 8. The relationship between the embryo and the oviduct in the viviparous placental onychophorans *Epiperipatus trinidadensis* Bouvier and *Macroperipatus torquatus* (Kennel) from Trinidad. *Phil. Trans. Roy. Soc. Lond.* (B)264:161-189. (with D.T. Anderson).

1973   The evolution of arthropodan locomotory mechanisms. 11. Habits, morphology, and evolution of the Uniramia (Onychophora, Myriapoda, Hexapoda) and comparisons with the Arachnida, together with a functional review of uniramian musculature. *Zool. J. Linn. Soc.* 53:257-375.
Arthropod phylogeny--a modern synthesis. *J. Zool.* 171:111-130.

1974   Segmentation in Symphyla, Chilopoda, and Pauropoda in relation to phylogeny. *Symp. Zool. Soc. Lond.* 32:163-190.

1977   *The Arthropoda. Habits, Functional Morphology, and Evolution.* Oxford: Clarendon Press.

1978   The Golgi-complex and inter-relationships of arthropods. *Nature* 276:849-850.
Habits, functional morphology, and evolution of pycnogonids. *Zool. J. Linn. Soc.* 63:1-21.

1979   *Colourpoint, Longhair, and Himalayan Cats*, 2nd edit. London: Ferendune Books.
Functional morphology and the evolution of hexapod classes. In A.P. Gupta (ed.), *Arthropod Phylogeny*, pp. 387-465. New York: Van Nostrand Reinhold.
Polyphyly and the evolution of arthropods. In M.R. House (ed.), *The Origin of Major Invertebrate Groups*, pp. 269-321. London: Academic Press.
Uniramian evolution with particular reference to the pleuron. In M. Camatini (ed.), *Myriapod Biology*, pp. 317-348. London: Academic Press.

REFERENCES

Attramadal, Y.G. 1981. On a non-existent ventral filtration current in *Hemimysis lamornae* and *Praunus flexuosus*. *Sarsia* 66:283-286.

Bonner, J.T. 1966. Introduction. In *On Growth and Form*, abridged edit. Cambridge: Cambridge U. Press

Borradaile, L.A. 1926. Notes upon crustacean limbs. *Ann. Mag. Nat. Hist.* (9)17:193-213, pls. 7-10.

Blower, J.G. 1989. Ralph Dennell (1907-1989). *The Linnean* 5(3):35-38.

Buckland, W. 1869. *Geology and Mineralogy as Exhibiting the Power, Wisdom, and Goodness of God*, 4th Edit. London: Bell & Daldy.

Calman, W.T. 1892. On certain new or rare rotifers from Forfarshire. *Ann. Scot. Nat. Hist.* 4:240-245, pl.8.

Calman, W.T. 1893. A new *Pedalion. Ann. Mag. Nat. Hist.* (6)11:332-333.

Calman, W.T. 1894. On *Julinia*, a new species of compound ascidians from the Antarctic Ocean. *Quart. J. Micro. Sci.* 37:1-17, pls. 1-3.

Calman, W.T. 1896. On the genus *Anaspides* and its affinities with certain fossil Crustacea. *Trans. Roy. Soc. Edinb.* 38:787-802, 2 pls.

Calman, W.T. 1899. On the characters of the crustacean genus *Bathynella. J. Linn. Soc. (Zool.)* 27:338-344, pl. 20.

Calman, W.J.(sic) 1902. *Uronectes* and *Anaspides. Zool. Anz.* 25:65-66.

Calman, W.T. 1904. On the classification of the Crustacea Malacostraca. *Ann. Mag. Nat. Hist.* (7)13:144-158.

Calman, W.T. 1908. On a stridulating organ in certain African river-crabs. *Ann. Mag. Nat. Hist.* (8)1:469-473.

Calman, W.T. 1909. *A Treatise on Zoology. Part 7, Appendiculata. Third Fascicle, Crustacea.* London: Adam & Charles Black.

Calman, W.T. 1911. *The Life of Crustacea.* London: Methuen & Co.

Calman, W.T. 1931. The taxonomic outlook in zoology. *Nature* 126:440-444.

Calman, W.T. 1934. Notes on *Uronectes* fimbriatus (Jordan), a fossil crustacean of the Division Syncarida. *Ann. Mag. Nat. Hist.* (10)13:321-330, pls. 12 & 13.

Calman, W.T. 1935. The meaning of biological classification. *Proc. Linn. Soc.* 1934-35:145-158.

Cannon, H.G. 1925. Ectodermal muscles in a crustacean. *Nature* 115:458-459.

Cannon, H.G. 1928. On the feeding mechanisms of the copepods *Calanus finmarchicus* and *Diaptomus gracilis. Brit. J. Exp. Biol.* 6:131-144.

Cannon, H.G., & F.M.C. Leak 1933. On the feeding mechanism of the Branchiopoda. *Phil. Trans. Roy. Soc. Lond.* (B)222:267-352.

Cannon, H.G. 1953. William Thomas Calman (1871-1952). *Obit. Not. Roy. Soc.* 8:355-372.

Cannon, H.G. 1956. An essay on evolution and modern genetics. *J. Linn. Soc. (Zool.)* 43:1-17.

Cannon, H.G. 1957. What Lamarck really said. *Proc. Linn. Soc., Lond.* 1956:70-85.

Cannon, H.G. 1959. *Lamarck and Modern Genetics.* Springfield: Charles C. Thomas Pub.

Cannon, H.G., & S.M. Manton 1927a. On the feeding mechanism of a mysid crustacean *Hemimysis lamornae. Trans. Roy. Soc. Edinb.* 55:219-253, pls. 1-4.

Cannon, H.G., & S.M. Manton 1927b. Notes on the segmental excretory organs of Crustacea. *J. Linn. Soc. (Zool.)* 36:439-456.

Cannon, H.G., & S.M. Manton 1929. On the feeding mechanism of the syncarid Crustacea. *Trans. Roy. Soc. Edinb.* 56:175-189.

Chambers, R. 1844. *Vestiges of the Natural History of Creation.* London: Churchill.

Chambers, R. 1949. Sir D'Arcy Wentworth Thompson, C.B., F.R.S. (1860-1948). *Science* 109:138-139,151.

Darwin, C. 1878. *On the Origin of Species*, 6th edit. London: John Murray.

Dennell, R. 1934. The feeding mechanism of the cumacean crustacean *Diastylus bradyi. Trans. Roy. Soc. Edinb.* 58:125-142.

Dennell, R. 1937. On the feeding mechanism of *Apseudes talpa* and the evolution of the peracaridan feeding mechanism. *Trans. Roy. Soc. Edinb.* 59:57-78.

Desmond, A. 1982. *Archetypes and Ancestors.* Chicago: Univ. Chicago Press.

di Gregorio, M.A. 1984. *T.H. Huxley's Place in Natural Science.* New Haven: Yale Univ. Press.

Doncaster, L. & H.G. Cannon 1920. On the spermatogenesis of the louse (*Pediculus corporis* and *P. capitis*) with some observations on the maturation of the egg. *Quart. J. Micro. Sci.* 64:303-328, pl. 15.

Fryer, G. 1980. Sidnie Milana Manton (1902-1979). *Biogr. Mem. Fellows Roy. Soc.* 26:327-356, portr.

Gordon, I. 1952. Dr. W.T. Calman, C.B., F.R.S. *Nature* 170:780-781.

Hansen, H.J. 1925. *Studies on Arthropoda II.* Copenhagen: Gyldendalske Boghandel.

Hansen, H.J. 1930. *Studies on Arthropoda III.* Copenhagen: Gyldendalske Boghandel.

Jordan, H. & H. von Meyer. 1854. Über die Crustaceen der Steinkohlenformation von Saarbrücken. *Paläontographica* 4:1-8.

Koehl, M.A.R., & J.R. Strickler 1981. Copepod feeding currents: food capture at low Reynolds number. *Limnol. Oceanogr.* 26:1062-1073.

Livingstone, D.N. 1986. *Nathaniel Southgate Shaler and the Culture of American Science*. Tuscaloosa: Univ. Alabama Press.

Lyell, C. 1830-1833. *Principles of Geology, Being an Attempt to Explain the Former Changes of the Earth's Surface by Reference to Causes Now in Operation*. London: John Murray.

Macbeth, N. 1971. *Darwin Retried*. Boston: Dell Publ.

Manton, S.M. 1928a. On the embryology of the mysid crustacean *Hemimysis lamornae*. *Phil. Trans. Roy. Soc., Lond.* (B)216:363-463.

Manton, S.M. 1928b. On some points in the anatomy and habits of the lophogastrid Crustacea. *Trans. Roy. Soc. Edinb.* 56:103-119.

Manton, S.M. 1929. Observations on the habits of some Tasmanian Crustacea. *Victorian Nat.* 45:298-300.

Manton, S.M. 1930. Notes on the habits and feeding mechanisms of *Anaspides* and *Paranaspides*. *Proc. Zool. Soc. Lond.* 1930:791-800.

Manton, S.M. & J.T. Saunders. 1931. *A Manual of Practical Vertebrate Morphology*. Oxford: Oxford Univ. Press.

Manton, S.M. 1934. On the embryology of the crustacean *Nebalia bipes*. *Phil. Trans. Roy. Soc., Lond.* (B) 223:163-238.

Manton, S.M. 1959. Functional morphology and the evolution of diagnostic characters of arthropodan groups. *Proc. 15th Int. Cong. Zool.*, pp. 390-393.

Manton, S.M. 1960. Concerning head development in the arthropods. *Biol. Rev.* 35:265-282.

Manton, S.M. 1961. Experimental zoology and the problems of arthropod evolution. In J.A. Ramsay & V.B. Wiggleworth (eds), *The Cell and the Organism*: pp. 234-255. Cambridge: Cambridge Univ. Press.

Manton, S.M. 1963. Jaw mechanisms of Arthropoda with particular reference to the evolution of Crustacea. In H.B. Whittington & W.D.I. Rolfe (eds), *Phylogeny and Evolution of Crustacea*: pp. 111-140. Cambridge: Mus. Comp. Zool.

Manton, S.M. 1964. Mandibular mechanisms and the evolution of arthropods. *Phil. Trans. Roy. Soc. Lond.* (B)247:1-183.

Manton, S.M. 1967. The polychaete *Spinther* and the origin of the Arthropoda. *J. Nat. Hist.* 1:1-22.

Manton, S.M. 1968. Terrestrial Arthropoda (II). In J. Gray (ed.), *Animal Locomotion*: pp. 333-376. London: Weidenfield & Nicholson.

Manton, S.M. 1969. Evolution and affinities of Onychophora, Myriapoda, Hexapoda, and Crustacea. In R.C. Moore (ed.), *Treatise on Invertebrate Paleontology* Part R, Arthropoda 4: pp. 15-56. Lawrence: Geol. Soc. Amer. & Univ. Kansas Press.

Manton, S.M. 1970. Arthropods: Introduction. In M. Florkin & B.T. Scheer (eds), *Chemical Zoology*, vol 5: pp. 1-34. London: Academic.

Manton, S.M. 1973. Arthropod phylogeny – a modern synthesis. *J. Zool., Lond.* 171:111-130.

Manton, S.M. 1977. *The Arthropoda. Habits, Functional Morphology, and Evolution*. Oxford: Clarendon Press.

Manton, S.M. 1979. Functional morphology and the evolution of hexapod classes. In A.P. Gupta, *Arthropod Phylogeny*: pp. 387-465. New York: Van Nostrand Reinhold.

Manton, S.M. & D.T. Anderson 1979. Polyphyly and the evolution of arthropods. M.R. House (ed.), *The Origin of Major Invertebrate Groups*: pp. 269-321.

Packard, A.S. 1885. The Syncarida, a group of Carboniferous Crustacea. *Amer. Nat.* 19:700-703.

Packard, A.S. 1886. On the Gampsonychidae, an undescribed family of fossil schizopod Crustacea. *Mem. Natl. Acad. Sci.* 3:123-128.

Paley, W. 1802. *Natural Theology*. London: R. Faulder.

Peacock, A.D. 1953. William Thomas Calman, C.B., D.Sc., LL.D., F.R.S., Hon. F.R.S.E. *Year Book Roy. Soc. Edib.* 1951-52:12-14.

Schram, F.R. 1984. Fossil Syncarida. *Trans. San Diego Soc. Nat. Hist.* 20:189-246.

Schram, F.R., & J.W. Hedgpeth 1978. Locomotory mechanisms in Antarctic pycnogonids. *Zool. J. Linn. Soc.* 63:145-169.

Smith, J.E. 1963. Herbert Graham Cannon (1897-1963). *Biogr. Mem. Roy. Soc., Lond.* 9:55-68, port.

Regan, T. 1930. The evolution of the Primates. *Ann. Mag. Nat. Hist.* (10)6:383-392.

Thompson, D.W. 1942. *On Growth and Form*, second edit. New York: Macmillan.

Thomson, G.M. 1893. Notes on Tasmanian Crustacea, with descriptions of new species. *Proc. Roy. Soc. Tasmania* 1892:45-76.

Thomson, G.M. 1894. On a freshwater schizopod from Tasmania. *Trans. Linn. Soc. (Zool.)* 6:285-303.

Tiegs, O.W. & S.M. Manton 1958. The Evolution of the Arthropoda. *Biol. Rev.* 33:255-337.

Yonge, C.M. 1963. Herbert Graham Cannon, M.A., Sc.D. (Cantab.), D.Sc. (Lond.), M.Sc. (Manch.), F.R.S., F.L.S., Hon. F.R.M.S. *Year Book Roy. Soc. Edinb.* 1962-63:14-15.

Darwin and cirripedology

William A. Newman
Scripps Institution of Oceanography, La Jolla, CA, USA

1 INTRODUCTION

1.1 *What was accomplished*

Charles Darwin (1809-1882) (Fig. 1) began his work on barnacles in 1846, about 20 years after they were recognized if not fully accepted as crustaceans. Needless to say, knowledge of their systematics and morphology was in an early stage of development. Thus Darwin took on a substantial challenge, applying what we now call the 'Darwinian Method' (Ghiselin 1969). He looked at barnacles from every vantage point including using the best dissecting and compound microscopes available at the time. Much of the work required delicate anatomical dissections on minuscule animals because embedding and cutting of thin sections would not come into use until the 1860's. His studies included larval development and metamorphosis as well as adult morphology. He became intensely interested in their sexuality, powers of burrowing, method of flotation, and fossil record, and he created a nomenclature of their soft as well as their hard parts in order to discuss them in sufficient detail. Darwin's four volumes on barnacles were divided two each between fossil and living Cirripedia (1851, 1852, 1854, and 1855; or FC1, LC1, LC2 and FC2, respectively, where cited herein). In these works, he not only elucidated barnacle crustacean affinities, functional morphology, and evolution, but he also established the basis for much of the classification we use today. When questions arose that would not yield to observation, he applied deduction that often led to novel hypotheses to fill the voids. Many of Darwin's deductions and ideas, today dismissed as totally bizarre, were not at all out of keeping with knowledge of the times.

Most of Darwin's anatomical errors and misconceptions quickly drew attention and were rectified. On the other hand, he was extremely cautious in recognizing species and higher taxa, and the problems involved are still being worked out today. For example, he placed all of the balanomorphs in a single family of relatively few genera and species. Many of these species he believed had numerous geographical races or varieties, especially in the wide ranging forms. With the underlying basis for *The Origin* (1859) constantly lurking in his mind, divergence and the variation associated with it were needed to supply the grist upon which natural selection could work. Recognition of varieties of supposed wide-ranging species, as well as the conservative taxonomy he produced, persisted pretty much intact for more than a 100 years after Darwin's barnacle monographs, largely, I believe, because of his great authority and loyal following.

349

Figure 1. Darwin in 1849 at the age of 40, three years after he began work on the barnacles (from Burkhardt & Smith 1985, with permission of F. Burkhardt).

One of the many advances Darwin made in our understanding of the evolution of the barnacles was his concept of the origin of the sessile barnacles[1] from a stalked or pedunculate ancestor. However, a number of relevant fossil and living forms have been discovered since Darwin, and current indications are that the origin of sessile barnacles was less direct and considerably more complicated than he had perceived. Nonetheless, his reasoning and the assumptions underlying his perception of the matter are still taken seriously.

Darwin did more than simply place cirripedology on a firm footing. He had already produced other meticulously detailed, insightful and unprecedented works in natural history, such as *Journal of Researches* (1839), *The Structure and Distribution of Coral Reefs* (1842) and two other volumes on geology (.1844 & 1846), and shortly after the barnacles (1851-55) came *The Origin of Species by Means of Natural Selection* (1859). From these and subsequent works (Fig. 2), it must be confessed, barnacles gained an aura enjoyed by few other invertebrate groups except perhaps earthworms.

1.2   *Sources*

Bookshelves literally groan under the weight of Darwinian biographies, monographs, and Ph.D. theses (Brackman 1980). In *Darwin and Evolution by Natural Selection*, de Beer (1963) noted that the literature on Darwin was immense, and therefore he restricted the bibliography in his book to a number of general works. Darwin himself penned an *Autobiography*, mainly written between May and August 1876 and first made available, although incomplete, by his

29	1839	Journal of researches.
32	1842	The structure and distribution of coral reefs.
34	1844	Geological observations on the volcanic islands.
36	1846	Geological observations on South America.
41	**1851**	**A monograph of the fossil Lepadidae, or, pedunculated cirripedes of Great Britain.**
42	**1852**	**A monograph on the sub-class Cirripedia, with figures of all the species. The Lepadidae; or, pedunculated cirripedes (1851).**
44	**1854**	**A monograph on the sub-class Cirripedia, with figures of all species. The Balanidae, Verrucidae &c.**
45	**1855**	**A monograph on the fossil Balanidae and Verrucidae of Great Britain (1854).**
49	1859	On the origin of species by means of natural selection.
52	1862	On the various contrivances by which British and foreign orchids are fertilized by insects.
55	1865	On the movements and habits of climbing plants.
58	1868	The variation of animals and plants under domestication.
61	1871	The descent of man, and selection in relation to sex.
62	1872	The expression of the emotions in man and animals.
65	1875	Insectivorous plants.
66	1876	The effect of cross and self fertilization in the vegetable kingdom.
67	1877	Different forms of flowers on plants of the same species.
70	1880	The power of movement of plants.
71	1881	The formation of vegetable mould, through the action of worms.

Figure 2. Position of the barnacle monographs (bold face) among 15 other principal works (numerals on the left indicate Darwin's approximate age at the time of publication).

son Francis in 1887, and finally published in its entirety by his granddaughter Nora Barlow in 1958.

In 1980, two stimulating books appeared: Brackman's *A delicate Arrangement*, a well documented and indexed query into why Darwin rather than Alfred Russel Wallace (Append. 1) received the most credit for origin of species by natural selection; and Stone's *The Origin*, an unindexed historical novel including a substantial 'Select Bibliography' condensed in eight and one half pages but making light of the barnacle years. In *The Growth of Biological Thought*, Mayr (1982) gives about a complete a review of the development of Darwin's science and ideas.

More recently a *Calendar of the Correspondence of Charles Darwin, 1812-1882* was put together by Burkhardt & Smith (1985a). In addition to brief abstracts of 13769 letters, this work contains a meticulous index, a bibliography of Darwin's publications, and a 'Biographical Resister' of the correspondents including an index to the letters they wrote or, more often than not, received. There are over 300 letters concerning barnacles and closely related matters, mostly (95%) from Darwin between 1847 and 1854, (avg. 43/yr.). This is out of a total of approximately 475 letters over the same period (avg. 68/yr.). Many of the barnacle letters involved administrative rather than scientific matters, the most intense of which (more than 22 letters between January 1850 and February 1851) involved James de Carle Sowerby and the illustrations for Darwin's his first volume on the fossil barnacles (FC1).

Five volumes of *The Correspondence of Charles Darwin* have been published (Burkhardt & Smith 1985, 1986, 1987, 1988, 1989; or C1, C2, C3, C4, and C5, respectively, where cited herein). They cover the correspondence up through 1855, or though the barnacle years. As in the *Calendar*, the letters are numbered as well as dated, but the numbers (with their corresponding dates) only appear at the beginning of each volume, the letters themselves being identified by date alone. This can be awkward, especially when the date of a letter has been emended. And sometimes letters that were not included in the *Calendar* have been added to the *Correspondence*; in which case, they are referred to by date alone[2.]

Each volume of the *Correspondence* has an introduction that gives an historical overview of the correspondence covered, and each has a number of appendices, bibliographies, biographical registers and excellent indexes. Burkhardt and Smith footnote many of Darwin's letters, and footnotes concerning barnacles are apparently primarily the work of the second author, the late Sydney Smith (Saint Catharine's College, Cambridge), who had a long standing interest in such matters. Not only do the notes frequently reference where the material appears in the barnacle volumes but often analyze related aspects or problems.

Thus it is now relatively easy to trace many of Darwin's interactions with colleagues and friends. However, what remains of the correspondence, especially during the barnacle years, is a little one-sided. While many of his letters to colleagues survive, it is obvious that Darwin had a habit of destroying letters he received. In light of this, the mystery of what happened to Wallace's early letters to Darwin (Brackman 1980: 3, 351) seems of no more consequence than why so little correspondence from Louis Agassiz, James Dana, Albany Hancock, John Henslow, Joseph Hooker, Charles Lyell, Henri Milne Edwards [3], and Johannes Steenstrup survived: Darwin routinely threw out old letters (Smith 1968).

2   THE BARNACLE YEARS: 1846-1855

In his autobiography, mainly written in 1876, Darwin wrote all too briefly of his barnacle work and concluded that, while it was of considerable use when he had to discuss the principles of a natural classification in *The Origin*, he doubted whether it was worth the consumption of so much time (Barlow 1958). Mayr (1982) noted that Darwin's autobiography was 'not at all reliable, not only because his memory occasionally let him down, but also because it was written with that exaggerated Victorian modesty that induced Darwin to belittle his own achievements.' It needs to be noted, however, that while Darwin was indeed a modest man, by 1876 he had suffered several crushing setbacks involving his anatomical work on organ systems in adult cirripedes, the first inklings of which came from Thomas Huxley in September 1855 (C-1757) and to which he replied, 'If these points are false, I shall never trust myself again.' That he was indeed wrong regarding much of the female reproductive system came to light in 1859 in a paper by the Russian anatomist August Krohn, about which Darwin followed up with a none too satisfactory reply in 1863 (see Sections 5.1-5.3).

On the heels of this came criticism of his work on larval structures and appendages, the first inklings of which came from earlier discussions with Dana, especially in late 1853, concerning the three pairs of naupliar limbs (C-1542). Published reverberations began with Krohn (1860) refuting Darwin's claim that the prehensile limbs of the cyprid were homologous with the frontolateral horns of the nauplius (see Section 5.4.1). This was followed by a letter from Anton Dohrn concerning the homologies of the remainder of the naupliar limbs, to which Darwin replied in 1867, 'I cannot yet quite persuade myself that the view I have taken...of the homologies of the appendages is wrong' (see Section 5.4). Furthermore, Darwin's analysis of the differences between *Cryptophialus* Darwin, 1854, and *Alcippe* Hancock, 1849 (= *Trypetesa* Norman, 1903, and usually referred to by this name hereafter), which had formed the basis for placing them in different orders, came under critical scrutiny in 1866 and 1872 by Gerstaecker and Noll respectively (Section 5.6). From these examples it is understandable why, when Darwin wrote his autobiography, he may have truly believed that much of his cirripede work had not been worth the consumption of so much time.

Darwin's contemporaries, notable Darwin scholars, and the cirripedologists that followed held his barnacle work in higher esteem than he did. For example, Huxley, who got to know Darwin through the barnacle work and was among the first to detect errors in some of his barnacle adult morphology, nonetheless consulted him regularly on such matters and became his 'bulldog' in defense of *The Origin*. Fritz Müller (see Section 5.4.1), who finished the job started by Krohn (1860) on larval limb homologies, honored him with his book, *Für Darwin* (1864); and Carl Claus, who benefited from the corrections that had been made involving larval appendages, dedicated his *Crustaceen-Systems* (1876) to Darwin (Fig. 3). Among historians, de Beer (1963), Ghiselin (1969), Crisp (1983a), Southward (1983), and Rachootin (1985) each devote 10 or more pages to Darwin's involvement with barnacles and conclude that his trials and tribulations were not only a significant part of self-training as a comparative anatomist and systematist, but also important in the development and testing of his ideas on evolution. More recently Burkhardt & Smith (C4:388-409) devote some 40 pages to 'Darwin's study of the Cirripedia', an essay conveniently tied in through footnotes with the meticulous indices of the volumes of *Correspondence* as well as with Darwin's barnacle monographs, and they reach much the same conclusions. Burkhardt & Smith noted that Darwin received the Royal Medal of the Royal Society of London in 1853 largely because of his first volumes on the fossil and living barnacles (FC1 and LC1), i.e., before his equally if not more important second volumes on living and fossil forms (LC2 and FC2) appeared. While a number of monographs and several

HERRN

CHARLES DARWIN

IN INNIGER VEREHRUNG

GEWIDMET.

Figure 3. Dedication of 'Crustaceen-Systems' (from Claus 1876).

thousand papers on barnacles appeared since Darwin, much of the anatomical and virtually all of the taxonomic portions of his monographs remain primary sources in cirripedology.

Darwin was in the prime of life during the barnacles years. Well before 1846 he had become widely recognized as a scientist through the collections he made (1832-1836) while on the Beagle [more than 40 species had been named after him by 1850 (Sherborn 1922)] and his publications resulting from the voyage of the Beagle narratives of the voyage, works on the geology of South America, coral reefs, and volcanic islands all dating from 1839-1846. These works consumed the decade following the Beagle voyage, and he was elected a fellow of both the Geological Society and the Royal Society in the middle of it (1839), well before he began on barnacles.

In addition to being firmly entrenched as a scientist, Darwin had a private income, a fine and devoted wife (his cousin, Emma Wedgwood), and a growing family comfortably ensconced at Down House in Kent (Atkins 1974). Of his 10 children, whom he and friends like Henslow referred to as little 'd's' 6 through 9 (Elizabeth 1847, Francis 1848, Leonard 1850, and Horace 1851) were born during the barnacle years. The barnacles were apparently a significant part of their lives for it has been said that when visiting, the 'd's' would often ask when their friend's father worked on his barnacles.

However, the barnacle years were marred by the deaths of Darwin's father, Robert Waring Darwin (1766-1848), and his second and 'favourite' child, Anne (1841-1851). He also lost 'about two years out of this time' to a chronic intestinal disorder, a lifetime illness that began before the voyage of the Beagle and for which in later life he gained relief by going to spas for water cures (Barlow 1958). On the other hand, he was still much more active than I had previously thought, going to meetings, visiting London, making additions to Down House, entertaining house guests, keeping detailed accounts, and carrying on a voluminous correspondence with family members and friends, as well as scientists. He evidently also had a social conscience; for example, in 1852 he wrote a second cousin and fellow beetle collector while at Cambridge, William Fox (C-1476), that he had joined a society to prosecute violators of the act against use of children in cleaning chimneys.

While Darwin did not begin work on barnacles in earnest until 1846, he knew about them from his college days. He became specifically interested in them when on the Beagle, largely because of a little form found burrowing into the exterior of the shell of the gastropod, *Concholepas*, from islands off the coast of Chile (see C3, facing p. 320, for illustrations Darwin made of this minute form during the voyage). In 1835, he wrote from Valparaiso to his Cambridge professor and friend, Henslow (C-272), about the new form he at first considered represented 'a genus in the family Balanidae, which has not a true case, but lives in minute cavities in the shells of *Concholepas*.' The form, referred to as 'Arthrobalanus' in most of his subsequent correspondence, became *Cryptophialus minutus* when finally described (LC2, 1854).

From the beginning Darwin considered that *Cryptophialus* held a special place in the Cirripedia. This was largely because it appeared well articulated. At the time, the cirripedes were considered an unarticulated 'inosculate' group, linking such groups as the mollusks and annelids (Smith 1968). Later, Darwin decided *C. minutus* formed an intermediate between the archetype cirripede and the Thoracica for which he erected the order Abdominalia (= Acrothoracica Gruvel, 1905; Section 5.6).

Darwin was also intensely interested in species, varieties, and their distributions. From the collections in the British Museum as well as from his own experiences, he knew that the barnacles were a readily available and commonly encountered group having a substantial fossil record. However, Mayr (1982) muses over Darwin's choice of an 'obscure group' like barnacles upon which to work. Obscurity, like beauty, is often in the eye of the beholder. Darwin had recently finished publishing on platyhelminths and chaetognaths (Section 2.1), invertebrates that are not only obscure, but also not all that easy to find. With barnacles, Darwin could write to a friend living by the seashore anywhere in the world and not only expect them to know what he was talking about but to collect some specimens for him. Furthermore, he could study their fossil record, which at the time ranged back into the Mesozoic. As invertebrates go, barnacles were an excellent choice for study. So, why did Darwin devote so much time to barnacles? While one can only agree with Mayr that Darwin must have felt he had a tiger by the tail, it is also true that becoming recognized as an accomplished systematist would add weight to his theorizing on variation and divergence of species.

2.1 *Getting started*

In October 1845, Darwin wrote Hooker (C-924): 'I hope this next summer to finish my S. American geology, then to get out a little zoology, and hurrah for my species work, in which, according to every law of probability, I shall stick and be confounded in the mud.' The 'species work' he referred to was, of course, the basis of *The Origin*, the search for variation and divergence among species upon which natural selection could work. A year later he noted in a letter to Hooker (C-1003): 'I am going to begin some papers on lower marine animals, which will last me some months, perhaps a year, and then I shall begin looking over my ten-year-long accumulation of notes on species and varieties which, with writing, I dare say will take me five years, and then when published, I dare say I shall stand infinitely low in the opinion of all sound naturalists – so this is my prospect for the future.' And again to Hooker a few weeks later (C-1012) he wrote, 'Your drawing (of *Cryptophialus*) is quite beautiful... I have been reading heaps of papers on Cirripedia, & your drawing is clearer than almost any of them. The more I read, the more singular does our little fellow appear, and as you say, looking at its natural size, a microscope is a most wonderful instrument.' And to Robert FitzRoy two days later (C-1014) he wrote, '(I have) for the last half month (been) daily hard at work in dissecting a little animal about the size of a pin's head from the Chonos Arch. (Chile) and I could spend another month on it, and daily see some more beautiful structure!'

Thus it appears that this last 'little zoology,' *Cryptophialus*, was to be simply another study of the more singular invertebrates from the voyage of the Beagle, the first having been on structure and propagation in the chaetognath *Sagitta*, and the second on some terrestrial and marine planarians (Darwin 1844a, 1844b, respectively). Smith (1968) infers that Darwin's initial interest in *Cryptophialus* arose because 'examined with care and drawn (text and drawings in the Beagle Zoology notes) from life this creature looked too much like an 'inosculating' form for the equanimity of a would-be transmutationist,' and 'for his own peace of mind over the *Species Theory* he had to be sure of his odd little *Cryptophialus*.' Perhaps Darwin's initial interest in *Cryptophialus* was the inosculation of the Cirripedia between, say, the annelids and the mollusks, as Lamarck (1809) and Cuvier (1817) had placed them. Barnacles were shelled and were thought to lack segmentation. While Cuvier (1834:137) recognized that their cirri were '...similar to the limbs observed under the tail of several of the crustacea;...', he considered them '...analogous to the articulate appendages of certain species of teredo...', and he placed the 'Cirrhopoda' as a sixth class of the Mollusca. Darwin did initially refer to *Cryptophialus* in letters as 'Arthrobalanus' implying that, unlike what was believed to be the case in the rest of the tribe, it was segmented (Rachootin 1985). It likely did not take him long to determine that the rest of the barnacles were also segmented. But normal barnacles had six thoracic segments, and he thought that *Cryptophialus* had eight just like another odd form he had discovered, *Proteolepas*, which he thought was a barnacle. Darwin reasoned the larger number of segments must be primitive, and therefore his archetype cirripede had to have eight thoracic segments. This agreed with Milne Edwards's archetype crustacean, based primarily it seems on malacostracans. Darwin pursued this idea in his analyses of cirripede larval forms (see Section 5.4.1-2). By December 1846 Darwin had written Hooker (C-1035): 'Thanks for the corallines; Heaven knows when I shall begin them; I have been nearly 3 months on Cirripedia and have done only 3 genera!!!'

2.2   *Microscopes*

The simple microscope Darwin took on the Beagle in 1831 was one Robert Brown recommended and, according to C4:126, n.3, 'it is almost certainly the Bancks microscope now at Down House, while the compound microscope at Down was donated by the Lubbock family.' When he began work on barnacles in 1846, Darwin considered the Bancks dissecting microscope inadequate for his needs and he went about looking for the best available at the time (Southward 1983). He ended up purchasing two microscopes for his personal use, a compound microscope in 1847 and a dissecting microscope in 1848 (C-1050, n. 1 and C-1345, n. 6).

The search for a suitable compound microscope apparently began in 1846. He wrote Hooker (C-1035) that he had been impressed by a good one, evidently belonging to an artist, Samuel Leonard, who he hoped to employ to illustrate some Beagle invertebrates (C-1001, n. 1) and who also illustrated a barnacle (C-1035, n. 4): 'When I was drawing with Leonard, I was so delighted with the appearance of the objects, especially with their perspective, as seen through the weak powers of a good compound microscope that I am going to order one: Indeed I often have structures, in which the 1/30th is not power enough.'

But the selection of a dissecting scope was another matter. Stone (1980) described how Darwin, after looking in London in vain for the scope he had in mind, came upon the plans of an advanced French design by Chevalier in Paris, which he modified and had manufactured by Smith and Beck, London. This may be, but I have not been able to document it in the references cited herein. While the dissecting scope Darwin took on the Beagle was a Bancks, Hooker's had been made by Charles Chevalier, Paris. In November 1846, Darwin wrote Hooker (C-1022): 'The alteration in my microscope (the Bancks), in accordance with your advice, has really been beyond value: The porcupine quills better than glass tubes; the Chutney Sauce capital, so that I have many daily memorials of you. N.B. I have cleverly invented two blocks of wood to support my wrists when dissecting under microscope a splendid invention.' A little more than a year later, when writing Henslow (C-1167) about progress with sexuality in cirripedes and his views that 'an instinct for truth, or knowledge or discovery...is reason enough for scientific researches without any practical results *ever* ensuing from them,' Darwin recommended Smith and Beck for a simple microscope *lately made for him* (Darwin's emphasis). He noted that it is 'partly from my own model and with hints from Hooker, wonderfully superior for coarse and fine dissections than any I ever before worked with' and 'if I had had it sooner, it would have saved me many an hour.' In other correspondence, he mentions that he was pleased with an arrangement he made for switching magnifications, a micrometer he fitted for making measurements, and the pair of wooden blocks arranged as hand rests for dissecting, a practice acquired on the Beagle. While both the compound and dissecting microscopes came from Smith and Beck, London, it was clearly the latter that was truly special; in 1852, Smith & Beck advertised

'Darwin's Improved Single (Microscope)' and a full description is to be found in Beck (1865:102-104; C-1166, n. 2).

Darwin spread the word of his new dissecting scope beyond his close friends, and to Richard Owen he wrote (C-1166), 'I have derived such infinitely great advantage from my new simple microscope, in comparison with the one, which I used on the Beagle... that I cannot forego the mere *chance* of advantage of urging this on you... Smith and Beck were so pleased with the simple microscope they made for me; that they have made another as a model: If you are consulted by any young Naturalist, do recommend them to look at this; I really feel quite a personal gratitude to this form of microscope & quite a hatred to my old one' (Darwin's emphasis).

His children took pleasure in looking through his microscopes and probably marveled at his fine dissecting scissors modeled after a pair belonging to George Newport (C-1445 and 1450) and made for him by Weiss & Co. Perhaps these experiences led one of his sons, Horace (1851-1928), into becoming an engineer of scientific instruments, which in turn led to the founding of Cambridge Instruments (C5:615, Anonymous 1991).

2.3   *The real work begins*

Shortly after Darwin began work on the curious cirripede he discovered burrowing into the shells of *Concholepas* from the coast of Chile, he was writing abroad as well as in the British Isles for fossil and Recent specimens: 'I had originally intended to have described only a single abnormal Cirripede, from the shores of South America, and was led, for the sake of comparison, to examine the internal parts of as many genera as I could procure' (LC2:566). Most of the material came from European and American collections as well as his own from the Beagle. He wrote to individuals for specimens of particularly interesting species, such as to Sven Lovén in Sweden for *Alepas (= Anelasma) squalicola* (1849, C-1269) parasitic on a dogfish, and to Ferdinand Krauss (1851, C-1465) in Germany for *Conia (= Tesseropora) rosea* from South Africa. Regarding the latter, he also wanted to know if Krauss had collected the specimen himself (which he did), or was it given to him, because he had a very similar form from Australia, and this surprised him. He also wrote to faraway places for general collections of specimens, such as to Syms Covington in Australia (1850, C-1370), his servant aboard the Beagle, who worked for him afterwards until he immigrated to Australia in 1839.

In February 1847, he wrote Ernst Dieffenbach (C-1059): 'I have for the *present* given up Geology, & am hard at work at pure Zoology & am dissecting various genera of Cirripedia, and am extremely interested in the subject. I always, however, keep on reading and observing on my favourite work on Variation or on Species, and shall in a year's time or so, commence & get my notes in order' (Darwin's emphasis). It is amusing to note, in searching for his barnacles from the Beagle, presumably stored at the Royal College of Surgeons, he had to write John Quekett (C-1114): 'Most mortifying to me to have lost my own Cirripedia, now that I am at work on them.' Shortly thereafter he mentioned in a letter to John Gray at the British Museum (C-1139) that he planned to follow up on Gray's suggestion and monograph the barnacles, and he acknowledged him for the suggestion (LC1:v). A year later Darwin also wrote Agassiz (C-1205) that his encouragement had helped him decide to monograph the barnacles.

Darwin apparently began corresponding with Milne Edwards late in 1847 (C-1139) and in June 1848 he wrote Edward Cresy (C-1228) that he planned to go to Paris about barnacles, but his chronic illness intervened. As it ended up, he borrowed the specimens from Paris and remained in England, mostly at Down, for the rest of his life.

One need only skim his monographs on the barnacles to be convinced that he labored mightily (see Appendix II for a table of contents). He kept up his spirits and sense of humor through his illnesses, as words written to his colleagues and friends reveal. In late October 1846, he wrote FitzRoy (C-1282 adjusted to C-1077 in C3) that he was enjoying dissecting *Cryptophialus*, but in May 1847 he mentioned to Hooker (C-1077) that, in addition to illness, he has 'been getting on wretchedly with the Barnacles, and have done only two other genera.' It is therefore refreshing to note a year later he advised Hooker (C-1174), 'I have been getting on well with my beloved Cirripedia, and got more skillful in dissection: I have worked out the nervous system pretty well in several genera, and made out their ears and nostrils, which were quite unknown (see Sections 5.2, 5.3). I have lately got a bisexual cirripede (*Ibla cumingi*), the male being microscopically small and parasitic within the sac of the female (see Section 5.4.3); I tell you this to boast of my species theory.... But I can hardly explain what I mean, and you will perhaps wish my Barnacles and species theory al Diabolo together. But I don't care what you say, my species theory is all gospel.' A month latter he was writing J. Gray (C-1187), at a time (June 28, 1848) when he thought he might split the Cirripedia into two volumes (systematic and anatomical): 'In truth never will a mountain of labor have brought forth such a mouse as my book on the Cirripedia: It is ridiculous the time each species takes me.' More than a year later in a letter to Lyell (C-1252) he confessed, 'It makes me groan to think that probably, I shall never again have the exquisite pleasure of making out some new district, of evoking geological light out of some troubled, dark region. So I must make the best of my Cirripedia.'

With much of the anatomical work over, the chores of species descriptions, synonymies, and attendant matters began taking their toll, and in 1850 he wrote Hooker (C-1300): 'I have now for a long time been at work at the fossil cirripedes, which take up more time even than the recent; confound and exterminate the whole tribe; I can see no end to my work.' He was anxious to get the fossil cirripedes out of the way, but illness intervened again, and early in March 1850 he wrote his illustrator, James Sowerby (C-1306): 'I have lost a good many days of late and now I do not think I shall have finished all the pedunculate fossil Cirripedia for 10 days or a fortnight.... If I were but in better health, I would work quicker.' A few days later he wrote to Lyell (C-1308): 'My cirripedial task is an eternal one. I make no perceptible progress I am sure that they belong to the Hour-hand, - and I groan under my task.' And to Hancock (C-1316) a month or so later he wrote, 'I mean now to continue at Systematic Part till I have finished, a period which will arrive, Heaven only knows when.' Encouragement often came, however, such as from Hooker who wrote from Calcutta in April of 1850 (C-1319) that he used to think Darwin was too prone to theoretical considerations about species and was therefore pleased that he had taken up a difficult group like barnacles. Part of the slowness (see below) involved Sowerby and getting the illustrations done (Section 4), but by June 1850 he wrote to Hooker (C-1339): 'At last I am going to press with a small first fruit of my confounded Cirripedia, viz the fossil pedunculated Cirripedia,' and 3 months later Owen (C-1355): 'I hope soon to go to press with my weariful fossil Cirripedia' (FC1, finally published in June 1851).

Darwin was optimistic about the future volumes and, in a letter to Edwin Lankester at the Ray Society in November 1850 (C-1367) he expresses hope that both volumes of the living Cirripedia (LC1 and LC2) would be published in 1851. However, while LC1 is dated 1851, it was not actually published until 1852 (see Section 3.1), and LC2 was even further delayed. In October 1852, he wrote Fox again (C-1489): 'I am at work on the second vol. of the Cirripedia (LC2), of which I am wonderfully tired: I hate a Barnacle as no man ever did before, not even a Sailor in a slow-sailing ship. My first vol. is out (LC1): The only part worth looking at is on the sexes of *Ibla* and *Scalpellum*; I hope next summer to have done with my tedious work.' By

March 1853 (C-1507) he was objecting to an early deadline for LC2 and the volume did not appear until 1854. As for the last volume (FC2), while dated the same year, it did not actually appear until May 1855 (Withers 1928).

Stone (1980) believed that Darwin's barnacles were a mindless task that caused no sleepless nights and he wrote, 'Though (Darwin was) dissecting, discovering, describing in minute detail, and cataloging a field which had been neglected, it was largely a mechanical operation, absorbing while he was at his microscope *but making no demands upon him once he closed the study door at the end of the day*' (Stone 1980: 492, emphases added). From Darwin's letters it was clearly a different matter. For example, the only letter he wrote between April 18 and June 5, 1849 was to Henslow (C-1241), while taking Dr. Gulley's cold water cure for 2 months at Malvern, noted, 'All last autumn and winter my health grew worse and worse; incessant sickness, tremulous hands and swimming head: I thought I was going the way of all flesh.... One most singular effect of the treatment is, that it induces in most people, and eminently in my case, the most complete stagnation of mind: I have ceased to think even of Barnacles!' However, he was soon back at it and in April 1853 he wrote Huxley (C-1514): 'I have become a man of one idea cirripedes morning and night.'

2.4 *Species work*

Next to barnacles during the barnacle years, the question of species and varieties were foremost in Darwin's mind. He had written a sketch and an essay on his species theories (1842 and 1844, respectively; F. Darwin 1887, de Beer 1963:150), which were recovered after his wife's death in 1896 when the cupboards of Down House were cleared (Smith 1968). In November 1845, he had sent Hooker (C-924) a copy of the sketch concerning natural selection and some time later pressed him for feedback on the matter (C-1058). However, despite warnings from friends that somebody else might come up with similar ideas, Darwin was dedicated to the barnacle task, certainly because he was interested in them, but apparently in good part because he wanted to establish himself as a systematist. This is reflected in a remark he made in September 1845 to Hooker (C-915): ' How painfully true is your remark that no one has hardly the right to examine the question of species who has not minutely described many.' So he labored on with the barnacles, whereby the 'hurrah,' alluded to in October 1845, would not appear until the glimmer in 1858 and the brilliant flash of 1859, some 4 years after the last of his four volumes on barnacles was published.

By 1856, Darwin's species work was well underway again; he sent a sketch to Asa Gray, and de Beer (1963:148) outlines his progress on the chapters for his big 'hurrah' on species up through April 1858. Darwin wrote Huxley (C-2143) in September 1857 that the 'time will come when we shall see true genealogical trees of each great kingdom of nature' and in November 1857 he wrote A. Gray (C-2176) that he found it difficult to avoid using the term 'natural selection' as an agent, and that he found the rule of large genera having the most varieties, as he observed in his barnacle work, holds good and regarded it as most important for his 'principle of divergence.'

Darwin began publishing routinely in journals in the 1830's, and he managed to average more than one article a year during the barnacle years. With completion of the barnacle volumes, he published 10 articles in journals in 1855, 2 in 1856, and 5 in 1857. The articles were on a variety of subjects but most concerned the germination of seeds, relating to his interests in the distribution of plants, especially on islands (Darwin 1855b). Darwin read Wallace's 1855 paper, 'On the law which has regulated the introduction of new species' (see Brackman 1980:

309-325 for a reprint), and Wallace wrote him about it in October 1856. Darwin was busy with his work on seeds and apparently didn't reply until May 1857 (C-2086) at which time he expressed his agreement with Wallace's conclusions and asked him for help with experiments on dispersal to oceanic islands. There was further correspondence on related matters from Darwin in 1857 (C-2145 and 2192). Then, on June 18, 1858, Darwin received Wallace's manuscript 'On the tendency of varieties to depart indefinitely from the original type' (see Brackman 1980:326-337 or de Beer 1958:258-279 for the text). On the same day, Darwin wrote Lyell (C-2285): 'Your words have come true with a vengeance that I should be forestalled... if Wallace had my MS sketch written out in 1842, he could not have made a better short abstract!'

Had Wallace's manuscript of 1858 arrived a decade or so earlier it seems likely that Hooker and Lyell, the friends who arranged for a short paper by Darwin and Wallace's paper to be read sequentially before the Linnean Society in 1858, would have prevailed in much the same way. If such had been the case, would Darwin have monographed the barnacles? Probably not, because what was in good part the major impetus, hands-on experience with variation in species, would have evaporated. If he had not so fully monographed the barnacles, one must agree with Crisp (1983a) that the depth and breadth of our knowledge about them would be significantly less than it is today. On the other hand, if following publication of his last volume on the barnacles Darwin had not been given principal credit for the theory of evolution by natural selection, to the extent that Wallace rather than Darwin was virtually a household word, I think it reasonable to suspect that cirripedology would have been less attractive to many young naturalists, except perhaps for ecologists, and consequently we would know less about barnacles today.

While more than 40 species of animals were named for Darwin (Sherborn 1922) as a result of his Beagle collections, no genera were named for him until after publication of the last volume of the Cirripedia. The first of these was a crustacean, *Darwinia*, described by C. Spence Bate in 1856, in honor of Darwin for his work on barnacles. It was not until 8 years later (1864) that the first of nine additional generic proposals appeared (Neave 1939; most as *Darwinia* and hence preoccupied), and these were I believe largely attributable to publication of *The Origin* (1859).

2.4.1 *Variation*

One of Darwin's big problems, as intimated earlier, was with species, varieties thereof, and to some extent with genus-group and higher taxa (see Section 5.8, 5.10). In June 1850, he wrote Hooker (C-1339): 'You ask what effect studying species has had on my variations theories; I do not think much; I have felt some difficulties more; on the other hand I have been struck (and probably unfairly from the class) with the variability of every part in some slight degree of every species: when the same organ is *rigorously* compared in many individuals I always find some slight variability, and consequently that the diagnosis of species from minute differences is always dangerous. I had thought the same parts, of the same species more resembled than they do anyhow in Cirripedia, objects cast in the same mould. Systematic work would be easy were it not for this confounded variation, which, however, is pleasant to me as a speculist though odious to me as a systematist' (Darwin's emphasis). Having gotten further into the details of balanomorph systematics, he wrote Hooker (C-1532) in September 1853: '(After) making them separate, and then making them one again, I have gnashed my teeth, cursed species, and ask what sin I have committed to be so punished: But I must confess, that perhaps nearly the same thing would have happened to me on any scheme of work.'

The problem of course was fundamental to his species work: If he could not decide whether he was dealing with a single wide-ranging and variable species or with several closely related species spatially separated over the same range, how could he trust the notion that wide ranging species were variable? He wrote to numerous colleagues about the problem in plants as well as in animals, and in barnacles he noted, 'I must express my deliberate conviction that it is hopeless to find in any species, *which has a wide range, and of which numerous specimens from different districts* are presented for examination, any one part or organ... absolutely invariable in form or structure' (LC2:155, Darwin's emphases).

A good example is found under *Balanus tintinnabulum*, a species in which he recognized 11 varieties (Henry & McLaughlin 1986; Section 5.8.3): 'The difficulty in determining whether or not the differences (between forms) are specific, is wonderfully increased by whole groups of individuals varying in exactly the same manner. I have seen three most distinct varieties taken from the bottom of a vessel, so that I did not at first entertain the least doubt that they were three distinct species' (LC2:197). Here three distinct forms without intergradation occurred together on a ship back from a long cruise. Why did Darwin end up deciding they were varieties rather than species? Because, he said, 'After going over the several immense collections of specimens placed at my disposal, I came to the conclusion that the above three, and several other forms presently to be described, were only varieties' (LC2:197). Darwin believed, while the three forms appeared distinct even when sympatric, they could be viewed as intergrading by characters found in related forms from different districts. As has already been noted (Southward 1983), and as will be seen shortly, Darwin's view of large genera of wide-ranging species of many varieties among the sessile barnacles did not hold up (Section 5.8). This was because his method was faulty and the results also led him astray, to some extent, when it came to barnacle biogeography (Newman 1982a, 1989a; Section 5.12).

2.4.2 *Biogeography and dispersal*

In addition to his concerns over species and his work on barnacles, Darwin probed into biogeographical problems, especially involving terrestrial biotas separated by vast reaches of open water. There was controversy at the time as to whether ancient land connections or dispersal by various means explained some situations, especially in plants. For example, Hooker wrote Darwin from New Zealand in November 1851 (C-1460) that he was reconsidering variability of insular species and was becoming convinced of the probability that the southern flora is the fragmentary remains of a great southern continent (eventually to become known as Gondwanaland). So Darwin decided to do some experiments to determine what seeds were present in mud taken from bird's feet. By placing the mud in a favorable place in his study for germination, he found that seeds of numerous families, genera, and species sprouted. But it occurred to him that there must be other means of dispersal, and he wondered for seeds that could float or be rafted how long would they remain viable when steeped in seawater? This work required writing to many colleagues in various parts of the world for seeds as well as mud and carried on after completion of the last cirripede volume (FC2, May 1855). By June 1855, he wearied of the task and wrote Hooker (C-1693): 'I begin to think they (seeds) are immortal and that the seed job will be another Barnacle job.' If he had not received Wallace's manuscript of June 1858, 'On the tendency of varieties to depart indefinitely from the original type,' seeds indeed might have become 'another Barnacle job.'

3 THE BARNACLE MONOGRAPHS

The completeness of Darwin's barnacle monographs is remarkable, especially when it is

TO

PROFESSOR H. MILNE EDWARDS,

DEAN OF THE FACULTY OF SCIENCES OF PARIS; PROFESSOR AT THE MUSEUM OF NATURAL HISTORY;
MEMBER OF THE INSTITUTE OF FRANCE;
FOREIGN MEMBER OF THE ROYAL SOCIETY OF LONDON, OF THE ACADEMIES OF BERLIN, STOCKHOLM,
ST. PETERSBURG, VIENNA, KONIGSBERG, MOSCOW, BRUSSELS, HAARLEM, BOSTON,
PHILADELPHIA, ETC.

THIS WORK IS DEDICATED,

WITH THE MOST SINCERE RESPECT,

AS THE ONLY, THOUGH VERY INADEQUATE ACKNOWLEDGMENT
WHICH THE AUTHOR CAN MAKE OF HIS GREAT AND
CONTINUED OBLIGATIONS TO THE

'HISTOIRE NATURELLE DES CRUSTACÉS,'

AND TO

THE OTHER MEMOIRS AND WORKS ON NATURAL HISTORY PUBLISHED BY
THIS ILLUSTRIOUS NATURALIST.

Figure 4. Dedication of 'Living Cirripedia 2' (from Darwin 1852).

realized how little was known about the anatomy as well as the systematics of most groups of invertebrates at the time. I am not aware of what work he may have relied upon for guidance in organizing the monographs, especially the two on living forms. Perhaps it was that of Milne Edwards to whom he dedicated LC1 (Fig. 4). However, while the volumes are well indexed, none has a table of contents per se, to give an idea of the depths probed as well as the overall organization, I have prepared a table of contents for LC1 and LC2 (Appendix II).

Darwin went about studying the cirripedes in a very thorough and well-organized manner. In fact, in my opinion if the monographs were revised today, as far as the format is concerned, other than having a reference section at the end, there is little I would consider changing. Two minor points of interest involve the practice of giving diagnoses in both Latin and English (translated by G.B. Sowerby Jr.) and not putting the author's name after a taxon. Darwin decided in favor of both languages for the diagnoses in all volumes, although in LC2 the Latin diagnoses appear in the Synopsis rather than along with the English at the appropriate places in the text. He relented when it came to attaching authors' names to taxa, which he considered more a matter of vanity than of value in classification (see Section 5.10): Author's names for orders through species, including those proposed by Darwin, are found in the Synopsis (LC2, see Appendix II).

It was Darwin's opinion, at least to his relative and friend W.D. Fox (C-1489), that the principal worth of LC1 was in the discovery of dwarf and complemental males in *Ibla* and *Scalpellum* s.l. Naturally, he thought more of the volume than that and was looking for the approval of others. In April 1853 (C-1514), Darwin wrote Huxley requesting that he review LC1 since 'it has been published a year, and no notice has been taken of it by any zoologist, except briefly by Dana (1852).' [Dana's review emphasized discovery of the complemental males, and Darwin informed Dana he was 'very much pleased' and that it gave him 'great satisfaction'(C-1492).] As already noted, publication of LC1 was largely responsible for Darwin receiving the Royal Society Medal in 1853, and it comes as a bit of a disappointment to find that, while his work on males has been verified and enlarged upon, much of the work upon which the award was based (C4:406-407) turned out to be wrong (Rachootin 1985; Sections 5.1-5.7).

3.1 *Dates of publication*

The publication dates of the barnacle monographs have long been considered as 1851a, 1854a and for FC1, LC1, LC2 and FC2, respectively (Newman et al. 1969; Burkhardt & Smith 1985, Southward 1987). Date of publication is of course important in matters of priority involving ideas as well as introductions of new taxa.

There would seem to be no problem with the publication date of FC1, for Darwin wrote Dana on September 9, 1851, that he had received copies (C-1453). On the other hand, Withers (1928) noted that FC2, dated 1854, was not actually published by the Palaeontographical Society until May 1855. The publication of LC1 has always been considered to have been 1851, but for some curious reason the French cirripede monographer, A. Gruvel (1905), cites LC2 as published by the Ray Society in 1853 rather than 1854, and the German cirripede monographer, P. Kruger (1940) followed suit.

My study confirms that FC1 was published in 1851. It is also clear that Withers was correct: FC2 was published in 1855 rather than 1854 and that Gruvel's and Kruger's date of 1853 rather than 1854 for LC2, was a slip of the pen. However, in the process of checking these points out through the *Calendar* a new problem arose. Although Darwin could have received 22 copies of LC1 from the Ray Society on or before December 19, 1851, (C-1464) and possibly could have distributed a few copies to colleagues late in 1851 (C-1453), the general distribution may have been in 1852. According to the International Code of Zoological Nomenclature, the possibility that some copies were distributed privately in 1851 would not have constituted publication that year. Thus, the situation needed to be carefully checked out.

It turns out that a letter written by Darwin on December 19 (C-1464), dated in the *Calendar* as 1851, had been changed to '1852 or 1854' in C5. The full text of this letter indicates that, in addition to having received 22 copies from the Ray Society, Darwin was anxious to know if the volume were for sale at the time and, if so, for how much? If this letter were '1852', then LC1 certainly was not published in 1851, but if '1854', then perhaps Darwin was worried about the publication date of LC2 rather than LC1. On August 13, 1854, Darwin wrote Joseph Bosquet (C-1578) that he would send him a copy of LC2 as soon as it was published, and he wrote Huxley on September 2 (C-1587), Hooker and Steenstrup on September 7 (C-1588 and 1589) and August Gould on September 9 (C-1591), of the same year, that his second volume on the living barnacles had been published. Thus, there is apparently no difficulty with the year of publication for LC2: It must have been published between September 8 (C-1487) and October 24 (C-1489) of 1854. Thus our quarry is still the publication date of LC1.

In C5:79, there is an unnumbered letter from Darwin to Lankester at the Ray Society, falling between C-1472 and 1473 and dated January 30, 1852: 'I have received my 20 copies and am much obliged to you for the same. I write now to say that the Binder by some wonderful blunder has bound the enclosed in all my copies (of LC1). Will you please order it to be pulled out. It belongs perhaps, to Leighton's Volume (also dated 1851). I know not the address of the Binder, otherwise I would not have troubled you' (interpolations from footnotes 1 and 2, C5:79). Therefore not only were the advanced copies of LC1 apparently received in January 1852 but they were unsuitable for distribution.

On February 15, 1852, Darwin wrote Dana (C-1473) that he hoped soon to be able to send him a copy of LC1 and on May 8 that he was glad to have Dana's opinion of LC1 (C-1481), so the copies for distribution must have become available sometime between late February and April 1852. In April 1853, Darwin wrote Huxley (C-1514) with the hope that he would review his book (LC1) which had been published for a year with no notice taken of it except by Dana in 1852. It seems clear that LC1, while dated '1851', was not published until 1852. While Darwin had some copies in January, removal of the 'wonderful Blunder' and rebinding apparently not only delayed the release of the advanced copies until April but also the release of the bulk of printing until late in 1852. On September 8, 1852, (C-1487) Darwin wrote George Waterhouse that 'The Ray Society has delayed distributing my first vol. *for the last nine months!*' (Darwin's emphases). And October 24, 1852, (C-1489) he wrote William Fox, 'My first volume is out.' Finally, it may be that American barnacle monographer Henry A. Pilsbry was aware of the 1852 publication date because, while in 1907 he cites LC1 as 1851, in 1916 (p.12) he wrote of Darwin's classification of '1852-1854.'

4 DARWIN'S BARNACLE ILLUSTRATORS

Darwin could make reasonable sketches, and he once wrote Huxley (C-1759), in compliment, that 'no one has a right to attempt to be a naturalist who cannot (sketch easily)' (C-1759). He also had firm opinions on how shells should be illustrated. Nonetheless, he employed artists to illustrate his works, and for the barnacle volumes he employed three who were also scientists. One was of course Samuel Leonard, who worked with Darwin briefly in 1846 and who introduced him to a good compound microscope (C-1035; see Section 2.2). The other two had the same surname, James de Carle Sowerby and G.B. Sowerby Jr., the brother and the eldest son of George Brettingham Sowerby, respectively. In the following, I have taken the liberty of distinguishing between the Sowerbys by their first names.

Both James and his nephew, George Jr., had been involved in George Sr's. *Genera of Recent and Fossil Shells* (1820-1824), 16 numbers of which included barnacles. When it came to Darwin's barnacles, James made drawings and woodcuts, and engraved the plates for FC1 while George Jr. did likewise for LC1, LC2, and FC2. George Jr. also translated Darwin's taxonomic diagnoses into Latin in all four volumes (C4:300, n. 1 & 2).

George Sr. and Darwin had correspondence over the Tertiary shells of South America, and George Sr. advised Darwin in 1846 (C-939) that, in his opinion, the Latin for 'Darwin' was 'Darvinius' since 'w' is not a Latin letter. In the same year, George Jr. worked on engravings of Cordillera shells (C-969), but for some reason, in August 1849 (C-1267) Darwin asked N.T. Wetherell's permission to have George Sr.'s brother, James, figure Wetherell's specimen of the fossil pedunculate barnacle, *Loricula*. This is curious because George Jr. had already illustrated and described *Loricula* for Wetherell (G.B. Sowerby Jr. 1843), but perhaps George Jr. was simply not available to do the cirripedes at the time and, since the specimen of *Loricula* had been further prepared by Wetherell, a new drawing was needed (C4:349, n. 3).

To get the full flavor of what followed, one needs to read all the letters in their entirety. Here I will briefly elucidate how tensions built between Darwin and James over the illustrations for FC1 to the point that James was effectively discharged and George Jr. was asked to do the drawings and make the engravings for LC1 and the subsequent two volumes. Since the correspondence is one sided (apparently none of James' letters to Darwin survived), what Darwin had to say reveals more of his than of James' personality, but Darwin was a man of strong ideas and in a rush; James' slowness was apparently primarily responsible for their incompatibility.

Near the beginning of the correspondence involving illustrations for the cirripede volumes (January 1850), Darwin wrote Robert Fitch (C-1290) and James Bowerbank (C-1294) for permission for James to draw some of their fossil pedunculate barnacles. In February (C-1303), he wrote James that because of ill health he would postpone coming to London until all drawings were finished, and he asked for a statement of the total owed as a guide to the future. Darwin wrote James again in March (C-1306) that he had lost a good many days to illness and would need another fortnight to finish the pedunculate fossil cirripedes. In an April letter to Fitch (April 13?, 1850; C-1315), he stated, 'I am ready to go to press' but implies that James is delaying the project: 'You may rely on it, I shall *urge* Mr. Sowerby on in his drawings' (Darwin's extra emphasis). By April Darwin had become exasperated (C-1336); he wanted the drawings harder, with lines of growth more distinct; no shading or similarity to lithography, which he thought harmed natural science: 'I do not care for artistic effect, but only for *hard rigid* accuracy. ...The inside drawings of the scuta ... are useless, from indistinctness and shading.... I must I am sorry to say give you the trouble of going over them again' (Darwin's extra emphases). The Palaeontographical Society agreed to publish the volume, but Darwin was subsequently informed that they would only allow one plate for foreign species. So, on May 4 (C-1324) he wrote James not to illustrate certain forms until he could see how many would 'pack in' to the plate. On the other hand, he must know what progress had been made. On May 26 (C-1333) Darwin pressed James again on how he was getting on, noting that he would be coming up to London soon with a few more specimens and some of the drawings for a few small corrections. Darwin suggested that, to speed things up, more simple drawings might be done directly on copper. In closing he states, 'My answer to everyone is that my M.S. is and has been for some time ready, & all depends on you.'

By June 1850 Darwin (C-1338) was pleased with James's drawings, but wanted a few corrections, and '*very soon*' (Darwin's emphases). A month later (C-1343) he acknowledged receiving the plates and gave instructions for the arrangement of engravings. But in August, he wrote James (C-1346) that Bowerbank, for the Palaeontographical Society, had asked (again) what progress had been made and to which he was forced to reply (again) that 'everything depended on you (James).' And later in August he advised James (C-1347) that two more minute specimens were being sent to be figured with the hope that they would not cause too much trouble in altering the arrangement, but 'pray observe how time slips by. Remember that before the Plates are all engraved, I must have drawings for the woodcuts, as these come in Introduction.'

In September of 1850, Darwin (C-1350) advised James that he received, to his 'grief be it confessed,' an enormous lot of Scanian and Copenhagen specimens, some of which were better than those sent earlier. So he suggested that James delay engraving foreign specimens until he had time to go through the new lot. Some 10 days later Darwin noted (C-1354) that the new specimens showed that he had placed two distinct species under one name. Darwin suggested new figures and deletions, and affirmed that he would come up to London when the plates are all engraved and he had the proofs. On September 12 (C-1358), Darwin sent corrections for the woodcuts, and on September 23 (C-1360) he wrote (again) that he would come to London to examine the proofs, when they were ready, enclosing a figure of a specimen from a newly arrived German lot for which he desired a woodcut be made.

I have heard it suggested that Darwin devoted his time to barnacles in order to stave off the day he would have to write up his species work for publication. One does not get this impression from reading his correspondence concerning the illustrations for FC1. To the contrary, he was a

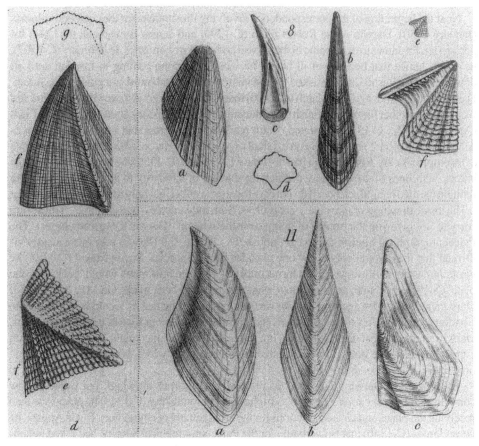

A

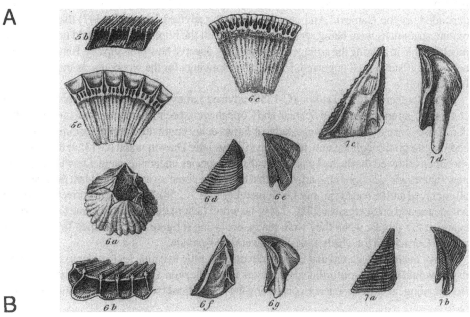

B

man in a big hurry. But he was also a perfectionist, and two such goals do not go easily together when other persons are involved.

While letters from Darwin to Fitch in November and December 1850 (C-1366, 1369, 1379) lamented the slow progress on the engravings for FC1, the decision for a change of illustrators had already been made. In his letter to Lankester at the Ray Society in October 1850 (C-1364), Darwin not only commented that the manuscript for LC1 would be ready in 2 weeks, and that he would like a decision from the Council on the number of plates, *but that the specimens should be sent to George Jr.* for an estimate on the price of the engravings. He was still dealing with James, as letters of January and February 1851 (C-1386, 1388, 1389) concerning the engravings of the fossils for FC1 confirm. While one can only guess the state of James's mind, he was certainly anxious if not distraught, and Darwin was clearly exasperated over the corrections he was trying to have made.

By February 1851 the artwork was finally finished, and Darwin wrote James a somewhat trite letter of appreciation (C-1391) in which he expressed concern that if James were not properly compensated by the Council then he would see to it. In the postscript to this last letter, Darwin asked James if he could borrow several named fossil Balani as they would probably be of great service to him. And that seems to have been it; James's nephew, George Jr., went on to illustrate FC2 as well as LC1 and LC2, and if there was much in the way of further correspondence between George Jr. and Darwin, nothing of significance has been recorded. Interestingly, when one compares the plates in FC1 and FC2, made by James and George Jr., respectively, those of James seem to come closer to fulfilling Darwin's goal of 'hard rigid accuracy' than do those of George Jr.; even James's name is more sharply inscribed (Fig. 5).

5 THE MONOGRAPHS IN LIGHT OF CONTEMPORARY KNOWLEDGE

Darwin's barnacle volumes are still often cited, primary sources. Winsor (1969) refered to the work as a 'definitive monograph of great detail and soundness,' and Crisp (1983a) suggested that '130 years later it remains a standard text.' The fact of the matter is, however, that use of the monographs is best left to experienced cirripedologists and carcinologists, because much of the anatomical as well as the systematic work is outdated. Study out of context with contemporary knowledge can only lead to serious misconceptions. Nonetheless, while most of Darwin's novel interpretations of cirripede anatomy turned out to be wrong, they 'were sufficiently in keeping with the times for Owen (1855) to accept all of them in the second edition of his lectures on invertebrates' (Rachootin 1985).

As the tables of contents for Darwin's volumes LC1 and LC2 (Appendix II) show, he systematically went through each organ system. Since there had been relatively little previous work, mistakes were inevitably made, and it is not only a little amusing how one mistake could

---

Figure 5. Comparison of illustrations by J. De C. Sowerby and George B. Sowerby Jr.: A. Sample of drawings of James De Carle Sowerby (from Darwin 1851, pl.1). Darwin had written James that he wanted the drawings harder, with growth lines more distinct, no shading or similarity to lithography which he thought harmed natural science; '...I do not care for artistic effect, but only for hard rigid accuracy.' B. A sample of drawings of George Brettingham Sowerby, Jr. (from Darwin 1855, pl. 1). Shortly before the completion of the drawings for 'Fossil Cirripedia 1' (1851) Darwin retained James' nephew, George, to do the rest of the illustrations. Compare and judge which Sowerby came closest to achieving Darwin's wishes (see Section 4 for discussion).

lead to another but how Darwin's fertile imagination came into play in explaining away difficulties that appeared. The principal problems, including those touched upon to varying degrees by Winsor (1969), Crisp (1983a), Southward (1983), and Rachootin (1985), will be briefly discussed.

5.1   *Cement glands and 'ovarian tubules'*

Darwin noted that the characteristic distinguishing barnacles from all other non-parasitic crustaceans was attachment throughout adult life, and he studied how this attachment was accomplished in considerable detail. He determined that, following attachment of the cyprid by its prehensile antennae and metamorphosis into a juvenile, glands near the base of the antennae continued to produce cement, and the antennae were buried in the cementing process (Fig. 6 A; see Walker & Yule 1984). Furthermore, he found that in the juveniles of some pedunculate barnacles, the cement continued to flow from the penultimate segments of the antennae (presumably the primitive condition), while in others the antennae were by-passed by distributional ducts. In sessile barnacles, he found a comparable but more elaborate system. Curiously, while the differences in the cementing apparatus discovered by Darwin are great, they were not used then, nor have they been used since, to elucidate the evolution of the barnacles.

Crisp (1983a) suggests Darwin's evolutionary viewpoint, viz., that new structures and organs were derived from preexisting structures having a different function, led him to make what we now consider the curious observation that there was a connection between the cement glands and ovarian tubules, and Burkhardt and Smith apparently agree with Crisp (C5:93, n. 5 and C5:444, n. 11). It is certainly true that once one gets an idea the search is then on for corroborating evidence. However, it is also true that ideas are not pulled out of thin air but stem from observations. Darwin did not imagine a connection between the cement glands and what he thought were ovarian tubules before he thought he saw evidence for it. To the contrary, he went as far as his observational and dissecting skills would take him before he wove the facts as he perceived them into a evolutionary hypothesis. At this point in his study of cirripede anatomy, he thought he saw an ontogenetic as well as a definitive anatomical connection between the cement glands and ovarian tubules. In light of these perceptions, he came up with the following speculation: 'I am well aware how extremely improbable it must appear, that part of an ovarian tube should be converted into a gland, in which cellular matter is modified, so that instead of aiding in the development of new beings, it forms itself into a tissue or substance, which leaves the body in order to fasten it to a foreign support' (LC1:37). This is fundamentally good science.

The broader evolutionary corollary to this speculation came several years later (1854): 'To conclude with an hypothesis, those naturalists who believe that all gaps in the chain of nature would be filled up, if the structure of every extinct and existing creature were known, will readily admit, that Cirripedes were once separated by scarcely sensible intervals from some other, unknown, Crustaceans. Should these intervening forms ever be discovered, I imagine they would prove to be Crustaceans, of not very low rank, with their oviducts opening at or near their second pair of antennae, and that their ova escaped, at a period of exuviation, invested with an adhesive substance or tissue, which served to cement them, together, probably, with the exuviae of the parent, to a supporting surface. In Cirripedes, we may suppose the cementing apparatus to have been retained; the parent herself, instead of the exuviae, being cemented down, whereas the ova have come to escape by a new and anomalous course' (LC2:151; see Section 5.2 and Fig. 9).

In September 1854 (C-1592), Darwin suggested to Huxley that he look into the connection

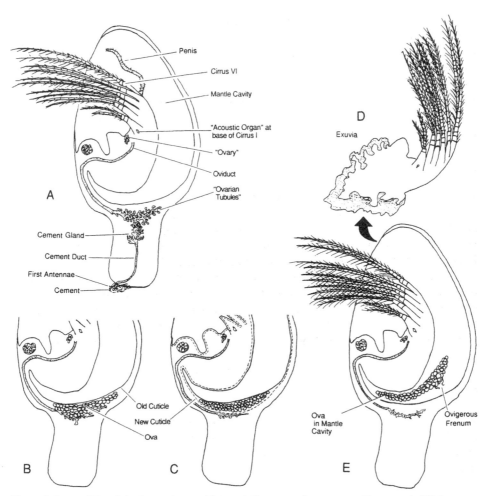

Figure 6. Some of Darwin's observations and interpretations proved erroneouos. For example (A), he saw a connection between the cement glands and the ovaries ('ovarian tubules'). He missed the connection between the oviduct and the genital aperture and oviducal gland ('acoustic organ'). He postulated a connection between the oviduct and the salivary gland ('ovarian gland'). Thus to Darwin the female genital system had no openings to the exterior. He explained this by postulating that eggs went from inside the body into the mantle cavity during a molt (B-E). (see Sections 5.1 and 5.2)

between the 'ovarian tubules' and the cement glands, and Huxley looked into the matter while at Tenby in southern Wales. By September 1855 Darwin (C-1757) was perplexed by Huxley's findings: 'I cannot conceive that I can have erred in (1st) so repeatedly tracing the branching ovarian tubes into the duct close to the gut-formed glands and (2d) and more especially in tracing the cement duct into the cement-gland. If these points are false, I shall never trust myself again.' A year later (December 1856, C-2017) he was still convinced he was correct.

It was a Russian anatomist, August Krohn (1859), who in 1857 discovered that what Darwin had taken for the connection between the cement glands and the ovarian tubules in *Conchoderma* was in fact the connection between the cement ducts and the cement glands. Darwin was more or less prepared for this by his correspondence with Huxley. Krohn, however, went further with the shocking announcement that 1) the cement glands intermingle with the ovaries in the peduncle, 2) that Darwin's so-called 'ovaries' were in fact the salivary glands of earlier authors, and 3) that his 'acoustic organs' were the female genital apertures and associated structures (see Section 5.2). Darwin was naturally chagrined. He wrote a letter to *Natural History Review* (1863) in which, while conceding that Krohn was possibly correct, he offered further observations, such as the formation of ovigerous lamellae in the feeble form, *Anelasma* (see Section 5.2, egg laying), that he felt were hard to reconcile with Krohn's observations.

5.2   *'Acoustic organs', salivary glands, and egg laying*

While at Tenby, Darwin observed that barnacles were sensitive to vibrations, and Crisp (1983a) suggests that the search for receptors led Darwin to mistake what are now known as the oviducal glands, associated with female genital apertures on the first thoracic segment, for acoustic or auditory organs. Furthermore, Burkhardt and Smith (C5:93, n. 5), as well as Crisp, suggest that Darwin's belief that the cement glands were modified ovarian tubules led him to misunderstand the route of the oviducts and the function of the so-called auditory sacs. The first suggestion may be partially correct, but the second is surely incorrect. Darwin thought he saw a connection between cement glands and 'ovarian tubules,' which often contain eggs in the peduncle (see Section 5.1), and he set out to trace a pair of ducts he found extending from them. If he had found that the ducts led to apertures near the bases of the first pair of cirri, which they do, he would have had it right. He came very close to discovering the connection, but he was distracted by glands near the buccal mass: '(The) two main unbranched ovarian ducts, followed up the peduncle, are seen to enter the body of the Cirripedia (close along side the great double peduncular nerves), and then separating, they sweep in a large curve along each flank of the prosoma, under the superficial muscles, *towards the bases of the first pair of cirri*; and then rising up, they run into two glandular masses. These latter (masses) rest on the upper edge of the stomach, and touch the caeca where such exist' (emphases added). Darwin goes on to note that these glandular masses were 'thought by Cuvier to be salivary glands,' but he reasoned they must be the ovaries (LC1:57; Fig. 6A).

Darwin just missed locating the connection of the oviducts with the female genital apertures at the bases of the first cirri. These relatively inconspicuous apertures must have some function, he reasoned, and he interpreted them and their associated oviducal glands as possibly acoustic organs. If he had succeeded in tracing the oviduct to its genital aperture, the auditory meatus of his acoustic organ, would he have interpreted the associated oviducal gland as acoustic? Likely not. On the other hand, even though he was 'not able to ascertain whether the two main ducts, coming from the peduncle, expanded to envelop them (ovaries according to Darwin), or what the precise connection was,' he was firmly convinced that these salivary glands of previous authors were the ovaries (LC1:57).

Crisp (1983a) muses over the fact that Darwin noted that the males of *Ibla* (Section 5.4.3) lacked 'auditory organs,' and he suggested that this should have alerted Darwin to the possibility that the supposed auditory organs were female structures. Perhaps, but the males of *Ibla* live within the protective confines of the mantle cavity of the female or hermaphrodite, and to Darwin the protection they received explained the loss of opercular plates and spiny armament, as well as reductions in the extent of the mantle and the number of cirri. Thus while Darwin did not address the absence of the so-called auditory sacs in males, there would have been no reason to assume that their absence was anything other than another set of reductions. After all, in keeping with Darwin's views on disuse and the economy of nature, views prevalent even today, the auditory sacs would no longer be needed to warn the male of an advancing predator.

The female reproductive system, as Darwin saw it, had no openings to the exterior (Fig. 6A). How then were the eggs laid? The eggs, or embryos in various stages of development, are found in the mantle cavity of the female or hermaphrodite held together by a secreted substance. In pedunculate barnacles, such as *Lepas*, the egg masses are formed into two relatively flat sheets (ovigerous lamellae of Steenstrup and others) which in turn are usually held in place by small attachment organs on the interior of the mantle lining, called ovigerous frena by Darwin (see Section 5.5.2). Darwin believed the clue as to how the eggs were laid was their appearing in flat or saucer-shaped sheets, and therefore he speculated, 'Immediately before one of the periods of exuviation, the ova burst forth from the ovarian tubes in the peduncle and round the sack and, carried along the open circulatory channels, are collected (by means unknown to me) beneath the chitin-tunic of the sack, in the corium, which is at this period remarkably spongy and full of cavities. The corium then forms or rather (as I believe) resolves itself into a very delicate membrane separately enveloping them together into two lamellae... As soon as this exuviation is effected, the tender ova, united into two lamellae, and adhering, as yet to the bottom of the sac, are exposed: as the membranes harden, the lamellae become detached from the bottom of the sack, and are attached to the ovigerous frena' (LC1:59; see Fig. 6 and Section 5.5.2; for recent research on the actual process, see Walley 1965, Walker 1983).

5.3 *Olfactory organs and excretory glands*

Darwin observed, 'In the outer (second) maxillae, at their bases where united together, ... there are, in all the genera, a pair of orifices.... It is impossible to behold these organs, and doubt that they are of high functional importance to the animal.' He went on to describe the internal structure and apparent innervation, and he could 'hardly avoid concluding that (it) is an organ of sense; and considering that the outer maxillae serve to carry the prey entangled by the cirri towards the maxillae and mandibles, the position seems so admirable adapted for an olfactory organ, whereby the animal could at once perceive the nature of any floating object thus caught, that I have ventured provisionally to designate the two orifices and sacks as olfactory' (LC1:52).

The preparation of specimens, by infiltration and embedding, for the cutting of thin sections was not developed until the 1860's, and the Dutch cirripedologist Paulus Hoek (1883) applied it to his anatomical studies of the barnacles of the Challenger Expedition. In the process, Hoek discovered that the provisionally identified 'olfactory organs' of Darwin were excretory organs, the maxillary glands, since discovered in other crustaceans.

5.4 *Larvae and related matters*

John Vaughan Thompson, a surgeon who practiced natural history in his spare time, published his first paper on larvae and the crustacean affinities of cirripedes in 1830 (see Winsor 1969). He discovered that curious bivalved crustaceans (subsequently known as cyprid larvae), netted from the plankton near Cork, Ireland, settled in the bottom of the jars in which they were held and metamorphosed into juvenile sessile barnacles (likely *Semibalanus balanoides*). At the time. Cuvier's (1817) classification placed cirripedes and most molluscs within Embranche-ment II, and of two other contemporary zoologists, Henri Milne Edwards did not fully accept cirripedes as crustaceans until 1852, and Richard Owen clung to his opinion that barnacles not be placed in the Crustacea until 1855, curiously recognizing the larvae of cirripedes as crustacean but denying such affinities in the sessile adults (Winsor 1969).

Thompson predicted that pedunculate barnacles would also have cyprid larvae, but when he got *Lepas* from the bottom of a ship, he found they spawned a different larva (subsequently known as the nauplius), and he concluded (1835) there was a larval difference between the two families. However, as Darwin noted (LC1:9 n. 1), J. Gray (1833) had already briefly described the first or nauplius larva of *Balanus*, although he mistook the anterior for posterior end, and Burmeister (1834) showed that *Lepas* passed through both naupliar and cyprid stages. So, unknown to Thompson, his original prediction proved correct.

Thompson went on to make another important discovery, viz, that barnacle nauplii were very similar to those obtained from the crab parasite *Sacculina* Thompson, 1836, a parasite belonging to what became known as the Rhizocephala Müller, 1862. Although Thomson demonstrated the cirripede as well as crustacean affinities of this parasite, if Darwin noted the matter, he ignored it.

Since the larvae of cirripedes where known by the time Darwin began work on them, he had no doubts that cirripedes were crustaceans. It is a little surprising though that biologists of that time had yet to determine the homologies of the naupliar appendages of any crustacean. So Darwin, in correspondence with Dana and C. Spence Bate, tried to work them out. Unfortunately, cirripede nauplii have frontal filaments and frontolateral horns, and Darwin mistook these for the anlagen of appendages. Furthermore, he let theory override his intuitions and, as a result, he mistook the homologies of the three pairs of naupliar limbs, and thus to some extent the limbs of adult Thoracica, Abdominalia, and the archetype cirripede (see Sections 5.4.1, 5.4.2, 5.6).

5.4.1   *The nauplius and the cyprid*

Darwin's contemporaries clarified the homologies of the larval limbs, and Darwin scholars worked out to various degrees why Darwin got it wrong. Winsor (1969), for example, briefly discusses Darwin's interpretations of adult segmentation, but she does not attempt to compare his homologies of the naupliar structures with those of cyprid and adult cirripedes. On the other hand, Burkhardt and Smith, while effectively ignoring segmentation, mistakenly reversed Darwin's view of the homologies of the two pairs of antennae and therefore lost the thread (C4:381, n. 4.; see LC1:9 paragraph 2 and LC2 Pl. 29 9c and 10c for a relatively clear statement of the Darwin's view of these homologies). The situation is further confused by Rachootin's (1985:235) table, which compared Darwin's homologies of the naupliar structure to how we see them today. While in a general way the table is correct, Rachootin neglected the eyes, which leaves the reader one segment short when it comes to Darwin's segment counts. Like Burkhardt and Smith (C4:404), Rachootin side-stepped Darwin's homologies of the three pairs of naupliar limbs by simply referring to them as three pairs of natatory limbs, which they are, but they also represent the first and second antennae and the mandibles (Fig. 7). Darwin ended up considering the first pair as belonging to the last cephalic segment and the second two pairs as being thoracic. The reasoning Darwin went through in reaching this erroneous conclusion has not been explored.

Cirripede nauplii differ from all other nauplii in having frontolateral horns, and from many other nauplii in having well developed frontal filaments (Fig. 7). It was the 1) presence of these structures, 2) the interpretation of the naupliar labrum as a proboscis with a terminal mouth, 3) the absence of mouth parts in the cyprid larva, 4) the interpretation of the first cirri of *Cryptophialus* as first maxillipedes, 5) the belief that *Proteolepas* was a barnacle, and 6) the theoretical considerations that led Darwin astray in his analysis of the homologies of naupliar limbs.

The situation is terribly confusing when one reads Darwin concerning the homologies of the

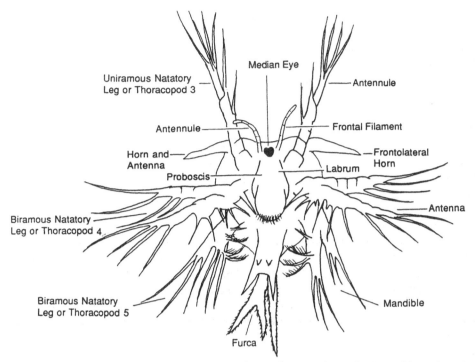

Figure 7. Darwin was the first to homolgize the barnacle naupliar appendages with those of the adult. Unfortunately the frontal filaments and unique frontolateral horns in the nauplius and the lack of mouthparts in the cyprid led him astray when it came to the homologies of the remaining three pair of appendages. In this figure, labels on the left are Darwin's identifications (from Darwin 1852, pl. 29,10), those on the right are as presently understood (see Sections 5.4-5.4.2 for explanation).

naupliar limbs because he interpreted these limbs differently in different parts of the text in LC1, and in the text and the caption for plate 29, 8-10 in LC2. No wonder Darwin scholars avoided getting deeply involved with the matter.

It is best to begin where Burkhardt and Smith went astray, with the frontal filaments and the naupliar frontolateral horns. Burmeister (1834) suggested that the frontal filaments had something to do with the prehensile limbs of the cyprid and this may be what initially led Darwin astray. Darwin thought he saw an articulated appendage within both the frontal filaments and the frontolateral horns[4] in a second stage nauplius of *Chthamalus stellatus*, and he noted (LC2) that Bate independently saw the same thing.

Darwin thought he found corroboration for the homology of the naupliar frontolateral horns with the prehensile antennae of the cyprid in his study of *Cryptophialus minutus*. The species had an abbreviated development, i.e., a recognizable naupliar stage is suppressed. Darwin illustrated a series of stages from an early egg-shaped embryo to the cyprid stage, which lacks thoracic limbs and is incapable of swimming (LC2, Pl. 24, 15-18). The embryo, which he interpreted as the nauplius, has a pair of more or less frontolateral projections that he thought were the frontolateral horns. This interpretation ignored the fact, of which he was well aware, that in normal development frontolateral horns are folded back during development through the first free naupliar stage.

In following the development of the 'horns' of *Cryptophialus* embryos through successive stages, Darwin found that the 'horns' contained the developing prehensile antennae of the cyprid. This is not surprising because they were actually the anlagen of the cyprid first antennae. But to Darwin they were homologous with the frontolateral horns of the nauplius, and thus the anlagen of the second rather than the first antennae (LC1:9, LC2:106 and Pl. 29, Figs 9, 10).

Darwin argued with Dana over whether the cyprid attachment limbs were the first or second antennae (unnumbered letter in C4:380 preceding C-1381, also C-1481, C-1493, C-1542). The correspondence that survived is primarily from Darwin to Dana, so the details of the argument are unknown, but it seems that both men were partially correct: Darwin thought the cyprid initially attached by its second antennae (incorrect) and the front of the head formed the peduncle (correct), while Dana thought the cyprid attached by its first antennae (correct) that also became the peduncle (incorrect). Both were mislead in trying to find homologies between the naupliar frontal filaments and frontolateral horns, and the antennae of the cyprid, but, as we now know, there are none.

Having satisfied himself regarding the homologies of the frontolateral filaments and frontolateral horns, Darwin went on to examine the naupliar limbs, and in LC2:107 he noted: 'With regard to the homologies of these three pairs of limbs, my first impression was that they were mandibles and the two pairs of maxillae in their earliest condition.' He dismissed this interpretation in LC1 because of 1) the wide interval between their bases and the mouth itself, 2) the somewhat variable position of the mouth with respect to the legs, and 3) the position that the legs occupy in the cyprid larva. There was a fourth and very important consideration implicit in his reasoning: since the mouth parts are clearly missing in the cyprid, which cannot feed, they must be missing in the nauplius, and therefore the three pairs of naupliar limbs, by two of which the nauplius feeds, must be something else. Perfectly reasonable, who would have thought mouth parts present in the nauplius would be lost in the cyprid and regained in the adult?

The first reason Darwin gave for considering that the three naupliar limbs were not mouth parts is puzzling. He noted that although the first naupliar limbs are uniramous and anterior to the mouth, and the second and third are biramous, behind the mouth, and provided with masticatory or prehensile inner spines. These spines could not 'but...serve some important end, namely, as organs of prehension for the larvae, like the mandibles and maxillae of mature Cirripedes, for seizing their prey, and conveying it to their movable mouths, conveniently seated for this purpose' (LC1:12). However, if these appendages became legs rather than mouth parts in the mature cirripede, Darwin needed to explain why they were so closely associated with the mouth in the nauplius. He reasoned, 'The highly remarkable position of the mouth in the larva, either between the bases of the two posterior pairs of legs, or at least posteriorly to the first pair, together with the probable functions of the spiny points springing from the basal segments of the two hinder pair of true thoracic limbs, forcibly bring to mind the anomalous structure of the mouth being situated in the middle of the under side of the thorax, in *Limulus*, that most ancient of crustaceans, and therefore one likely to exhibit a structure now embryonic in other orders' (LC1:12). While *Limulus* is now known to be more closely allied to eurypterids and scorpions than to the crustaceans, Darwin was correct concerning their gnathobasic limbs. While they appear to surround the mouth, they are in fact fundamentally post-oral, and their position and function in trilobitomorphs as well as *Limulus* and crustacean nauplius larvae may be more than coincidental (Hessler & Newman 1975).

The second reason Darwin considered that the three naupliar limbs were not true mouth parts was the variable position of the mouth, which he thought was at the distal end of a proboscis (now known to be the labrum). The third reason, the position he inferred the naupliar limbs occupy in the cyprid larva, stemmed from an erroneous observation of Burmeister (1834) that the 'second larva' (early cyprid; LC2, Pl. 30, 1) had only three pairs of thoracic limbs, the remaining three pairs of thoracic limbs being added when the second larvae went to the 'third' or so-called 'pupal stage.' While ascothoracicans can pass through several cyprid stages (Grygier 1984, 1987), we now know there is only one in barnacles and rhizocephalans, and in all three the anlagen of the six thoracic limbs appear during the last few naupliar stages while the three naupliar appendages are still present.

Having decided that the three pairs of naupliar limbs were not the mouthparts, Darwin initially considered them equivalent to the 'outer' or third pair of maxillipede of higher crustaceans followed by 'the first two pairs of thoracic limbs' (LC1:12; Fig. 8B)[5]. This brought him to a 'theoretical consideration': while the adult thoracican cirripede has six pairs of 'thoracic' limbs, in light of his Apoda and Abdominalia, the archetype cirripede would have to have eight. Therefore, two pairs of thoracic limbs must have been lost between the archetype cirripede and the Thoracica, and there is some correspondence with Dana concerning whether these limbs were lost from the proximal or distal end of the thorax. Darwin concluded that the first pair of naupliar limbs must be equivalent to the third maxillipedes of higher crustaceans (LC2:107), but perhaps by a slip of the pen he designated the first three trunk limbs of the cyprid as the 'second, third and fourth thoracic limbs' (LC2:108). It becomes clear in reading the complex discussion a few pages later (LC2:111) that he meant the third, fourth and fifth, the third being the segment of the third maxillipedes and hence his conclusion that the first two pairs of maxillipedes had been lost (Fig. 8B).

In support of his view that the first two pairs of 'thoracic' limbs were missing in thoracican cirripedes (LC2:108; see Fig. 8), Darwin noted that in the structure and origin of their nerves the six pairs of cirri resembled the third maxillipedes and five pairs of ambulatory legs in certain decapods. He also observed that (LC2:111) 'between the mouth of the pupa (the normal cyprid) and the first pair of natatory legs, there is a space of membrane, equaling when stretched out, the three succeeding thoracic segments in length and breadth; this interspace, I conceive, must have some homological significance; here then we have at least an appearance of the abortion of appendages; whereas, at the posterior end of the cephalo-thorax, no such appearance is present. Moreover this interspace of membrane is divided nearly in the middle by a most conspicuous fold, which, on the view here adopted, would mark the separation of the seventh (cephalic) from the eighth (thoracic) segment; and the interspace and fold are thus simply explained.' So, Darwin concluded that the supposedly missing appendages were the first two pairs of 'thoracic' limbs (LC2:112). Since the archetype *had* to have eight thoracic segments (see Note 5 and Section 5.4.2), Darwin concluded that his orders Apoda and Abdominalia as well as the Thoracica could be included in the Cirripedia.

Although now it is thought unlikely that anterior thoracic segments were lost in the evolution of the archetype cirripede (see Newman 1987; Boxshall & Huys 1989), Darwin's consideration of the matter was driven largely by the 8-segmented thorax of *Proteolepas* (LC2, Pl. 25, 7) and the inferred 8-segmented thorax, including that segment bearing the 'rudimentary first maxillipedes,' of *Cryptophialus* (LC2, Pl. 22, 5h) [These are actually the rudimentary first cirri and not too dissimilar in form and position to those of *Trypetesa*, which he had already identified as such (LC2).] Darwin's interpretation of *Proteolepas* and *Cryptophialus* dictated an archetype cirripede with 17 rather than 15 segments (LC2:111 and 122; see Fig. 8A).

As noted above, Krohn (1860) questioned Darwin's interpretation of the frontolateral horns as the anlagen of the cyprid antennae, but if the two corresponded over the matter there is no record. On the other hand, in August 1865, Darwin wrote Müller (C-4881) that he admired his

A

```
1)  1  2  3  4  5  6  7  8  9  10 11 12 13 14 15 16 17 18 19 20 21
          Head                    Thorax              (Abdomen)

2)  1  2  3  4  5  6  -  -  3  4  5  6  7  8  (1  2  3)

3)  A' A" M  MX' MX" 1  2  3  4  5  6  (1  2  3)
```

B

	NAUPLIUS			CYPRID				JUVENILE
	LC1:9&12	LC2:107	LC2:107	LC2:107	LC2:108	LC2:111	Meta.	Present
1	eyes	eyes	eyes	eyes	eyes	eyes	eyes	
2	FF	A'	A'	A'	A'	A'	--	
3	FLH	A"	A"	A"	A"	A"	A"	eyes 0
4	--	--	M	--	--	--	M	A' 1
5	--	--	MX'	--	--	--	MX'	-- 2
6	--	--	MX"	--	--	--	MX"	M 3
7	--	--	--	MXPED1	--	--	--	MX' 4
8	--	--	--	MXPED2	--	--	--	MX" 5
9	1	MXPED3		MXPED3	T2	T3	T3	T1 6
10	2	T1			T3	T4	T4	T2 7
11	3	T2			T4	T5	T5	T3 8
12					T5	T6	T6	T4 9
13					T6	T7	T7	T5 10
14					T7	T8	T8	T6 11
15						a1	--	-- 12
16						a2	--	-- 13
17						a3	--	-- 14
18							--	
19							--	
20							--	
21							--	

Figure 8. The archetype and cirripede segmentation: A. The archetype crustacean of Milne Edwards (A, 1) consisted of 21 segments (incl. eyes). Darwin believed (A, 2) that the archetype cirripede had at least 17 segments (incl. eyes), 2 more than Thoracica, where the anterior 2 pair of thoracopods and their segments were supposedly lost. The cirripedes as presently known (A, 3) includes 14 segments (incl. 3 abd. seg. in the cyprid larva, but excluding eyes (Section 5.4.2).

B. The history of Darwin's shifting analysis of larval and juvenile limb homologies: The left column includes the 21 segs of an archetype. The third, fourth, & fifth columns list the naupliar homologies Darwin surmised (see Fig. 7). He initially homologized the 3 limb pairs with maxillipede 3 and thoracopods 1 and 2, or possibly (column 4) the mandible and maxillae 1 and 2. What followed is confusing (columns 5-8, see Section 5.4.2 and Note 5). Columns 9 and 10 give the limb homologies and segment numbers as presently known.

work on crustaceans and requested that he make some dissections of cirripedes to confirm some of his (Darwin's) observations. On November 29, 1867, Darwin wrote Anton Dohrn (C-5698, letter available in its entirety in Groeben 1982) thanking him for a paper on 'The morphology of the Arthropoda' and adding: 'I cannot yet persuade myself that the view which I have taken in my volume on Balanidae page 105, of the homologies of the appendages is wrong. I still believe (though Fritz Müller writes to me that he thinks I am mistaken) that I saw within the anterior-lateral horns of the carapace the prehensile antennae in the process of developing.' Ostensibly Müller voiced his doubts to Darwin after making the dissections Darwin requested in August 1865 (C-4881); when this occurred or what Müller actually said is not reported in the *Calendar*.

The short paper on arthropod morphology that Dohrn sent Darwin had been read before the British Association in September 1867; it was subsequently published in the *Journal of Anatomy and Physiology* in 1868. The paper announced the results of an investigation into the homologies of crustaceans and insects submitted to the same journal. In May 1868, Dohrn became aware that his results were based on incorrect observations and withdrew the publication (Groeben 1982:19, n. 1). However, in July 1868, he wrote Huxley: 'Unfortunately (I) cannot devour all the copies of the Journal of Anat. and Physiol.' announcing it (Groeben 1982:19, n. 1). Apparently Dohrn's application of Darwin's theory of evolution to the Crustacea was at odds with what Müller (1864) reported, and Dohrn acknowleged this in a letter to Darwin on 30 December 1869 (letter available in its entirety in Groeben 1982). I see no evidence that Darwin was fully aware of what happened with Dohrn, who had a nervous breakdown caused by this scientific failure (Groeben 1982:90, n. 10).

5.4.2 *Segmentation and the archetype or urcirripede*

Darwin observed 1) that *Proteolepas*, which turned out to be a malacostracan rather than a cirripede (see Section 5.7), had eight thoracic and three abdominal segments, 2) that the trunk of *Cryptophialus* also appeared to be comprised of eight thoracic and three abdominal segments (see Section 5.6), 3) that the thoracican cyprid had six thoracic and three abdominal segments, and 4) that the adult thoracican had completely lost the abdomen. Thus he envisaged a series based on the reduction of a malacostracan crustacean (like *Lucifer*, LC1:28; Fig. 9) to the archetype cirripede plan, followed by reductions from the archetype cirripede to the thoracican plan.

Darwin's choice of a malacostracan as a model from which the cirripedes could be derived was influenced by Milne Edwards's views on crustacean segmentation (see Section 2.1). His choice of *Lucifer* was apparently influenced by 1) the elongate cephalon (potentially the peduncle), 2) the cirral-like thoracopods, 3) the reduction of at least the last pleopods, and 4) the carapace divided into plates by linea and arranged in tiers. He wrote (LC2:133): 'I stated that I believed that the carapace of the Cirripedes presented more real resemblance with the carapaces of ...higher Crustacea, than with those of the lower Crustacea, though in mere shape they more nearly resemble the latter.' In light of the ascothoracids, this 'mere' resemblance to 'lower' crustaceans turned out to be prophetic.

One contemporary view of segmentation in thoracican and acrothoracican cirripedes is 5-6-(3) (not including eyes; see Newman et al. 1969; Newman 1983), the three abdominal segments being seen only in the cyprid. However, if the penis is considered to be thoracic rather than abdominal, then the cirripede plan would be 5-7-(2), and 5-7-4 in their closest living relatives, the ascothoracids (see Newman 1982b, Grygier 1983). These two subclasses are in the Class Thecostraca Gruvel, 1905, (sensu Grygier 1984), Superclass Maxillopoda, which, in light of the Skaracarida and Tantulocarida (Müller & Walossek 1988 and Boxshall & Lincoln 1983) has a fundamental 5-7-5 or 17 segmented plan (Boxshall & Huys 1989, Newman 1992). Thus, considerations of how reduction led to the maxillopodan and hence the cirripede plan, implicit in Darwin's thinking, is of more than simply historical interest.

5.4.3 *Males*

One of Darwin's remarkable discoveries was the existence of males in several distantly related barnacles, at a time when, except for the work of Goodsir (1843), barnacles were thought to be

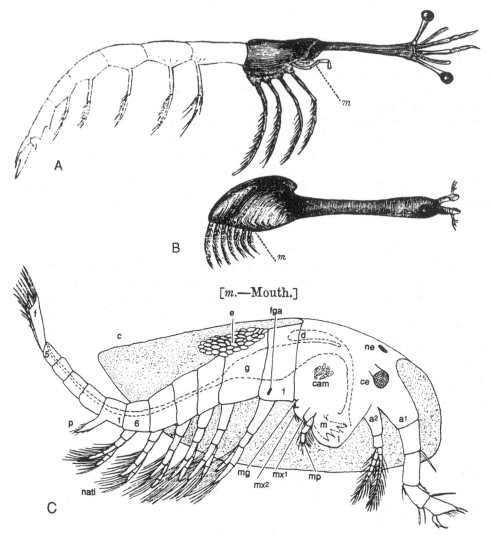

Figure 9. The origin of Cirripedia: Darwin took the aberrant sergestid, *Lucifer* (A, from Darwin 1852 text-fig. after Milne Edwards), as a model archetype, which he compared to *Lepas* (B, see Section 5.4.2). The discovery of parasitic ascothoracidans, like *Synagoga* (see Fig. 21A) negated his view that the barnacle carapace resembled more the carapaces of higher crustaceans than lower forms (see Sections 5.4.2 and 5.9). The ancestral cirripede (C) is currently viewed as a free-living ascothoracican (a^1 = 1st ant., a^2 = 2nd ant., c = carapace, cam = carapace adductor muscle, ce = cmpd. eye, d = gut diverticulum, e = eggs, f = caudal rami, fga = female genital aperture, g = gut, l = labrum, m = mandibles, mg = max. gland opening, mp = mand. palp, mx^1 = maxillule, mx^2 = maxilla, natl = natatory limb, ne = naupliar eye, p = penis, 1-6 = thrx. limbs, 1-5 = abd. segs. (after Newman et al. 1967).

exclusively hermaphroditic. Burkhardt and Smith (C3:xvii) implied that the first male Darwin discovered was that of the minute, highly reduced 'Arthrobalanus' (= *Cryptophialus minutus*), the species he collected and studied in 1836 when on the Beagle, and that the fact that this barnacle had separate sexes was one of the peculiarities that launched Darwin into a survey of the entire subclass. Although it is true that Darwin began his studies of the barnacles with *Cryptophialus* and had by October 1, 1846, sent a short manuscript on this form to Owen for comments (C4:xvii), Darwin had not yet found the males. The males of *Cryptophialus*, like those of *Trypetesa*, are not only the most reduced of those Darwin discovered, but they are connected around the attachment disc, outside the female. That he would have seen them before the males of *Ibla* and *Scalpellum* seems unlikely, not only because of their size and position, but also because by at least as late as November 1846, he thought the two median-dorsal filamentary appendages on what turned out to be the female of *Cryptophialus* were penes (C-1022; LC2:574); that is, Darwin initially thought *Cryptophialus* was a hermaphrodite! Study of the correspondence reveals the order in which all the males were discovered, and it is evident that his opinion that *Cryptophialus*, and *Trypetesa* for that matter, were hermaphroditic (without even dwarf males) did not change until he was nearly finished with LC2 in 1853.

As expected, Darwin discovered the males of barnacles essentially in the order of their decreasing size and, concomitantly, reduced structure. The largest and fairly generalized males residing in the mantle cavity of *Ibla cumingi* were discovered in April 1848, and Darwin wrote Henslow (C-1167): 'I must tell you a curious case I have just these few last days made out: All the Cirripedia are bisexual (hermaphroditic), except one genus, and in this the female has the ordinary appearance, where as the male has no one part of its body like the female & is microscopically minute; but here comes the odd fact, the male or sometimes two males, at the instant they cease being locomotive larvae become parasitic within the sack of the female, and thus fixed & half embedded in the flesh of their wives they pass their whole lives & can never move again. Is it not strange that nature should have made this one genus unisexual....' He subsequently discovered that *Ibla quadrivalvis* (Cuvier), while hermaphroditic, had similar males, and these he dubbed 'complemental.' The series from moderately to fairly reduced males residing in pits along the inner margins of the scuta of *Scalpellum* was discovered next, in May 1848 (C-1174).

It may be a little puzzling why Darwin went to such lengths in explaining why he was sure that the supposed males of *Ibla* were not some crustacean parasite (LC1), until it is recalled that it was he who refuted Goodsir's (1843) claim of males in *Balanus*. In November 1847, Darwin wrote Milne Edwards (C-1136) of his doubts concerning Goodsir's findings, before Darwin's discovery of males in *Ibla*. He also wrote, in 1850, detailed accounts to Dana (C-1305) of what he believed Goodsir saw, and to Hancock: '(Goodsir's) *male Balanus* is a *female* crustacean allied to *Bopyrus* and his parasite is the male of this female. But now comes the odd case, I have found two genera of cirripedes (*Ibla* and *Scalpellum*) with males separate and parasitic on the females; in these cases I am sure there can be no mistake, though I will not take up your time with details' (C-1311, Darwin's emphases). When Darwin found the little 'parasite' fastened in the mantle cavity of a hermaphrodite of *Ibla*, he had every reason to be cautious since there was no obvious reason why it should be a male. Like Goodsir's form, it too could have been simply a parasite. It was the fact that forms such as *Ibla cumingi* were female that clinched it. If the accompanying 'parasite' were not the male in this species, reproduction would have to have been parthenogenetic.

Having discovered males in *Ibla*, when he came to *Scalpellum* Darwin was prepared to find males, and he did. Some were, like those of *Ibla*, clearly little barnacles with reduced armament yet capable of feeding. However, others males were reduced to little spheres having traces of capitular armament but lacking feeding and digestive organs, and were embedded in small pockets in the scuta of each side. In his letter to Hooker of May 1848 (C-1174), Darwin noted that it was his species theory that made it possible for him to understand the reductional series seen in the males he had discovered first in *Ibla* and then in *Scalpellum*: 'I tell you this to boast of my species theory, for the nearest and closely allied genus to it (*Ibla*) is (*Scalpellum* and), as usual, hermaphrodite, but I had observed some minute parasites adhering to it (*Scalpellum*), and these parasites, I now can show, are supplemental males, the male organs in the hermaphrodite being unusually small, though perfect and containing zoosperms: so we have almost a polygamous animal, simple females alone being wanting.' This turned out to be a nice intermediate form between purely hermaphroditic species and those having separate sexes but dwarf males, as in one of six species of *Scalpellum*; and Darwin was able to discuss coevolution between the sexes (LC1:285).

In Darwin's letters to Dana and Hancock in February 1850 (C-1305, C-1311) and Steenstrup in May 1850 (C-1330), he mentioned finding males in species from only two genera (*Ibla* and *Scalpellum*). So clearly, Darwin was unaware of males in *Trypetesa* at this time, and the only previous mention of sexuality in *Cryptophialus* was that it had a pair of median penes (see above). I can find no evidence that Darwin discovered the minute and extremely reduced males of *Trypetesa* any earlier than February 1853 (C-1500), or the similar males of *Cryptophialus* before March 1853 (C-1509). In addition to being minute, males of both *Trypetesa* and *Cryptophialus* are virtually transparent and occur attached to the outside of the female on and among the shreds of the horny attachment disk of the rostral region.

There were half a dozen letters on the sexuality of cirripedes between 1850 and 1853, especially following publication of LC1 in 1852, and these dealt with the males of *Ibla* and *Scalpellum*. Then in January 1853 Darwin wrote hurriedly to Hancock (C-1499): 'I am almost driven mad by its (*Trypetesa* Hancock, 1849) generative system and I write to ask whether you have any (specimens) you could send me.' (Darwin expected to pay the post if heavy, and he would be willing to hire a fisherman to collect if necessary.) While Darwin made no mention of males, interestingly he noted he no longer thought, as he had in 1849 (C-1253), that *Trypetesa* was allied to his South American form, *Cryptophialus*. However, less than 2 weeks later another letter to Hancock (C-1500) excitedly announced that he had found males on specimens of *Trypetesa* already on hand: 'You may imagine how peculiar the appearance of the male *Trypetesa* is, when I mention that, though having had experience how diverse an aspect the males put on, I now know that I looked at a Male, during the first day or two, and never dreamed it was a cirripede!'

His first mention of males in *Cryptophialus*, which are quite similar to those of *Trypetesa*, came a little more than a month later (March 30, 1853) in yet another letter to Hancock (C-1509): 'I have now finished with my S. American boring cirripede and this has utterly confounded my previous confusion how to rank *Trypetesa*; for they present some most remarkable similarity, for instance they are both bisexual, with the males remarkably alike.' Although it is possible, I see nothing to suggest Darwin discovered the males of *Cryptophialus* just before writing Hancock for fresh specimens of *Trypetesa* (C-1499) and held back informing Hancock of them until he had a chance to see if *Trypetesa* had them too. Therefore I conclude that Darwin discovered the males of *Cryptophialus* not first, but last.

Darwin was proud of his discovery of a diversity of males in barnacles, and it much influenced his future thoughts on sexual selection. Although he only published specifically on barnacles twice after completing his barnacle volumes, one of these (1873) was a letter to *Nature* in response to Wyville Thomson's (1873) 'interesting account of rudimentary males of *Scalpellum regium*' dredged in the Atlantic abyss by the Challenger. [The other (1863) was a response to Krohn's (1859) criticism of Darwin's 'auditory sacs' and related matters (see Section 5.2).] Darwin had been 'most anxious that some competent naturalist should re-examine them (the males); more especially as a German, without apparently having taken the trouble to look at any specimens, has spoken of (Darwin's) descriptions as a fantastic dream.' Thomson's article gave Darwin the opportunity to review and update his views on what were the possible conditions that first favored a separation of sexes in attached and normally hermaphroditic species. Darwin discussed in a general way how the inherited effects of disuse and the principle of the economy of growth could come into play when a species finds its resources in short supply. An essential factor unknown at the time (1873) was that cirripede hermaphrodites tend to be protandric, and therefore have a proclivity for developing males when conditions place a premium on them (Ghiselin 1974; Dayton et al. 1982; Charnov 1987).

While Darwin discovered males in the acrothoracicans, and in *Ibla* and *Scalpellum* s.l. among the lepadomorphans, he missed them in two balanomorphs he otherwise studied in detail. One was an obligate commensal of gorgonians, *Conopea galeata* (Henry 1965; Henry & McLaughlin 1967; McLaughlin & Henry 1972). The genus *Chelonibia*[6] was the other (Crisp 1983b, pers. obs.).

5.5   *Minor items; penes, frena and branchiae, and calcareous bases*

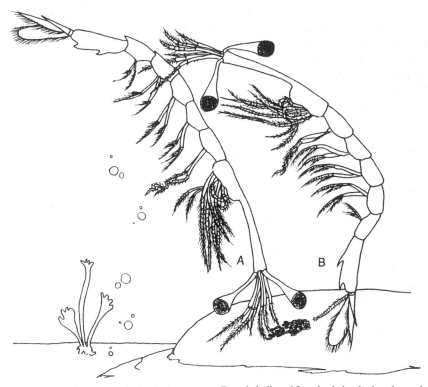

Figure 10. Mating and egg laying in the ancestor: Darwin believed female cirripedes lost the genital openings. He thought this related to a change in function of cement from affixing eggs to substrates (A) to affixing adults themselves (Section 5.2). He also thought 1) *Proteolepas* had a rudimentary penis at the end of the abdomen and 2) *Cryptophialus* females had an abdomen while males had a terminal penis, and 3) he observed prosciform penes of thoracicans were also terminal (see Section 5.5). So he concluded the penis was terminal in the ancestral male (B) and was used to fertilize eggs deposited and affixed to substrates by the female (A).

5.5.1   *Penes*

Failing to find the female genital apertures in barnacles, Darwin inferred that the ancestral cirripede laid and cemented its eggs to the substratum via the antennae, and that the advent of permanent attachment of the adult via the antennae required the development of a new method of egg laying (see Section 5.2). He got similarly, but considerably less, enmeshed when it came to deducing the ancestral position of the penis. His study of *Cryptophialus* suggested that the form had a three-segmented abdomen and that its males, like those of *Trypetesa*, had a terminal penis. While the abdomen is missing in adult thoracicans, the penis is terminal, and therefore Darwin concluded, 'From the attachment of the penis at the posterior end and on the under side of the anus, from the position of the caudal appendages (where such occur) over the anus, from the position of these same appendages in the pupa, and lastly, from the position of the papillae-like penis in the abnormal *Proteolepas*, I infer that, homologically, the penis is situated at the apex of the abdomen, on its ventral surface; and that, consequently, this organ cannot be considered as the abdomen itself in a modified condition' (LC2:100).

To those steeped in contemporary carcinology, a terminal penis would seem an odd inference, but the field was in its infancy in the 1850's. On the other hand, it was known then that the gonopores in both sexes of opisthogoniate insects emptied on the end of the abdomen, so a crustacean having a terminal penis apparently did not strike carcinologists as implausible. So, Darwin's ancestral cirripede (Fig. 10) would not only have had female genital apertures opening on, or in the vicinity of, the antennae, but also a penis at the posterior end of the abdomen, both conditions without precedent in crustaceans known then or today. While the connection between the oviducts and the female genital apertures on the first thoracic segment was established by Krohn (1859), it took the discovery of ascothoracicans by Lacaze-Duthiers (1880; Append. III,1) to demonstrate that the original position of the cirripede penis was on the first abdominal segment, the seventh postcephalic or last thoracic segment of some investigators.

5.5.2   *Frena and branchiae*

Darwin observed that in many pedunculate barnacles, eggs were brooded as 'ovigerous lamellae' in the mantle cavity, where they may be attached and held in place by 'ovigerous frena' (Section 5.2; Figs 6, 11). While he thought he had found structures in *Trypetesa* that might be frena that had lost their function, frena are absent in sessile barnacles, the eggs simply being packed into the bottom of the mantle cavity as in verrucids (= Verrucomorpha) and in all other acrothoracicans. Instead he found 'branchiae' in balanids, which, like the frena, are outgrowths of the interior of the mantle lining. Since pedunculate barnacles gave rise to the sessile barnacles, he reasoned that while the frena were lost in verrucomorphs and perhaps in acrothoracicans, they were converted into branchiae in balanids.

To Huxley in September 1854 (C-1590) Darwin wrote, 'I am *much pleased* to hear what you say about Branchiae of *Balanus*: It is exactly my view, the ovigerous frena having been metamorphosed to this new function, and the glands having become aborted. And the case strikes me as rather pretty' (Darwin's emphases). Unfortunately Huxley's letter did not survive, but since he queried Darwin over the latter's observations on the function of the 'auditory sacs' and the supposed connection between the cement glands and ovarian tubules, Darwin would naturally have been pleased over anything favorable Huxley said regarding his inferred transformation of frena.

Darwin's hypothesis that the frena had been transformed into branchiae survived for nearly 130 years, until Walker (1983) published a detailed study on the branchiae and frena of thoracicans in which he reported that a relatively primitive sessile barnacle from Australia, *Catomerus*

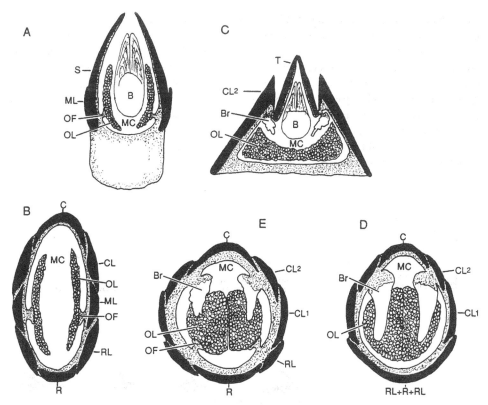

Figure 11. Ovigerous frena and branchiae: Darwin observed most pedunculates have 'ovigerous frena,' mantle outgrowths that hold the egg masses (A, frontal, and B, plan schematic of a scalpellomorph). He also saw similar outgrowths in balanomorphs that appeared to be respiratory, 'branchiae' (C, frontal, D, plan schematic of a balanoid). He inferred branchiae were frena with a new function. However, *Catomerus polymerus*, a primitive balanomorph, has both frena and branchiae (E). (Br = branchiae, CL = carinolatus, CL^1 = 1st or primary carinolatus, CL^2 = secondary carinolatus, MC = mantle cavity, ML = median latus, OF = ovigerous frenum, OL = ovigerous lamella, r = rostrum, RL = rostrolatus, RL+R+RL = compound rostrum, S = scutum, T = tergum; modified from Walker 1983.)

polymerus had both sets of organs in the same individual. This was the deathblow to Darwin's frena and branchiae hypothesis: The frena disappeared without a trace in most balanomorphs as well as in the verrucomorphs, and the branchiae apparently appeared de novo in balanomorphs.

5.5.3 *Calcareous bases*

The basis of a sessile barnacle, the bottom of the frustum formed by the wall plates cemented to the substrate, can be either membranous or calcareous, or a combination of the two. Darwin initially thought it plausible that the calcareous basis evolved from a fusion of peduncular scales of a pedunculate ancestor, potentially another transformation hypothesis. However, it soon became apparent the membranous basis was more prevalent in primitive than in higher sessile barnacles, so he did not pursue the notion further. On the other hand, when Darwin encountered a calcareous basis in the large species *Chthamalus hembeli* (= *Euraphia hembeli*)[7], a member of his relatively primitive Chthamalinae, he considered it a 'false' basis because he thought it was formed by an inflection of the basal margin of the wall (Fig. 12): 'The most remarkable character is, that all these old specimens had a flat, wide, calcareous basis, which is absolutely continuous with the inner lamina of the parietes, whereas in the younger specimen there was no appearance of any tendency in the parietes thus to grow inflected. There can be hardly any doubt that in a series of specimens some would be found with the parietes first forming a flat narrow ledge round the true basal membrane (as in the following species, *C. intertextus*); and that in others, this ledge would be wider and wider, till its edges met in the middle, and coalesced into a continuous plate' (LC2:466).

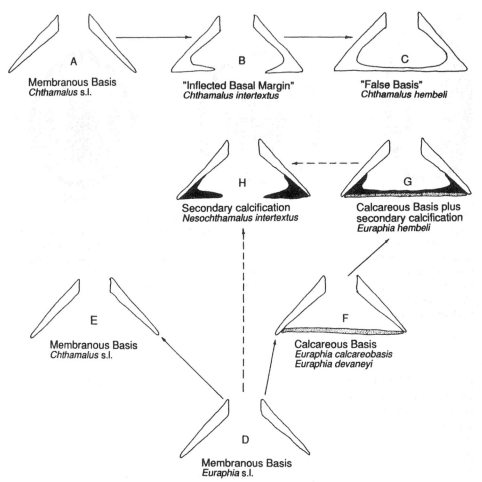

Figure 12. The basis in sessile barnacles: Darwin observed several kinds of bases, the most common being either membranous (A, D & E), or calcareous (F). Two other kinds of calcareous bases Darwin thought formed by an an 'inflection of the basal margin of the wall' (B) or by the inflections meeting in the center to form a 'false basis' (C), with an evolutionary series from A to C. However, we now know that the inflected margin and the false basis are formed by secondary clacification (G & H), where H could evolve from G either by loss of the basis or independently. (True basis stippled, secondary calcifications black; after Newman 1961, see Section 5.5.3.)

To Darwin, *Chthamalus intertextus (= Nesochthamalus intertextus*[7] displayed the missing stage in the hembeli series, viz., the encircling calcareous basis between no calcareous basis in the juvenile and the full calcareous but false basis of the adults. This highly plausible scenario prevailed until it was discovered 1) that the juveniles of *C. hembeli* actually had a normal but thin calcareous basis (it commonly remains with the substrate when the animal is removed and, since Darwin did not collect the specimens himself, this may account for why he did not see it), and 2) that the so-called inflected wall of large individuals did not develop until after full size had been reached, i.e., the basis was neither 'false' nor was there an intermediate, incomplete basis during ontogeny in this species (Newman 1961). What then was the 'inflected wall' Darwin saw in both *hembeli* and *intertextus*?

Darwin determined how sessile barnacles generally grow, 'diametrically', i.e., growth increments being made 1) between adjacent plates enabling them to increase the diameter of the frustum, from top to bottom, and 2) along the basal margin enabling the frustum to increase in height. Diametric growth is possible as long as growth takes place between adjacent plates and between the plates and the basis. The orifice surrounding the opercular valves also enlarges with diametric growth. However, Darwin explained how, when growth between the plates ceases or the plates become concrescent early in ontogeny, overall growth becomes 'monometric,' i.e., while the frustum can only grow higher and wider by increments around its base, the orifice remains the same sized unless its upper margin is gradually eroded away, as he observed was the case in some species of *Tetraclita* (LC2:324).

Although Darwin discovered and explained the relationships involving diametric and monometric growth in sessile barnacles, he apparently did not appreciate that these same relationships were at work when the so-called inflected wall developed in *hembeli* and *intertextus*. In both, diametric and monometric growth are precluded by 'the inflected basal margin of the wall,' whereby the frustum can no longer enlarge from top to bottom by growth between plates, nor can it increase in height by additions around the basal margin. The successive layers of secondary calcification are applied from within to both the interior of the wall and the substratum. This makes the entire structure monolithic, i.e., capable of neither monometric nor diametric growth. While a body chamber increases in volume with monometric as well as diametric growth, it is decreased by the sort of secondary internal calcification seen in these two species. Since the ultimate decrease can be substantial, the capacity of the body chamber, prior to the onset of the secondary internal calcification, must exceed that needed later in life (Newman 1961).

5.6 *Alcippe (= Trypetesa) and Cryptophialus*

It is difficult in hindsight to conceive of how Darwin could have had such difficulties with the affinities of *Trypetesa lampas* (Hancock 1849) and *Cryptophialus minutus* Darwin, 1854, and how, despite their morphological and sexual similarities, he ended up making *Cryptophialus* the sole representative of his order, the Abdominalia (= Acrothoracica). The similarities include not just general form, size, armament, burrowing habitat and the deployment and number of limbs, but the same sexual system and very similar males. As for the four pairs of limbs in *Trypetesa* and *Cryptophialus*, Darwin interpreted the first pair in *Trypetesa* as the first cirri (which they are) and in *Cryptophialus* as the first maxillipedes; and rather than the fourth, fifth, and sixth cirri as in normal cirripedes, he interpreted the terminal three pairs as the fifth and sixth cirri plus the caudal appendages in *Trypetesa* and as three pairs of abdominal cirri in *Cryp-*

tophialus. Because of these supposed differences between the two forms, Darwin felt compelled to create a new order for *Cryptophialus*.

A good part of Darwin's problem was that the two forms represented the tips of two divergent branches of the Acrothoracica, branches that would eventually become known as the suborders Apygophora and Pygophora Berndt, 1907. When Darwin first wrote Hancock in 1849 (C-1253) he told him he had a form from South America, allied to Hancock's form from England, which he thought constituted a new family. The reductions not only made that analysis difficult but he got involved in supposed differences in segmentation (Section 5.4.2). Even after discovering in 1853 that both forms had separate sexes, and that their males were not only extremely similar to each other but were very different from those of *Ibla* and *Scalpellum* (Section 5.4.3), he placed Hancock's form in the Thoracica and proposed a new order, the Abdominalia, for his *Cryptophialus* (CL2).

Darwin's contemporaries had their suspicions. Gerstaecker (1866), with nothing new to go on, argued for close affinities between *Trypetesa* and *Cryptophialus*, and Noll (1872), who described the first less specialized allied genus, *Kochlorine*, followed suite. Lack of knowledge of even more generalized allies was short-lived; the two most primitive of the seven pygophoran genera, *Lithoglyptes* with six pairs of cirri and *Weltneria* with six pairs of cirri plus caudal appendages, were soon discovered. Thus the basic segmentation of the Abdominalia was like that of ordinary cirripedes. Therefore Berndt (1903:436) concurred with Noll's recommendation that the Cryptophialidae Gerstaecker, 1866, as well as the Alcippidae Hancock, 1849 (= Trypetesidae Stebbing, 1910) be assigned to the Abdominalia. However, in light of the lack of an abdomen, Gruvel (1905) changed the name from Abdominalia to Acrothoracica, to contrast with Darwin's Thoracica (acro-, referring to all but the first pair of cirri being clumped at the posterior end of the thorax), rather than adopt Hancock's name 'Cryptosomata,' which had been intended to conform with Leach's classification of 1825 (see Fig. 26).

If Darwin had followed his intuitions rather than doggedly adhering to his perception of certain homologies (see Section 5.4.2), he would have had it right for he wrote (LC1:565, footnote): 'I may add that I have several times tried to persuade myself, with no success, into the belief that I have somehow misunderstood the homologies of the thoracic segments and cirri of *Trypetesa* and *Cryptophialus*; for if this were so, the two genera could be brought into much closer relationship...'

5.7   *Proteolepas, sole representative of Darwin's order, the Apoda*

In LC1, Darwin mentioned a new and primitive genus, *Proteolepas*, and he described it at some length in LC2 (p.589) as the sole representative of his new order, the Apoda. He had but a single specimen, a parasite taken from the mantle cavity of a pedunculate barnacle he described from the Caribbean (LC1:165). The sole specimen of the 'legless' parasite was attached to the host by two slender filaments each terminating in what he believed were typical cyprid 'antennae' (Fig. 13). In a letter to Agassiz in October 1848 (C-1205), he mentioned that this was the only character indicating that *Proteolepas* was a cirripede. Darwin dealt with the fact that the supposed antennae emanated from the 'dorsal surface' on the second thoracic segment as 'monstrous and incredible an inversion of the laws of nature, as those fabulous half-human monsters, with an eye seated in the middle of their stomachs' (LC2:602), and he went about explaining how this disconcerting anomaly may have evolved. There was however a second anomaly: The specimen had an additional two thoracic segments, bringing the number to eight (see Section 5.4.2). From this Darwin inferred that it was closer to the archetype than were the remainder of

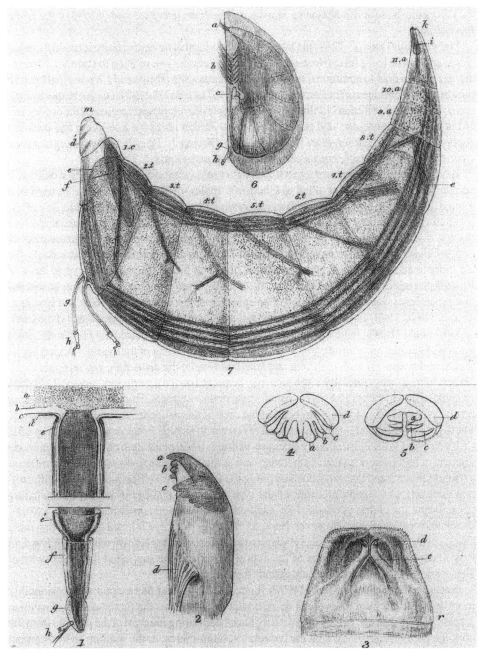

Figure 13. *Proteolepas bivincta*, sole represenative of the order Apoda: The single specimen (7), a parasite from a pedunculate barnacle, had three pairs of muscles running to each of the pair of gnathites (d), and 11 trunk segments (1 cephalic, 2-8 thoracic, 9-11 abdominal). A limb pair stemming from the 2nd trunk segment was used for attachment, the distal portion of which (1f, g, h 7h) appeared to Darwin identical to the attachment organ seen in cyprids (see Fig. 14); 2, one pair of jaws inferred to include the mand. plus 2 max. (a, b, c); 3, the orientation of the jaws to the mouth field; 4-5, comparison of the mouth field of ordinary cirripedes and the inferred fusion of mouthparts and the rotation of their common support; and 6, its hypothetical cyprid. *Proteolepas* is now recognized as an epicaridean isopod (from pls. 24, 25, Darwin 1854; see Section 5.7).

the Cirripedia, and for this reason he considered the form primitive and worthy of the name *Proteolepas*.

Darwin wrote Dana (C-1259) in October 1849, saying that he agreed cirripedes were crustaceans, and that he had a form (*Proteolepas*) that bore 'about same relation to common Cirripedia, as *Lernaea* does to common Crustacea.' To Linneaus, the affinities of *Lernaea* had been as puzzling as the affinities of the barnacles (*Vermes mollusca* and *Vermes testacea*, respectively), and it was not until 1832 that Nordmann's researches on development confirmed the crustacean affinities of the Lernaeidae, and not until 1854 that Zenker allied the free-living and parasitic copepods under the Entomostraca (Calman 1909). Obviously, Darwin was working at a time when the nature of parasitic crustaceans was not well understood.

If only Darwin had possessed a specimen of *Proteolepas* brooding embryos, he would have known it was an epicaridean allied to Goodsir's 'males' of *Balanus*, which he himself had refuted (Section 5.4.3). This single observation would have saved him the considerable effort required to explain how the rest of the cirripedes jibed with an 8-segmented archetype (see Section 5.4.2). It might also have saved him 1) from misinterpreting the segmentation of *Cryptophialus*, which led to its separation from *Trypetesa* and the misnomer Abdominalia for the order to accommodate it, and 2) from the bother of illustrating what the cyprid larva of *Proteolepas* might have looked like (Fig. 13.6). Since Darwin's time investigators suspected that *Proteolepas* was either a copepod or an epicaridean (Newman et al. 1969); and Bocquet-Védrine (1972, 1979) and Bocquet-Védrine & Bocquet (1972) demonstrated that it is the latter.

How could Darwin have made such a colossal error? He had placed his faith in the distal portion of the attachment limb, which he thought identical to that of the cyprid: 'Perceiving its very singular nature, I took such care and length of time in the dissection, and repeated every observation so many times that... reliance may be placed on the description...given' (LC2:589, footnote). Darwin thought he saw a 'penultimate or disc segment... by which the antennae are firmly cemented down to the rostral end of the sack of the cirripede...on which it is parasitic.' The disc-like penultimate segment compelled Darwin to include *Proteolepas* in the Cirripedia.

If one inspects the illustration by Bocquet-Védrine (1972) of the limb on the second thoracic segment of the epicaridean that is otherwise very similar to Darwin's *Proteolepas*, it is difficult to believe that she and Darwin were describing the same structure. Bocquet-Védrine's illustration is clearly of a subchelate limb of the type seen on a number of peracarids including epicarideans, not of the cyprid-like first antenna Darwin carefully described and had cursorily illustrated for *Proteolepas*. However, Bocquet-Védrine (1972:2147) noted that if the first three articles of the attachment limb (dactyl-carpus-propodus) were broken off, what remains would look much like the first antennae of the cyprid larvae and she suggested this must be what Darwin saw. Her figure does not substantiate this however.

Bocquet-Védrine & Bocquet's (1972) figures of the second thoracopod are considerably more detailed, and from them it can be ascertained that if either the first two or the first three articles were broken off, as they might easily have been during removal of the parasite from the host, the portions remaining with the parasite (ischium-merus, or the ischium-merus-carpus) both bear a striking resemblance to *cyprid* antennules, including the penultimate segment or attachment disc (Fig. 14). Whatever shreds of chitin might have remained hanging from the 'attachment disc' in Darwin's specimen could easily have been interpreted as the cement he thought had held the parasite to the host. This must account for Darwin's mistake: He actually saw something that looked very much like cyprid first antennae.

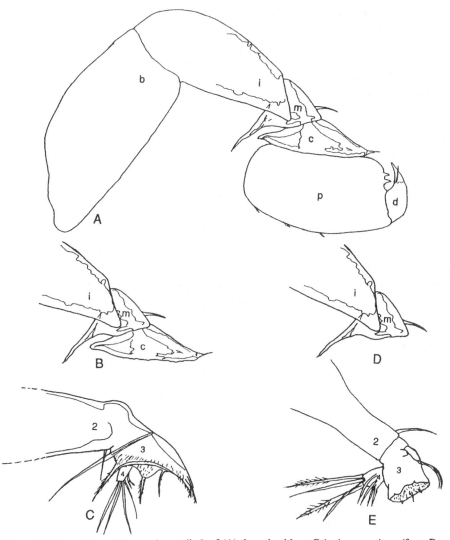

Figure 14. Comparison of the attachment limb of (A) the epicaridean *Crioniscus equitans* (from Bocquet-Védrine & Bocquet 1972) with the antennule of cyprids of (C) *Ornatoscalpellum gibberum* (from Nilsson-Cantell 1921) and (E) *Semibalanus balanoides* (modified from Nott & Foster 1969). If the dactyl and propodus (B), or these and the carpus (D), were broken off when the parasite was removed from the host, then the remnant looks like cyprid anntennules. (compare B-C and D-E; see Section 5.7; b = basis, c = carpus, d = dactyl, i = ischium, m = merus, p = propodus.)

5.8 *Genera of few species, many varieties and related matters*

Darwin struggled with the problem of distinguishing between varieties and species and, as mentioned above (Section 2.4), he ended up with relatively few genera, some with few species but many varieties (subspecies). He worried about this for he wrote (LC2:190): 'I hope that, owing to having examined a vast number of specimens of the most varying species, I have not fallen into very many errors. I have endeavoured to err on the side of making too few species instead of too many species... I would have gladly divided this genus (*Balanus*), already including 45 species, into smaller genera; but...I have been compelled to form my sections on characters not absolutely invariable, and far from obvious.' He divided *Balanus* into seven sections (effectively subgenera) and his authority on the matter prevailed for many years. In order to adhere to this arrangement, a cumbersome quadrinomial system developed in some species groups, such as in *Balanus tintinnabulum* in his Section A of *Balanus*, and in *Tetraclita squamosa* (Sections 5.8.3, 5.8.4).

Most of Darwin's varieties or subspecies have, in relatively recent years, become recognized as species, and numerous new species have been assigned to their respective species groups. Where Darwin's species groups were well defined, most were eventually designated as subgenera before being elevated to genera. As a simple illustration, consider two forms from California recognized as subspecies of Darwin's 'wide-ranging' species, *Tetraclita squamosa* and *Balanus tintinnabulum*. The Californian subspecies were designated as *Tetraclita (Tetraclita) squamosa rubescens* Darwin, 1854, and *Balanus (Megabalanus) tintinnabulum californicus* Pilsbry, 1916, by Pilsbry (1916) and referred to as such in the handbook, *The Intertidal Invertebrates of the Central California Coast*, as recently as 1975 (Newman 1975). Today these two taxa are known simply as *Tetraclita rubescens* Darwin and *Megabalanus californicus* Pilsbry (Newman & Abbott 1980). While it is certainly true that some of the delay in making these changes was due to lack of ecological as well as morphological evidence needed to demonstrate that one was dealing with well defined species groups and species, it was also due to faithful adherence to the system Darwin developed.

Not all workers have been happy with the apparent progress. Crisp & Fogg (1988), for example, voiced concerns over generic name changes among the barnacles and suggested, for the sake of stability, a return to Darwin's system. They claimed that the 'current half-life of many generic names is only about 30 years,' but this is certainly not true as far as the barnacles are concerned. Prior to Darwin there had been no in-depth attempt to revise the barnacles, and one of Darwin's numerous contributions to the field was the working out of synonymies among genera as well as species, a complex and tedious task. He ended up with 102 extant species of balanomorph barnacles divided between 14 genera and two subgenera (*Acasta* as a subgenus of *Balanus* (LC2:303), and *Creusia* as a subgenus of *Pyrgoma*), and for the last 100 years, or so, all 16 of these taxa have been recognized as full genera.

There are now over 530 species in the Balanomorpha, more than five times as many as were known to Darwin, but the additions have not been uniform. For 12 of the genera there has been little or no change in the number of species, and for 3 there have been increases: *Balanus* from 45 to some 135 species (but to more than 275 before subdivision of the genus by Newman & Ross 1976), *Acasta* from 9 to 30 species, and *Chthamalus* from 8 to 17 species (Section 5.8.6). Only one, *Pyrgoma*, has suffered a substantial loss, from nine to one species as a result of compelling evidence for convergence among its members (Ross & Newman 1973; Section 5.8.5). Thus, Crisp & Fogg's (1988) allegation applied to sessile barnacles – that since Darwin there has been a 30 year half-life among the genera – is simply not true, and the same can be said for pedunculate and acrothoracican barnacles. The principal changes made in Darwin's systematic work have involved 1) distinguishing species from varieties (Section 5.8) and 2) the formal recognition of his species groups as genera and of higher categories to accommodate them (Section 5.10).

The following examples taken from Darwin's Balanidae illustrate the extent to which the ascribing differences between populations to geographical variation within a species biased

Darwin's perception of variability in natural populations. Although this has been pointed out with a few examples by Southward (1983), a fuller rendition of examples will illustrate the enormity of the matter and the extent to which Darwin's conservative view of species- and genus-group taxa blurred his perception of the biogeography of the cirripedes (Section 5.12).

On the other hand, the basis for any classification is the degree of relatedness between species. Therefore, the following examples are instructive when it comes to understanding that, in spite of his incorrect interpretation of evidence for variation within 'wide ranging' species, Darwin's awareness of natural groups (in the form of informal numerical and alphabetical sections as well as genera and subfamilies in the his Balanidae) formed the basis for the current hierarchial levels in the Balanomorpha (Section 5.10.3).

5.8.1 *Varieties of Balanus amphitrite*
Pilsbry (1916: 93-94) pointed out that the 'The definition of the subspecies of *Balanus amphitrite* is a very intricate problem (and that) first of all the Darwin collection must be restudied and type localities for his varieties selected'. This was accomplished by Harding (1962) who made a major advance in defining species groups within Darwin's *amphitrite* complex. However, Harding left us with a conflicting description of the nominotypical subspecies, *Balanus amphitrite amphitrite*. This was because Darwin confounded two species under *Balanus amphitrite communis* (= *Balanus amphitrite* s.s.), and this was discovered by Utinomi (1967) and confirmed by Southward (1975). Henry & McLaughlin (1975) further revised the *amphitrite* complex. They recognized some 19 species and a few subspecies, but none of the latter were those recognized by Darwin. Today, more than 100 years after Darwin, of the nine varieties of *Balanus amphitrite* he described, one has been recognized as two species, three have been elevated to full species, and the remaining ones have been distributed between them (Fig. 15).

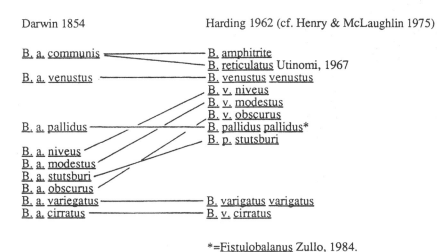

Darwin 1854   Harding 1962 (cf. Henry & McLaughlin 1975)

B. a. communis — B. amphitrite / B. reticulatus Utinomi, 1967
B. a. venustus — B. venustus venustus
B. v. niveus
B. v. modestus
B. v. obscurus
B. a. pallidus — B. pallidus pallidus*
B. p. stutsburi
B. a. niveus
B. a. modestus
B. a. stutsburi
B. a. obscurus
B. a. variegatus — B. varigatus varigatus
B. a. cirratus — B. v. cirratus

*=Fistulobalanus Zullo, 1984.

Figure 15. Varieties of *Balanus amphitrite*: Observe that Darwin's 9 varieties are now distributed between 5 species, and one of these is now in its own genus. It took over 100 years to discover that most of the variation in *B.a. communis* (= *B. amphitrite* s.s.) was due to Darwin confounding 2 species under 1 name. (Utinomi 1967; see Section 5.8.1.)

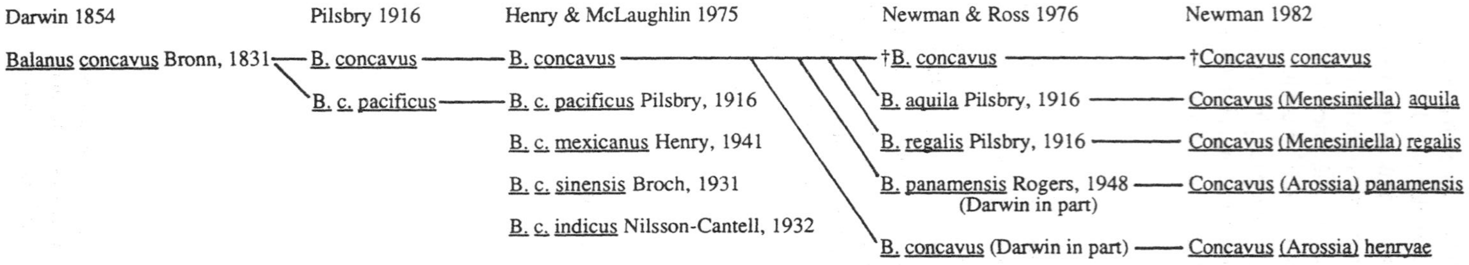

Figure 16. Variation in *Balanus concavus*: Darwin did not specify varieties, but he considered *B. concavus* another example of the high variability found in wide ranging species. However, he had mixed 2 species groups, *B. concavus* s.s. and *B. pacificus. Concavus* includes numerous fossils species, and only 4 extant species divided between 2 subgenera and Darwin had studied 2 if not 3 of them (*C. panamensis, henryae,* and perhaps *aquila*; see Section 5.8.2).

5.8.2 *Confounding of several species under Balanus concavus Bronn, 1831*

Darwin discusses geographical variation in *Balanus concavus* Bronn, 1831 at some length. The species was originally described from the Tertiary of Europe, and Darwin believed it was represented by extant populations in the Philippines and Australia as well in the eastern Pacific. Under this one, variable species, he observed two extremes: Varieties that were moderate in size and similar to certain varieties of *B. amphitrite*, and varieties similar to the great fossil Coralline Craig specimens with their ribbed walls, very oblique radii, and coarsely striated scuta.

From the onset, *Balanus concavus* gave Darwin trouble: 'The recent specimens... from Panama and California, which, though differing greatly in colour, resemble each other in their scuta (and by) characters which first appeared of high specific value; but I soon found other specimens from Panama in which these peculiarities were barely developed. I then examined a single specimen from the Philippine Archipelago, resembling in external appearance one of the Panama varieties, but differing (in a few characters)...I therefore considered this as a distinct species. I then examined a single white rugged specimen from...Peru, which differed from the Philippine specimen in the shape of well-defined denticulations...and some other trifling respects...with considerable doubt, I also named this as distinct. But when I came to examine a large series of fossil specimens from the Coralline Crag of England, from northern Italy, from Portugal, and from the southern United States, I at once discovered that the form of the denticuli on the scuta was a quite worthless character... Hence I have been compelled to throw all of these forms...into one species...

'I enter into the above particulars, on account of, in the first place, its offering an excellent example how hopeless it is in most cases to make out the species (and) as a good instance of the amount of variation which seems to occur in most of the species which have very extensive ranges' (LC2:235-236).

The large fossil varieties with ribbed walls turned out to be represented by four extant species in the eastern Pacific that are related to certain fossils in eastern North America and Europe, and these have been equally divided between two subgenera of *Concavus* Newman, 1982 (Fig. 16). Darwin's other group within *Balanus concavus*, including those less robust species with similarities to certain varieties of *Balanus amphitrite*, is also represented in the eastern Pacific by the extant *Balanus pacificus* Pilsbry, 1916, and they will be included in a new genus (Zullo, 1992a).

The existence of the Australian and Philippine populations related by Darwin has never been confirmed, but I suspect that if the localities were correct the specimens were of the *Balanus pacificus* form rather than the *concavus* form. Pilsbry (1916) described the former as *B. concavus pacificus* from California. Darwin actually had a specimen, apparently from Panama, of the very closely related form described as *B. c. mexicanus* by Henry (1941) from West Mexico. In fact, all three of the Panamanian varieties of *B. concavus* studied by Darwin turned out to be distinct species, the other two were *B. panamensis* Rogers, 1948 (= *B. eyerdami* Henry, 1960) and a new species, *Concavus* (*Arossia*) *henryae* Newman, 1982. However, the sole specimen of *Concavus henryae*, while labeled as from Panama on the pallet Darwin studied, was associated with another barnacle, *Notobalanus flosculus* (Darwin), on a gastropod shell, *Trochita trochiformis*, collected by Hugh Cuming who also collected the 'Philippine' specimen. Both *N. flosculus* and the gastropod are known only from Peru and Chile, and Darwin (1854:240) mentions having a 'Peru specimen (of *B. concavus*)... associated with *B.* (= *Notobalanus*) *flosculus*'. Therefore the type locality of *C. henryae* must have been in Peru or Chile, and the species is now known from as far south as Chiloe (J.-C. Castilla, pers. comm.).

If it seems that Darwin became confused, I think it fair to say that it was in part due to an apparent mix-up of a number of Cuming's specimens. However, even under the best of circumstances, the problem of distinguishing between geographical races and distinct species by morphology alone can be a difficult matter. Fortunately today, much of the difficulty has been overcome by the application of ecological, isotopic, and biochemical techniques (Dando & Southward 1980, Section 5.8.6; Newman & Killingley 1985).

5.8.3   Section A of Balanus and the varieties of Balanus tintinnabulum

Darwin divided *Balanus* into 6 alphabetical sections, and Section A included 8 species, 1 of which, *Balanus tintinnabulum*, had 11 varieties (Henry & McLaughlin 1986; Fig. 17). Darwin's 11 varieties were treated as subspecies and, so the nomenclature would formally distinguish the section's members from those of the other five sections of *Balanus*, Hoek (1913) proposed the subgenus *Megabalanus* to accomodate them. Until recently, 10 of Darwin's 11 extant varieties of *Megabalanus* were considered subspecies, and at least 8 additional new subspecies were described under this quadrinomial system, the most recent being *Balanus (Megabalanus) tintinnabulum linzei* Foster, 1979.

That most if not all of the so-called varieties or subspecies of *B. (M.) tintinnabulum* were good species came under scrutiny when three East Pacific subspecies, *B. (M.) t. californicus, peninsularis*, and *concopoma*, were encountered living side-by-side on a fishing boat in the Gulf of California (Newman & Ross 1976), a circumstance essentially identical to one Darwin

Darwin 1854	Present
Balaninae ———————————	Balanidae
Section A ———————————	Megabalaninae Newman, 1979
	Megabalanus Hoek, 1913
Balanus tintinnabulum communis Darwin	M. tintinnabulum (Linneaus, 1757)
B. t. coccopoma Darwin	M. coccopoma (Darwin)
B. t. concinnus Darwin	M. concinnus (Darwin)
B. t. crispatus (Schöter, 1786)	M. crispatus (Schöter)
B. t. dorbignii Chenu, 1843	M. dorbignii (Chenu)
B. t. intermedius Darwin	=M. tintinnabulum (L.)
B. t. occator Darwin	M. occator (Darwin)
B. t. spinosus Brugière, 1789	M. spinosus (Brugière)
B. t. validus Darwin	M. validus (Darwin)
B. t. vesiculosus Darwin	M. vesiculosus (Darwin)
B. t. zebra Darwin	M. zebra (Darwin)
B. ajax Darwin	M. ajax (Darwin)
B. echinata (Spengler, 1790)	=M. spinosus (Brugière)
B. tulipiformis Darwin	M. tulipiformis (Darwin)
B. vinaceus Darwin	M. vinaceous (Darwin)
	M. stultus (Darwin) (*from sect. B)
B. cylindricus (Gmelin, 1790)	Austromegabalanus cylindricus (Gmelin)
B. nigrescens (Lamarck, 1818)	A. nigrescens (Lamarck)
B. psittacus (Molina, 1782)	A. psittacus (Molina)
B. decorus (Darwin)	Notomegabalanus decorus (Darwin)

Figure 17. Darwin's section A and varieties of *Balanus tintinnabulum*: Section A is now recognized as a family-group taxon of several genera mostly confined to the southern hemisphere. His varieties of *Balanus tintinnabulum* now are recognized as good species of a distinct genus *Megabalanus*, primarily confined to the tropics and including the only obligate coral-inhabiting form, *M. stultus*, from Darwin's Section B (see Section 5.8.3).

had encountered (see LC2:197). Since they could not be ecotypic variants, since intergrades were unknown, and because Darwin's belief that geographic variants may not interbreed when provided the opportunity was no longer acceptable, the alternative was that these three forms were good species. Furthermore, these three species plus the others of Section A formed an exceptionally well defined species group. Therefore, Hoek's subgenus *Megabalanus* was raised to genus (Newman & Ross 1976).

With further revision, Darwin's Section A became a subfamily, and his original 8 species plus all varieties recognized as species were divided between 3 genera including 23 extant species (Henry & McLaughlin 1986; Fig. 17). The zoogeographic significance of the group also became evident (Newman 1979a, 1991; Section 5.12.1).

5.8.4 *Darwin's varieties of Tetraclita porosa (= squamosa), and other forms now mostly attributed to the Tetraclitidae*

The history of Darwin's varieties of *Tetraclita porosa* (Gmelin, 1789) [= *squamosa* (Brugière, 1789)] is similar to that for those of *Balanus amphitrite* and *Balanus tintinnabulum* (Fig. 18). Of the original seven subspecies, five are recognized as distinct species today (Newman & Ross 1976, Yamaguchi 1987; Fig. 18). However, of the original eight species, only two, *T. serrata* Darwin and *T. squamosa*, are still included in *Tetraclita*. *Tetraclita rosea* (Krauss, 1848) is now the type for the genus *Tesseropora* Pilsbry, 1916, subfamily Tetraclitinae; *T. purpurascens*

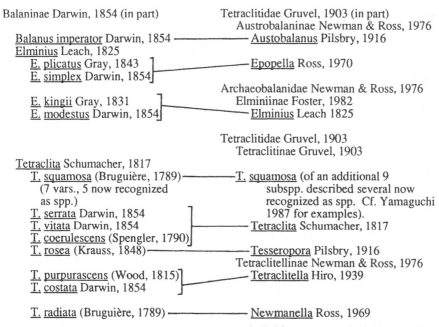

Figure 18. Varieties of *Tetraclita porosa (= squamosa)* and allied forms now are divided between 3 subfamiles of Tertraclitidae and 1 subfamily of Archaeobalanidae. Of Darwin's original 7 varieties of *T. squamosa*, 5 are now species and additional species have been described. Darwin did note similarities between *B. imperator* and *Tetraclita*, but he chose to include it within *Balanus*. However, it is now known as a 6-plated tetraclitid which, along with 2 of 4 species of *Elminius* studied by Darwin, is assigned to the subfamily Austrobalaninae (see Section 5.8.4).

(Wood, 1815) is the type for the genus *Tetraclitella* Hiro, 1939; and *T. radiata* (Brugière, 1789) is the type for the genus *Newmanella* Ross, 1969. While *Tetraclita* and *Tesseropora* are in the subfamily Tetraclitinae Gruvel, 1903, *Tetraclitella* and *Newmanella* were placed in the Tetraclitellinae (Newman & Ross 1976).

Darwin (LC2:42) inferred that the four-plated wall of *Tetraclita* was likely derived from a six-plated form comparable to *Balanus* because the rostrum is already compound; and all that was needed was to loose the carinolatera, either by fusion to adjacent plates or, more simply since they appear late in ontogeny, by abortion (see Section 5.9.1). In light of this view, it is interesting to note that when he described *Balanus imperator* he observed (LC2:290): 'In the nature of the basis; in the structure, to a certain extent, of the walls of the shell; in the narrowness of the carino-lateral compartments; in the elongation of the third pair of cirri; in the crests for the rostral and lateral scutal depressors, *B. imperator* comes nearer to the genus *Tetraclita* than does any other species of *Balanus*.' Indeed, Darwin effectively had the ancestral tetraclitid in his hand, and had he found it suitable to recognize a subfamily for *Tetraclita*, perhaps he would have proposed a new genus for *imperator* and placed it there.

Pilsbry (1916:218) was likewise struck with the uniqueness of Darwin's *Balanus imperator*, and for it he erected the subgenus *Austrobalanus* (Fig. 18) in which he included two other but admittedly disparate Darwinian species from the Southern Hemisphere, *Balanus flosculus* and *B. vestitus*. Pilsbry, unlike Darwin, was apparently unaware of the tetraclitid affinities of *B. imperator*, but as it has turned out, Darwin's *Austrobalanus imperator* has been moved into the Tetraclitidae, under the subfamily Austrobalaninae Newman & Ross, 1976. *B. flosculus* and *B. vestitus* do not have these affinities, and they have since been placed in an austral archaeobalanid genus, *Notobalanus*.

There was another four-plated genus, *Elminius* Leach, 1825 that included two species described by Gray (1831, 1843) and to it Darwin added two new species (Fig. 18); all these species were from the Southern Hemisphere. Since then, it has been determined that *E. kingii* Gray, 1831 and *E. modestus* Darwin, 1854 are archaeobalanids. Foster (1982) erected the Elminiinae to accommodate them. *E. plicatus* Gray, 1843, and *E. simplex* Darwin, 1854 have been placed in the genus *Epopella* Ross, 1970 and, along with *Austrobalanus imperator*, in the tetraclitid subfamily Austrobalaninae. Further adjustments are likely as our knowledge increases.

5.8.5   *The coral balanids, Pyrgoma and the varieties of Pyrgoma (Creusia)*

Before Darwin (1854), a number of genera of coral barnacles had been described, and the group was reviewed by Gray in 1825. Gray recognized six genera, one of which was divided into two subgenera. Darwin reduced these to one genus of two subgenera by excluding all six-plated forms (*Acasta* and *Conopea*), and by placing all one-plated forms in the *Pyrgoma (Pyrgoma)* and all four-plated forms in *Pyrgoma (Creusia)*. *Pyrgoma (Pyrgoma)* consisted of 9 species, some with varieties, while the *P. (Cresuia)* consisted of 1 species and 11 varieties (Fig. 19). Cirripedologists of the day were not pleased; *Creusia* was soon returned to full generic status, and over the years, not only was it concluded that most of Darwin's varieties were good species but many new species were added. Furthermore, with time and study, species groups were identified among the 4-plated and 1-plated forms, and for most generic names were to be found among those placed in synonymy by Darwin. This activity by several authors, mainly Annandale (1924), Nilsson-Cantell (1938) and Hiro (1938), precipitated the revision of Ross & Newman (1973) and the classification of Newman & Ross (1976), and the coral barnacles are now divided between more than the nine genera noted here (Fig. 19; see Anderson 1992).

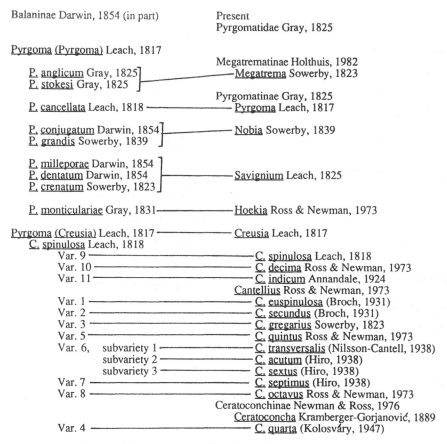

Figure 19. Darwin lumped the numerous species and several genera of coral-inhabiting balanids into a single genus that he divided into 2 subgenera. He recognized 9 species in the subgenus *Pyrgoma (Pyrgoma)* and a single species with 10 varieties and 2 subvarieties in *Pyrgoma (Creusia)*. These are now divided among 3 subfamilies and 8 genera, and other genera and species have been described (see Section 5.8.5).

In no other group of barnacles did Darwin's judgments lead to such an over-simplification, which is puzzling because he had already determined that 4-platedness was convergent in distantly related genera. In addition to the 4-plated coral barnacles there was *Tetraclita* Schumacher, 1817, and *Elminius* Leach, 1825, and Darwin himself described (1854) the 4-plated genus *Chamaesipho* and the 4-plated form of *Pachylasma*. It is puzzling why he lumped some rather divergent 4-plated forms of coral barnacles under varieties of a species that is now distributed among four genera. Equally puzzling is that he also lumped 1-plate, coral barnacles under a subgenus, but they are now distributed among six genera that apparently evolved independently from 4-plated ancestors at least two if not three times (Ross & Newman 1976).

5.8.6 *The Chthamalinae s.s. and the varieties of Chthamalus stellatus*
Darwin divided his Balanidae (=Balanomorpha) into the Balaninae and the Chthamalinae. The foregoing examples belonged to the Balaninae and are, for the most part, medium to relatively large, low intertidal and subtidal forms. The majority of the Chthamalinae s.s. (Fig. 20) are for

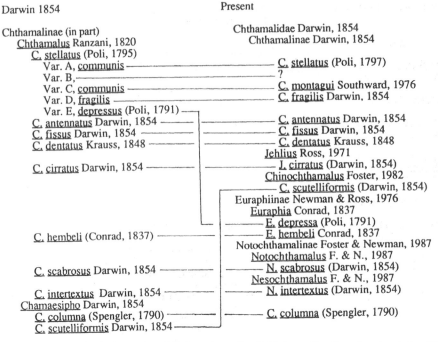

Figure 20. Darwin studied 8 species of *Chthamalus* and he divided one, *C.stellatus* into 5 varieties. Of these, 4 varieties are now good species and 1 of them has been transferred to *Euraphia*. Of the 7 other species he assigned to *Chthamalus* 4 have been divided among 4 genera in 2 subfamilies. One of the species of *Chamaesipho* he described turned out to have closer affinities with *Chthamalus* than with *Chamaesipho* (B. Foster, pers. comm.), and a new genus, *Chinochthamalus*, was erected to accomodate it (see Section 5.8.6).

the most part relatively small, high intertidal forms. There is little in the way of field characters to distinguish most species of the species-rich genus, *Chthamalus*, and even species that are in quite different sections of the genus, like *C. dalli* and *C. fissus* from the shores of western North America, are often difficult to tell apart even upon dissection (Newman 1975). However, despite their small size and nondescript general appearance, Darwin not only recognized several new species of *Chthamalus*, but he identified five varieties of *Chthamalus stellatus* (Poli, 1795). In the process, he confounded a previously described species, *C. punctuatus* (Montagu, 1803) (= *C. montagui* Southward, 1976) with *C. stellatus* and suppressed the genus *Euraphia* Conrad, 1837 (see Fig. 20 for the current status of Darwin's varieties of *Chthamalus*).

There are some 20 species of *Chthamalus* s.s. recognized today (Stanley & Newman 1980; Newman & Stanley 1981). These include four of eight species and four of five varieties originally recognized by Darwin; however, one of the species and one of the varieties are in different genera. His remaining four species and one variety are in different genera. The fate of one of his species of *Chamaesipho* is also interesting (Fig. 20).

5.9   Homologies of the plates

Many recently discovered forms have helped us clarify the identity and relationship of the barnacle plates (e.g., see Fig. 21). Without access to such information, Darwin established the

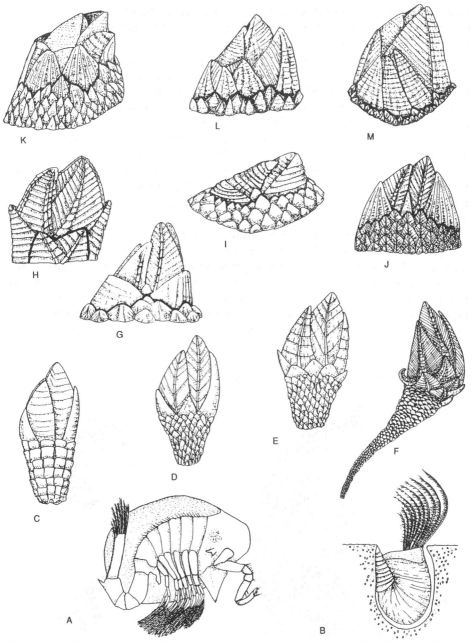

Figure 21. Genera discovered since Darwin (1855) relevant to our understanding the origin and evolution of burrowing and sessile forms (see Appendix III). A. *Synagoga* (after Newman 1974), B. *Weltneria* (after Newman 1982b), C. *Eolepas* (or *Archaeolepas*) (after Withers 1928), D. *Neolepas* (after Newman 1979b), E. *Scillaelepas* (after Newman 1980), F. *Newmanilepas* (after Zevina & Yakontova 1987), G. *Neoverruca* (after Newman & Hessler 1989), H. *Eoverruca* (after Withers 1935), I. *Neobrachylepas* (after Newman & Yamaguchi, in prep.), J. *Brachylepas* (after Woodward 1901), K. *Eochionelasmus* (after Yamaguchi & Newman 1990), L. *Chionelasmus* (after Nilsson-Cantell 1928), M. *Waikalasma* (after Buckeridge & Newman 1992).

nomenclature of shell plates of barnacles (Figs 22, 23, and 24) and determined the homologies of the plates between pedunculate and sessile barnacles. He went about it like an experienced comparative anatomist, utilizing ontogenetic changes and the presence and position of primordial valves as well as the shape and position of valves and musculature in adult forms.

The scalpellomorphs (Fig. 22), treated in his first two volumes (FC1 and LC1), have the most complex calcareous capitular armament of the pedunculate barnacles, and he determined their manner of growth, explaining why their plates have apical umbones. He named the scuta and terga (S-T), the median latera (L, but known in descriptive works as the upper latera (UL) when lower latera (LL) are present), and the surrounding whorl often comprising the rostrum, rostrolatera, carinolatera, and the carina (R-RL-CL-C). He also named the margins and angles of each plate as well as prominent features such as ridges, furrows, pits, and apodemes. In a few genera, such as *Pollicipes*, there are whorls of imbricating plates produced at the capitulo-peduncular junction that grow upward and overlap the major capitular whorl, but Darwin named only the subrostrum and subcarina (SR and SC).

The peduncle in scalpellomorphs is also armed with imbricating whorls of calcareous plates. Darwin explained that, like those of the capitulum, they generally appeared during development along the growth zone at the capitulo-peduncular junction, but once formed they were generally carried basally rather than apically with growth (LC1). The number of peduncular plates or scales per whorl can be rather large, at least in adults and, as with the plates of the minor capitular whorls, he did not apply names to them either. However, a system for keeping track of these minor plates on both the capitulum and the peduncle, at least during early ontogenetic

NOMENCLATURE OF THE VALVES.

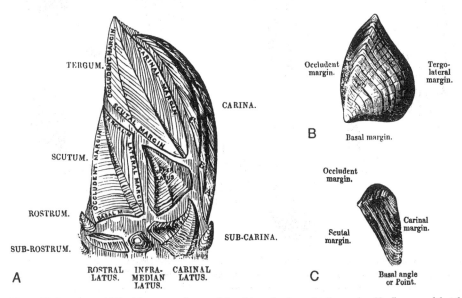

Figure 22. Darwin established the nomenclature of the plates of pedunculate barnacles. He discovered that the 5 principal plates in juveniles had 'primordial' valves, also found in verrucids and perhaps 1 chthamaline known at the time. These, plus muscle insertions and general topography, helped establish plate homologies between his families (from text-fig. in Darwin 1851, see Section 5.9). A. Capitulum of a scalpellomorph, B. Scutum of *Lepas*, C. Tergum of *Lepas*.

NOMENCLATURE OF THE SHELL OF A SESSILE CIRRIPEDE.

SHELL. Fig. 1.

Orifice of shell, surrounded by the *sheath.* *Sheath* formed by the *ala* (*a—a.*) and by portions of the upper and inner surfaces of the *parietes* (*p—p.*)

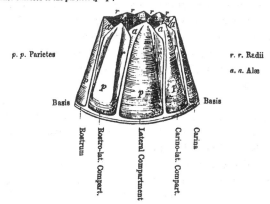

p. p. Parietes

r. r. Radii

a. a. Alæ

Basis

Basis

Rostrum

Rostro-lat. Compart.

Lateral Compartment

Carino-lat. Compart.

Carina

N.B. In Balanus, and many other genera, the Rostrum and Rostro-lateral compartments are confluent, and hence the Rostrum has the structure of Fig. 2.

COMPARTMENTS.

Fig. 2. Fig. 3. Fig. 4.

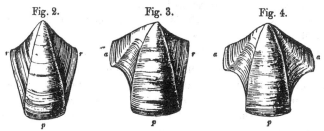

Fig. 2. Compartment with two radii, serving either as a Rostrum or Rostro-lateral compartment.
Fig. 3. serves as a Lateral and Carino-lateral Compartment. Fig. 4. serves as a Carina or Rostrum.

OPERCULAR VALVES.

Fig. 5. Scutum (internal view of). Fig. 6. Tergum (external view).

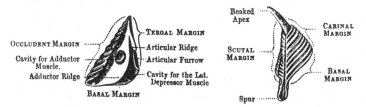

OCCLUDENT MARGIN

Cavity for Adductor Muscle.

Adductor Ridge

BASAL MARGIN

TERGAL MARGIN
Articular Ridge
Articular Furrow
Cavity for the Lat. Depressor Muscle.

Beaked Apex

SCUTAL MARGIN

CARINAL MARGIN

BASAL MARGIN

Spur

Fig. 7. Tergum (internal view).

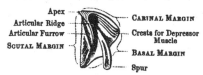

Apex
Articular Ridge
Articular Furrow
SCUTAL MARGIN

CARINAL MARGIN
Crests for Depressor Muscle
BASAL MARGIN
Spur

Figure 23. Darwin established the nomenclature of the plates of balanids, observing 8 as the maximum number of wall plates (1). These he noted were of 3 types: with radii on both sides (2), with an ala on one side and a radius on the other (3), or with alae on both sides (4), and he named them. He also named the opercular valves (5 and 6) and the details related to their construction (5-7). He noted the 'beaked apex' and deduced how it formed (text-fig. from Darwin 1855; see Section 5.9).

stages, has recently been found necessary (Newman 1987, 1989b; Yamaguchi & Newman 1990).

Darwin found most of the names he gave the plates of the capitulum of his Lepadidae applicable to the plates of his Verrucidae and Balanidae (Figs. 23, 24). *Verruca*, the sole genus of the Verrucidae known to Darwin, is bilaterally asymmetrical, and he wrote, 'At first glance it appeared hopelessly difficult to identify, in a homological sense, these six valves, with those of ordinary cirripedes, but the difficulty soon quite vanished' (LC2:497) While the difficulty vanished, one can only marvel at the reasoning he went through in making the determinations; it still takes study to understand the homologies of these plates (e.g., see extant *Neoverruca*, Fig. 21, Appendix III, 12). In *Verruca*, the wall has four plates attached to the substratum: a fixed scutum and tergum (FS and FT) making up most of one side and a rostrum and carina (R and C) making up the opposite side and both ends. The lid or operculum comprises two plates, the movable scutum and tergum (MS and MT). [Having the tergum and scutum of one side immovably incorporated into the wall produces the asymmetry.] The asymmetry can be either left or right sided and the arrangement of plates on the movable side reflects the arrangement of plates on both sides of the symmetrical ancestor. Contrary to the conclusion of Ghiselin & Jaffe (1973), Darwin could not decide whether the verrucomorphs were closer to the lepadomorphs or balanomorphs, and no wonder since it now appears that the ancestor was a brachylepado- morph (e.g., see extant *Neobrachylepas*, Fig. 21, Appendix III, 14).

The Brachylepadomorpha were unknown to Darwin. His Balanidae included all the sym- metrical sessile barnacles and the homologies of the plates with those of the pedunculate barnacles were considered quite straightforward. As in the verrucids, he identified the plates of the operculum as the paired scuta and terga (S and T), but those of the primary wall in primitive forms such as *Catophragmus* were more numerous: Rostrolatera, median latera, and carino- latera in addition to the rostrum and the carina (R-RL-L-CL-C). While Darwin based these names on presumed homologies with the plates in scalpellomorphs, he did not actually specify that the balanid 'Latus' (L) was homologous with the upper latera or lower latera (UL or LL) of the scalpellomorph pedunculate barnacles (Newman 1987). Perhaps he was being careful in not being specific, but he went on to write (LC2:486): 'The Chthamalinae...fill up the interval between the Balaninae and Lepadidae; and *Catophragmus* forms, in a very remarkable manner, the transitional link, for it is impossible not to be struck with the resemblance of its shell with the capitulum of *Pollicipes*. In *Pollicipes*, at least in certain species [particularly that known as *Pollicipes (= Capitulum) mitella*], the scuta and terga are articulated together, the carina, rostrum and *three pairs of latera*, making altogether eight inner valves, are considerably larger than those of the outer whorls; the arrangement of the latter, their manner of growth and union, all are as in *Catophragmus*' (emphases added). Thus Darwin considered that the 8-plated wall of balanomorphans consisted of the rostrum, three pairs of latera, and the carina (R-RL-L-CL- C), all inferred to be homologous with and to have descended from those found in the scalpello- morphs. He neither broached the question of how L, associated with S-T in scalpellomorphs, became associated with the wall (R-RL-L-CL-C) in balanomorphs, nor how it came to overlap CL in the process, and subsequent authors pretty much followed suit (Newman et al. 1969; Anderson 1983).

If sessile barnacles more primitive than *Catophragmus* had not been discovered (Section 5.9, Appendix III), Darwin's work probably would be the state of our knowledge today. The first was a Cretaceous form, *Brachylepas* Woodward, 1901, from the Norwich Chalk (Fig. 21). Shells Darwin had attributed to a pedunculate fossil species from the Norwich Chalk, collected by Fitch (C-1288), *Pollicipes fallax* (FC1), were subsequently shown to be those of a *Brachy-*

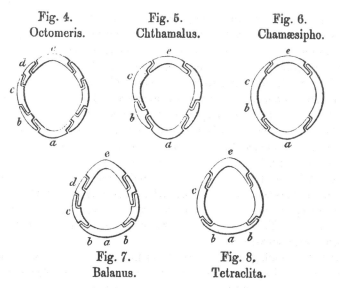

Fig. 4.
Octomeris.

Fig. 5.
Chthamalus.

Fig. 6.
Chamæsipho.

Fig. 7.
Balanus.

Fig. 8,
Tetraclita.

a, Rostrum; *b*, Rostro-lateral, *c*, Lateral, *d*, Carino-lateral compartment; *e*, Carina.

Horizontal sections through the Shells of the principal genera of Balanidæ, showing the arrangement of the Compartments. Genera 4, 5, and 6 belong to the Chthamalinæ; 7 and 8 to the Balaninæ.

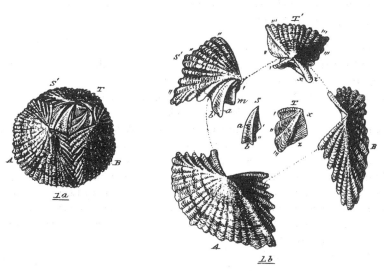

Figure 24. Grades of wall construction in 5 genera of Darwin's balaníds, and the organization of the veruccid wall: Darwin observed that the 8-plated wall of *Octomeris* was the model from which other forms could be derived by reduction and/or fusion of plates. Thus 6-plated forms could arise by loss of the carinolatera, as in *Chthamalus*, or by fusion of the rostrolaterals with the rostrum, as in *Balanus*. These types of 6-plated forms could give rise to 4-plated forms, as in *Chamaesipho* and *Tetraclita* (text-fig. from Darwin 1854; see Section 5.9). (a = rostrum, b = rostrolateral, bab = fused rostrum and rostrolatera (tripartitie rostrum, c = lateral, d = carinolateral, e = carina). Darwin also determined the puzzling identities of the plates in the asymmetrical wall of *Verucca* by using the position of the primordia valves, origins and insertions of muscles, and position and shape of plates (from Darwin 1854, pl. 29; see Section 5.9). (A = rostrum, B = carina, S = scutum, S' = fixed scutum, T = tergum, T' = fixed tergum).

lepas, B. fallax (Darwin 1851) (Woodward 1906; Withers 1914a, 1935). However, Darwin had also compared the shells of his *P. fallax* to those of *Verruca*, and he concluded that in certain details they were very similar. Thus Darwin unwittingly provided evidence allying the Verrucomorpha to the Brachylepadomorpha (Newman 1987; Newman & Hessler 1989).

Darwin had implied that the median latus of scalpellomorph *Pollicipes* was homologous with that of balanomorph *Catophragmus* (see above). In *Pollicipes*, L is closely associated with the scuta and terga, but in *Catophragmus* its presumed homolog is dedicated to the wall. The brachylepadomorphs are the same as *Pollicipes* in this regard: Woodward (1906) inferred, Withers (1912) concurred, and modern *Neobrachylepas* (Fig. 21) proved that L was associated with the operculum (Newman & Yamaguchi, in prep.). This must have been the case in *Catophragmus* (Newman 1987), and it was instructive to learn that it was also the case in *Neoverruca*, the most primitive of the known verrucomorphs (Newman & Hessler 1989). Thus if L of scalpellomorphs, brachylepadomorphs and verrucomorphs is not homologous with the median latus of the balanomorph wall, the lower latus (LL or l^1) would, by default, seem the obvious candidate for the homology (Newman 1987).

The difficulty with this homology is that the median latus would overlap both RL and CL, and therefore it has been considered unlikely (Newman & Hessler 1989; Newman 1989b). Fortunately there is an alternative explanation: if the balanomorphan median latus is not homologous with L or l^1, it must be the homologue of CL (Yamaguchi & Newman 1990). This is a significant departure from the Darwinian hypothesis involving the origin of the latera of the balanomorph wall, and it follows that, primitively, the wall was 6- rather than 8-plated, as in chionelasmatines (see *Eochionelasma*, Section 5.9, Fig. 21, Appendix III, 13). If so, where did the third pair of latera in balanomorphs come from?

5.9.1   *Ontogeny and the origin of the third pair of latera in Balanomorpha*
Darwin observed that, shortly after metamorphosis of the cyprid to the juvenile balanomorph, the balanomorph wall was composed of four plates, R-'L'-C; the rostrolatera and 'carinolatera' had not yet formed (LC2:130; Fig. 25G-I). In his Balaninae, the rostrum was compound (RL-R-RL), so only CL needed to be added to complete ontogeny. On the other hand, in the relatively primitive Chthamalinae, the rostrolatera are usually free, and they are apparently added to the wall at about the same time as the carinolatera (Newman 1987, Fig 5B). However, there is an important difference in the way they are added. While the carinolatera become intercalated between the 'latera' and the carina, overlapping the former and underlapping the latter, the rostrolatera simply grow up and overlap the 'latera' and rostrum. Furthermore, according to the observations of Runnström (1925), the intercalation of the 'carinolatera' in *Semibalanus balanoides* begins by their being budded off the 'latera,' i.e., they appear to be replicas of that plate (Fig. 25H-I).

The chionelasmatine wall is 6 rather than 8-plated as it is in *Catophragmus*; that is, it includes two rather than three pairs of latera. For 8-plated balanomorphs to evolve from a chionelasmatine-like ancestor, a third pair of latera has to be added. What had previously been considered the latus (L) in *Chionelasmus* begins development at the carinolateral position in the youngest known juvenile, which happened to be a male (Yamaguchi & Newman 1990). In light of this, the 4-plated ontogenetic stage in balanomorphs is apparently R-CL-C, rather than R-L-C as Darwin and subsequent workers had supposed. By addition of the RL pair, the level of wall development seen in adult chionelasmatines is attained (R-RL-CL-C), and by replication of CL, the third pair of latera is added in 8-plated balanomorphs (R-RL-CL^1-CL^2-C) (Yamaguchi & Newman 1990; Buckeridge & Newman 1992; Newman & Yamaguchi, in press; Appendix III 6,10,14). What evidences is there for the addition of plates by replication?

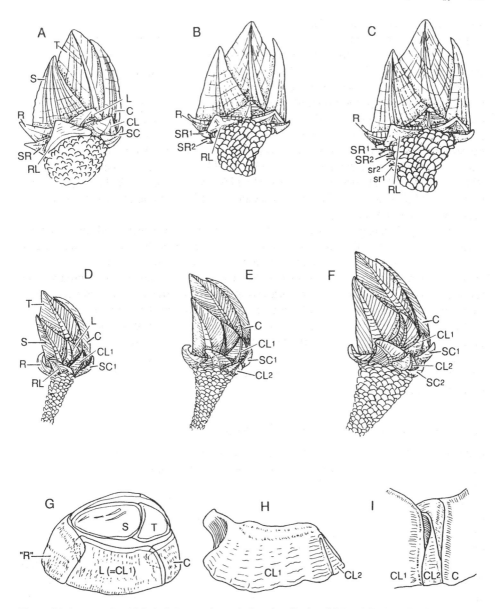

Figure 25. Ontogenetic addition of plates to the capitulum (see Section 5.9.1 and 2). A, *Scillaelepas (Aurivil-lialepas)* with 1 subrostrum, B, *Scillaelepas (Gruvellialepas)* which generally has 2 subrostra. C, occassionally during ontogeny in *Gruvellialepas* subrostra destined for the capitulum remain in the peduncle (sr^2 and sr^1; after Newman 1980). D, *Newmanilepas* has the same number of plates of *Scillaelepas* s.s. when young, E, during ontogeny an extra carinolatus and F, an extra subcarina are added by replication at the same sites as the original plates (CL_2 and SR_2; after Zevina & Yakhontova 1987). G-I. Darwin knew the 'carinolatus' was added last in balanomorph ontogeny, but he did not try to explain why it underlapped rather than overlapped the 'latus,' not knowing it was apparently a replicate of this plate (*Semibalanus balanoides*, after Runnström 1925) (C = carina, CL = carinolatus = CL^1, CL^2 = replicate of CL^1, L (in A-F) = median latus, L (in G) = CL^1, R = rostrum, RL = rostrolatus, S = scutum, SR = capitular subrostrum = SR^1, SR^2 = replicate of SR^1, sr^1 = peduncular subrostrum, sr^2 = replicate of sr^1, T = tergum).

5.9.2   *Addition of capitular plates by transfer and by replication*

The idea that, in an evolutionary sense, plates can be transferred from the peduncle to the capitulum in scalpellomorphs stems back to Darwin's (CL2) observations on plate production at the capitulo-peduncular junction in *Pollicipes* s.l. (Fig. 23B) and, as far as I am aware, the evolutionary possibilities were first alluded to by Broch (1922). Darwin noted that there were normally two types of plates produced at the capitulo-peduncular junction, those being added to the capitulum and those being added to the peduncle. Sometimes plates normally destined for the peduncle can instead be transferred to the capitulum, and apparently largely by this process, the primitive and simply armored scalpellomorphs like *Eolepas* (Appendix III 2; Fig. 21C) gave rise to higher, more complexly armored forms like the calanticines and pollicipedines (Newman 1979b, 1982b, 1987). The ontogenetic addition of plates, in scalpellomorphs such as *Pollicipes* and *Capitulum*, is effectively a recapitulation of their evolutionary acquisition from a calanticine-like ancestor.

A capitulo-peduncular junction exists between the principal wall plates and the imbricating whorls in neoverrucids and brachylepadomorphs (Newman 1989b; Newman & Yamaguchi, in prep.). However, both these groups have a second growth zone below this junction that, following metamorphosis from the pedunculate to the sessile stage during ontogeny, produces the basis. Only the basal growth zone persists in eoverrucine and verrucine verrucomorphs, and in balanomorphs, where it is responsible, in addition to growth of the basis, for the growth of the wall plates and, when present, the addition of basal imbricating plates. Thus, a significant evolutionary step between lower and higher forms in both the Verrucomorpha and Balanomorpha was the assimilation of the capitulo-peduncular growth zone into the basal growth zone, i.e, the failure of the two zones to separate during ontogeny. Not until this was accomplished could lateral plates be incorporated into the wall.

5.10   *Systematics and classification of the Cirripedia s.s.*

Darwin had a long standing interest in systematics, stemming from college days, and by February 1842 he had been appointed by the Council of the British Association for the Advancement of Science to a committee, chaired by Hugh Strickland, to make recommendations on the subject of zoological nomenclature (C-619, n. 1). By that May he had read Strickland's revised manuscript of the rules, resulting from the committee's first meeting and subsequent comments from other naturalists. One of Darwin's comments (C-630) concerning changing erroneous names was that changes should be limited to geographical names and otherwise to glaring errors. To illustrate what was not glaring he thought it well to provide examples such as the following: 'An animal if called 'alpinus' and afterwards found to inhabit a low as well as an alpine country, ought not, I presume, to have its name changed: an animal called 'albus' and afterwards found to be white only in the female or young state ought not, I presume, to have its name changed. All authors, if left quite to themselves without *rule* or *examples*,...will think errors of their *own discovery* 'glaring' ones, and will think themselves justified in changing the name, coining another and attaching their own name...after it.' In the same letter, Darwin discussed the 'authority for a species' when the genus has been changed and concludes that it should remain with the original author since this will 'send the inquirer direct to the original description.' The revised rules based on the recommendations of this committee were published by the Association the same year.

In 1849, when he was well into the barnacles, Darwin wrote Stickland (C-1215): 'I have found the rules very useful... something to rest on in the turbulent ocean of nomenclature,...

though I find it very difficult to obey always. Here is a case (among the barnacles) *Coronula, Cineras* and *Otion* are names adopted by Cuvier, Lamarck, Owen and almost *every* well-known writer, but I find that all 3 names were anticipated by a German: Now I believe if I were to follow strict rule of priority more harm would be done than good and more especially as I feel sure the newly fished up names would not be adopted. I have almost made up my mind to reject rule of priority in this case: Would you grudge the trouble to send me your opinion.' Darwin then added: 'I have come to a fixed opinion that the plan of the first describer's name being appended for perpetuity to species has been the greatest curse to natural History... I feel sure as long as species-mongers have their vanity tickled by seeing their own names appended to a species, because they first miserably described it, in two or three lines, we shall have the same *vast* amount of bad work as at present... I find every genus of Cirripedia has half a dozen names and not one careful description of any one species in any one genus' (Darwin's emphasis).

Strickland's long and thoughtful replies in two letters (C-1216 and 1223) apparently encouraged Darwin to follow the rule of priority in the aforementioned case (Darwin used *Conchoderma* Olfers. 1814, rather one of the then currently popular names *Otion* or *Cineras* Leach, 1817). However, further along in LC1 he selects *Pollicipes* Leach, 1817, rather than *Mitella* Oken, 1815, not only because it had been universally adopted throughout Europe and North America and had been extensively used in geological works, but because Oken's work was little known and displayed entire ignorance regarding the Cirripedia (LC1:293, footnote).

While poor Strickland (1811-1853) barely lived to see Darwin's brazen departure from the rules, he likely would have been pleased with Pilsbry (1907) who, in following the rules, reinstated *Mitella*: 'It is much to be regretted that Darwin allowed the general use of the name *Pollicipes* to influence his course in rejecting the earliest generic name for this group, contrary, as he writes, to the rules of the British Association. Had he accepted the earliest name, it would long ago have become the generic term *Mitella*, which Darwin himself showed to be prior to *Pollicipes*.' So after 1907 most cirripedologists switched to *Mitella*, but the switch did not last long. Some 50 years later all of Oken's molluscan names were suppressed and some cirripedologists were consulted as to what to do in the case of *Mitella* and *Pollicipes*. They voted to again suppress *Mitella* (ICZN Ruling 417), which quite naturally caused some temporary confusion among non-systematists.

Darwin allowed a matter of opinion to be taken into consideration involving priority. If he had followed the rule, the stability the rule is intended to preserve would have been properly served. Furthermore, despite Strickland's well-reasoned arguments, Darwin objected to having an author's name attached to a taxon because he believed to do so placed a premium on hasty and careless work, despite the fact, as he argued in 1842 (C-630), there is often need to get back to the original description even though it may not be the best one available. While he carried his objection over into the main parts of his barnacle monographs (this can be inconvenient to readers), he later relented and added author's names, including his own, to the taxa covered in his Synopsis at the end of the volume (LC2:606-640).

5.10.1 *Classification of the Cirripedia and its roots in Darwin's classification*

Darwin (1854) practically ignored the classifications of the thoracican barnacles of Lamarck (1818), Leach (1825), and Gray (1825). However, while Darwin's classification essentially forms the basis for the present classification (Figs 26, 27), Lamarck's and Leach's division of the thoracican cirripedes into pedunculate and sessile forms, ignored by Gray as well as Darwin, has been resurrected (Section 5.10.2), as has Leach's and Gray's distinction between the balanoid and coronuloid barnacles. Furthermore, family-group taxa have been added, espe-

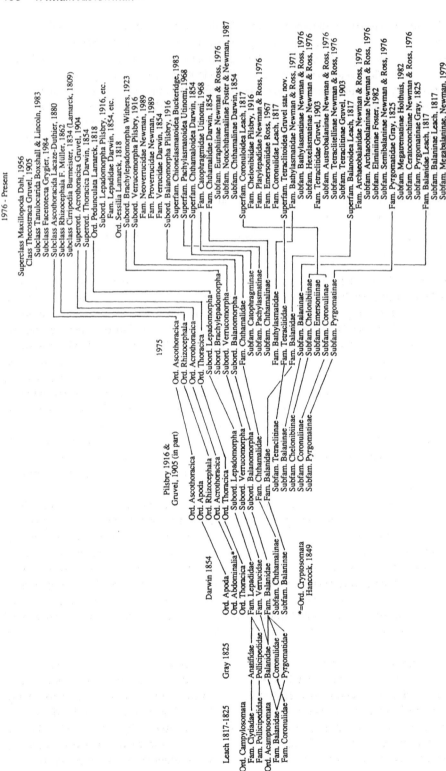

Figure 26. History of classification of the Cirripedia Burmeister 1834 (= Cirrhipedes Lamarck 1806). For a fuller rendition of Leach's (1825) classification see Pilsbry (1916:12) and for the suborders of Pedunculata (see Fig. 27 and Section 5.10.1; modified from Newman & Ross 1976).

cially among the Balanomorpha. However, this has been accomplished within the confines of the Darwinian framework because two unnamed hierarchial levels, fully defined and utilized in his analysis of natural groups within the Balanidae (= Balanomorpha), were given taxonomic status in subsequent classifications (Section 5.10.3).

Darwin neither reviewed previous classifications nor explained his rejection of the ordinal and most of the familial designations of earlier workers, such as those of Leach (1825) and Gray (1825). Such may have been customary, however, at the time because Gray to some extent ignored Leach, and all three authors ignored Lamarck (1818). While Leach (1825) divided the families of barnacles between the Campylostomata and Acamptostomata (= Pedunculata and Sessilia Lamarck, 1818), both Gray (1825) and Darwin (1854) ignored this division. Darwin's reasons are taken up in Section 5.10.2.

Darwin's Order Apoda disappeared before 1975, after it was discovered that *Proteolepas bivincta* Darwin, 1854, was not a cirripede (Section 5.7); his ordinal name Abdominalia (= Cryptosomata Hancock, 1849, and subsequently divided into two suborders, Section 5.6), was replaced by the superorder Acrothoracica Gruvel, 1905; and his order Thoracica has expanded dramatically (Fig. 26). Two important parts of the expansion of the Thoracica are based on Darwin: 1) his having divided the Balanidae of Gray into the Balanidae and Verrucidae which, together with Darwin's Lepadidae, have since become suborders, and 2) his having split the Chthamalinae off from the Balanidae s.s. and thereby forming the basis for the superfamily Chthamaloidea. Thus the expansion in the present classification between the Acrothoracica and the Chthamalinae has occurred pretty much within the frame work of Darwin's original classification, although much of it has also resulted from the recognition of coordinate taxa, an example of which can be seen in the subdivision of the Chthamalinae s.s. (Section 5.8.6).

The Coronuloidea, following the Chthamaloidea (Fig. 26), stems from Leach, 1817, but the grouping was ignored by Darwin. Prior to the present classification, the Coronuloidea included the Tetraclitidae. While this taxon shares a number of primitive features with the coronuloids it is nonetheless distinct at a coordinate level, and it is therefore treated here as a superfamily, Tetraclitoidea Gruvel, stat. nov. The Tetraclitoidea also shares a number of characteristics with the Balanoidea (see Section 5.8.4 for the present status of Darwin's species of *Tetraclita* and its allies in which the reassignment *Balanus imperator* and two of the species of *Elminius* more sharply defined the Tetraclitidae). On the other hand, the more highly evolved Balanoidea Leach, 1817, includes the coral inhabiting Pyrgomatidae of Gray, 1825 (Section 5.8.5).

Casual inspection of Figure 26 might suggest that there has not only been an expansion of the higher classification of the barnacles, e.g., more subfamilies and families, but also that there has been an increase in hierarchial levels over those utilized by Darwin, e.g., superfamilies, suborders, and superorders. However, only one level has been added, an ordinal-level category involving the Pedunculata and Sessilia of Lamarck, a category abandoned by Gray, 1825, and Darwin (see Sections 5.10.2, 5.10.3). Although for the most part Darwinian, the present classification includes underlying concepts of all these early workers.

5.10.2 *Darwin's suppression of the Pedunculata and Sessilia Lamarck, 1818*
Lamarck (1818) divided the ordinary barnacles into two groups, Pedunculata and Sessilia (Fig. 27). Leach (1825) recognized the same division but coined his own names, Campylostomata and Acamptostomata, respectively. Darwin (LC2) ignored the orders of Lamarck and Leach and divided the ordinary barnacles between three families, the Lepadidae, Verrucidae, and Balanidae. These he lumped without explanation under his new order Thoracica

Leach and Gray lumped the asymmetrical sessile barnacles, *Verruca*, with the symmetrical

Lamarck 1818 Thompson 1830	Darwin 1852-54 (Agassiz 1846) [Woodward 1901]	Gruvel 1902-05 (Calman 1909) [Hoek 1913]	Pilsbry 1916 Nilsson-Cantell 1921 (Withers 1923)	Newman 1987
Order Pedunculata ——	Order Thoracica ——([Order Pedunculata]) Family Lepadidae	Order Thoracica ——Suborder Pedunculata	Order Thoracica ——Suborder Lepadomorpha	Superorder Thoracica ——Order Pedunculata Suborder Cyprilepadomorpha Suborder Iblomorpha Suborder Heteralepadomorpha Suborder Praelepadomorpha Suborder Lepadomorpha Suborder Scalpellomorpha
Order Sessilia ——	(Order Sessilia) [Order Operculata] [Family Brachylepadidae] Family Verrucidae Family Balanidae	[Suborder Sessilia] Suborder Operculata ——Tribe Asymmetrica (Verrucidae) ——Tribe Symmetrica (Balanidae)	(Suborder Brachylepadomorpha) Suborder Verrucomorpha ——Suborder Balanomorpha	——Order Sessilia Suborder Brachylepadomorpha Suborder Verrucomorpha Suborder Balanomorpha

Figure 27. The history of the higher classification of the Thoracica demonstrating Darwin's suppression of Lamarck's (1818) Pedunculata and Sessilia (= Camptylostomata and Acamptostomata Leach, 1825). (Modified from Newman 1987; see Section 5.10.2).

sessile barnacles, *Balanus* and its allies, under the Balanidae. However, Darwin decided these forms represented distinct families, the Verrucidae and Balanidae, because, upon looking into the affinities he could not closely ally *Verruca* with either the Lepadidae or the Balanidae. From his assessment of shell structure, muscular attachments, and mouthparts and other appendages, Darwin found that *Verruca* shared some characters with the balanids, especially the chthamalines, but also shared a number of primitive features with the lepadids. Because he could not commit himself one way or the other, he apparently chose to ignored the Sessilia as a taxon uniting the verrucids and balanids.

However, for the next 70 years or so, notable authors, e.g., Woodward (1901), Gruvel (1905), Calman (1909), and Hoek (1913), continued to recognized the Sessilia and Pedunculata of Lamarck (Fig. 27). Whether they simply failed to notice that Darwin had omitted the taxon, or they actually disagreed with him is not known. I suspect it went unnoticed, for otherwise one would have expected some discussion over a monophyletic versus a diphyletic origin of the sessile barnacles. In any event, it was Withers (1914) who concluded, following his description of a primitive fossil verrucid, *Proverruca*, that the verrucids could readily have been derived from a pedunculate ancestor like *Scillaelepas* (see Newman et al. 1969, Fig. 113). Thus he allied the verrucids to the pedunculate rather than the symmetrical sessile barnacles. Apparently because of this, subsequent classifications, beginning with that of Pilsbry (1916), abandoned the Sessilia which, in effect, returned us to the system of Darwin. However, Withers never gave up Woodward's view that *Brachylepas* Woodward, 1901 (Brachylepadomorpha Withers, 1923, a suborder based on forms seen but not comprehended by Darwin), was a primitive sessile barnacle from which higher sessile barnacles could have evolved, and this view was accepted and enlarged upon by Newman (1987), in returning to Lamarck's system (Figs 26, 27).

5.10.3 *Congruency between hierarchial levels of Darwin in the present classification of the Balanomorpha*

Darwin recognized natural groups among and within genera of his subfamily Balaninae, and these he designated 'Sections' (Appendix II; LC2:175-397). These were subsequently incorporated into the formal classification of the Balanomorpha, with the elevation of some ranks (Fig. 28). How do Darwin's categories line up within the present classification, especially with regard to his species-rich Balaninae?

Darwin divided the Balanidae into two subfamilies, the Balaninae and Chthamalinae (LC2; Figs 26, 28). The Balaninae was ranked more highly evolved than the Chthamalinae by characteristics of their appendages as well as hard parts, but both contained a range of relatively primitive to advanced forms. Darwin divided the Balaninae into Sections 1 and 2, Section 2 equaling the Coronulidae of Leach and Gray (cf. Figs 26, 28). Section 1 he divided into six subsections, A through F (Fig. 28). Essentially, these subsections ranged from higher to lower groups in the characteristics of their hard parts, i.e., advanced forms were more readily grouped and defined by their specializations (apomorphies) and he separated them out first.

Most of these sections represented natural groups that subsequently became better defined, and both sections 1 and 2, and several of the alphabetical subsections of section 1, were given formal taxonomic ranks by subsequent workers. Some of the new genera were broken into subgenera, and some of these were subsequently elevated to generic status. As the process continued, new family-group taxa were required (Fig. 26), and the discovery of new groups led to further changes in the system. However, the hierarchy remains essentially congruent with Darwin's original classification, from the Balanomorpha down to his alphabetical sections

A

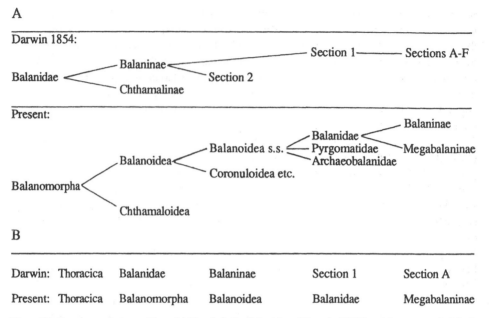

B

| Darwin: | Thoracica | Balanidae | Balaninae | Section 1 | Section A |
| Present: | Thoracica | Balanomorpha | Balanoidea | Balanidae | Megabalaninae |

Figure 28. Congruency between hierarchial levels in the Balanidae of Darwin (1854) and the present: A. Much of the infrastructure of the present taxonomy has roots in Darwin's numerical and alphabetic sections. B. From Thoracica to Megabalaninae no hierarchial levels are needed beyond those provided by Darwin (Section 5.10.3). A dramatic expansion at the subordinal level, however, took place in Peduculata because of the discovery of Paleozoic fossils and deep-sea forms unknown to Darwin (see Fig. 27).

(Fig. 28). That this complex part of Darwin's system has been so adaptable is a testimonial to his remarkable understanding of natural relationships among the Balanomorpha.

5.11 *Geological History*

Darwin summarized his conclusions concerning the geological history of the pedunculate barnacles (FC1) in a paragraph in LC1:172, and they are summarized even more briefly here. Although the occurrence of pedunculate barnacles in the Paleozoic had yet to be discovered, and the conclusion that balanines did not radiate until the Cenozoic was admittedly based on negative evidence, Darwin's deduction that pedunculate barnacles were morphologically more primitive and geologically older has been essentially borne out by subsequent findings. Relatively unarmored pedunculate barnacles have since been discovered in the Paleozoic (Collins & Rudkin 1982; Schram 1986). The commonest fossil forms were scalpellomorphs, and he concluded these had reached their culmination during the Late Mesozoic (Cretaceous), at which time a scalpellomorph such as *Capitulum* gave rise to balanomorphs.

Darwin's perception of the culmination of shallow-water scalpellomorphs in the Late Mesozoic, is much the way we see it today (Withers 1935; Newman et al. 1969). The intertidal forms, *Pollicipes* and *Capitulum*, are but a paltry representation of what shallow-water scalpellomorphs were in the past. However, deep-sea explorations, beginning with the Challenger Expedition of the 1870's, revealed a substantial diversity of scalpellomorphs; in fact, judging from Zevina (1981), about 250 of the more than 280 known living species are from the deep sea, and some of these such as *Scillaelepas*, *Newmanilepas*, and the hydrothermal vent barnacles are relicts of largely Late Mesozoic shallow-water radiations unknown to Darwin.

Nonetheless, Darwin (LC2) observed that, compared to the scalpellomorphs, the history of the balanomorphs has been a brief one. Other than the bogus record of a Cretaceous *Chthamalus* by Bosquet, Darwin reported no Mesozoic forms. Since then the catophragmoid *Pachydiadema* has been discovered from the Upper Cretaceous of Sweden (Withers 1935), but otherwise Darwin's general view that the great radiation of the balanomorphs was in the Cenozoic remains valid.

Darwin believed that the naming of fossil species based on incomplete material (lacking opercular valves for example) was the reason for so many nominal species, thus making it appear that Tertiary seas abounded with species of *Balanus*. Nonetheless, according to the compilation of Buckeridge (1983), the Miocene was richer in barnacle genera, on a world-wide basis, than was the Pliocene and Pleistocene, and there was a greater diversity of coral barnacles in the Atlantic in the Miocene than there is there today (Newman & Ladd 1974; Zullo & Portell in prep.). A similar trend has been observed among mollusks (Stanley 1986). Therefore, the number of valid species in the fossil record may not be as exaggerated as Darwin feared.

5.12 *Biogeography*

Darwin was intensely interested in biogeography (see Section 2.4.2), especially the fauna and flora of oceanic islands such as the Galapagos and Madeiras, and the similarities between biotas of continents, such as between Northern Eurasia and North America, and Australia and South Africa. His interests in bipolarity (amphitropicality), the biotic affinities between the Northern and Southern hemispheres, discussed in *The Origin*, seem to have arisen shortly after the barnacles, for in 1856 he asked Huxley (C-1914) about bipolarity in ascidians and in 1857, asked Dana (C-1925 and 2072) about the phenomenon in crustaceans.

He not only discussed the geographic distribution of barnacles at length (LC2:159-171), but he later devoted nearly 40 pages to general biogeography in *The Origin* (1859 (1872:278-317)) In LC2, Darwin compares his ideas on the distribution of barnacles with Dana's views on the distribution of other crustaceans. In *The Origin*, he discusses the distribution of biotas, including such topics as dispersal, bipolarity, and the importance of barriers.

Despite Darwin's obvious interests in biogeography, his conclusions on the distribution of barnacles have, to my knowledge, never been reviewed. Indeed, Eckman (1953) did not cite Darwin at all. This is perhaps because Darwin discourages the reader before he has gotten into the interesting aspects of the subject.

5.12.1 *Sessile barnacles and biogeography*
A short section in LC1, 'Geographical range; Habitats,' deals with geographic distribution of genera, but Darwin reached no conclusions. In his introductory comments concerning the geographic distribution of barnacles in LC2 he wrote, 'With respect to range, the results arrived at have no particular interest, for the species are not sufficiently numerous; and what is still more adverse, the genera, with unimportant exceptions, range over the world; so that there is no scale of differences, and it cannot be said that these two regions differ in their genera, and these two only in their species.' Yet he goes on to devote a dozen pages to the geographical distribution of primarily sessile barnacles, relating it where possible to ecology; he defines the four biogeo-

graphical provinces (Section 5.12.2.) and he is responsible for the following seven major conclusions:

1) Darwin observed that the Malay-Australian region, despite its representing a relatively small portion of the globe, had the greatest total number as well as the greatest number of peculiar (endemic) sessile barnacle species. This is true today for many groups of marine invertebrates, and for sessile barnacles Darwin suggested it might be due to the broken nature of the land affording a diversity of habitats and to much of the coast being rocky. Another important aspect, unknown at the time, was that in addition to cooling at the poles, the tropics of the earth apparently have warmed since the Miocene (Shackleton 1984; Valentine 1984). The concomitant extinctions (that cirripedes seemed to have been more diverse in the Miocene and Pliocene was discussed by Darwin; see Section 5.12) largely related to these two factors, apparently included paramphitropical and amphitropical forms, especially in the Northern Hemisphere. Survivors of these extinctions contribute to the relative species richness and high degree of endemism among marine invertebrates in the Austro-Malayan region and in the Southern Hemisphere in general (Newman 1991; see conclusion 4, below).

2) Darwin noted that coral reefs were not favorable for cirripedes, and consequently few were known to inhabit islands of the Pacific Ocean. However, where they can live, such as on the shores of western South America, England, and North America, individuals of but a few species abounded in infinite numbers. It is true that coral reefs are unfavorable for ordinary barnacles, and this has been attributed largely to the grazing of limestone-rasping fish (Newman 1960). However, there is a relatively high diversity of tropical barnacles when symbiotic and burrowing forms are taken into consideration.

It is instructive to observe that the tropics per se are not unfavorable for ordinary forms; to the contrary, where coral growth is limited to offshore islands, such as in the Gulf of Panama, the rocky coastal shores, which experience a tidal range in the neighborhood of 7 m, are covered with sessile barnacles representing half a dozen genera and numerous species (Southward & Newman 1979; Laguna 1990). I doubt that there is another place on earth where there is a greater diversity and abundance of intertidal balanomorph barnacles than on the Pacific coast of Panama.

3) Darwin concluded that there were no genera of sessile barnacles, having more than one species, confined to the tropics. He concluded this largely because he had lumped the coral barnacles into one genus (Section 5.8.5), whereby some species on ahermatypic corals were extratropical. If he had included pedunculate barnacles (e.g., *Lithotrya*), this would not have been true at the time and, in light of subsequent revisions of sessile barnacles, it is no longer true for them. Good examples of the latter are found among non-commensal tetraclitids and chthamalids, as well as the coral balanoids (Newman & Ross 1977; Foster & Newman 1987; Sections 5.8.4, 5.8.6).

4) Darwin noted that two genera, *Elminius* and *Chamaesipho*, were found in the Southern Hemisphere, although until recently it was not certain that a species of *Chamaesipho* did not also occur in the Northern Hemisphere (*Chinachthamalus scutelliformis* (Darwin, 1854) Foster 1982; Fig. 20). Darwin's observation was the first indication that there were endemics among the barnacles of the Southern Hemisphere, but he chose to ignore it.

The involvement of Darwin's *Elminius* in Southern Hemisphere endemism expanded when it was discovered that it was not a monophyletic taxon (see *Epopella*, Fig. 18). Furthermore, revisions of the sessile barnacles have increased the number of genera recognized from the Southern Hemisphere. It is now known that at least one genus among the megabalanines (Fig.

17) became a Southern Hemisphere endemic through extinction in the Northern Hemisphere (Newman 1991; conclusion 1 above).

5) Returning to the effects of temperature, Darwin noted that Dana's conclusion concerning the distribution of other crustaceans, that there are more species in the tropics than in the temperate region, is in keeping with the greater area of the tropics. However, taking all orders of cirripedes into consideration, Darwin pointed out that of a relevant 140 species only 46 were found exclusively in the torrid zone (tropics). This left the temperate regions more diverse even though they included considerably less area (about a sixth less according to Dana's calculations). Darwin explained why his result differed from that of Dana: The temperate regions are separated by the great width of the tropics and thus, being isolated from each other, they would be expected to have more, perhaps doubly more species than the single tropical area.

There is undoubtedly some truth in this explanation, but the data base has changed. While the number of barnacle species known from the temperate regions has increased since Darwin, it has increased more dramatically in the tropics. Some of the increase in the tropics is due to Darwin's numerous varieties of coral and other barnacles now being recognized as species, but it is also due to the discovery of many tropical forms in cryptic habitats. For example, while Darwin recognized only two acrothoracican species (*Trypetesa lampas* and *Cryptophialus minutus*; Section 5.6), neither was from the coralline seas. Now there are more than 45 extant species of acrothoracicans, most of which are known from the tropics (see Tomlinson 1969, 1987, Baird et al. 1990). Also, Darwin recognized few barnacles from the Galapagos, but 18 extant species are known from there today (Zullo 1991).

6) Darwin reviewed the sizes acquired by different species of sessile barnacles in relation to temperature and concluded that *Balanus* s.l. did not require the temperature of the tropics for its largest development, an observation that holds today. On the other hand, it must be considered that both a very large and a very small size can afford protection from predators (Paine 1981). (Apparently the relatively large and internally thickened shell of the tropical chthamalid *Euraphia hembeli* (Section 5.5.3) deters predators.) Furthermore, while in general the ambient temperatures for deep-sea scalpellomorphs are low, the sizes of the species range from minute to gigantic, further indicating that biological rather than physical parameters are important in determining the size of barnacle species.

7) With regard to wide-ranging species, Darwin noted that that ships could account for the wide range of a number of species, such as *Balanus* (= *Megabalanus*) *tintinnabulum* s.l. (Linnaeus), *B. amphitrite* Darwin, *B. improvisus* Darwin, and *B. trigonus* Darwin. Interestingly, some species needed the advent of high-speed ships, and perhaps ballast water for transport of their larvae, to increase their range (see Carlton 1985); *B. amphitrite* s.s. apparently did not arrive on the west coast of North America until 1939 (Zullo et al. 1972) or in Europe until World War II, where it was followed by an unexpected form, *Elminius modestus* Darwin, from Australia and New Zealand (Bishop & Crisp 1957). On the other hand, Zullo (1992b) has suggested that *B. trigonus*, while long known from the Atlantic, was known to Darwin only from the Pacific and was likely introduced from there.

Darwin found it surprising that *Balanus* (= *Austromegabalanus*) *psittacus* and *Balanus eburneus* should still be confined to southern South America and the western Atlantic, respectively, since they commonly foul ships. *Austromegabalanus psittacus* has yet to become established elsewhere. However, one of its relatives in the same west wind drift complex, *Notomegabalanus algicola*, was introduced from South Africa to southeast Australia, while *Elminius modestus*, mentioned above as introduce to Europe, was also introduced to South Africa (see Newman 1979a). However, Darwin's query concerning *Balanus eburneus* was short lived: it has become established in the Mediterranean, Japan, and the Marshall and Hawaiian Islands (Pilsbry 1916; Matsui et al. 1964; Utinomi 1966; Zullo et al. 1972). Curiously however, other than having been found in the locks at the Pacific terminus of the Panama Canal (Spivey 1976), *B. eburneus* is not known to have become established on the west coast of the Americas (Laguna 1990). Clearly, as Darwin likely would be the first to admit, there is more to becoming an introduced species than simply being transported.

Darwin noted that many introduced forms were estuarine, at least in their home waters, and apparently the introduction of a coastal intertidal species was unusual. This rule apparently holds for the coastal species such as *Balanus glandula* from the west coast of North America introduced near La Plata, Argentina (Spivak, pers. comm.; erroneously reported from Rio De Janeiro by Newman & Ross 1976), and *Chirona amaryllis* from the Indo-Pacific introduced in the Rio Parnaba estuary in Brazil (Young 1989).

5.12.2 *Darwin's biogeographical provinces based on barnacles*

Purposely disregarding all probabilities or conclusions deduced from other animals, Darwin (LC2:164) concluded that the barnacles, including all orders and families, but excluding species attached to floating or swimming objects, fell into four provinces, one subprovince, and one or perhaps two additional subprovinces. The provinces were based on the proportion of peculiar (endemic) to wide-ranging species (approximately 2/3 endemic), and their affinities with other provinces or regions by shared species. The ranges of most species were not sufficiently known for the breaks between provinces to be preceived, even along latitudinal gradients; if they had been, Darwin likely would have recognized more species and fewer varieties (Section 5.8).

1) Darwin's first or North Atlantic province ranged from the Arctic to 30°N, and held affinities with the West Indies south to Rio Plata on the west and the Mediterranean and equatorial Africa on the east side of the Atlantic. While he noted that a North Atlantic species, *Balanus crenatus* Brugière, 1789, was known from the Bering Straits, he was not aware that *B. porcatus (= balanus)* and *Semibalanus balanoides* also occurred in the North Pacific, and he did not discuss this interesting amphipolar distrubution pattern (see Flowerdew 1983 regarding *Semibalanus balanoides* and Høeg & Lutzen 1985 regarding rhizocephalans).

Darwin's single subprovince, having principal affinities with Europe by virtue of share species, was the southern extremity of Africa. It no longer holds that the tropical Atlantic coasts of South America and Africa are remarkably barren in cirripedes (see Stubbings 1967; Young 1988, 1990), and some of the uniqueness Darwin reported for the South African province has disappeared with more knowledge of the ranges of species previously thought endemic there. Yet, South Africa is clearly identifiable as a transition zone between the southern Atlantic and Indian oceans (Ekman 1953; Newman 1991). The short-range transition zone endemics include the relict, *Octomeris angulosa, Balanus capensis (= Austromegabalanus cylindricus)* and *Notomegabalanus algicola*. The last two are apparently derivatives from the west wind drift, their closest affinities being with species from New Zealand and South America, Ascension Island, Saint Peter and Saint Paul Islands, and southern Australia (Newman 1979a, 1991); that is, they are apparently relict populations taking advantage of refugia afforded by these southern outposts.

2) Darwin's second province encompasses the west coast of North and South America, from the Bering Strait to Tierra del Fuego, an enormous coast line for which only 22 species were known at the time. For Darwin it was the least prolific of the provinces, but as noted below there are more than three times as many species of ordinary barnacles known there today. This certainly cannot be said of the North Atlantic province.

Of the 15 endemic species known to Darwin from the west coast of the Americas, four were not known south and 8 were not known north of the tropics. Thus Darwin considered that the province might be divided into two subprovinces. However, because 8 species were found on both sides of the tropics, he declined to make the separation. As it turned out, one of these species, *Pollicipes elegans*, is paramphitropical (Newman & Foster 1987) and *Balanus concavus* (Section 5.8.2.) represents a latitudinal series of four species ranging across the tropics into the temperate zones of both hemispheresof the eastern Pacific.

There is no question today that, based on barnacles as well as other organisms, this extensive coastline is divisible not just into two but into six or more provinces. Such divisions were made possible by accumulating data on the actual ranges of individual species, first begun on the Pacific coast of North America by Henry (1942). When ranges are plotted, natural breaks between faunal regions become evident, especially where the plots are reduced to a set of curves utilizing the 'end-point' method developed in 1948 by Newell (see Newman 1979c).

Laguna (1990) applied the end-point method to the nearly 70 barnacle species now known from between latitude 32°N and 8°S on the west coast of the Americas and thus illustrated the existence of at least three major provinces between them. While comparable work has not been carried out farther south, by simply plotting ranges Knox (1960) showed a break comparable to that at 35°N near 35°S. Thus, Darwin's west coast province of North and South America, from the Bering Strait to Tierra del Fuego, has been divided into approximately seven major provinces (Valentine 1966).

This recognition today of more provinces than in Darwin's day is partly due to Darwin delineating provinces based on a criterion of at least 2/3 of the species being endemic to a large region rather than on the 1/2 to 1/3 value used today. However, it is also due to the higher resolving power gained by advances in taxonomy involving the recognition of species and species-group taxa, a better data base on the ranges of these species, and the use of the end-point method in determining where one province ends and another begins. The provincial boundaries at this level, at least along latitudinal gradients, are relatively subtle discontinuities in the hydrographic environment; whereas in Darwin's scheme they were the wide expanses of the tropics separating the northern and southern temperate zones.

3) Darwin called his third province the East Indian Archipelago, a region encompassing the Philippines south to New Guinea and west to the coast of India, in which 24 of the 37 species were endemic. He noted that he had seen no species from Madagascar or the east coast of Africa and few from India and China. While his suspicions that few species existed in any of these regions turned out to be false, true to his predictions this third province came to include the entire Indian Ocean, and it is close to what is known as the Indo-West Pacific province today.

4) Darwin's fourth province included Australia and New Zealand, in which 21 of the 30 species were endemic, and the uniqueness of the region has become more generally recognized and better documented (Foster 1979; Jones et al. 1990). While Darwin listed the species, he did not discuss endemism at the generic level; i.e., *Elminius* and *Chamaesipho* (see Sections 5.8.4 and 5.8.6). Today it is endemism at the generic level that distinguishes Southern Australia and New Zealand, and other Southern Hemisphere outposts (southern South America and South Africa) from the rest of the world (Newman 1979a). As with other marine groups, much of this endemism has affinities with a largely extinct extratropical cirripede fauna of the Northern Hemisphere (Newman & Foster 1987; Newman 1991).

5.12.3 *Darwin's comparison of the distribution of barnacles and other crustaceans*

Darwin concluded that the facts showed his four provinces were natural divisions of the world, as far as cirripedes were concerned. Since cirripedes were crustaceans, Darwin considered it worthwhile to compare his results with those of Dana (1853), especially because the results were very different. Darwin noted that the differences might be due to his provinces being formed merely on the proportion of endemic to shared species (his barnacle genera generally being wide ranging); whereas Dana's were founded mainly on endemism at the generic level.

Dana divided the surface of the globe into three great sections or provinces: 1) Africo-European, 2) Occidental (both coasts of the Americas), and 3) Oriental (Indian and Pacific Oceans and Australia). Thus he considered the eastern shores of North America to be distinct from Europe, and this was not the case with the cirripedes at the generic and to some extent at the species level. Furthermore, Darwin had placed the east and west shores of the Americas into two distinct cirripede provinces. And finally, Dana's conclusions would have united Darwin's East Indian Archipelago and Australian provinces, and made New Zealand a sub-province apparently distinct from Australia. Darwin noted in closing that the provinces deduced from cirripedes were more in accord with those of the molluscan provinces than those Dana had deduced for the rest of the crustaceans, and by and large, this is true today.

6 CONCLUSIONS

Charles Darwin's work on cirripedes was truly remarkable and all encompassing. That he made several errors in the course of this work is more a reflection of the state of technology and scientific knowledge of his times than of any defects in his own mental capacities. The imagination that he exhibited in developing explanations for the problems he encountered truly displays his genius.

ACKNOWLEDGMENTS

Thanks are due F.R. Schram, for much editorial advice and relentless encouragement, and similarly F.M. Truesdale, who clarified the circumstances surrounding Anton Dohrn's withdrawl of a paper on arthropod evolution. I also need to thank B.A. Foster for information on the affinities of *Chinochthamalus*, L.B. Holthuis concerning authorship of 'Cirripedia,' E. Mayr for a discussion that improved my understanding of Darwin's species concept at the time he was studying barnacles and of what was wrong with his methods when it came to varieties, M. Boudrias for alerting me to and R. Shadwick for providing a copy of an article noting the connection between one of Darwin's sons and the founding of Cambridge Instruments (Annonymous 1991), M. Ghiselin for clarifying several points and providing a copy of his Darwin bibliography and V.A. Zullo for tweaking my curiosity over 30 years ago as to why Darwin changed illustrators in midstream. I would also like to thank the following for gift copies: D. Souerbry for Stone's *The Origin*, R. Lewin for Groeben's *Charles Darwin, Anton Dohrn: Correspondence*, and M.J. Grygier for Atkin's *Down: The Home of the Darwins*. I am grateful to Darwin not only for laying the foundations and methods for studying barnacles, but also for a natural and ethical philosophy that has helped make studying them so satisfying. This paper is dedicated to the late Brian A. Foster (1942-1992), a friend, colleague, and Darwin disciple whose untimely death leaves a substantial void in cirripedology.

NOTES

1 Darwin (LC2) was not certain of the affinities between the Balanidae (Balanomorpha) and the Verrucidae (Verrucomorpha). Therefore, he did not divide the Thoracica into the orders, Pedunculata and Sessilia of Lamarck, 1818 (Section 5.10.2). While the relationships have yet to be fully resolved, since Darwin, the Brachylepadomorpha have intervened between the Lepadomorpha s.l. and the Verrucomorpha as well as the Balanomorpha, and therefore it is argued the sessile barnacles are monophyletic (Newman 1987; Newman & Hessler 1989; Yamaguchi & Newman 1990), all three suborders being included under the order Sessilia Lamarck, 1818 (Section 5.10, 2).

2 The search for letters concerning barnacles began in the *Calendar* and therefore they are cited herein by their number plus the prefixed 'C-', unless identifiable only by date in the *Correspondence*. Quotations from letters may be verbatim or slightly condensed. Other than for words that preserve the British 'flavour', spellings have been modernized, and words Darwin was fond of abbreviating, such as 'c^d' for 'could' and 'w^d' for 'would', have been spelled out. And lastly, generic and specific names have been italicized and proper names capitalized where appropriate.

3 The surname name, 'Milne Edwards' is commonly hyphenated, as it is in the Burkhardt and Smith's volumes. However, Darwin (1854:v, see Fig. 4) did not hyphenate it, nor did Henri Milne Edwards; therefore it is not hyphenated herein.

4 The frontolateral horns of cirripede nauplii have gland cells at their bases, and perforate tips, but their function is yet unknown. Taylor (1970) thought they might serve as antipredation devices, but his experiments were inconclusive. Walker (1973), on the other hand, suggested that the secretions they apparently produced might be used in feeding, but frontolateral horns are present and apparently functional in nonfeeding nauplii of Cirripedia and Rhizocephala.

5 In Darwin's time the division between the cephalon and the thorax in decapod crustaceans was considered to be between the first and second maxillipedes (seventh and eighth segments). In recognition of this, Darwin considered the homolog of the first maxillipedes of the archetype cirripede to be cephalic, rather than thoracic as were the second and third. So in Darwin's view the first thoracic segment of the archetype supported the second maxillipedes and this leads to much confusion, both for Darwin and the reader (Section 5.4.2), and therefore it is important that the problem be recognized here. Note (Fig. 8B) that between LC2:107 and LC2:108 the name of the first naupliar limb was changed from maxillipede 3 (maxped3) to thoracic limb 2 (T2), and then on LC2:111 it became thoracic limb 3 (T3).

6 The original spelling for a genus of turtle barnacles was *Chelonibia* Leach, 1817. Darwin corrected the spelling to *Chelonobia* in 1854, presumably on the advice of George B. Sowerby Jr. A number of authors including Gruvel (1905) followed Darwin on the matter, but in adherence to the rules Pilsbry (1916) returned to the original spelling. There is no quarrel with Crisp (1983b), Darwin's spelling is correct. However, according to Article 32, c, ii, of the International Code of Zoological Nomenclature (Ride et al. 1985), an inappropriate connecting vowel is not an inadvertent error (Frazier & Margarioulis 1990) and, therefore, Leach's original spelling is retained here.

7 While Darwin considered *Chthamalus hembeli* and *intertextus* congeneric, they now reside in two distinct genera, *Euraphia* and *Nesochthamalus*, respectively (Section 5.8.6). It is therefore interesting to note that while the two species most closely related to the Indo-West Pacific species, *Euraphia hembeli, E. calcareobasis*(Henry, 1957) and *E. devaneyi* Foster & Newman, 1987, from the Southeast Pacific, also have a normal calcareous basis, but they apparently forego secondary calcification throughout life.

REFERENCES

Sources for most taxa in text described before 1969, and not entered below, may be found in Newman et al. (1969).

Anderson, D.T. 1983. *Catomerus polymerus* and the evolution of the balanomorph form in barnacles (Cirripedia). *Memoir Aust. Mus.* 18:7-20.

420   *William A. Newman*

Anderson, D.T. 1992. Structure, function and phylogeny of coral-inhabiting barnacles (Cirripedia, Balanoidea). *Zool. J. Linn. Soc.* 106:277-339.

Annandale, N. 1924. Cirripedes associated with Indian corals of the families Astraeidae and Fungidae. *Mem. Indian Mus., Calcutta* 8:61-68 + pl. 12.

Anonymous 1991. Darwin's company: Cambridge Instruments. *Microscopion News* 11:15. Leica Heerbrugg AG: Switzerland.

Atkins, H. 1974. *Down, the home of the Darwins.* London: The Curwen Press Ltd.

Aurivillius, C.W.S. 1892. Neue Cirripeden aus dem Atlantishchen, Indischen und Stillen Ocean. Öfver. Kongl. Vetensk.-Akad. Fîrhandl. Stockholm 3:123-134.

Aurivillius, C.W.S. 1894. Studien über Cirripeden. *Kongl. Svenska Vetensk.-Akad.* Handl. 26(7):5-107.

Baird, G.C., C. E. Brett & J.T. Tomlinson 1990. Host-specific acrothoracid barnacles on Middle Devonian platyceratid gastropods. *Hist. Biol.* 4:221-244.

Barlow, N. ed. 1958. *The Autobiography of Charles Darwin.* Harcourt, Brace & Co.: New York.

Beck, R. 1865. *A Treatise of the Construction, Proper Use, and Capabilities of Smith, and Beck's Achromatic Microscopes.* London.

Berndt, W. 1903. Die anatomie von *Cryptophialus striatus n.* sp. *Sb. Ges. Naturf. Fr. Berlin* 1903:436-444.

Berndt, W. 1907. Öber das System der Acrothoracica. *Arch. Naturg.* 73:287-289.

Bishop, M.W.H. & D.J. Crisp 1957. The Australian barnacle, *Elminius modestus,* in France. *Nature* 179:482-483.

Bocquet-Védrine, J. 1972. Suppression de l'order des Apodes (Crustacés Cirripèdes) et rattachement de son unique représentant à Crinoniscidae (Crustacés Isopodes, Cryptonisciens). *C.R. Acad. Sci., Paris.* (D) 275:2145-2148.

Bocquet-Védrine, J. 1979. Interprétation actuelle de la description de *Proteolepas bivincta* Darwin, 1854 (Représentant unique de l'ancien ordre des Cirripédes Apodes). *Crustaceana* 37:153-164.

Bocquet-Védrine, J. & C. Bocquet 1972. La ceinture d'attache de la femelle juvénile de *Crinoniscus equitans* Pérez (Isopode Cryptoniscien) et son importance adaptative. *C.R. Acad. Sci., Paris.* (D) 275:2235-2238.

Boxshall, G.A. & R. Huys. 1989. New tantulocarid, *Stygotantulus stocki,* parasitic on harpacticoid copepods, with an analysis of the phylogenetic relationships within the Maxillopoda. *J. Crust. Biol.* 9:126-140.

Boxshall, G.A. & R.J. Lincoln 1983. Tantulocarida, a new class of Crustacea ectoparasitic on other Crustacea. *J. Crust. Biol.* 3:1-16.

Brackman, A.C. 1980. *A delicate arrangement.* New York: Times Books.

Broch, H. 1922. Studies on Pacific cirripeds. In *Papers from Dr. Th. Mortensen's Pacific Expedition 1914-1916. X. Vidensk. meddel. fra Dansk Naturh. For., København* 73:215-358.

Bronn, H.G. 1831. *Italiens Tertiär-Gebilde und deren organische Einschlüsse.* Heidelberg: K. Groos.

Buckeridge, J.S. 1983. Fossil barnacles (Cirripedia: Thoracica) of New Zealand and Australia. New Zealand Geological Survey *Paleontol. Bull.* 50:1-151 + pls. 1-13.

Buckeridge, J.S. & J.A. Grant-Mackie 1985. A new scalpellid barnacle (Cirripedia: Thoracica) from the Lower Jurrasic of New Caledonia. *Géol. de la France* 1:77-80.

Buckeridge, J.S. & W.A. Newman 1992. A reexamination of *Waikalasma* (Cirripedia: Thoracica) and its significance in balanomorph phylogeny. *J. Paleo.* 66:341-345.

Burkhardt, F. & S. Smith (eds) 1985. *A calendar of the correspondence of Charles Darwin, 1821-1882.* Garland Publishing, Inc.: New York & London.

Burkhardt, F. & S. Smith (eds). 1985-1989. *The correspondence of Charles Darwin.* Cambridge University Press: Cambridge. Vols. 1, 1985 (1821-36); 2, 1986 (1837-43); 3, 1987 (1844-46); 4, 1988 (1847-50); 5, 1989 (1851-55).

Burmeister, H. 1834. *Beiträge zur Naturgeschichte der Rankenfüsser (Cirripedia).* Berlin.

Calman, W.T. 1909. Crustacea. In R. Lankester, (ed.), *A Treatise on Zoology* 7(3):1-346. London: Adam & Charles Black.

Carlton, J.T, 1985. Transoceanic and interoceanic dispersal of coastal marine organisms: The biology of ballast water. *Oceanogr. Mar. Biol. Rev.* 23:313-371.

Charnov, E.L. 1987. Sexuality and hermaphroditism in barnacles: A natural selection approach. *Crustacean Issues* 5:89-103.

Claus, C. 1876. *Untersuchungen zur Erforschung der Genealogischen Grundlage des Crustaceen-Systems.* Wien.

Collins, D.D.M. Rudkin 1981. *Priscansermarinus barnetti*, a probable lepadomorph barnacle from the Middle Cambrian Burgess Shale of British Columbia. *J. Paleo.* 55:1006-1015.

Crisp, D.J. 1983a. Extending Darwin's investigations on the barnacle life-history. *Biol. J. Linn. Soc.* 20:73-83.

Crisp, D.J. 1983b. *Chelonobia patula* (Ranzani), a pointer to the evolution of the complemental male. *Mar. Biol. Letters* 4:281-294.

Crisp, D.J. & G.E. Fogg 1988. Taxonomic instability continues to irritate. *Nature* 335:120-121.

Cuvier, G. 1817. *Le Règne Animal Distribué après son Organisation.* Paris.

Cuvier, G. 1834. *The Animal Kingdom, volume 12; the Mollusca and Radiate. Ciirhopoda*, pp.134-138 + plates 2, 7, 17. London: Whittaker & Co.

Dana, J.D. 1852. (Review of Darwin's first volume on the living Cirripedia) *Amer. J. Sci. Arts,* (2) 14:125-127.

Dana, J.D. 1853. *On the classification and geographical distribution of Crustacea*; from the report on Crustacea of the United States Exploring Expedition, under Captain Charles Wilkes, USN., during the years 1838-1842. Philadelphia.

Dando, P.R. & A.J. Southward 1980. A new species of *Chthamalus* (Crustacea: Cirripedia) characterized by enzyme electrophoresis and shell morphology; with a revision of other species of *Chthamalus* from the western shores of the Atlantic Ocean. *J. Mar. Biol. Assc. UK* 60:787-831.

Darwin, C. 1839. *Journal of Researches into the Geology and Natural History of the Various Countries Visited by H.M.S. Beagle, under the command of Captain FitzRoy, R.N., during the Years 1832-36.* London: Colburn.

Darwin C. 1842. *The structure and distribution of coral reefs.* London: Smith, Elder.

Darwin, C. 1844a. Observations on the structure and propagation of the genus *Sagitta. Ann. Mag. Nat. Hist.* 13:1-6.

Darwin, C. 1844b. Brief descriptions of several terrestrial *Planariae*, and some remarkable marine species, with an account of their habits. *Ann. Mag. Nat. Hist.* 14:241-251.

Darwin, C. 1844c. *Geological observations on the volcanic islands visited during the voyage of H.M.S. Beagle, together with some brief notices of the geology of Australia and the Cape of Good Hope. Being the second part of the geology of the voyage of the Beagle, under the command of Capt. Fitz-Roy, R.N., during the years 1832 to 1836.* London: Smith, Elder.

Darwin, C. 1846. *Geological observations on South America. Being the third part of the geology of the voyage of the Beagle, under the command of Capt. Fitz-Roy, R.N., during the years 1832-1836.* London: Smith, Elder.

Darwin, C. 1849 (Remarks on Cirripedia). *Athenaeum. J. of Lit., Sci., & Fine Arts.* 1143:966.

Darwin, C. 1850. On British fossil Lepadidae. *Quart. J. Geol. Soc., Lond., Proc. Geol. Soc.* 6:439-440.

Darwin, C. 1851. *A monograph of the fossil Lepadidae, or, pedunculated cirripedes of Great Britain.* Pp. 1-88 + pls. 1-5. London: Palaeontographical Society.

Darwin, C. 1852. *A monograph on the sub-class Cirripedia, with figures of all the species. The Lepadidae; or, pedunculated cirripedes.* pp. 1-400 + pls. 1-10. London: Ray Society (1851).

Darwin, C. 1854. *A monograph on the sub-class Cirripedia, with figures of all species. The Balanidae, Verrucidae*, etc. pp. 1-684 + pls. 1-30. London: Ray Society.

Darwin, C. 1855a. *A monograph on the fossil Balanidae and Verrucidae of Great Britain.* pp. 1-44 + pls. 1-2. London: Palaeontographical Society (1854).

Darwin, C. 1855b. Does sea-water kill seeds? *Gardener's Chronicle & Agricultural Gazette* 15:242.

Darwin, C. 1858. On the tendency of species to form varieties and on the perpetuation of varieties and species by natural means of selection. *J. Proc. Linn. Soc. (Zoology)* 3:45-53 (reprints can be found in de Beer 1958 & Brackman 1980).

Darwin, C. 1859. *The Origin of Species by Means of Natural Selection* (Reprint of 6th edition (1872), pp. 386. New York: Modern Library).

Darwin, C. 1862. *On the various contrivances by which British and foreign orchids are fertilised by insects, and on the good effects of intercrossing.* London: Murray.

Darwin, C. 1863. On the so-called auditory sac of cirripedes. *Nat. Hist. Rev.* 3(9):115-116.

Darwin, C. 1865. On the movement and habits of climbing plants. *J. Linn. Soc. London* (Bot.) 9: 1-118.

Darwin, C. 1868. *The variation of animals and plants under domestication*, 2 Vols. London: Murray.

Darwin, C. 1871. *The descent of man, and selection in relation to sex*, 2 Vols. London: Murray.

Darwin, C. 1872. *The expression of the emotions in man and animals*, 1st ed. London: Murray.

Darwin, C. 1873. On the males and complemental males of certain cirripedes, and on rudimentary structures. *Nature* 8:431-432 (Sept. 25).

Darwin, C. 1875. *Insectivorous Plants*. London: Murray.

Darwin, C. 1876. *The effects of cross and self fertilisation in the vegetable kingdom*. London: Murray.

Darwin, C. 1877. *The different forms of flowers on plants of the same species*, 1st ed. London: Murray.

Darwin, C. 1880. *The power of movement in plants*. London: Murray.

Darwin, C. 1881. *The formation of vegetable mould, through the actions of worms, with observations on their habits*. London: Murray.

Darwin, F. (ed.) 1887. *The Life and Letters of Charles Darwin*. London: Murray.

Dayton, P.K., W.A. Newman & J. Oliver 1982. The vertical zonation of the deep-sea Antarctic acorn barnacle, *Bathylasma corolliforme* (Hoek): Experimental transplants from the shelf into shallow water. *J. Biogeogr.* 9:95-109.

de Beer, G. 1958. *Evolution by Natural Selection. Charles Darwin and Alfred Russel Wallace*. Cambridge: Cambridge University Press.

de Beer, G. 1963. *Charles Darwin; evolution by natural selection*. London: T. Nelson & Sons.

Dohrn, A. 1868. On the morphology of the Arthropoda. *J. Anat. Physio.* 2:80-86.

Ekman, S. 1953. *Zoogeography of the sea*. London: Sidgwick & Jackson Ltd.

Flowerdew, M.W. 1983. The circumboreal barnacle *Balanus balanoides* (L.) and its subpopulations. In G.S. Oxford & D. Rollinson (eds), *Protein polymorphism: adaptive and taxonomic significance*. Systematic Association Special Volume 24. London: Academic Press.

Foster, B.A. 1979. The Marine Fauna of New Zealand; Barnacles (Cirripedia: Thoracica). *N. Z. Oceanogr. Inst. Memoir* 69:1-159.

Foster, B.A. 1982. Shallow water barnacles from Hong Kong. In B.S. Morton & H. Tseng (eds) *Proceedings of the first international marine biological workshop on the marine flora and fauna of Hong Kong and southern China* 1:207-232. Hong Kong: Hong Kong University Press.

Foster, B.A. 1983. Complemental males in the barnacle *Bathylasma alearum* (Cirripedia: Pachylasmatidae). In J.K. Lowry (ed.)., Proceedings from the Conference on Biology and Evolution of Crustacea. *Aust. Mus. Memoir* 18:133-139.

Foster B.A. & W.A. Newman 1987. Chthamalid barnacles of Easter Island, Southeast Pacific: Peripheral isolation of Notochthamalinae subfam. nov. and the Hembeli-group of Euraphiinae (Cirripedia: Chthamaloidea). *Bull. Mar. Sci.* 41:322-336.

Frazier, J.G. & D. Margaritoulis 1990. The occurrence of the barnacle, *Chelonibia patula* (Ranzani, 1818), on an inanimate substratum (Cirripedia, Thoracica). *Crustaceana* 59:213-218.

Gerstaecker, A. 1866-1879. Arthropoda. In *H.G. Bronns Klassen und Ordungen des Tierreichs 5, 1 Abt. Crustacea, 1. Hefte, Rankenfussler: Cirripedia Burm.* pp. 406-589, 1 Abb., Taf. I-VI. C.F. Winter: Leipzig u. Heidelberg.

Ghiselin, M.T. 1969. *The triumph of the Darwinian method*. University of California Press, Berkeley.

Ghiselin, M.T. & L. Jaffe 1973. Phylogenetic classification in Darwin's monograph on the sub-class Cirripedia. *Syst. Zool.* 22:132-140.

Ghiselin, M.T. 1974. *The economy of nature and the evolution of sex*. Berkeley: University of California Press.

Goodsir, H.D.S. 1843. On the sexes, organs of reproduction, and mode of development, of the cirripeds. *Edinb. New Phil. J.* 35:88-104 + pls. 3 & 4.

Gray, J.E. 1825. A synopsis of the genera of cirripedes arranged in natural families, with a description of some new species. *Ann. Phil.* (n.s.) 10(2):97-107.

Groeben, D. (ed.), 1982. *Charles Darwin, 1809-1882, Anton Dohrn 1840-1909: Correspondence*. Macchiaroli, Naples. 188 pp.

Gruvel, A. 1905. *Monographie des Cirrhipèdes ou Thécostracés*. Masson Cie: Paris (reprinted 1965). Amsterdam: A. Asher & Co.

Grygier, M.J. 1983. Ascothoracida and the unity of the Maxillopoda. *Crust. Issues* 1: 73-104.

Grygier, M.J. 1984. Comparative morphology and ontogeny of the Ascothoracida, a step toward a phylogeny of the Maxillopoda. Doctoral Dissertation. Univ. California, San Diego.

Grygier, M.J. 1987. Nauplii, antennular ontogeny, and the position of the Ascothoracida within the Maxillopoda. *J. Crust. Biol.* 7:87-104.

Hancock, A. 1849. Notice of the occurrence on the British coast of a Burrowing Barnacle belonging to the new Order of the Class Cirripedia. *Ann. Mag. Nat. Hist.*(2) 23:305-314 + 2 pls.

Harding, J.P. 1962. Darwin's type specimens of varieties of *Balanus amphitrite*. *Bull. Brit. Mus. (N. H.), Zool.* 9:273-296 + 10 pls.

Henry, D.P. 1941. Notes on some sessile barnacles from Lower California and the west coast of Mexico. *Proc. New England Zool. Club* 18:99-106.

Henry, D.P. 1942. Studies on the sessile Cirripedia of the Pacific Coast of North America. *Univ. Wash. Publ. Oceanogr.* 4:95-134.

Henry, D.P. 1957, Some littoral barnacles from the Tuamotu, Marshall, and Caroline Islands. *Proc. US Natl. Mus.* 107:25-38.

Henry, D.P. 1960. Thoracic Cirripedia of the Gulf of California. *Univ. Wash. Publ. Oceanogr.* 4:135-158.

Henry, D.P. 1965. Unique occurrence of complemental males in a sessile barnacle. *Nature* 207:1107-1108.

Henry, D.P. & P.A. McLaughlin 1967. A revision of the subgenus *Solidobalanus* Hoek (Cirripedia Thoracica) including a description of a new species with complemental males. *Crustaceana* 12:43-58.

Henry, D.P. & P.A. McLaughlin 1975. The barnacles of the *Balanus amphitrite* complex (Cirripedia, Thoracica). *Zool. Verhand.* 141:203-212.

Henry, D.P. & P.A. McLaughlin 1987. The Recent species of *Megabalanus* (Cirripedia: Balanomorpha) with special emphasis on *Balanus tintinnabulum* (Linnaeus) sensu lato. *Zool. Verhand.* 235:1-69.

Hessler, R.R. & W.A. Newman 1975. A trilobitomorph origin for the Crustacea. *Fossils and Strata* 4:437-459.

Hiro, F. 1938. Studies on animals inhabiting reef corals. II. Cirripeds of the genera *Creusia* and *Pyrgoma*. *Palaeo. Trop. Biol. Stn.* Stud. no. 3:391-416.

Hiro, F. 1939. Studies on the cirripedian fauna of Japan. IV. Cirripeds of Formosa (Taiwan), with some geographical and ecological remarks of the littoral forms. *Mem. Coll. Sci., Kyoto Imp. U.* (B)15:245-284.

Høeg, J.T. & J. Lützen. 1985. Crustacea Rhizocephala. *Marine Invertebrates of Scandinavia* 6:1-90.

Hoek, P.P.C. 1883. Report on the Cirripedia collected by H.M.S. Challenger during the years 1873-1876. *Repts. Sci. Res., Zool.* 8(25):1-169.

Hoek, P.P.C. 1907 and 1913. Cirripedia of the Siboga-Expedition: Cirripedia Pedunculata. *Siboga-Expeditie* 21:1-127 + pls. 1- 10 & Cirripedia Sessilia. *Ibid.* 21:129-275 + pls. 11-27. Leiden: Brill (sic).

Hui, E. & J. Moyse 1984. Complemental males of the primitive balanomorph barnacle, *Chionelasmus darwini*. *J. Mar. Biol. Assc. UK* 64: 91-97.

Jones, D.S., J.T. Anderson & D.T. Anderson 1990. Checklist of the Australian Cirripedia. *Tech. Repts. Aust. Mus.* 3:1-38.

Knox, G.A. 1960. Littoral ecology and biogeography of the southern ocean. *Proc. Roy. Soc., Lond.* 152B:429-677.

Kolbasov, G. 1992. Two new species of the genus *Acasta* from the southwestern part of the Indian Ocean. *Akad. Nauk Zool. Zh.* 1:140-145.

Krohn, A. 1859. Beobachtungen über den Cementapparat und die weiblichen Zeugungsorgane einiger Cirripedien. *Archiv f. Naturgesch.* 1:355-364.

Krohn, A. 1860. Observations on the development of the Cirripedia. *Ann. Mag. Nat. Hist.* (3) :423-428.

Krüger, P. 1940. Cirripedia, Ascothoracida, and Apoda. In *H.G. Bronns Klassen und Ordungen des Tierreichs.* 5,1,3,3:1-560, Leipzig.

Lacaze-Duthiers, H. de 1865. Mémoire sur un mode nouveau de parasitisme observé chez un animal non décrit. *C.R. Hebd. Scéan. Acad. Sci. Paris* 61:838-841.

Lacaze-Duthiers, H. de 1880 and 1883. Histoire de la *Laura gerardiae*: type nouveau de Crustacé parasite. *Arch. Zool. Exp. Gen.* (1) 8:537-581, and *Mém. Acad. Sci. Inst. Fr.* (2)42(2):1-160.

Laguna, J.E. 1990. Shore barnacles (Cirripedia, Thoracica) and a revision of their provincialism and transition zones in the tropical eastern Pacific. *Bull. Mar. Sci.* 46:406-424.

Lamarck, J.B. de 1809. *Philosophie Zoologique*. Paris.

Lamarck, J.B. de 1818. *Histoire naturelle des animaux sans vertèbres* 5:382-383 (cirrhipèdes sessiles), 5:401 (Cirrhipèdes pédonculées).

Leach, W.E. 1817. Distribution systématique de la classe Cirrhipèdes. *J. Phys. Chim. Hist. Nat. Arts.* 85:67-69.

Leach, W.E. 1825. A tabular review of the cirripedes, with descriptions of the species of *Otion, Cineras* and *Clypta. Zool. J., Lond.* 2:208-215.

Linnaeus, C. 1767. *Systema Naturae*. Ed. 12. Stockholm.

Matsui, T., G. Shane & W.A. Newman 1964. On *Balanus eburneus* Gould (Cirripedia, Thoracica) in Hawaii. *Crustaceana* 7:141-145.

Mayr, E. 1982. *The growth of biological thought.* Cambridge: Harvard Univ. Press.

McLaughlin, P.A., & D.P. Henry 1972. Comparative morphology of complemental males in four species of *Balanus* (Cirripedia: Thoracica). *Crustaceana* 22:13-30.

Milne Edwards, H. 1852. Observations sur les affinités zoologiques et les classification naturelle des Crustacés. *Ann. Sci. Nat.* (3)18:109-166.

Müller, F. 1862. Die Rhizocephalen, eine neue Gruppe schmarotzender Kruster. *Arch. Naturgesch.* 28:1-9 + pl. 1.

Müller, F. 1864. *Für Darwin.* Leipzig (Translated by W.S. Dallas, 1869, as *Facts and Arguments for Darwin.* London: John Murray.)

Müller, K.J. & D. Walossek 1988. External morphology and larval development of the Upper Cambrian maxillopod *Bredocaris admirabilis. Fossils and Strata* 23:3-70.

Neave, S.A. 1939. *Nomenclator Zoologicus* 2:14-15. London: Zoological Society.

Newman, W.A. 1960. On the paucity of intertidal barnacles in the tropical Western Pacific. *Veliger* 2:89-94.

Newman, W.A. 1961. On the nature of the basis in certain species of the *Hembeli* section of *Chthamalus* (Cirripedia, Thoracica). *Crustaceana* 2:142-150.

Newman, W.A. 1974. Two new deep-sea Cirripedia (Ascothoracica and Acrothoracica) from the Atlantic. *J. Mar. Biol. Assc. UK* 54:437-456.

Newman, W.A. 1975. Cirripedia. In R.I. Smith & J.T. Carlton (eds) *Intertidal Invertebrates of the Central California Coast*, pp. 259-269. Berkeley: Univ. California Press.

Newman, W.A. 1979a. On the biogeography of balanomorph barnacles of the southern ocean including new balanid taxa; a subfamily, two genera and three species. In *Proc. International Symposium on Marine Biogeography and Evolution in the Southern Hemisphere. N. Z. Dept. Sci. Indust. Res. Info.* (137)1:279-306.

Newman, W.A. 1979b. A new scalpellid (Cirripedia); a Mesozoic relic living near an abyssal hydrothermal spring. *Trans. San Diego Soc. Nat. Hist.* 19:153-167.

Newman, W. A. 1979c. Californian transition zone: Significance of short-range endemics. In J. Gray and A. J. Boucot (eds), *Historical Biogeography, Plate Tectonics and the changing Environments*, pp. 339-416. Oregon State Univ. Press: Corvallis.

Newman, W.A. 1980. A review of extant *Scillilepas* (Cirripedia:Scalpellidae) including recognition of new species from the North Atlantic, western Indian Ocean, and New Zealand. *Tethys* 9:379-398.

Newman, W.A. 1982a. A review of extant taxa of the 'Group of *Balanus concavus*' (Cirripedia; Thoracica) and a proposal for genus-group ranks. *Crustaceana* 43:25-36.

Newman, W.A. 1982b. Cirripedia. In L. Abele (ed.), *The Biology of Crustacea* 1:197-221. New York: Academic Press.

Newman, W.A. 1983. Origin of the Maxillopoda; urmalacostracan ontogeny and progenesis. *Crust. Issues* 1:105-119.

Newman, W.A. 1987. Evolution of cirripedes and their major groups. *Crust.* Issues 5:3-42.

Newman, W.A. 1989a. Barnacle taxonomy. *Nature* 337:23-24.

Newman, W.A. 1989b. Juvenile ontogeny and metamorphosis in the most primitive living sessile barnacles, *Neoverruca*, from an abyssal hydrothermal spring. *Bull. Mar. Sci.* 45:467-477.

Newman, W.A. 1991. Origins of southern hemisphere endemism, especially among marine Crustacea. *Mem. Queensland Mus.* 31:51-76.

Newman, W.A. 1992. Origin of the Maxillopoda. Acta Zoologica (Stockholm) 73: 317-322.

Newman, W.A. & D.P. Abbott 1980. Cirripedia. In R.H. Morris, D.P. Abbott E.C. Haderlie (eds), *Intertidal Invertebrates of California*, pp. 504-535 + pl. 20.1-20.36. Stanford Univ. Press: Palo Alto.

Newman, W.A. & B.A. Foster 1987. The highly endemic barnacle fauna of the southern hemisphere (Crustacea; Cirripedia): Explained in part by extinction of northern hemisphere members of previously amphitropical taxa? *Bull. Mar. Sci.* 41:361-377.

Newman, W.A. & R.R. Hessler 1989. A new abyssal hydrothermal verrucomorphan (Cirripedia; Sessilia): The most primitive living sessile barnacle. *Trans. San Diego Soc. Nat. Hist.* 21:259-273.

Newman W.A. & J.S. Killingley 1985. The north-east Pacific intertidal barnacle *Pollicipes polymerus* in India? A biogeographical enigma elucidated by [18]O fractionation in barnacle calcite. *J. Nat. Hist.* 19:1191-1196.

Newman, W.A. & H.S. Ladd 1974. Origin of coral-inhabiting balanids (Cirripedia, Thoracica). *Verhandl. Naturf. Ges. Basel* 84:381-396.

Newman, W.A. & A. Ross 1971. Antarctic Cirripedia. *Antarctic Research Series* 14:1-257. Amer. Geophys. Union: Wash., D.C.

Newman, W.A. & A. Ross 1976. Revision of the balanomorph barnacles; including a catalogue of the species. *Memoirs San Diego Soc. Nat. Hist.* 9:1-108.

Newman, W.A. & A. Ross 1977. A living *Tesseropora* (Cirripedia: Balanomorpha) from Bermuda and the Azores: First records from the Atlantic since the Oligocene. *Trans. San Diego Soc. Nat. Hist.* 18:207-216.

Newman, W.A. & S.M. Stanley 1981. Competition wins out overall: Reply to Paine. *Paleobiol.* 7:561-569.

Newman, W.A. & T. Yamaguchi. in press. A living brachylepadomorph from the Lau Basin; the first record since the Miocene.

Newman, W. A., V. A. Zullo & T.H. Withers 1969. Cirripedia. In R.C. Moore (ed.), *Treatise on Invertebrate Paleontology, Part R, Arthropoda 4*, 1:R206-295. Univ. Kansas & Geol. Soc. Amer.: Lawrence.

Nilsson-Cantell, C.A. 1921. Cirripeden Studien. Zur Kenntnis der Biologie, Anatomie, und Systematik dieser Gruppe. *Zool. Bidrag Uppsala* 7:75-390.

Nilsson-Cantell, C.A. 1928. The cirripede *Chionelasmus* (Pilsbry) and a discussion of its phylogeny. *Ann. Mag. Nat. Hist.* (10)2:445-455.

Nilsson-Cantell, C.A. 1938. Cirripedes from the Indian Ocean in the collection of the Indian Museum, Calcutta. *Mem. Indian Mus.* 13:1-81, pls. 1-3.

Noll, F.C. 1872. *Kochlorine hamata* N., ein bohrender Cirripede. *Bericht über die Senckenbergische naturf. Gesellsch., 1871-1872* 4:50-58.

Norman, A.M. 1888. Report on the occupation of the table. *Brit. Assc. Advan. Sci. Rept. 57th Meeting.* 1887:85-86.

Nott, J.A. & B.A. Foster. 1969. On the structure of the antennular attachment organ of the cypris larva of *Balanus balanoides*. *Phil. Trans. Roy. Soc. Lond.* (B)256:1215-134.

Owen, R. 1855. *Lectures on the Comparative Anatomy and Physiology of the Invertebrate Animals.* Second Edition. London.

Paine, R.T. 1981. Barnacle ecology. Is competition important? The forgotten roles of disturbance and predation. *Paleobiol.* 7:31-44.

Pilsbry, H.A. 1907. The barnacles (Cirripedia) contained in the collections of the US National Museum. *Bull. US Nat. Mus.* 60:1-122, pls. 1-11.

Pilsbry, H.A. 1916. The sessile barnacles (Cirripedia) contained in the collections of the US National Museum. *Bull. US Nat. Mus.* 93:1-366, pls. 1-76.

Rachootin, S.P. 1985. Darwin's embryology. Ph.D. Dissertation, Yale University (1984). Univ. Microflims: Ann Arbor.

Ride, W.D.L., C.W. Sabrosky, G. Bernardi & R.V. Melville (eds) 1985. *International code of zoological nomenclature* (3rd ed.). Internatl. Trust Zool. Nomen., Brit. Mus. (N. H.): London.

Ross, A. 1969. Studies on the Tetraclitidae (Cirripedia: Thoracica): Revision of *Tetraclita. Trans. San Diego Soc. Nat. Hist.* 15:237-251.

Ross, A. 1970. Studies on the Tetraclittidae (Cirripedia: Thoracica): A proposed new genus for the austral species *Tetraclita purpurascens brevisscutum. Trans. San Diego Soc. Nat. Hist.* 16:1-11.

Ross, A. & W.A. Newman 1973. Revision of the coral-inhabiting barnacles (Cirripedia: Balanidae). *Trans. San Diego Soc. Nat. Hist.* 17:137-174.

Runnström, S. 1925. Zur biologie und entwicklung von *Balanus balanoides* (Linné). *Bergens Mus. Aarbuk* 5:1-46.

Schram, F.R. 1986. *Crustacea.* New York: Oxford Univ. Press.

Shackleton, N.J. 1984. Oxygen isotope evidence for Cenozoic climatic change. In P. Brenchley (ed.), *Fossils and Climate*, pp. 27-34. New York: John Wiley.

Sherborn, C.D. 1922. *Index Animalius 1801-1850.* Parts 6-10:1786-1787. Cambridge: Cambridge Univ. Press.

Smith, S. 1968. The Darwin collection at Cambridge with one example of its use; Charles Darwin and Cirripedes. *Coll. Trav. Acad. Internat. Hist. Sci. Paris.* 11:96-100.

Southward, A.J. 1975. Intertidal and shallow water Cirripedia of the Caribbean. *Studies on the Fauna of Curaçao and other Caribbean Islands* 46:1-53 + 5 plates.

Southward, A.J. 1976. On the taxonomic status and distribution of *Chthamalus stellatus* (Cirripedia) in the North-East Atlantic Region: With a key to the common intertidal barnacles of Britain. *J. Mar. Biol. Assc. UK* 56:1007-1028.

Southward, A.J. 1983. A new look at variation in Darwin's species of acorn barnacles. *Biol. J. Linn. Soc., Lond.* 20:59-72.

Southward, A.J. (ed.) 1987. Barnacle Biology. *Crust. Issues* 5. Rotterdam: Balkema.

Southward, A.J. & W.A. Newman 1977. Aspects of the ecology and biogeography of the intertidal and shallow-water balanomorph Cirripedia of the Caribbean and adjacent sea-areas. *FAO Fish. Rept.* 200:407-425.

Sowerby, G.B. 1820-1825 and 1833. *The genera of recent and fossil shells, for use of students in Conchology and Geology.* London: E.J. Sterling.

Sowerby Jr., G.B. 1843. Description of a new fossil cirripede from the Upper Chalk near Rochester. *Ann. Mag. Nat. Hist.* (2)12:260-261.

Spivey, H.R. 1976. The cirripeds of the Panama Canal. *Corrosion Marine-Fouling* 1(1):43-50.

Stanley, S.M. 1986. Anatomy of a regional mass extinction: Plio-Pleistocene decimation of the Western Atlantic bivalve fauna. *Palaios* 1:17-36.

Stanley, S.M. & W.A. Newman 1980. Competitive exclusion in evolutionary time: the case of the acorn barnacles. *Paleobiol.* 6:173-183.

Stebbing, T.R.R. 1910. General catalogue of South African crustaceans. *Ann. S. Afr. Mus.* 6:563-575.

Stone, I. 1980. *The origin, a biographical novel of Charles Darwin.* New York: Doubleday.

Stubbings, H.G. 1967. The cirriped fauna of tropical West Africa. *Bull. Brit. Mus. (N.H.), Zool.* 15:229-319.

Taylor, P.A. 1970. Observations on the function of the frontolateral horns and horn glands of barnacle nauplii (Cirripedia). *Biol. Bull.* 138:211-218.

Thompson, J. V. 1830. On the cirripedes or barnacles; demonstrating their deceptive character; the extraordinary metamorphosis they undergo, and the class of animals to which they indisputably belong. *Zoological Researches, and Illustrations; or, Natural History of nondescript or imperfectly known animals.* Vol. 1, Part 1, Memoir IV:69-82 + Plates IX & X, 87-88. Cork. (Sherborn Fund Facsimile No. 2, 1968, Soc. Bibliog. Nat. Hist.: London).

Thompson, J.V. 1835. Discovery of the metamorphosis in the second type of the cirripedes, viz. the Lepades, completing the natural history of these singular animals, and confirming their affinity with the Crustacea. *Phil. Trans. Roy. Soc., Lond.* :355-358.

Thompson, J.V. 1836. Natural history and metamorphosis of an anomalous crustaceous parasite of *Carcinus maenas*, and the *Sacculina carcini*. *Entomol. Mag., Lond.* 3:452-456.

Thomson, (C.) Wyville 1873. Notes from the 'Challenger' VI. *Nature*, pp. 347-349 (Aug. 28, 1873).

Tomlinson, J.T. 1987. The burrowing barnacles (Acrothoracica). *Crust. Issues* 5:63-71. Rotterdam: Balkema.

Utinomi, H. 1966. Recent immigration of two foreign barnacles into Japanese waters. *Proc. Japan. Soc. Syst. Zool.* 2:36-39.

Utinomi, H. 1967. Comments on some new and already known cirripeds with emended taxa, with special reference to the parietal structure. *Publ. Seto Mar. Biol. Lab.* 15:199-237.

Valentine, J.W. 1966. Numerical analysis of marine molluscan ranges on the extratropical northeastern Pacific shelf. *Limnol. Oceanogr.* 11:198-211.

Valentine, J.W. 1984. Neogene marine climate trends; Implications for biogeography and evolution of the shallow-sea biota. *Geology* 12:647-650.

Walker, G. 1973. Frontal horns and associated gland cells of the nauplii of the barnacles, *Balanus hameri*, *Balanus balanoides* and *Elminius modestus* (Cirripedia: Crustacea). *J. Mar. Biol. Assc. UK* 53:455-463.

Walker, G. 1983. A study of the ovigerous fraena of banacles. *Proc. Roy. Soc., Lond.* (B)218:425-442.

Walker, G. & A.B. Yule 1984. Temporary adhesion of the barnacle cyprid; The existence of an antennular adhesive secretion. *J. Mar. Biol. Assc. UK* 64:679-686.

Wallace, A.F. 1855. On the law which has regulated the introduction of new species. *Ann. Mag. Nat. Hist.* (2)16:184-196.

Wallace, A.F. 1858. On the tendency of varieties to depart indefinitely from the original type. *J. Proc. Linn. Soc. (Zool.)* 3:53-62.

Walley, L.J. 1965. The development and function of the oviducal gland in *Balanus balanoides*. *J. Mar. Biol. Assc. UK* 45:115-128.

Winsor, M.P. 1969. Barnacle larvae in the nineteenth century: A case study in taxonomic theory. *J. Hist. Medicine and Allied Sci.* 24:294-309.

Withers, T.H. 1912a. The cirripede *Brachylepas cretacea* H. Woodward. *Geol. Mag., Lond.* 9:321-326.

Withers, T.H. 1912b. The cirripede *Brachylepas cretacea* H. Woodward. *Geol. Mag., Lond.* 9:353-359 + pl. 20.

Withers, T.H. 1914a. Some Cretaceous and Tertiary cirripedes referred to *Pollicipes*. *Ann. Mag. Nat. Hist.* (8)14:167-206 + pls. 7 & 8.

Withers, T.H. 1914b. A remarkable new cirripede from the chalk of Surrey and Hertfordshire. *Proc. Zool. Soc. Lond.*: 945-953.

Withers, T.H. 1923. Die Cirripedien der Kreide Rügens. *Abh. geol.-palaeont. Inst. Greifswald* 3:1-54 + 3 pls.

Withers, T.H. 1928. *Catalogue of fossil Cirripedia in the Department of Geology. 1 (Triassic and Jurassic).* Brit. Mus. (N.H.): London.

Withers, T.H. 1935. *Catalogue of fossil Cirripedia in the Department of Geology. 2 (Cretaceous).* Brit. Mus. (N.H.): London.

Withers, T.H. 1953. *Catalogue of Fossil Cirripedia in the Department of Geology* 3(Tertiary): Brit. Mus. (N.H.): London.

Woodward, H. 1901. On *Pyrgoma cretacea*, a cirripede from the Upper Chalk of Norwich and Margate. *Geol. Mag., Lond.* (n.s) (4)8:145-152 + pl. 8 (erratum:240; additional note:528).

Woodward, H. 1906. Cirripedes from the Trimmingham Chalk and other localities in Norfolk. *Geol. Mag., Lond.* (n.s.) (5)3:337-353.

Yamaguchi, T. 1987. Changes in the barnacle fauna since the Miocene and the infraspecific structure of *Tetraclita* in Japan (Cirripedia: Balanomorpha). *Bull. Mar. Sci.* 41:337-350.

Yamaguchi, T. & W.A. Newman 1990. A new and primitive barnacle (Cirripedia: Balanomropha) from the North Fiji Basin abyssal hydrothermal field, and its evolutionary implications. *Pac. Sci.* 44:135-155.

Young, P.S. 1988. Recent cnidarian-associated barnacles (Cirripedia, Balanomorpha) from the Brazilian coast. *Rev. Brasil. Zool.* 5:353-369.

Young, P.S. 1989. Establishment of an Indo-Pacific Barnacle (Cirripedia, Thoracica) in Brazil. *Crustaceana* 56:212-214.

Young, P.S. 1990. Lepadomorph cirripeds from the Brazilian coast. I. Families Lepadidae, Poecilasmatidae and Heteralepadidae. *Bull. Mar. Sci.* 47:641-655.

Zevina, G.B. 1976. Abyssal species of barnacles (Cirripedia: Thoracica) of the North Atlantic. *Zool. Zh. Akad. Nauk SSSR* 55:1149-1156.

Zevina, G.B. 1981. Barnacles of the suborder Lepadomorpha (Cirripedia, Thoracica) of the world ocean. Part 1: Family Scalpellidae. *Guides to the Fauna of the USSR, 127:406 pp. Zool. Inst. Acad. Sci. USSR*: Leningrad.

Zevina, G.B. & I.V. Yakhontova 1987. A new barnacle of the family Scalpelidae from the North Atlantic. *Zool. Zh. Akad. Nauk SSSR Zool. Zh.* 66: 1261-1264.

Zullo, V.A. 1984. New genera and species of balanoid barnacles from the Oligocene and Miocene of North Carolina. *J. Paleo.* 58:1313-1338.

Zullo, V.A. 1991. Zoogeography of the shallow-water cirriped fauna of the Galapagos Islands and adjacent regions in the tropical eastern Pacific. In M.J. James (ed.), *Galapagos Marine Invertebrates. Taxonomy, Biogeography, and Evolution in Darwin's Islands*, pp. 173-192. New York: Plenum Publ.

Zullo, V.A. 1992a. Revision of the balanid barnacle genus *Concavus* Newman, 1982, with the description of a new subfamily, two new genera, and eight new species. Paleontological Society Memoir 27: 1-46.

Zullo, V.A. 1992b. *Balanus trigonus* in the Atlantic basin; an introduced species? *Bull. Mar. Sci.* 50:66-74.

Zullo, V.A. & R.W. Portell (in prep.). Revision of the coral- inhabiting barnacles, Pyrgomatidae (Cirripedia, Thoracica) with special reference to the tropical Atlantic.

Zullo, V.A., D.B. Beach & J.T. Carlton 1972. New barnacle records (Cirripedia, Thoracica). *Proc. California Acad. Sci.*. Fourth Series, 39(6):65-74.

APPENDIX I: BIOGRAPHIES OF CONTEMPORARIES OF DARWIN MENTIONED IN THIS WORK

Agassiz, Louis (1807-1873). Swiss geologist and zoologist, professor at Harvard from 1847 and founder of the Museum of Comparative Zoology, sent Darwin specimens and encouraged him to monograph cirripedes, became most influencial opponent of Darwinian evolution in US.

Bate, Charles Spence (1819-1889). Dentist, carcinologist, counseled Darwin on the cirripede nauplius.

Bosquet, Joseph Augustin Hubert (1814-1880). Belgian invertebrate paleontologist, counseled Darwin on plates of *Verruca*, misleads Darwin by 'discovery' of *Chthamalus* in the Cretaceous, since recognized as part of a kitchen midden (Southward 1983).

Bowerbank, James Scott (1779-1877). London geologist, provided Darwin with specimens of living and fossil cirripeds and administered publication of Darwin's volumes on fossil barnacles by the Palaeontographical Society, devoted later career to study of sponges.

Brown, Robert (1772-1858). Keeper of botanical collections, British Museum. FRS 1811.

Claus, Carl Fredrich Wilhelm (1835-1899). German carcinologist, followed up F. Müller's work with larval cirripeds, admired Darwin's evolutionary work and dedicated a book to him.

Covington, Syms (1816?-1861). Became Darwin's servant aboard the Beagle in 1833 and remained his assistant until 1839 when he emigrated to Australia, sent Darwin barnacles from Australia.

Cresy, Edward (1792-1858). Architect and civil engineer, a neighbor in Down who Darwin told he planned to go to Paris to study barnacles.

Cuming, Hugh (1791-1865). Naturalist, collected a number of Indo-Pacific barnacles described by Darwin, occasionally specimens got mixed up so that some localities were wrong.

Cuvier, Georges (1769-1832), French systematist and paleontologist, Professor of Natural History (from 1800) and Comparative Anatomy (from 1802) at the Collège de France and the Muséum d'Histoire Naturelle respectively, secretary of the Académie des Sciences (from 1803), and FRS (1806).

Dana, James Dwight (1813-1895). Geologist and carcinologist, professor at Yale from 1856, naturalist on the Wilkes Expedition (1838-1842), editor of *American Journal of Science* from 1846, corresponded with Darwin over coral reefs, cirriped larvae, and biogeography.

Dieffenbach, Ernst (1811-1855). German naturalist and geologist, translated Darwin's *Voyage of the Beagle* into German, corresponded with Darwin on geology of islands until Darwin wrote him he had embarked on barnacles.

Dohrn, (Felix) Anton (1840-1909). German zoologist, established the Zoological Station at Naples and was its first director, got involved with Darwin over larval work and administrative matters involving the laboratory.

Fitch, Robert (1802-1895). Pharmacist and geologist, loaned Darwin fossil cirripedes from Norwich Castle (Cretaceous; see Woodward).

FitzRoy, Robert (1805-1865). Vice-admiral, hydrographer and meteorologist, Commander of the Beagle (1828-1836). Darwin's friend on the voyage with whom he corresponded until he got into barnacles.

Fox, William Darwin (1805-1880). Darwin's second cousin, close friend and fellow beetle collector at Cambridge. Rector of Delamere, Cheshire (1838-1873).

Goodsir, Henry D.S. (d. 1845). Physician and zoologist whose males of *Balanus* (1843) were shown to be female epicarideans by Darwin (LC1, 2). Goodsir perished with the Franklin Expedition in a search for the northwest passage in 1845, and Darwin wrote J.C. Ross in December 1847 (C-1140) to ask if he would collect Arctic cirripedes while on the search for Sir John Franklin and his party.

Gould, August Addison (1805-1866). American physician and zoologist, expert on mollusks, provided Darwin with barnacle specimens, described *Balanus ebruneus* Gould, 1841.

Gray, Asa (1810-1888). American botanist, professor at Harvard from 1842, corresponded with Darwin on biogeographical matters involving plants, received a sketch of Darwin's species theory in 1856 and became leading supporter of Darwinian evolution in the US.

Gray, John Edward (1800-1875). Keeper of Zoology, British Museum (1840-1874), published a revision of the barnacles in 1825, encouraged Darwin to monographic cirripedes and supplied specimens.

Hancock, Albany (1806-1873). Zoologist, morphology of mollusks, described *Alcippe* (= *Trypetesa*) and provided Darwin with specimens in which Darwin discovered males.

Henslow, John Stevens (1796-1861). Botanist and clergyman, professor of botany, Cambridge (1827-1861), Darwin's teacher and friend with whom he corresponded about barnacles, salting seeds, and problems concerning dispersal, up to the year before Henslow died.

Hoek, Paulus Peronius Cato (1851-1914). Dutch zoologist, published reports on the cirripedes of the Challenger Expedition (1883) and the Saboga Expedition (1907 and 1913) emphasizing anatomy of adults and males in the former (probably because of the influence of Wyville Thomson as well as Darwin, and systematics in the latter. Discovered Darwin's olfactory organs were maxillary glands. Corresponded briefly with Darwin in connection with some earlier papers on barnacles he had sent him.

Hooker, Joseph Dalton (1817-1911). Botanist, assistant director and director, Kew. Over 1300 known communications with Darwin over some 40 years (1843-1882), thus on the average more than 30 communications a year. Advised Darwin on virtually every subject he was interested in, arranged with Lyell to have Darwin's and Wallace's papers read before the Linnaean Society and published in 1858.

Huxley, Thomas Henry (1825-1895). Zoologist, lecturer, Royal School of Mines from 1854, sometimes professor at the Royal College of Surgeons and the Royal Institution, powerful supporter of Darwinism. Some 300 communications between him and Darwin between 1851 and 1882, beginning with questions involving Darwin's anatomical work on barnacles.

Krohn, August Davin (1803-1891). Russian anatomist and embryologist, corrected Darwin's observations on the female reproductive system of barnacles.

Krauss, (Christian) Ferdinand (Friedrich von) (1812-1890). German naturalist and museum keeper who travelled in S. Africa, described *Tesseropora rosea* (Krauss, 1848).

Lankester, Edwin (1814-1874). Physician, botanist, Professor of Natural History, New College, London, administratively involved with Darwin over barnacle volumes published by the Ray Society.

Leonard, Samuel William. (no dates available) Artist, member of the Microscopical Society of London from 1840, introduced Darwin to a good compound microscope.

Lovén, Sven (1809-1859). Swedish marine biologist, worked on mollusks and echinoids, and described *Alepas* (= *Anelasma*) *squalicola*.

Lyell, Charles (1797-1875). Uniformitarian geologist who discredited the catastrophist school, over 260 communications with Darwin between 1836 and 1874. Arranged with Hooker to have Darwin's and Wallace's papers read before the Linnaean society in 1858.

Milne Edwards, Henri (1800-1885). French zoologist who specialized in invertebrates, especially crustaceans. His writings greatly influenced Darwin's thinking on the crustacean organization plan, but the trip Darwin planned to visit him in Paris in 1848 was cancelled due to Darwin's health.

Müller, Fritz (1822-1897). German naturalist and carcinologist, emigrated to Brazil in 1852, found flaws in Darwin's larval work, honored Darwin with *Für Darwin* (1869). Proposed the Rhizocephala (1862) to encompass the crustacean parasite *Sacculina* and its allies, used *Anelasma* as a model (1869) to explain their origin from cirripeds.

Newport, George (1803-1854). Entomologist, provided dissecting scissors Weiss & Company used as a model in fabricating a pair for Darwin.

Owen, Richard (1804-1892). Leading comparative anatomist, Hunterian Professor, Royal College of Surgeons, about 35 communications with Darwin.

Quekett, John Thomas (1815-1861). Histologist and microscopist, in charge of collections at the College of Surgeons where the cirirpedes of the Beagle were apparently stored.

Sowerby, George Brettingham (1788-1854). Conchologist and scientific illustrator, author of many genera of barnacles.

Sowerby Jr., George Brettingham (1812-1884). Conchologist and scientific illustrator, son of G.B. Sowerby, illustrated LC1, LC2, and FC2, and translated Darwin's diagnoses into Latin.

Sowerby, James de Carle (1787-1871). Naturalist and scientific illustrator, older brother of G.B. Sowerby, illustrated FC1.

Steenstrup, Johannes (1813-1897). Danish zoologist, described *Xenobalanus* in 1851, arranged for loans of Scandinavian living and fossil cirripedes for Darwin.

Strickland, Hugh Edwin (1811-1853). Geologist, zoologist, and author of major reforms in zoological nomenclature, chair of a committee of the British Association to which Darwin was appointed, and Darwin subsequently consulted with him over nomenclatural problems.

Thompson, John Vaughan (1779-1847). Surgeon and amateur naturalist, discovered that cyprids, and independently confirmed that nauplii, were cirripede larvae, discovered that *Sacculina*, a nondescript parasite of a crab, was related to cirripedes by its nauplii. Darwin held great admiration for his work but they did not correspond.

Thomson, (Charles) Wyville (1830-1882). Naturalist interested in oceanography, Professor of Natural History, Edinburgh; civilian director of Challenger Expedition. His publication on the males of cirripedes (1873) stimulated Darwin to write his last paper on barnacles the same year.

Wallace, Alfred Russel (1823-1913). Collector of tropical biota, natural historian, biogeographer. Independently discovered natural selection. FRS (1893).

Waterhouse, George Robert (1810-1888). Naturalist, Curator of the London Zoological Society. Described Darwin's entomological specimens from the Beagle.

Woodward, Henry (1832-1921). Geologist and paleontologist, founder and editor of the *Geological Magazine*, corresponded with Darwin in 1880 seeking a testimonial for the position of Keepership of Geology at the British Museum, discovered the sessile nature of *Brachylepas* (1901) and noted that some material Darwin had described as a pedunculate was actually a brachylepadomorph.

APPENDIX II

Contents of Living Cirripedia 1, 1852 (1851)

Summary of content on p. 8, based mainly on subheadings:

Metamorphosis: larva, first stage, pp. 9-12; larva, second stage, p. 13; larva, last stage, p. 14; its carapace ib.; acoustic organs, p. 15; antennae, ib.; eyes, p. 16; mouth, p. 17; thorax and limbs, p. 18; abdomen, p. 19; viscera, ib.; immature cirripede, p. 20; homologies of parts, p. 25.

Description of mature Lepadidae: capitulum, p. 28, peduncle, p. 31; attachment, p. 33; filamentary appendages, p. 38; shape of body, musculature system, p. 39; mouth (and parts), ib.; cirri, p. 42; caudal appendages, p. 43; alimentary canal, p. 44; circulatory system, p. 46; nervous system, ib.; eyes, p. 49; olfactory organs, p. 52; acoustic (?) organs, p. 53; male sexual organs, p. 55; female organs, p. 56; ovigerous lamellae, p. 58; ovigerous fraena, ib.; rate of growth, p. 61; size, p. ib.; affinities of family, p. 64; range and habitats, p. 65; geological history (refers to Fossil Cirripedia 1 for details), p. 66.

Genus 1. *Lepas* Linnaeus, 1767.

Diagnosis, p.67; description, ib.; scuta, p.68; terga, p. 69; peduncle,p. 70; size and color, ib.; filamentary appendages, ib.; mouth, p. 71; cirri, ib.; caudal appendages, ib.; distribution, p. 72; general remarks and affinities, ib. Sp. 1., p. 73, diagnosis, p. 73; varieties, ib.; general appearance, ib.; filamentary appendages, p. 75; mouth, ib.; size, ib.; colors, ib.; general remarks, p. 76. Spp. 2-6, same format as first species (+ cause of buoyancy p. 95), 77-99.

Genera 2-11 and their included species are treated similarly:

Genus 2. *Poecilasma* Darwin, 1852, spp. 1-4, pp. 99-115.

Genus 3. *Dichelaspis* Darwin, 1852, spp. 1-5, pp. 115-133.

Genus 4. *Oxynaspis* Darwin, 1852, sp. 1, pp. 133-136.

Genus 5. *Conchoderma* Olfers, 1814, spp.1-3, pp. 136-156.

Genus 6. *Alepas* Sanger Rang, 1829, spp. 1-4, pp. 156-169.

Genus 7. *Anelasma* Darwin, 1852, sp. 1, pp. 169-180.

Genus 8. *Ibla* Leach, 1825, spp. 1-2 + males, pp. 180-214.

Genus 9. *Scalpellum* Leach, 1817, spp. 1-6 + males, pp., 215-293. Spp. 1-3 without subcarina; spp. 4-6 with subcarina.

Genus 10. *Pollicipes* Darwin, 1852, spp. 1-6, pp. 293-331.

Genus 11. *Lithotrya* Sowerby, 1822, spp. 1-6 and burrowing, pp. 332-373.

Contents of Living Cirripedia 2, 1854

APPENDIX III: PRINCIPAL TAXA DISCOVERED SINCE DARWIN'S BARNACLE WORK RELEVANT TO THE ORIGIN AND RADIATION OF THE CIRRIPEDIA

Some 14 genera that have added substantially to our understanding of the origin and evolution of the Cirripedia have been discovered since Darwin (see Fig. 21). These are briefly discussed here for historical purposes in the order of their discovery rather than in their presently accepted systematic order. Genera relevant to the origin and radiation of the sessile barnacles are identified by an asterisk, and the homologies of the plates are discussed in Section 5.9 and Fig. 25.

1) *Laura* Lacaze-Duthiers, 1865: First known representative of a large group of parasitic crustaceans, the Ascothoracida Lacaze-Duthiers, 1880, once included in the Cirripedia but now included at the coordinate level in the Thecostraca Gruvel, 1905 (Grygier 1984; Boxshall & Huys 1989; Newman 1992; Section 5.10.1). *Synagoga* Norman 1888, is the most primitive living genus. Ascothoracids are important in considerations of cirripede origins, and it is unfortunate that Darwin did not known about them.

2) *Archaeolepas* Zittle, 1884, and *Eolepas* Withers, 1928: Upper Triassic-Jurassic. The most primitive known scalpellomorphs (Eolepadinae Buckeridge, 1983); simple 6-plated capitulum (S-T/R-C; see below) and, in *Archaeolepas*, a peduncle having imbricating plates in whorls of 8 tiers (sr-rl-l-cl-sc). The L, first appearing in *Neolepas* (S-L-T/R-C), is considered homologous with 1, and RL-CL and SC first appearing in *Scillaelepas* s.s. (S-L-T/R-RL-CL-C-SC) are considered homologous with rl-cl and sc.

3) *Scillaelepas* Seguenza, 1876: Upper Cretaceous- Recent. A scalpellomorph that, according to Withers (1953), was illustrated by Scilla in 1670 and therefore was the first Mesozoic pedunculate barnacle known. However, it was subsequently found living at 2000 m off Greenland (Aurivillius 1894). The genus forms the basis for the Darwinian and Aurivillian model for the origin of the Balanomorpha (see Newman et al. 1969, Fig. 90) and the Withersian model for the origin of the Verrucomorpha (Newman et al. 1969, Fig. 113). The capitulum is basically 13-plated (S-L-T/R-RL-CL-C-SC) and the peduncle is clothed with numerous whorls of relatively

large plates. Two of the three subgenera, *Gruvelialepas* Newman, 1980, and *Aurivillialepas* Newman, 1980, develop capitular subrostra (SR) from peduncular subrostra (sr) that, in the former, can remain either with the peduncle or the capitulum following formation and growth (Section 5.9.2; Fig. 25C).

4) *Brachylepas cretacea* (Woodward, 1901): Upper Cretaceous. The type specimen, housed in the British Museum (Natural History), represents the most complete fossil brachylepadomorph (Cretaceous-Recent) known. Darwin had described some brachylepadomorphs as species of *Pollicipes,* and at least until recently (see *Neobrachylepas* below), without Woodward's specimen, the sessile nature of the Brachylepadomorpha might not have been recognized. The wall consists of R-C supporting a 6-plated operculum (S-L-T) with in its aperture and surrounded by numerous whorls of imbricating plates. Those filling the gap between R and C on each side were called 'sublatera' by Woodward (1906), and he inferred and Withers demonstrated that the operculum included L as well as S-T. The order in which the basal imbricating whorls of latera were added was thought to be the same as in pollicipedines and lower balanomorphs (Newman 1987), but the discovery of *Neoverruca* demonstrated that this was not the case (Newman 1989b). Brachylepadomorphs are important in models involving the origin of the Verrucomorpha and Balanomorpha (see *Neobrachylepas* below and *Verruca*, Section 5.10).

5) *Kochlorine* Noll, 1872, *Lithoglyptes* Aurivillius, 1892, and *Weltneria* Berndt, 1907; Acrothoracica, Lower Devonian. As discussed in Section 5.6, Darwin ended up believing that the two highly reduced forms he studued (*Trypetesa* & *Cryptophialus*) were in different orders, in part because he thought they had different appendages. However, these turned out to be the rudimentary first cirri and three pairs of terminal cirri (1+3) in both species. The subsequent discovery of the foregoing three genera assured that this interpretation was correct; namely, in order of their discovery and the reverse of their phylogeny, their appendages counts are 1+3+ caudal furca, 1+4+c.f., and 1+5+c.f. The last, *Weltnaria*, has the full complement of limbs seen in the Thoracica, and it is being debated whether the Acrothoracica descended from the Thoracica, or if they shared a common ancestor (Newman 1982, 1987).

6) *Chionelasmus* Pilsbry, 1911; see Nilsson-Cantell (1928) for full description and Newman (1987) for ontogeny. An Indo-West Pacific balanomorph from bathyal depths and the Eocene of Tonga (Stanley & Newman 1980). The 6-plated wall, originally considered R-RL-L-C rather than R-RL-CL-C (Yamaguchi & Newman 1990), is surrounded by numerous whorls of immbricating plates of three types more or less condensed into a single whorl (see *Eochionelasmus* below for ancestral arrangement). The Chionelasmatinae has recently replaced *Catophragmus* as the primary candidate in considerarions of the origin of the Balanomorpha (Newman 1987; Yamaguchi & Newman 1990; Buckeridge & Newman 1992).

7) *Proverruca* Withers, 1914 and *Eoverruca* Withers, 1935: Upper Cretaceous. Verrucomorphans having, in addition to the basic plan of the Verrucinae (MS-MT/R-FS-FT-C), another pair of latera (RL-CL) closing the gap between R and C on the movable side. Withers thought that the proverrucine plan helped bridge the gap between *Scillaelepas* and the verrucines, but the neoverrucines were unknown to him and the proverrucine plan is now known to fit between them (see below) and verrucines (Newman & Hessler 1989).

8) *Capitulum* Gray, 1825 (Foster 1978). Darwin considered the pedunculate genus *Pollicipes* Leach, 1817, particularly *P. mitella* (Linnaeus 1767), to be an ideal model from which the primitive sessile barnacle *Catophragmus* could have been derived (Section 5.9). Foster (1978), in concurrence with Withers (1928:32), noted that *P. mitella* is very different from the three extant species presently attributed to *Pollicipes*. Foster has recommended that Gray's genus *Capitulum*, which was abandoned by Darwin, be resurrected to accommodate it, and this has been accepted by recent workers (Newman & Killingley 1985; Newman 1987; not in Fig. 21).

9) *Neolepas* Newman, 1979, ?L. Jurassic-Recent (Buckeridge & Grant-Mackie 1985). This is the most primitive living scalpellomoprh (S-L-T/R-C). It was discovered living near abyssal hydrothermal vents in the eastern Pacific (Newman 1979b), but it is now known by more-or-less closely related forms from vents in the western Pacific. A *Neolepas*-like rather than a *Scillaelepas*- or *Pollicipes*-like ancestor has been considered a more likely candidate for the origin in the Brachylepadomorpha (Newman 1982, Newman & Yamaguchi in prep.).

10) *Waikalasma* Buckeridge (1983): A unique balanomorph from the Late Miocene having an 8-plated wall interpreted as R-RL-CL^1-CL^2-C and surrounded basally by several whorls of small imbricating plates (Buckeridge & Newman 1992). Cl^1 and CL^2 are nearly identical in size and shape, but of the three pairs of latera only Cl^1 enters the sheath. While RL does not enter the sheath in *Pachylasma* (LC^2) *Chionelasmus* (Newman 1987), all three latera enter the sheath in higher balanomorphs such as *Catophragmus*, *Octomeris* and *Chelonibia*. The unintegrated condition of RL and CL^2 in *Waikalasma* supports the interpretation that CL^1 preceded RL and CL^2

in the formation of the sheath in higher forms (Buckeridge & Newman 1992).

11) *Newmanilepas* Zevina & Yakhontova, 1987 is a scalpellomorph from a depth of 3000-4000 m, the species of which, *N. mirifica*, was originally attributed to *Scillaelepas* by Zevina (1976). While young specimens are similar to *Scillaelepas*, mature individuals are similar to the Cretaceous *Cretiscalpellum* Withers, 1922. Zevina & Yakhontova (1987) describe several sizes that show that ontogeny is unusually slow; e.g., plates (SC^2 & CL^2) are added to the capitulum by replication after what might appear to be the adult form has been achieved (Section 5.9.2, Fig. 25D-F).

12) *Neoverruca* Newman (in Newman & Hessler 1989) is the most primitive known verrucomorph. It and related forms are from deep-water hydrothermal vents of the western Pacific. *Neoverruca* has the asymmetry characteristic of verrucomorphans (Section 5.9), but the unmodified side has the basic organization of a brachy-lepadomorph (S-L-T/R-C + whorls of imbricating plates). On the opposite side L has been lost, and T and S are fixed between R and C, whereby the wall is R-FS-FT-C, the hallmark of the Verrucomorpha. It is the perfect intermediate between the two suborders. The ontogeny of *Neoverruca* (Newman 1989b) includes numerous pedunculate stages followed by an abrupt metamorphosis into the sessile mode (a pedunculate stage was seen in *Semibalanus balanoides* by Darwin (LC2)). Ontogeny also reveals that the whorls of imbricating plates are added in the reverse order to that observed in *Pollicipes* (Darwin 1851; Broch 1922; Newman 1989b), *Cato-phragmus* (Darwin 1854; Newman 1987), *Chionelasmus* (Newman 1987), and as had been assumed in *Brachy-lepas* (Newman 1987). It now seems inescapable that brachylepadomorphs also add the imbricating plates in the same way as in neoverrucines (Newman 1989b).

13) *Eochionelasmus* Yamaguchi (in Yamaguchi & Newman 1990) is from abyssal hydrothermal vents in the western Pacific. It is more primitive than *Chionelasmus* by virtue of its basal imbricating whorls made up of unmodified rather than modified plates of three types (Nilsson-Cantell 1928). The study of these two genera led to the conclusion that the original balanomorph wall was R-RL-CL-C rather R-RL-L-C.

14) *Neobrachylepas* Newman & Yamaguchi, in prep., is represented by a single specimen recovered from near a hydrothermal vent in the West Pacific. It is the only known living representative of the Brachylepadomor-pha, a suborder that appeared in the Jurassic and was thought extinct since the Miocene (Newman et al. 1969). *Neobrachylepas* provides the first incontrovertible evidence that the operculum of brachylepadomorphs in-cluded L, and that the whorls of imbricating plates are added between the wall and the previous whorl, all as in *Neoverruca* except for the symmmetric opercular plates. The trophi of *Neobrachylepas* are modified in the same unique manner as in the other hydrothermal barnacles, *Neolepas, Neoverruca* and *Eochionelasmus*, for feeding on extremely fine particles, presumably bacteria.

There is a recent biography on *Darwin* (Desmond, A. & J. Moore 1992. Warner Books: New York) and I thank E.L. Winterer for alerting me to it. Unlike Stone's *The Origin*, this conveniently indexed and meticulous volume treats Darwin's involvement with barnacles in considerable depth. Some details are bound to be misleading or incorrect in a work of this size, such as saying '... all letters were salted away' by Darwin, mixing up of males of *Ibla* and *Scalpellum*, not recognizing that there were two George B. Sowerbys, father and son, and perpetuating a myth that genealogical trees of life '... were common in Darwin's day.' On the other hand, the order in which the males were discovered is eminently clear. How Darwin went about working on barnacles, his accomplish-ments with them, and the relationship of the work to his species work are well conveyed. Furthermore, the coverage goes well beyond the confines of Darwin's study and family life; it is enriched by an illuminating expose of the social, religious, and political turmoil of the times. The bottom line? In producing an exhaustive work on barnacles, Darwin became recognized as an accomplished systematist. This, in his mind's eye, gave him the authority to publicly theorize on the origin of species, which he then set out to do.

Index

T - #0316 - 101024 - C0 - 254/178/25 [27] - CB - 9789054101376 - Gloss Lamination